AF324856

# THE
# MULTIFACETED SKYRMION

# THE MULTIFACETED SKYRMION

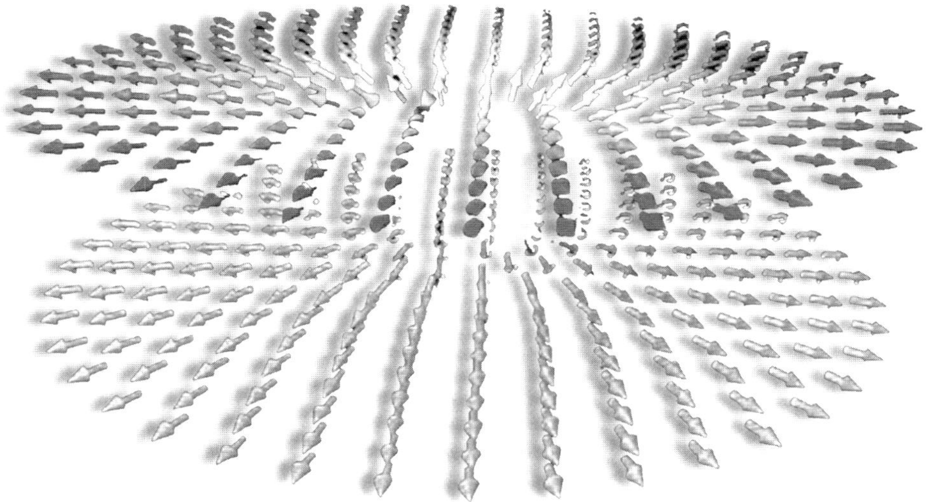

*Editors*

**Gerald E Brown**
State University of New York, USA

**Mannque Rho**
DSM-CEA Saclay, France and Hanyang University, Korea

**World Scientific**

NEW JERSEY · LONDON · SINGAPORE · BEIJING · SHANGHAI · HONG KONG · TAIPEI · CHENNAI

*Published by*

World Scientific Publishing Co. Pte. Ltd.

5 Toh Tuck Link, Singapore 596224

*USA office:* 27 Warren Street, Suite 401-402, Hackensack, NJ 07601

*UK office:* 57 Shelton Street, Covent Garden, London WC2H 9HE

**British Library Cataloguing-in-Publication Data**
A catalogue record for this book is available from the British Library.

**THE MULTIFACETED SKYRMION**



ISBN-13 978-981-4280-69-3
ISBN-10 981-4280-69-0

*Printed in Singapore by Mainland Press Pte Ltd*

# Preface

Two path-breaking developments took place consecutively in physics in the years 1983 and 1984: First in nuclear physics with the rediscovery of Skyrme's seminal idea on the structure of baryons and then a 'revolution' in string theory in the following year. One of us (Gerald E. Brown) edited a volume entitled *Selected Papers, with Commentary, of Tony Hilton Royle Skyrme* in 1994, recounting how at that time the most unconventional idea of Skyrme that fermionic baryons could emerge as topological solitons from $\pi$-meson cloud was confirmed in the context of quantum chromodynamics (QCD) in the large number-of-color ($N_c$) limit. It also confirmed how the solitonic structure of baryons, in particular, the nucleons, reconciled nuclear physics — which had been making an impressive progress phenomenologically, aided mostly by experiments — with QCD, the fundamental theory of strong interactions. Immediately after the rediscovery of what is now generically called 'skyrmion' came the first string theory revolution which then took most of the principal actors who played the dominant role in reviving the skyrmion picture away from that problem and swept them into the mainstream of string theory reaching out to a much higher energy scale. This was in some sense unfortunate for the skyrmion model *per se* but fortunate for nuclear physics, for it was then mostly nuclear theorists who picked up what was left behind in the wake of the celebrated string revolution and proceeded to uncover fascinating novel aspects of nuclear structure which otherwise would have eluded physicists, notably concepts such as the 'Cheshire Cat phenomenon' in hadronic dynamics.

What has taken place since 1983 is a beautiful story in physics. It has not only profoundly influenced nuclear physics — which was Skyrme's original aim — but also brought to light hitherto unforseen phenomena in other areas of physics, such as condensed matter physics, astrophysics and string theory.

The objective of this volume is to illustrate, with a few selected contributions from leading researchers, how profound and path-breaking the notion of skyrmion has turned out to be in various different areas of physics.

The first volume on Skyrme in 1994 contained his seminal articles dating from the late 1950's and early 1960's and a few selected articles that played a pivotal role in the 1980's in resurrecting, in the context of QCD, Skyrme's supremely original idea that had been slumbering in total obscurity for more than two decades. These articles were presented with the Editor's personal anecdotes on and about Skyrme and Skyrme's papers, supplemented with the Editor's commentaries on how the skyrmion picture fit in with what was then in vogue at the time of the rediscovery in the effort of modeling QCD, such as quark confinement and asymptotic freedom à la MIT bag, spontaneous breaking of chiral symmetry necessitating Nambu–Goldstone bosons etc. This volume picks up from what has taken place since then. In surveying the developments that have taken place in the past two and half decades, what's most significant of all is that the notion of skyrmion has found to be uncannily pervasive and universal, figuring in nearly all branches of physics and manifesting in a variety of different facets, from which came the title 'The Multifaceted Skyrmion.'

What was particularly appealing to nuclear physicists in the rediscovery of the skyrmion picture was that the highly successful standard nuclear physics approach to nuclear dynamics where nucleons, pions, vector mesons and other low-lying hadrons are treated as *the* relevant degrees of freedom could be naturally accommodated in the framework of quantum chromodynamics (QCD). That the nucleon emerges as a soliton made of coherent states of Nambu–Goldstone bosons — pions — rendered *natural* the standard Yukawa interactions between nucleons that had been taken for granted. This volume contains articles that support (sometimes very accurately) this expectation in several different aspects as well as those which make predictions that are accessible neither by QCD proper nor by the standard nuclear physics approach. The skyrmion approach both supplements what has already been established before and furthermore allows one to probe the regimes difficult to access, i.e., hadrons under extreme conditions as at high temperature and high density.

The intricate way the skyrmion notion pervades in nature is manifested the most beautifully in condensed matter systems where there is clear-cut evidence for topological excitations. In fact, it first surfaced, having nothing to do with QCD, in condensed matter physics at about the same time the 1994 volume appeared. The concept has become so familiar to the workers in the field that while the term 'skyrmion' is mentioned very frequently, Skyrme's original papers are rarely cited as one would notice in the contributions to this volume. It is not our aim here to give a broad overview of the development — for which we cannot claim to be sufficiently qualified — but to illustrate our principal thesis, namely, that skyrmions are *universal*. We focus on two most extensively studied low-dimensional strongly correlated condensed matter systems, namely, quantum Hall and high temperature superconductor. In these systems as well as in certain quantum critical phenomena, one of which is described in this volume, both skyrmions with integer charges and half-skyrmions (or merons) with half-integer charges that emerge as topological excitations in $(2+1)$ dimensions constitute *the* relevant physical degrees of freedom. In contrast to the current situation in strong-interaction systems in $(3+1)$ dimensions where fractionalized skyrmions also do appear, here both skyrmions and half-skyrmions are well exhibited and scrutinized both experimentally and theoretically.

The recent new development which makes up the last part of the volume is the re-emergence of skyrmions in string theory. To string theorists, this may be neither unexpected nor overly exciting: It brings string theory back to its initial objective of the 1960's when it was invented to address hadronic physics. However, for modern hadron physics in the QCD era, this development could signify a promising novel direction that will reveal surprises. The gravity/gauge holographic duality endows an extra dimension to hadron structure which makes the soliton for the baryon an instanton in (4+1) dimensions or a skyrmion in (3+1) dimensions in the *presence* of an infinite tower of hidden local gauge fields. The important aspect of this development is a possible new structure implied in nucleon as well as nuclear dynamics. While the original skyrmion was formed as a coherent state of pions, the instanton structure depicts the baryon as a coherent state of *both* pions *and* an infinite tower of vector mesons with hidden local symmetry. How the presence of this fifth dimension will influence nuclear dynamics in such extreme conditions as at high density and/or high temperature is an entirely open problem for the future.

What underlies the multifaceted nature of skyrmion(s) may very well be reflecting a deep principle in nature. As explained in the introductory section, Parts I and III are almost certainly connected by a string/gauge duality. The current development in understanding strongly-correlated phenomena in condensed matter systems also indicates the possible role of the string-gauge duality. It is therefore appealing to conjecture that all three parts are likewise intricately connected.

This volume was conceived when both of the authors were visiting the Korean Institute for Advanced Study in 2003. It was completed when one of us (Mannque Rho) was participating in Spring 2009 in the World Class University Program (R33-2008-000-10087-0) of the Korean Ministry of Education, Science and Technology at Hanyang University in Seoul, Korea. We are most grateful to all the contributors for their excellent exposés, reviews and essays and not least, for their generous help in our editing job.

*Gerald E. Brown*
*Mannque Rho*

# Contents


Preface     v

Introduction     xiii

**Hadrons and Nuclear Matter**     **1**

1. Skyrmions and Nuclei     3

   *R.A. Battye, N.S. Manton and P.M. Sutcliffe*

2. Electromagnetic Form Factors of the Nucleon in Chiral Soliton Models     41

   *G. Holzworth*

3. Exotic Baryon Resonances in the Skyrme Model     57

   *D. Diakonov and V. Petrov*

4. Heavy-Quark Skyrmions     91

   *N.N. Scoccola*

5. Skyrmion Approach to Finite Density and Temperature     115

   *B.-Y. Park and V. Vento*

6. Half-Skyrmion Hadronic Matter at High Density     147

   *H.K. Lee and M. Rho*

7.   Superqualitons: Baryons in Dense QCD      165
*D.K. Hong*

8.   Rotational Symmetry Breaking in Baby Skyrme Models      179
*M. Karliner and I. Hen*

**Condensed Matter**      **215**

9.   Spin and Isospin: Exotic Order in Quantum Hall Ferromagnets      217
*S.M. Girvin*

10.   Noncommutative Skyrmions in Quantum Hall Systems      233
*Z.F. Ezawa and G. Tsitsishvili*

11.   Skyrmions and Merons in Bilayer Quantum Hall System      269
*K. Moon*

12.   Spin and Pseudospin Textures in Quantum Hall Systems      291
*H.A. Fertig and L. Brey*

13.   Half-Skyrmion Theory for High-Temperature Superconductivity      311
*T. Morinari*

14.   Deconfined Quantum Critical Points      333
*T. Senthil, A. Vishwanath, L. Balents, S. Sachdev and M.P.A. Fisher*

**String Theory**      **345**

15.   Skyrmion and String Theory      347
*S. Sugimoto*

16.   Holographic Baryons      367
*P. Yi*

17.   The Cheshire Cat Principle from Holography      393
*H.B. Nielsen and I. Zahed*

18.   Baryon Physics in a Five-Dimensional Model of Hadrons          403
      *A. Pomarol and A. Wulzer*

Author Index          435

Subject Index          437

# Introduction

What is the most remarkable and significant of Skyrme's bold idea is that it figures ubiquitously in practically *all* branches of physics. This volume illustrates, with a few selected articles, how pervasive this idea is in a large variety of physical phenomena in particle/nuclear physics (Part I), condensed matter physics (Part II) and string theory (Part III). The topics included in this volume cover the more recent developments, leaving out those that can be found in the available reviews and books. In this introductory section, we give our personal assessment of how the basic idea figures in, and connects, these three seemingly different disciplines. The order in which the contributions are presented reflects this objective.

To begin with, we clarify where in the present state of strong interaction physics the original idea of Skyrme stands, and then proceed to treat the matter presented in the volume roughly in the order of evolution from the original formulation both in concepts and in practical applications.

It is widely accepted that at very low energies (or momenta) $E \ll \Lambda$ where $\Lambda$ is a scale set by QCD, strong interactions are accurately captured by current algebras in terms of low-energy theorems involving Nambu–Goldstone (or Goldstone for short) bosons, namely the pions (in the chiral limit where the quark masses are ignored or rather pseudo-Goldstone bosons with the light quark masses are taken into account), of chiral symmetry in the Goldstone mode. The effective theory as $E \to 0$ is then encapsulated in the chiral Lagrangian

$$\mathcal{L} = \frac{f^2}{4}\mathrm{Tr}(\partial^\mu U \partial_\mu U^\dagger) + \cdots \tag{1}$$

where $U$ represents the chiral field $U = exp(2i\pi/f)$ with $\pi$ the Goldstone boson ('pion') field (triplet for two flavors and octet for three flavors), $f$ is a mass-dimension-1 constant related to the pion decay constant and the ellipsis represents terms that become difficult to ignore as one departs from zero energy, which we will denote in what follows generically as $\mathcal{L}_{ho}$. The Lagrangian (1) encodes theorems that are rigorously valid at very low energy. Now given (1) at near zero energy, how does one go up in energy scale and probe physics up to near the scale $\Lambda$? This is the question that currently preoccupies many hadron/nuclear physicists.

At present, there are broadly two approaches to tackle the above task, both involving, in the absence of tractable QCD techniques in the nonperturbative regime, effective field theories in the spirit of a f(olk)-theorem largely attributed to

S. Weinberg, which amounts to saying that "the most general theory one can write down is the one with the most general possible Lagrangian consistent with the principles and symmetries of the theory," here QCD.

One modern strategy is to exploit hidden gauge symmetries and bring in massive gauge particles by having them 'emerge' from low-energy theories. The idea is to exploit a redundancy present in the chiral field $U(x)$. Valued in the algebra of the spontaneously broken chiral symmetry $SU(N_f)_L \times SU(N_f)_R \rightarrow SU(N_f)_{V=L+R}$, the chiral field $U$ can be written as a product $\xi_L^\dagger(x)\xi_R(x)$ with $\xi_{L,R} \in SU(N_f)_{L,R}$ transforming $\xi_{L,R} \rightarrow h(x)\xi_{L,R}g_{L,R}$ under rigid chiral rotation $g_{L,R} \in SU(N_f)_{L,R}$ and local hidden local transformation $h(x) \in SU(N_f)_V$. Here $N_f$ is the number of flavors which is typically three including the strangeness in nuclear physics. This redundancy, intrinsic in the way the chiral field is written, can be elevated to a gauge symmetry with the set of $SU(N_f)_V$ gauge fields identified with the low-lying vector and/or axial-vector mesons seen in nature. This procedure can be suitably utilized to elevate the energy scale to the mass of the vector mesons, $\lesssim 1$ GeV, and allows us to write an effective Lagrangian that accounts for, via hidden local vector fields, the terms represented by the ellipsis of Eq. (1), $\mathcal{L}_{ho}$. One can make this procedure consistent with QCD by suitably matching the correlators of the effective theory to those of QCD at a scale near $\Lambda$. Clearly this procedure is not limited to only one set of vector mesons; in fact, one can readily generalize it to an infinite number of hidden gauge fields in an effective Lagrangian. In so doing, it turns out that a fifth dimension is 'deconstructed' in a (4+1)-dimensional (or 5D) Yang–Mills type form. We will see in Part III that such a structure arises, top-down, in string theory.

An alternative but more microscopic approach, perhaps in a closer contact with QCD, to elevate the energy scale is to introduce explicit quark–gluon fields suitably coupled to the nonperturbative sector involving the Goldstone bosons (pions). How this can be done in a systematic way can be explained in terms of what is known as 'chiral quark model' (Chapter 3). There the pion mean-field — and also vector mean fields if incorporated — provides the background for nonperturbative properties of quarks. In this description, the skyrmion can be considered as the mean chiral field that *binds* the quarks. Now when the quarks are deeply bound by the strong mean field, the baryon charge winds up entirely in the soliton, and the system becomes the pure skyrmion baryon. The chiral quark soliton model plays the role of interpolating between the (constituent) quark description and the soliton description. How the two pictures are manifested in nature depends on the condition in which the system is probed and on which meson fields participate in the mean field for the process. A simpler but equivalent picture is given in terms of what's called the 'chiral bag' which was touched on by the editor in the 1994 volume. There quarks and gluons, weakly interacting in accordance with asymptotic freedom, are confined in a 'bag' of radius $R$ coupled to pions and other meson fields at the boundary with their mean fields absorbing the fractionized baryon charge, thereby conserving the total baryon charge exactly and other static properties, albeit approximately. There the bag radius is a gauge degree of freedom and plays no physical role. This means that one ultimately winds up with an effective Lagrangian of the type (1). Remarkably, this picture, dubbed as 'Cheshire Cat Mechanism,' is found to re-emerge in holographic dual QCD from string theory in the last part of the volume.

Skyrme's idea was that given an effective Lagrangian built entirely in meson fields as in (1), baryons — as fermions — could emerge from this Lagrangian as solitons. This was a totally original and unconventional concept that was largely unappreciated in the early years of the 1960's and had remained so until the idea was resurrected in the 1980's. Since the soliton would be unstable with only the first term of (1), namely the current algebra term, a stabilizing term — subsumed in the ellipsis — was needed, and Skyrme took for it the simplest possible form, i.e., a quartic term of the form $\sim \mathrm{Tr}[U^\dagger \partial_\mu U, U^\dagger \partial_\nu U]^2$. This term is known in the literature as the 'Skyrme term.' The soliton, so constructed with the Skyrme term for $\mathcal{L}_{ho}$, is referred to as the 'Skyrme model.'[1] In the modern development described in Part III, such a quartic term will be seen to play no significant role in the presence of both the tower of vector mesons and chiral anomalies.

An important aspect of the skyrmion picture, generalized from the original Skyrme model as understood now, is that it is a description of baryons in the limit that the number of colors $N_c$ — which is three in nature — is taken to be very large. In that limit, it is shown to be equivalent to the non-relativistic quark model, with the baryon mass scaling as $\mathcal{O}(N_c)$. Leading corrections, via moduli-space quantization, to the large $N_c$ limit give appropriate quantum numbers to the solitons allowing them to be identified as physical baryons. Accounting for systematic higher order $1/N_c$ corrections is a difficult problem and still remains to be worked out. Nonetheless what comes out, when computed to the manageable order, is surprisingly good. Even with the simplest Skyrme model, not only static properties of the nucleon but also the structure of finite nuclei can be described well.

The Skyrme model and its generalized version with low-lying vector fields are applied not only to systems with nucleons, finite nuclei and dense nuclear matter, but also to exotic baryons. In Chapter 1, the skyrmion in its simplest form, i.e., the Skyrme model, is shown to be capable of describing fairly well both the 'elementary' nucleons and finite nuclei with mass number up to $\sim 22$, with predictions of certain ground state properties that have not been revealed by the standard many-body approaches developed in nuclear theory. Although quantization has not yet been fully implemented in the model, and hence a detailed quantitative comparison with experiments is not feasible, it promises an exciting novel domain of nuclear structure physics to be explored. With a minimal implementation of vector meson degrees of freedom, the model can fairly accurately reproduce nucleon electromag-

---

[1]Unless otherwise specified, we will understand by 'skyrmion' in hadron/nuclear physics both the Skyrme model and generalized models that include both pions and vector mesons (either the lowest members or the infinite tower).

netic form factors up to large momentum transfers entering into the regime where asymptotic freedom is operative. This is discussed in some detail in Chapter 2. This result illustrates clearly the need for heavier degrees of freedom than pions in the nucleon structure, presaging *the* vector dominance involving an infinite tower of vector mesons discussed in Part III.

So far we have dealt with two light quark — up and down — flavors figuring crucially in nucleon structure. Heavier flavors do also give rise to skyrmions. In Chapters 3 and 4 'exotic' baryons and heavy-quark baryons are described, respectively, in terms of (generalized) skyrmions. Chapter 3 details how the controversial pentaquark was predicted in the Skyrme model and why it could have thus far escaped clear-cut experimental detection. Whether or not this prediction is viable is a highly disputed issue and will ultimately be settled by further experiments, but the merit of the approach adopted in Chapter 3 is that it indicates — in terms of chiral quark structure or hidden gauge fields — the limited validity of the skyrmion model with pion field only and how to improve on it. In Chapter 4, we discussed how one can reliably describe baryons that contain both heavy and light quarks by combining heavy-quark symmetry and light-quark chiral symmetry. Surprisingly, the strange quark can be approached from either the heavy-quark limit or the light-quark limit, accounting for its ambidextrous property. We should note that both the heavy quark symmetry and the solitonic baryon are anchored on 'heaviness,' so skeptics could argue that what one is doing for the hyperon when the 'heavy' kaon is bound to a skyrmion (e.g., the Callan–Klebanov bound-state model) is like "a tail wagging a dog."

Systematic analytic application of the skyrmion model to heavy nuclei and nuclear matter has proven to be difficult. The only analytic treatment available in the literature was the mean-field type prediction for in-medium scaling of light quark hadron masses (and coupling parameters) in temperature and/or density, known as 'Brown–Rho scaling.' There are, however, numerical simulations with skyrmions put on crystal lattice. Chapter 5 addresses dense nuclear matter in terms of skyrmions constructed with the pions together and with the lowest vector mesons — $\rho$ and $\omega$ — put on an FCC lattice. It is predicted by symmetry that when skyrmions on crystal lattice are squeezed to high density, a half-skyrmion matter should be energetically favored over the full skyrmion state. The half-skyrmion state is characterized by a vanishing chiral order parameter, that is, the quark condensate $\langle \bar{q}q \rangle = 0$, which would formally imply that chiral symmetry is restored. However, what differentiates this state from the standard chiral restoration is that the pion decay constant $f \equiv f_\pi$ can in general be nonzero. This phase — which is predicted to occur at a density near chiral restoration — is interpreted in Chapter 5 to be an analog to the pseudogap phase conjectured to be associated with high-temperature superconductivity discussed in Part II. Phrasing in terms of hidden local symmetry (HLS) theory, Chapter 6 identifies this half-skyrmion phase as an emergent 'vector symmetry' first discussed by H. Georgi in the large $N_c$ limit.

We note that this identification presages the chiral-symmetry-restored but confined 'quarkyonic phase' conjectured in large $N_c$ QCD. The half-skyrmions in HLS theory can be viewed as fractionized components in the chiral field, $U = \xi_L^\dagger \xi_R$, which has hidden gauge invariance referred to above (this will have an analogy in condensed matter where an abelian gauge degree of freedom emerges and figures in splitting a skyrmion into two half-skyrmions). This phase is argued in Chapter 6 to be relevant for describing compact stars, i.e., neutron stars and black holes, including the Brown–Bethe maximum neutron star mass.

Perhaps less appreciated but equally remarkable is that the skyrmion description can also be applied to color-flavor locked superconducting dense baryonic matter, providing a baryonic version of 'quark-hadron continuity' at high density. This is described in Chapter 7. When color in $SU(3)_c$ and flavor in $SU(3)_f$ are locked at high density, the symmetry $SU(3)_c \times SU(3)_L \times SU(3)_R$ is broken spontaneously by diquark condensate to $SU(3)_{c+L+R}$, the dynamics of which can be written in terms of octet Goldstone pseudo-scalar fields and octet vector fields in a form identical to the HLS Lagrangian encountered at low density. Here the vector fields arise Higgs-ed from the gluon fields; hence they are *not hidden, but explicit* gauge fields. In Chapter 7, it is seen how octet baryons can arise from this mesonic Lagrangian as skyrmions, called 'superqualitons,' which can be mapped one-to-one to the low-density baryons. The Fermi sea formed with superqualitons in dense matter could be identified as a Q-ball matter. Note, however, that the color-flavor locking must take place — if at all — at superhigh density, so it may not be physically relevant even for compact stars. It nonetheless is an interesting theoretical object that exemplifies the pervasive nature of the skyrmion structure.

There are strong compelling indications that heavier meson fields, in particular, in an infinite tower of vector mesons, could play a crucially important role, not only for elementary baryons but also in many-baryon systems and dense matter discussed in Chapter 5. This is not unexpected. Even to the leading order in $N_c$, there are an infinite number of terms in the ellipsis in (1). Since the solitonic baryon is built as a coherent superposition of mean fields, the construction of effective field theories at increasing energy scales must therefore involve all relevant fields in the tower. Their important role is clearly seen phenomenologically already in nucleon electromagnetic form factors (Chapter 2). However, at present, there is no systematic study on this issue from the point of view of effective field theories. The reason is simply that unguided by first principle theory or by experiments, there are too many undetermined parameters as the number of terms increases. In this connection, the recent holographic dual QCD could prove to be an invaluable guide. While the conventional treatment of the skyrmion involves four dimensions, holographic dual descriptions involve one extra dimension that represents spread in energy scale. This brings in new features that are discussed in Part III.

Before we go to the (4+1)-dimensional (or 5D for short) case that arises in string theory, we describe in Part II a few (2+1)-dimensional (or 3D) systems met in condensed matter physics. It is in condensed matter that the notion of skyrmion turns out to be the most successful in confronting nature, manifesting itself conspicuously in various experimental observables. It should be stressed that here skyrmion emerges in a setting totally unrelated to QCD. On a crystal lattice where many-skyrmion systems are simulated, one observes a close analogy between the 3D and (3+1)-dimensional (or 4D) systems. In 3D, the soliton, called 'baby skyrmion,' involves spin density — which is the analog to the isospin density in hadronic skyrmions in 4D. The skyrmion here is a coherent excitation of spins instead of isospins as in the case of baryons. The target manifold for the baby skyrmion is a unit three-dimensional vector field $\hat{n}$ which has an analogous topological structure as the chiral soliton while involving one dimension less. In contrast to the 4D object which carries no electric charge (both the proton which is charged and the neutron which is uncharged are skyrmions), the 3D soliton is electrically charged, quantized proportionally to the topological charge. The common element in the 4D and 3D systems is the leading term of a non-linear sigma model, i.e., the current algebra term. In addition, in 3D systems, a potential term is typically required for the soliton stability in contrast to the 4D case where there is no need for potentials (as far as stability is concerned) once there is the Skyrme term. All that is required of the potential is, however, that it vanishes at infinity for a given vacuum field, but otherwise it is arbitrary. This arbitrariness gives rise to a rich variety of baby-skyrmion models realized and observed in nature. This feature is discussed in detail in Chapter 8 where baby skyrmions are studied in flat as well as in curved spaces and also on crystal lattice, with focus on rotational symmetry breaking. A close parallel made in this chapter between 4D and 3D skyrmions, in particular on multi-skyrmion structure, provides a valuable and as yet unexplored bridge between the physics of Part I (hadronic matter) and that of Part II (condensed matter).

Both skyrmions and half-skyrmions, the latter also known as 'merons,' figure in a wide variety of different condensed matter systems. In this volume, to illustrate our principal theme — i.e., the multifaceted nature of skyrmions — we have picked, among others, a few selective articles on ferromagnetic quantum Hall, high T superconductivity and deconfined quantum critical phenomena. Other related matters such as fractionalization of quantum dots into merons are left out.

To give a general overview of what's happening in quantum Hall ferromagnets, we reproduced in Chapter 9 a review article from *Physics Today* whereby a language accessible to non-experts was employed. This article beautifully illustrates, with the help of several specific experiments such as NMR and various optical and transport measurements, how the topological description works. In Chapters 10, 11 and 12, this subject matter is taken up in detail and at different levels of rigor, i.e., both microscopically and phenomenologically, by the leading workers in the field. Particularly notable are the roles of the pseudospin degree of freedom in the bilayer quantum Hall structure involving 'pseudospin skyrmions' and half-skyrmions (i.e., merons with half electron charge) that constitute the bona fide quasiparticle degrees of freedom. In a simple term, one can say that a skyrmion is just a deformed bound two meron excitation. A number of remarkable features observed in experiments can also be understood in terms of merons made unbound by disorder.

How half-skyrmions could also figure in high-temperature superconductivity is discussed in Chapter 13. The rich phase diagram of high-T superconductors is believed to be controlled by one parameter, i.e., the doped hole contribution. In Chapter 13, we described how the doped hole can carry half of the topological charge, the half-skyrmion number. There a simple example is given in terms of a single hole embedded in an antiferromagnetic long-range ordered state. In the $CP^1$ representation with the spin vector field $\mathbf{n} = z^\dagger \sigma z$, the doped hole is argued to carry a topological charge — which is 1/2 of the skyrmion — represented by a gauge flux of the hidden gauge symmetry of the $CP^1$ representation. (Note the parallel between this argument and the analogous argument made in Chapter 6 for the hadronic 1/2-skyrmion where nonabelian hidden gauge fields figured.) How a pseudogap structure and d-wave superconductivity can arise is discussed in this chapter.

Perhaps intricately related to quantum Hall and high-T superconductivity phenomena is the role played by skyrmions and half-skyrmions in the Néel magnet-VBS (valence bond solid) paramagnet transition (and related transitions) described in Chapter 14. There the skyrmion texture present in the Néel magnet splits into two half-skyrmions at the phase transition, with the magnetic monopole of the $U(1)$ gauge field 'emerging' in the $CP^1$ representation and a Berry phase associated with the lattice structure playing key roles. The important point to note here is that the relevant degrees of freedom for the quantum critical phenomenon in between two phases are half-skyrmions. The phase transition involving the deconfinement of a single skyrmion into two unbound half-skyrmions, possessing no common order parameters, is said to belong to a class outside of the Ginzburg–Landau–Wilson paradigm.

While the problems treated in Parts I and II have a rather long history, the emergence of the skyrmion structure in the holographic description of baryons in string theory is quite recent and hence much less explored. We noted above that when one goes up in energy scale from the current algebra scale, there emerges generically, bottom-up, a 'deconstructed' fifth dimension that accounts for the multitude of scales involved. In terms of hidden gauge structure, the pertinent low-energy dynamics can be captured by a 5D Yang–Mills (YM) action — with the scale defined by a cut-off mass $\tilde{M}$ — of the form,

$$S_{YM} = - \int dx^4 dw \, \frac{1}{2e^2(w)} \, \mathrm{Tr} F_{AB} F^{AB} + \cdots , \qquad (2)$$

with $(A, B) = 0, 1, 2, 3, w$ where $w$ is the fifth dimension. Here $e(w)$ is a $w$-dependent effective constant that reflects the curved space encoding the complex background and the ellipsis stands for higher derivative terms and possible fields other than YM.

The action (2) must be supplemented by the Chern–Simons term that accounts for quantum anomalies associated with chiral symmetry.

An interesting modern development is that the 5D action of the form (2) naturally arises 'top-down' in certain limits from string theory. In a model constructed by Sakai and Sugimoto which correctly implements chiral symmetry of QCD in the chiral limit (called Sakai–Sugimoto — SS for short — model),[2] certain properties of hadrons can be addressed simply in the large $N_c$ limit, $N_c \to \infty$ and large 't Hooft limit, $\lambda \equiv g_{YM}^2 N_c \to \infty$, with only one additional parameter $M_{KK} \sim \tilde{M}$. When viewed in 4D, the 5D action comprises an infinite tower of vector and axial-vector mesons. In Chapters 15, 16 and 17, we discussed how baryons arise as instantons in the four-dimensional $(\mathbf{x}, w)$ space in the SS model. With $N_c = 3$, the physical pion decay constant $f_\pi = 93$ MeV and the parameter $M_{KK}$ fixed by the $\rho$-meson mass, $m_\rho = 770$ MeV, the model comes out to describe — unexpectedly well — low-energy properties of both mesons and baryons, in particular those properties reliably described in quenched lattice QCD simulations. In Chapter 15, the soliton is quantized in the same way as in the standard skyrmion (collective coordinate or moduli space) quantization employed in Part I, whereas in Chapter 16, an effective field theory involving *explicit* baryon fields in addition to the pion field and the infinite tower of vector mesons is formulated. The two approaches, presumably equivalent in the sense of the f-theorem mentioned above, give essentially the same results. The latter can be viewed as a holographic analog to heavy-baryon chiral perturbation theory in the large $N_c$ limit. In applying it to many-instanton systems, the former would then correspond to what's done with skyrmions in Chapters 5 and 6, while the latter would lend itself to a Walecka-type mean field theory familiar in nuclear physics, with, however, the infinite tower of vector meson fields — and not just the lowest — intervening in 4D. Application to many-body systems has, however, not yet been performed.

One of the most noticeable results of this holographic model is the *first derivation* of vector dominance (VD) that holds *both* for mesons and for baryons. It has been somewhat of an oddity and a puzzle that Sakurai's vector dominance — with the lowest vector mesons $\rho$ and $\omega$ — which held very well for pionic form factors at low momentum transfers famously failed for nucleon form factors. In this holographic model, the VD comes out automatically for both the pion and the nucleon provided that the infinite tower is included. While the VD for the pion with the infinite tower is not surprising given the successful Sakurai VD, that the VD holds also for the nucleons is highly nontrivial. In the large $\lambda$ limit in which the model is justified, the soliton — instanton — is point-like, but with $1/\lambda$ corrections added, it should develop a non-negligible size. Indeed, in the usual skyrmion picture described in Part I, the intrinsic skyrmion size accounts largely for the physical hadronic size as

---

[2]There are other holographic constructions for strong interaction dynamics, but at present, the SS construction is the only one that has the chiral symmetry property possessed by the QCD proper that we are interested in.

seen by EM probes even in the presence of vector mesons (see Chapter 2): In the Skyrme model, the size is in fact entirely given by the skyrmion size. The complete VD in the nucleon form factor means that the instanton size does not figure in the physical baryon size. How this comes about is explained in Chapter 17. It turns out to be a consequence of a holographic Cheshire Cat phenomenon, namely that the instanton size is not physical and can be 'gauged away' as was the case with the bag radius in Part I.

An alternative bottom-up approach to holographic dual model for baryons is described in Chapter 18. The effective 5D model treated in this chapter is the action of the form (1) with the effect of the energy scale in the $w$ coordinate encoded in a compact warp factor. Instead of descending from string theory in the specified limits, here the 5D action is interpreted à la AdS/CFT holographic correspondence in terms of a 4D QCD-like theory on the boundary with relevant symmetries. In the large $N_c$ limit, there are again three parameters, two (in the chiral limit) holographically related to $f_\pi$ and $m_\rho$ and the third, the cutoff which is fixed for given $N_c$, $\tilde{M} \sim 2$ GeV. This approach enjoys more flexibility than the top-down approach, so it could be made more versatile phenomenologically, though perhaps somewhat *ad hoc*. The results discussed in Chapter 18 differ in certain aspects from the SS model results, with the Cheshire Cat property missing therein as the 't Hooft constant $\lambda$ plays no visible role there. Otherwise the results are broadly similar including the vector dominance, with the agreement with experiments being in the same ball park. What this is indicating is that independent of how it is arrived at, top-down or bottom-up, the 5D structure (2) is a generic feature in strong interaction physics. Application to nuclear and dense matter within this approach again remains to be made.

In closing this introduction, we should mention that an extensive mathematical development existing in the literature on the skyrmion model and its variants in various dimensions has been left out in this volume. This is because our focus was principally on the phenomenological side of the development. With the advent of the instanton picture in the holographic approaches, however, such omission may no longer be warranted. As in gauge theories where mathematics and physics have invaluably helped each other, mathematics may also become more influential and conducive to breakthroughs in skyrmion physics by closely connecting various different branches of physics.

# PART 1
# Hadrons and Nuclear Matter

# Chapter 1

# Skyrmions and Nuclei

R.A. Battye[*], N.S. Manton[†] and P.M. Sutcliffe[‡]

*Jodrell Bank Centre for Astrophysics, University of Manchester,
Manchester M13 9PL, UK*

†*Department of Applied Mathematics and Theoretical Physics,
University of Cambridge, Wilberforce Road, Cambridge, CB3 0WA, UK*

‡*Department of Mathematical Sciences, Durham University,
Durham DH1 3LE, UK*

We review recent work on the modelling of atomic nuclei as quantized Skyrmions, using Skyrme's original model with pion fields only. Skyrmions are topological soliton solutions whose conserved topological charge $B$ is identified with the baryon number of a nucleus. Apart from an energy and length scale, the Skyrme model has just one dimensionless parameter $m$ proportional to the pion mass. It has been found that a good fit to experimental nuclear data requires $m$ to be of order 1. The Skyrmions for $B$ up to 7 have been known for some time, and are qualitatively insensitive to whether $m$ is zero or of order 1. However, for baryon numbers $B = 8$ and above, the Skyrmions have quite a compact structure for $m$ of order 1, rather than the hollow polyhedral structure found when $m = 0$. One finds that for baryon numbers which are multiples of four, the Skyrmions are composed of $B = 4$ sub-units, as in the $\alpha$-particle model of nuclei.

The rational map ansatz gives a useful approximation to the Skyrmion solutions for all baryon numbers when $m = 0$. For $m$ of order 1, it gives a good approximation for baryon numbers up to 7, and generalisations of this ansatz are helpful for higher baryon numbers.

We briefly review the work from the 1980s and 90s on the semiclassical rigid-body quantization of Skyrmions for $B = 1$, 2, 3 and 4. We then discuss more recent work extending this method to $B = 6$, 7, 8, 10 and 12. We determine the quantum states of the Skyrmions, finding their spins, isospins and parities, and compare with the experimental data on the ground and excited states of nuclei up to mass number 12.

---

[*]Richard.Battye@manchester.ac.uk
[†]N.S.Manton@damtp.cam.ac.uk
[‡]P.M.Sutcliffe@durham.ac.uk

## Contents


1.1  Introduction . . . . . . . . . . . . . . . . . . . . . . . . . . . . . . . . .  4
1.2  Skyrmions . . . . . . . . . . . . . . . . . . . . . . . . . . . . . . . . . . .  8
1.3  The Rational Map Ansatz . . . . . . . . . . . . . . . . . . . . . . . . . .  11
1.4  Skyrmions and $\alpha$-Particles . . . . . . . . . . . . . . . . . . . . . .  15
    1.4.1  $B = 4$ . . . . . . . . . . . . . . . . . . . . . . . . . . . . . . . . . .  15
    1.4.2  $B = 8$ . . . . . . . . . . . . . . . . . . . . . . . . . . . . . . . . . .  15
    1.4.3  $B = 10$ . . . . . . . . . . . . . . . . . . . . . . . . . . . . . . . . .  17
    1.4.4  $B = 12$ . . . . . . . . . . . . . . . . . . . . . . . . . . . . . . . . .  18
    1.4.5  $B = 16$ . . . . . . . . . . . . . . . . . . . . . . . . . . . . . . . . .  19
    1.4.6  $B = 32$ . . . . . . . . . . . . . . . . . . . . . . . . . . . . . . . . .  21
1.5  Quantization . . . . . . . . . . . . . . . . . . . . . . . . . . . . . . . . .  22
    1.5.1  $B = 4$ . . . . . . . . . . . . . . . . . . . . . . . . . . . . . . . . . .  25
    1.5.2  $B = 6$ . . . . . . . . . . . . . . . . . . . . . . . . . . . . . . . . . .  25
    1.5.3  $B = 8$ . . . . . . . . . . . . . . . . . . . . . . . . . . . . . . . . . .  26
    1.5.4  $B = 10$ . . . . . . . . . . . . . . . . . . . . . . . . . . . . . . . . .  27
    1.5.5  $B = 12$ . . . . . . . . . . . . . . . . . . . . . . . . . . . . . . . . .  30
1.6  Calibration and Energy Levels . . . . . . . . . . . . . . . . . . . . . . . .  32
1.7  Conclusion . . . . . . . . . . . . . . . . . . . . . . . . . . . . . . . . . .  36
References . . . . . . . . . . . . . . . . . . . . . . . . . . . . . . . . . . . . .  37


## 1.1. Introduction

The Skyrme model is a field theoretic description of nucleons and nuclei.[1,2] It is intermediate between the traditional models with point nucleons interacting through a potential, and a complete description based on quarks and gluons, as should emerge from Quantum Chromodynamics (QCD). The model captures the key feature of low-energy QCD with light up and down quarks, namely that of a broken chiral symmetry with light almost-Goldstone bosons. These bosons are the three pions. The unbroken internal symmetry is isospin symmetry.

The simplest and original Skyrme model, which is all we shall discuss, has an $SU(2)$-valued field $U$, constructed nonlinearly from the three pion fields, and the dynamics is determined by a Lagrangian with three terms — a kinetic term quadratic in field derivatives, a Skyrme term quartic in derivatives, and an explicit pion mass term, which is a field potential energy term. No dynamical electromagnetic effects are built in to the simplest model, as these appear to be unimportant for nuclear structure until one reaches nuclei beyond $^{40}$Ca, larger than anything we shall discuss. There are just three parameters, two of which set the mass and length scale of nuclear physics (the proton mass and proton size). There is one remaining dimensionless parameter, proportional to the pion mass. It is an attractive aspect of the Skyrme model that it has essentially no adjustable parameters, but a consequence is that its predictions are not as refined as those of other models.

The basic perturbative physics of the Skyrme field theory is that of interacting pions, but in addition, there are non-perturbative topological soliton solutions. The solitons have a conserved integer charge $B$, identified with baryon number. (In conventional nuclear physics, this is the mass number, or atomic number, and denoted $A$.) The classical solitons of minimal energy for each baryon number are called Skyrmions. They are static, but they can also acquire kinetic energy and be in translational or rotational motion. The field equation is not integrable, and no Skyrmion solution is known in closed form. The Skyrmions are determined following a substantial numerical search. They are found to have an interesting geometrical and physical structure which is now quite well understood, and is used to guide the search for the numerical solutions. Each solution has a smooth topological charge density and energy density localized in a region of physical size comparable with that of a nucleus. Very few Skyrmions, in fact only those with $B = 1$ and $B = 2$, have any continuous rotational symmetry, but almost all of them have some discrete symmetry, either a symmetry of one of the platonic solids, or a smaller cyclic or dihedral symmetry. Several Skyrmions are illustrated below. Since three pion fields are involved, we show a selected energy or baryon density contour (isosurface), sometimes with a colour scheme which indicates where each pion field is large.

The Skyrmion solutions approach the vacuum at infinity through a linearized pion tail. From the tail structure, one can calculate (most easily in the massless pion case) the interactions between two well-separated Skyrmions. These forces depend on the relative orientations of the Skyrmions in both space and isospace, and in almost all cases one can show that for some suitable orientations the Skyrmions attract, and hence if one minimizes the energy, the Skyrmions should merge, forming a new Skyrmion whose baryon number is the sum of the baryon numbers of the initial, separated Skyrmions. This physical argument suggests an approach to proving that Skyrmions of any non-zero baryon number rigorously exist, but so far such a proof has been elusive.[3,4] Numerical evidence shows without doubt that Skyrmions do exist for a large range of baryon numbers, and for a range of pion masses, and that they are all smooth. Mathematical proof that Skyrmions are smooth is also elusive.

Since Skyrmion matter is rather incompressible, the volume of the core region where the energy density of a Skyrmion is significantly different from zero tends to increase linearly with the baryon number. Consequently, in Skyrmions of higher baryon number, the lower baryon number constituents only partially merge, and some of the structure of the constituents remains visible. Nevertheless — and this is important — one cannot identify within a Skyrmion of baryon number $B$ a set of $B$ points that are centres of $B = 1$ Skyrmions. Because of this, it is almost impossible to compare step-by-step the Skyrme model and point nucleon models. They have different degrees of freedom. For example, the kinetic energy of nucleons in nuclei is significant, so nucleon spatial correlations are rather weak, and the intrinsic spatial arrangement of nucleons within a nucleus rather meaningless. The corresponding field kinetic energy in the Skyrme model is not really related to this, and the intrinsic shape of a Skyrmion is vital.

Classical Skyrmion solutions are not nuclei, since they have no spin or isospin quantum numbers. To obtain quantum states of a nucleus in the Skyrme model, one should in principle quantize the field fluctuations around a Skyrmion of the required baryon number. This is in practice too difficult, so we follow the lead of Skyrme, of Adkins, Nappi and Witten,[5] and of Braaten and Carson,[6] and quantize just the zero modes or collective coordinates of each Skyrmion. This means that we regard each Skyrmion as a rigid body that can translate and rotate in both space and isospace. The translational motion is rather trivial, so we concentrate on the six rotational and isorotational degrees of freedom. For baryon numbers 1 and 2, there are fewer degrees of freedom (respectively three and five), because of the continuous symmetries. Quantum states are tensor products of rigid-body states in space and isospace, or linear combinations of these. To determine the energy levels of the ground and excited states, one needs to know the $6 \times 6$ inertia tensor of the Skyrmion. The inertia tensors have been known for Skyrmions of small baryon number for some time, but have been accurately computed for most baryon numbers up to $B = 12$ only recently.[7]

The spectrum of a quantized Skyrmion is strongly constrained by its discrete symmetries. Schematically, if there is a $C_n$ cyclic symmetry around some axis, then a $2\pi/n$ rotation about the axis maps the Skyrmion into itself, and one can expect that only states which are invariant under this rotation are allowed. Therefore, only states with angular momentum component $0 \bmod n$ about this axis are allowed. In practice, this kind of constraint usually acts on combined spin and isospin states. A further complication is the fact that a $2\pi$ rotation in space or in isospace is not always represented by 1 on the quantum states. For Skyrmions of odd baryon number it is represented by $-1$. This ensures that a quantized Skyrmion of odd baryon number has half integer spin and isospin. The detailed quantization rules, related to the topology of the Skyrme field configuration space, were elucidated by Finkelstein and Rubinstein,[8] and we explain in detail how to impose the correct constraints below. A consequence is that every symmetry of a Skyrmion is represented by either $\pm 1$ on states. Recently, Krusch found a very useful formula for determining these Finkelstein–Rubinstein signs.[9] One finds, perhaps surprisingly, that for Skyrmions with even baryon number, some symmetry operations are represented by $-1$. A consequence is that the ground state of such a Skyrmion may not be allowed to have spin zero and isospin zero, which is fortunate, since, for example, the isospin zero nucleus $^6$Li has spin 1 in its ground state.

The spectrum that emerges from this quantization is rather different from what appears to be discussed in most of the experimental and theoretical nuclear physics literature, since there is a complete unification of spin and isospin excitations. This is worth some comment.

It appears to be no longer controversial to classify at least some states of small nuclei into rotational bands. For this to work, one needs a model of a nucleus with a non-spherical intrinsic shape. (For a review, see Ref. 10) The modern shell model description of a nucleus like $^8$Be seems to require a non-spherical potential well, or mean field. We have not looked into this, but we have looked more closely at cluster models of nuclei, which appear in some respects closer to the Skyrmion point of view. In particular, our recent work on Skyrmions has been influenced by the $\alpha$-particle model, which is used to model nuclei that have equal numbers of protons and neutrons, and baryon number a multiple of 4. Here, $^8$Be and $^{12}$C are viewed as molecules of $\alpha$-particles, a dimer in the first case, and with the shape of an equilateral triangle in the second. Strong evidence for these cluster models comes from the binding energy data, and from their consistency with the clear rotational bands observed among the low-lying nuclear states. The triangular symmetry, for example, implies a rotational band of $^{12}$C states of spin/parity $0^+$, $2^+$, $3^-$, $4^-$ and $4^+$, with characteristic energy spacings.

Many Skyrmion solutions are consistent with these intrinsic structures, and the Skyrme model (to the extent one believes it) gives a deeper understanding of them. First of all, the forces leading to these intrinsic structures need not be postulated, but are a consequence of the Skyrme field equation. Second, the $\alpha$-particle is no longer modelled as a structureless point, but is instead a $B = 4$ Skyrmion substructure (slightly deformed). Indeed, a given Skyrmion might be seen as made up of substructures in more than one way, and not necessarily all of baryon number 4, although these are energetically favoured. Consequently, the interpretation of Skyrmions as bound clusters of smaller Skyrmions applies to baryon numbers that need not be multiples of 4. For example, one can recognise $\alpha$-particle and nucleon substructures in the $B = 10$ Skyrmion.

Finally, and we think this is the most important difference from the traditional $\alpha$-particle models, Skyrmion quantization gives a spectrum of isospin excitations together with spin excitations. This is because Skyrmions have classical pion fields which have an intrinsic shape in isospace as well as ordinary space. More precisely, at each point in space, the pion fields of a Skyrmion have definite classical values. This, we believe, was Skyrme's vision, that the interior of a nucleus is a non-uniform pion condensate. In contrast, in standard nuclear physics, isospin is never regarded classically. Instead, nucleons are quantized as having isospin half from the start. In other words, it is not postulated that there can be a condensate or coherent state in isospace which spontaneously breaks isospin symmetry at each point. The classical pion field configurations of Skyrmions have this feature, so that the isospin symmetry needs to be restored by collective coordinate quantization. To get close to the quantized Skyrmion picture in conventional nuclear physics language, one would need to accept that nucleons in close proximity are quite strongly and coherently mixed with delta resonances and higher isospin objects.

There are several papers on the spectrum of $^{12}$C going beyond the rigid-body picture of a triangle of $\alpha$-particles, and explaining more of the spectrum than we shall be able to here. But none of these papers seems to treat the isospin triplet of $^{12}$B, $^{12}$C and $^{12}$N as a collective isorotational excitation of an intrinsic shape in isospace. Presumably, these latter states are usually interpreted as arising from one of the $\alpha$-particles being broken up, through a change of a proton into a neutron or vice versa. In the Skyrme model picture of these nuclei, the excitation is collective and involves all three $\alpha$-particles symmetrically. The classical Skyrmion solution is the same one, with triangular symmetry, that is quantized to give $^{12}$C in its ground state with isospin zero.

We shall show that the Skyrme model gives quite a good account of isospin excitations, and of the non-trivial constraints linking allowed spin and isospin states. Experimental data is available up to isospin 2 or 3 for the baryon numbers of interest. The reader will be left to judge how successful the model is in this regard. The basic energy scales come out right, with spin energies of order 1 MeV and isospin energies of order 10 MeV for the nuclei we consider. This is because rotational inertias increase quadratically with baryon number, whereas isospin inertias increase linearly. As with all Skyrme model predictions, the quantitative errors can easily be of the order of tens of percent.

The structure of this review is as follows. We briefly describe the Skyrme model and its solutions for low baryon numbers. Then we review the rational map ansatz, which has turned out to be the most useful mathematical approximation to Skyrmion solutions, helping us understand their symmetries and the Finkelstein–Rubinstein constraints on the quantum states of Skyrmions. We then describe some of the Skyrmion solutions, found fairly recently, that are clusters of $B = 4$ Skyrmions, the Skyrmion version of $\alpha$-particles. The heart of this review is the discussion of the allowed states of quantized Skyrmions with baryon numbers up to $B = 12$. Some of this is based on very recent work, partly done by O.V. Manko and S.W. Wood, students of the second author. Qualitatively, the results are encouraging, but no single calibration of the Skyrme model's three parameters matches the predicted spectra with the experimental data very well.

We end with a summary of the Skyrme model's successes and limitations as a model of nuclei of small and moderate size, and an indication of directions for further research.

For a rather more detailed review, especially of the classical Skyrmion solutions in the massless pion case, see Ref. 11.


## 1.2. Skyrmions

T.H.R. Skyrme[1,2] proposed that the interior of a nucleus is dominated by a nonlinear semiclassical medium formed from the three pion fields, and he introduced the Skyrme model, a Lorentz invariant, nonlinear sigma model, in which the pion fields $\pi = (\pi_1, \pi_2, \pi_3)$ are combined into an $SU(2)$-valued scalar field

$$U(\mathbf{x}) = (1 - \pi(\mathbf{x}) \cdot \pi(\mathbf{x}))^{1/2}\mathbf{1} + i\pi(\mathbf{x}) \cdot \tau, \qquad (1.2.1)$$

where $\tau$ are the Pauli matrices. (The possible time-dependence of $U$ is here suppressed.) There is an associated current, taking values in $\mathfrak{su}(2)$ (the Lie algebra of $SU(2)$), with spatial components $R_i = (\partial_i U)U^\dagger$. For static fields, the energy in the

Skyrme model is given by

$$E = \int \left\{ -\frac{1}{2}\mathrm{Tr}(R_i R_i) - \frac{1}{16}\mathrm{Tr}([R_i, R_j][R_i, R_j]) + m^2 \mathrm{Tr}(\mathbf{1} - U) \right\} d^3x\,, \qquad (1.2.2)$$

and the vacuum is $U = \mathbf{1}$. $E$ is invariant under translations and rotations in $\mathbb{R}^3$ and also under $SO(3)$ isospin rotations given by the conjugation

$$U(\mathbf{x}) \mapsto \mathcal{A}U(\mathbf{x})\mathcal{A}^\dagger, \qquad \mathcal{A} \in SU(2)\,. \qquad (1.2.3)$$

This rotates the pion fields among themselves. Stationary points of $E$ satisfy the Skyrme field equation, and we shall mostly consider minima of $E$. The Lorentz invariant extension of this energy function gives a dynamical Lagrangian and field equation.

Without the final, pion mass term, there would be a chiral symmetry $U(\mathbf{x}) \mapsto \mathcal{A}U(\mathbf{x})\mathcal{A}'^\dagger$, with $\mathcal{A}$ and $\mathcal{A}'$ independent elements of $SU(2)$, but this is broken by the mass term, and even without it by the vacuum boundary condition.

The expression (1.2.2) is in "Skyrme units" and $m$ is a dimensionless pion mass parameter. We will discuss below the calibration of the energy and length units by comparison with physical data. Traditionally $m$ has been given a value of approximately 0.5,[12] but recent work suggests a higher value, $m \approx 1$[13–15] or $m = 1.125$.[16] The physical pion mass is proportional to $m$, but also depends on the length unit.

The model has a conserved, integer-valued topological charge $B$, the baryon number. This is the degree of the map $U : \mathbb{R}^3 \to SU(2)$, which is well-defined because $U \to \mathbf{1}$ at spatial infinity. $B$ is the integral of the baryon density

$$B = -\frac{1}{24\pi^2}\epsilon_{ijk}\mathrm{Tr}(R_i R_j R_k)\,, \qquad (1.2.4)$$

which is proportional to the Jacobian of the map $U$. In Skyrme units there is the Faddeev–Bogomolny energy bound, $E \geq 12\pi^2|B|$, although equality is not attained for any field configurations with non-zero $B$. The minimal energy solutions for each $B$ are called Skyrmions, and their energy $E$ is identified with their mass, $\mathcal{M}_B$. (More loosely, local minima and saddle points of $E$ with nearby energies are also sometimes called Skyrmions.)

The $B = 1$ Skyrmion has the spherically symmetric, hedgehog form

$$U(\mathbf{x}) = \exp\left\{if(r)\hat{\mathbf{x}} \cdot \tau\right\} = \cos f(r)\mathbf{1} + i\sin f(r)\hat{\mathbf{x}} \cdot \tau\,. \qquad (1.2.5)$$

$f$ is a radial profile function obeying an ODE with the boundary conditions $f(0) = \pi$ and $f(\infty) = 0$. Skyrmions with baryon numbers greater than 1 all have interesting shapes (see Fig. 1.1); they are not spherical like the basic $B = 1$ Skyrmion. The $B = 2$ Skyrmion is toroidal, and the $B = 3$ Skyrmion tetrahedral. The $B = 4$ Skyrmion is cubic and can be obtained by bringing together two $B = 2$ toroids along their common axis. The $B = 6$ solution has $D_{4d}$ symmetry and can be formed from three $B = 2$ toroids stacked one above the other, and the $B = 7$ Skyrmion has icosahedral symmetry.

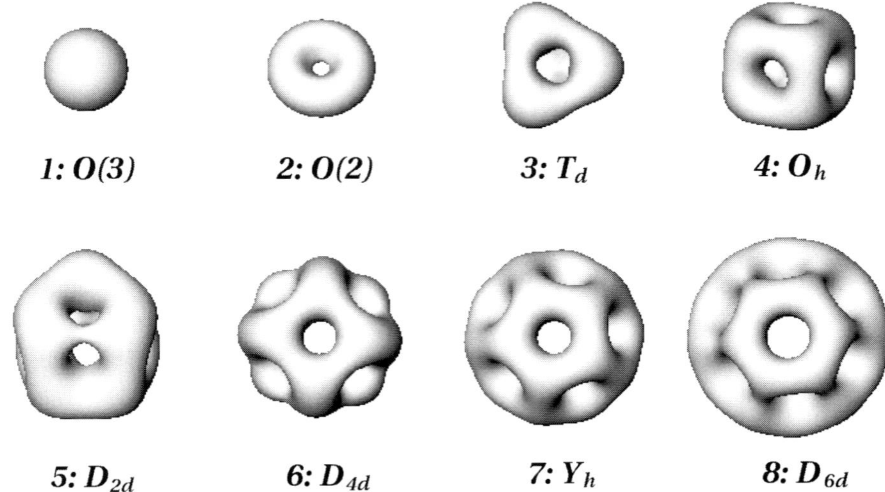

Fig. 1.1. Skyrmions for $1 \leq B \leq 8$, with $m = 0$. A surface of constant baryon density is shown, together with the baryon number and symmetry.

Table 1.1 presents, for $m = 0$, the symmetries and energies of the Skyrmions, computed from numerically obtained minima of the Skyrme energy.[17–19]

Table 1.1. The symmetry $K$, and normalized energy per baryon $E/12\pi^2 B$, for numerically computed Skyrmions with $m = 0$.

| $B$ | $K$ | $E/12\pi^2 B$ |
|---|---|---|
| 1 | $O(3)$ | 1.2322 |
| 2 | $D_{\infty h}$ | 1.1791 |
| 3 | $T_d$ | 1.1462 |
| 4 | $O_h$ | 1.1201 |
| 5 | $D_{2d}$ | 1.1172 |
| 6 | $D_{4d}$ | 1.1079 |
| 7 | $Y_h$ | 1.0947 |
| 8 | $D_{6d}$ | 1.0960 |

The toroidal structure of the $B = 2$ Skyrmion has some phenomenological support from nuclear physics,[20] since the particle density has a toroidal shape in models of the deuteron as a bound state of point-particle nucleons. This is because of the tensor forces. Recall that the deuteron has isospin zero and spin 1. When the spin component along the 3-axis is zero, then the particle density is concentrated in a torus whose symmetry axis is the 3-axis. If the spin component is $\pm 1$, then the density is the more familiar dumbbell, but this can be interpreted as a torus tipped through $90°$ and spinning about the 3-axis.

From Fig. 1.1, one sees that in Skyrmions of higher baryon number, approximately toroidal structures are ubiquitous. They surround every hole in the baryon density. In each case, along a circuit enclosing a hole, the Skyrme field winds twice around some axis in isospace. Therefore in each region around a hole there are two units of baryon number. For example, the $B = 4$ Skyrmion, if sliced in half, gives two slightly distorted $B = 2$ tori, and this can be done in three independent ways. The structure of Skyrmion solutions is therefore consistent with the fairly recent observation that if pairs of nucleons which are initially close together are knocked out of any nucleus, they are found to be usually rather strongly correlated as a proton-neutron pair, that is, as an isospin zero state.[21] The spin of the pair is not shown experimentally to be 1, but a theoretical understanding relies again on tensor forces.

The Skyrme model seems therefore to capture, at a classical level, some of the structural aspects of the many-body quantum states in nuclei.

The inclusion of the third term in the energy density, which involves the pion mass, has a significant effect on the shapes and symmetries of the Skyrmion solutions, the effect being more marked for larger values of $B$. For zero pion mass, the Skyrmions with $B$ up to 22 and beyond resemble hollow polyhedra. Their baryon density is concentrated in a shell of roughly constant thickness, with $2B - 2$ holes, surrounding a region in which the baryon density is very small.[19] This disagrees with the approximately uniform baryon density observed in the interior of real nuclei. Fortunately, it has been established that the hollow polyhedral solutions for $B \geq 8$ do not remain stable when the pion mass parameter $m$ is set at a physically reasonable value, of order 1.[14,15] This is because in the interior of the hollow polyhedra the Skyrme field is very close to $U = -\mathbf{1}$, and here the pion mass term gives the field a maximal potential energy, and hence instability. This instability results in the interior region splitting into separate smaller subregions. The stable Skyrmion solutions are found to exhibit clustering: small Skyrmion solutions, such as the cubically symmetric $B = 4$ solution, appear as substructures within larger solutions.[22] This is encouraging, as it has been believed for some time that $\alpha$-particles exist as stable substructures inside heavier nuclei.

Some further Skyrmion solutions for $10 \leq B \leq 16$, of minimal or close-to-minimal energy, have a planar, layered character.[15] One may interpret these solutions as fragments of an infinite crystalline sheet with hexagonal (or in some cases, square) symmetry, a two-layer version of the one-layer crystalline sheet presented in Ref. 23, which by itself has the wrong boundary conditions.

## 1.3. The Rational Map Ansatz

Skyrmions and $SU(2)$ Yang–Mills–Higgs monopoles are both examples of topological solitons in three dimensions, with an integer-valued topological charge. So-called BPS monopoles, satisfying the Bogomolny equation, have been constructed with very similar symmetries to Skyrmions with the corresponding charges,[24–26] and there has been, historically, an interesting interplay between the discovery of symmetric monopoles and symmetric Skyrmions.

It is known that there is a precise 1–1 correspondence between charge $N$ monopoles and degree $N$ rational maps between Riemann spheres, the Jarvis rational maps.[27] The observed similarity between Skyrmions and monopoles leads to an approximate construction of Skyrmions using rational maps. This is the rational map ansatz of Houghton, Manton and Sutcliffe,[28] which separates the angular from the radial dependence of the Skyrme field $U$. One introduces a complex (Riemann sphere) coordinate $z = \tan \frac{\theta}{2} e^{i\phi}$, where $\theta$ and $\phi$ are the usual spherical polar coordinates, and constructs the Skyrme field from a rational function of $z$,

$$R(z) = \frac{p(z)}{q(z)}, \tag{1.3.6}$$

where $p$ and $q$ are polynomials with no common root. One also needs a radial profile function $f(r)$ satisfying $f(0) = \pi$ and $f(\infty) = 0$. One should think of $R$ as a smooth map from a 2-sphere in space (at a given radius) to a 2-sphere in the target $SU(2)$ (at a given distance from the identity). By standard stereographic projection, the point $z$ corresponds to the Cartesian unit vector

$$\mathbf{n}_z = \frac{1}{1 + |z|^2} (z + \bar{z}, \, i(\bar{z} - z), \, 1 - |z|^2), \tag{1.3.7}$$

and conversely

$$z_{\mathbf{n}} = \frac{(\mathbf{n})_1 + i(\mathbf{n})_2}{1 + (\mathbf{n})_3}. \tag{1.3.8}$$

Similarly, an image point $R$ can be expressed as a unit vector

$$\mathbf{n}_R = \frac{1}{1 + |R|^2} (R + \bar{R}, \, i(\bar{R} - R), \, 1 - |R|^2). \tag{1.3.9}$$

The rational map ansatz for the Skyrme field is

$$U(r, z) = \exp\{if(r)\mathbf{n}_{R(z)} \cdot \tau\} = \cos f(r)\mathbf{1} + i \sin f(r)\mathbf{n}_{R(z)} \cdot \tau, \tag{1.3.10}$$

generalising the hedgehog formula (1.2.5). The baryon number $B$ of this Skyrme field equals the topological degree of the rational map $R : S^2 \to S^2$, and this is the higher of the algebraic degrees of the polynomials $p$ and $q$.

An $SU(2)$ Möbius transformation on the domain $S^2$ of the rational map corresponds to a spatial rotation, whereas an $SU(2)$ Möbius transformation on the target $S^2$ corresponds to a rotation of $\mathbf{n}_R$, and hence to an isospin rotation of the Skyrme field. Thus if a rational map $R$ has some symmetry (i.e. a rotation of the domain can be compensated by a rotation of the target), then the resulting Skyrme field has that symmetry (i.e. a spatial rotation can be compensated by an isospin rotation).

An important feature of the rational map ansatz is that, when one substitutes it into the Skyrme energy function (1.2.2), the angular and radial parts decouple. The energy simplifies to

$$E = 4\pi \int_0^\infty \left( r^2 f'^2 + 2B(f'^2 + 1)\sin^2 f + \mathcal{I}\,\frac{\sin^4 f}{r^2} + 2m^2 r^2 (1 - \cos f) \right) dr \,, \quad (1.3.11)$$

where $\mathcal{I}$ denotes the angular integral

$$\mathcal{I} = \frac{1}{4\pi} \int \left( \frac{1 + |z|^2}{1 + |R|^2} \left| \frac{dR}{dz} \right| \right)^4 \frac{2i\,dz d\bar{z}}{(1 + |z|^2)^2} \,, \quad (1.3.12)$$

which only depends on the rational map $R(z)$. $\mathcal{I}$ is an interesting function on the space of rational maps. To minimize the energy (for given $B$), it is sufficient to first minimize $\mathcal{I}$ with respect to the coefficients occurring in the rational map, and then to solve an ODE for $f(r)$ whose coefficients depend on the rational map only through $B$ and the minimized $\mathcal{I}$. Optimal rational maps, and the associated profile functions, have been found for many values of $B$, and often have a high degree of symmetry. The optimized fields within the rational map ansatz are good approximations to Skyrmions, and they are also used as starting points for numerical relaxations to true Skyrmion solutions (which almost always have the same symmetry, but no exact separation of the angular and radial dependence of $U$).

The simplest degree 1 rational map is $R(z) = z$, which is spherically symmetric. The ansatz (1.3.10) then reduces to the hedgehog field (1.2.5). For $B = 2, 3, 4, 7$ the symmetry groups of the numerically computed Skyrmions are $D_{\infty h}$, $T_d$, $O_h$, $Y_h$ respectively. In each of these cases there is a unique rational map with this symmetry, up to rotations and isorotations, namely

$$R(z) = z^2 \,, \quad R(z) = \frac{z^3 - \sqrt{3}iz}{\sqrt{3}iz^2 - 1} \,, \quad R(z) = \frac{z^4 + 2\sqrt{3}iz^2 + 1}{z^4 - 2\sqrt{3}iz^2 + 1} \,,$$

$$R(z) = \frac{z^7 - 7z^5 - 7z^2 - 1}{z^7 + 7z^5 - 7z^2 + 1} \,, \quad (1.3.13)$$

and these also minimize $\mathcal{I}$. For $B = 5, 6$ and $8$, rational maps with dihedral symmetries are required, and these involve one or two coefficients that need to be determined numerically. Table 1.2 lists the energies of the approximate solutions obtained using the rational map ansatz, together with the values of $\mathcal{I}$, again for $m = 0$.[18,19]

The Wronskian of a rational map $R(z) = p(z)/q(z)$ of degree $B$ is the polynomial

$$W(z) = p'(z)q(z) - q'(z)p(z) \quad (1.3.14)$$

of degree $2B - 2$. Where $W$ is zero, the derivative $dR/dz$ is zero, so only the radial derivative of $U$ is non-vanishing. The baryon density therefore vanishes along the entire radial half-line in the direction of a zero of $W$ (exactly within the rational map ansatz, and approximately for the true Skyrmions), and the energy density is also low. This explains why the Skyrmion baryon density contours look like

Table 1.2. Symmetry group $K$, the value of the angular integral $\mathcal{I}$, and the energy per baryon $E/12\pi^2 B$ of the approximate Skyrmions obtained using the rational map ansatz with $m = 0$.

| $B$ | $K$ | $\mathcal{I}$ | $E/12\pi^2 B$ |
|---|---|---|---|
| 1 | $O(3)$ | 1.0 | 1.232 |
| 2 | $D_{\infty h}$ | 5.8 | 1.208 |
| 3 | $T_d$ | 13.6 | 1.184 |
| 4 | $O_h$ | 20.7 | 1.137 |
| 5 | $D_{2d}$ | 35.8 | 1.147 |
| 6 | $D_{4d}$ | 50.8 | 1.137 |
| 7 | $Y_h$ | 60.9 | 1.107 |
| 8 | $D_{6d}$ | 85.6 | 1.118 |

polyhedra with holes in the directions given by the zeros of $W$, and why there are $2B - 2$ such holes, precisely the structures seen in Fig. 1. As an example, the icosahedrally-symmetric degree 7 map in (1.3.13) has Wronskian

$$W(z) = 28z(z^{10} + 11z^5 - 1),\qquad (1.3.15)$$

which is proportional to one of the icosahedral Klein polynomials, and vanishes at the twelve face centres of a regular dodecahedron (including $z = \infty$).

The solutions we have described so far are for $m = 0$, but it is found that qualitatively similar solutions with 10% to 20% higher energy exist for $m$ up to 1 and beyond, provided $B \leq 7$. There is, however, a qualitative change for Skyrmions with $B \geq 8$, as we will see in the next section.

For these Skyrmions of higher baryon number, it is sometimes helpful to use a generalisation of the rational map ansatz, called the double rational map ansatz.[29] This uses two rational maps $R^{\mathrm{in}}(z)$ and $R^{\mathrm{out}}(z)$, with a profile function $f(r)$ satisfying $f(0) = 2\pi$ and $f(\infty) = 0$, and decreasing monotonically as $r$ increases, passing through $\pi$ at a radius $r_0$. The ansatz for the Skyrme field is again (1.3.10), with $R(z) = R^{\mathrm{in}}(z)$ for $r \leq r_0$, and $R(z) = R^{\mathrm{out}}(z)$ for $r > r_0$. Notice now that $U = \mathbf{1}$ both at the origin and at spatial infinity, and $U = -\mathbf{1}$ at $r = r_0$. The total baryon number is the sum of the degrees of the maps $R^{\mathrm{in}}$ and $R^{\mathrm{out}}$. The ansatz is optimized by adjusting the coefficients of both maps, allowing variations of $r_0$, and solving for $f(r)$. All this is quite hard, but easier if $R^{\mathrm{in}}$ and $R^{\mathrm{out}}$ share a substantial symmetry. The double rational map ansatz is a special case of Skyrme's product ansatz,[2] in which a non-trivial Skyrme field $U_1(\mathbf{x})$ with baryon number $B_1$ defined inside radius $r_0$, is multiplied by $U_2(\mathbf{x})$ with baryon number $B_2$ defined outside, giving the field $U(\mathbf{x}) = U_1(\mathbf{x})U_2(\mathbf{x})$ with baryon number $B_1 + B_2$. Here, $U_1(\mathbf{x}) = \mathbf{1}$ outside radius $r_0$, and $U_2(\mathbf{x}) = -\mathbf{1}$ inside $r_0$.

There is a problem here, since for true Skyrmions, $U$ does not take the value $-\mathbf{1}$ on the entire sphere at radius $r_0$. This problem is avoided if one just takes the product of two fields $U_1$ and $U_2$ defined by the original rational map ansatz, with rational maps $R_1$ and $R_2$, and with profiles $f_1$ and $f_2$ decreasing freely from $\pi$ at $r = 0$ to 0 at $r = \infty$, without further constraint at an intermediate radius $r_0$. The product $U_1 U_2$ still preserves the joint rotational symmetries of $U_1$ and $U_2$, but not any inversion or reflection symmetries. Along a generic radial line the field no longer passes through $U = -\mathbf{1}$, but rather takes a short-cut, reducing the radial derivative of $U$ without a significant increase in the angular derivatives, and also reducing the potential energy. The non-generic lines are those for which $R_1(z) = R_2(z)$, and there are $B$ of these, counted with multiplicity. Therefore $U = -\mathbf{1}$ at $B$ points, the number expected topologically. Their distance $r$ from the origin is where $f_1(r) + f_2(r) = \pi$. A detailed investigation of this type of product field has not been made, but would be worthwhile.

## 1.4. Skyrmions and $\alpha$-Particles

The $^4$He nucleus, or $\alpha$-particle, is particularly stable and can be regarded as a building block for nuclei with baryon number a multiple of four and having equal numbers of protons and neutrons. The $\alpha$-particle model[10,30–32] has considerable success describing the nuclei $^8$Be, $^{12}$C, $^{16}$O etc. as "molecules" of pointlike $\alpha$-particles. For $m = 1$, Skyrmion solutions with baryon number a multiple of four have been found, which make contact with the $\alpha$-particle model.[22] These solutions are clusters of cubic $B = 4$ Skyrmions, and for $B \geq 12$ they are energetically more stable than the hollow polyhedral Skyrmions, the effect being marginal for $B = 8$.

The planar solutions mentioned earlier can also be thought of as made up of $B = 4$ Skyrmions, with one or two $B = 1$ Skyrmions added or removed. In particular, the solution for $B = 10$ can be thought of this way.

### 1.4.1. $B = 4$

In order to understand the interaction of several $B = 4$ cubic Skyrmions it is useful to introduce a colour scheme that represents the direction in isospace of the associated pion fields. For regions in space where at least one of the pion fields does not vanish, the normalized pion field $\hat{\pi}$ can be defined, and takes values in the unit sphere. We colour this sphere by making a region close to the north pole white and a region close to the south pole black. On an equatorial band, where $\hat{\pi}_3$ is small, we divide the sphere into three segments and colour these as red, blue and green.

A baryon density isosurface for the $B = 4$ Skyrmion is diplayed in Fig. 1.2, using this colour scheme. It can be seen that opposite faces share the same colour and vertices alternate between black and white.

### 1.4.2. $B = 8$

We saw that when $m = 0$, the $B = 8$ Skyrmion is a hollow polyhedron with $D_{6d}$ symmetry, with no obvious relation to a pair of cubic $B = 4$ Skyrmions. Motivated

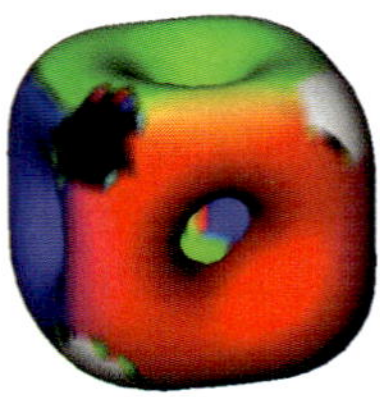

Fig. 1.2.   Surface of constant baryon density for the $B = 4$ Skyrmion. Different colours indicate different directions of the pion fields.

by the $\alpha$-particle model, one expects that for $m$ sufficiently large, the lowest energy solution is a dimer of two cubic, $B = 4$ Skyrmions. Two such Skyrmions, placed initially in the same orientation and next to each other, have a weak quadrupole-quadrupole attraction.[4,33] Because of a significant short-range octupole interaction in the single pion field component that has no quadrupole moment, it is best to also twist one cube by $90^0$ relative to the other around the axis joining them (Fig. 1.3).

The reason this twist is favourable is clear from the colour representation in Fig. 1.3, since the $90^0$ rotation allows the vertices of one cube to be close to vertices of the same colour on the other cube. Without the rotation the vertices that are close would be of opposite colours and this results in a significant gradient energy, since black and white points are antipodal on the sphere of pion field directions.

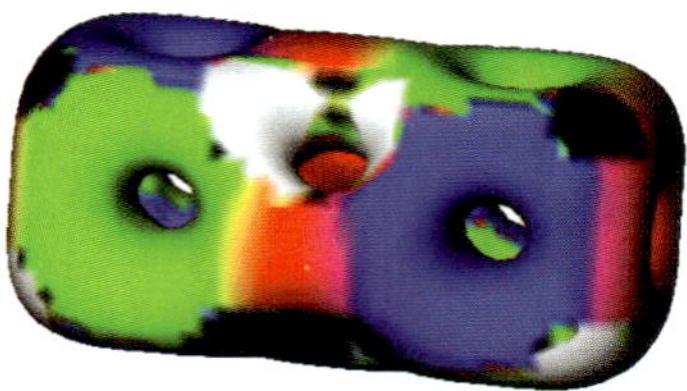

Fig. 1.3.   Surface of constant baryon density for two $B = 4$ cubes, with one of the cubes rotated by $90°$. The colour scheme indicates the direction of the pions fields.

For comparison with other solutions, the configuration displayed in Fig. 1.3 is reproduced in Fig. 1.4(a) without the colour scheme. Another suitable starting configuration has the shape of a truncated octahedron and is obtained using the rational map ansatz with an $O_h$-symmetric degree 8 map (Fig. 1.4(b)).

Numerical relaxation from either starting point (including a symmetry breaking perturbation for the truncated octahedron) produces the stable solution displayed in Fig. 1.4(c), which has $D_{4h}$ symmetry. There are still 14 holes in the baryon density. For $m = 1$, the energy per baryon of this new Skyrmion and also of the old $D_{6d}$-symmetric Skyrmion is $E/12\pi^2 B = 1.294$. The change of structure therefore has a marginal effect in this case, but one expects that for $m > 1$ and for larger

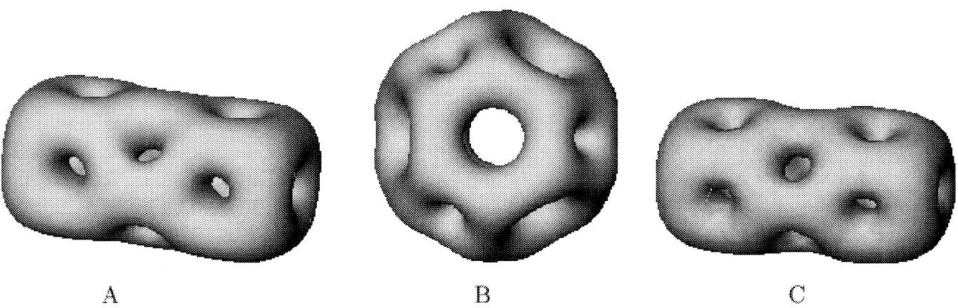



Fig. 1.4.   Baryon density contours for (a) two $B = 4$ cubes with one of the cubes rotated by $90°$ around the line joining them; (b) the $B = 8$ truncated octahedron; (c) the relaxed $B = 8$ Skyrmion with $m = 1$.

$B$, clusters of $B = 4$ cubes will be the more stable solutions. Note that there is a definite attraction between $B = 4$ cubes, because the energy per baryon of the $B = 4$ cube is $E/12\pi^2 B = 1.307$ for $m = 1$. Numerical errors are estimated as $0.5\%$ or less.

### 1.4.3.   $B = 10$

For $m = 1$, the $B = 10$ Skyrmion has $D_{2h}$ symmetry and may be viewed as a pair of $B = 4$ cubes with two single Skyrmions between them. This interpretation is suggested by the baryon density plot in Fig. 1.5, where two deformed cubes are visible at the two ends. This interpretation is also consistent with the distribution of points in space where $U = -\mathbf{1}$, which are grouped into two sets of four and two single points.

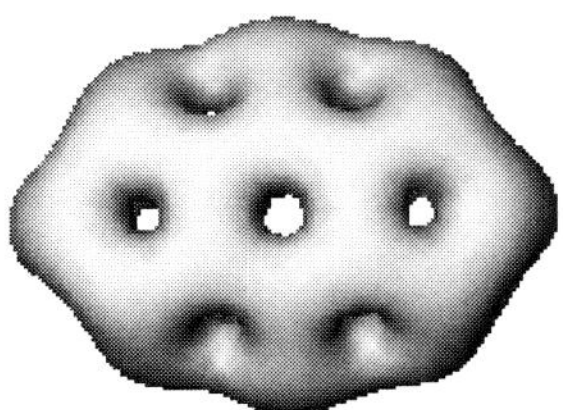

Fig. 1.5.   Baryon density isosurface for the $B = 10$ Skyrmion with $m = 1$.

A suitable rational map is

$$R(z) = \frac{a + bz^2 + cz^4 + dz^6 + ez^8 + z^{10}}{1 + ez^2 + dz^4 + cz^6 + bz^8 + az^{10}}, \tag{1.4.16}$$

with $a = 0.28$, $b = -9.37$, $c = 14.83$, $d = 4.98$ and $e = 3.02$. The $D_2$ rotation group is generated by $180°$ rotations about the spatial 3-axis and spatial 1-axis, under which the rational map has the symmetries

$$R(-z) = R(z), \quad R(1/z) = 1/R(z).  \tag{1.4.17}$$

The true $B = 10$ solution has the same symmetries as this rational map, but is less spherical.

### 1.4.4.  $B = 12$

In the $\alpha$-particle model, three $\alpha$-particles form an equilateral triangle. This motivates the search for a triangular $B = 12$ solution in the Skyrme model, composed of three $B = 4$ cubes. A configuration with approximate $D_{3h}$ symmetry can be obtained with each cube related to its neighbour by a spatial rotation through $120°$ combined with an isorotation by $120°$. The isorotation cyclically permutes the values of the pion fields on the faces of the cube, so that these values match on touching faces, and the cubes attract. It is fairly easy to see that around the centre of the triangle the field has a winding equivalent to a $B = 1$ Skyrmion. From this starting configuration, numerical relaxation leads to the true $B = 12$ Skyrmion.

If a configuration of the above form is constructed using the product ansatz then it has only an approximate $D_{3h}$ symmetry. However, it looks similar to the $B = 11$ Skyrmion, whose baryon density has 20 holes, which suggests that the initial arrangement of three cubes can also be viewed as a $B = 11$ Skyrmion with a $B = 1$ Skyrmion placed inside at the origin. Such a field configuration can be constructed with exact $D_{3h}$ symmetry using the double rational map ansatz. This involves a $D_{3h}$-symmetric outer map of degree 11, $R^{\text{out}}$, and a spherically-symmetric degree 1 inner map, $R^{\text{in}}$, together with an overall radial profile function. The maps are

$$R^{\text{out}}(z) = \frac{z^2(1 + az^3 + bz^6 + cz^9)}{c + bz^3 + az^6 + z^9}  \tag{1.4.18}$$

$$R^{\text{in}}(z) = -\frac{1}{z},  \tag{1.4.19}$$

where $a = -2.47$, $b = -0.84$ and $c = -0.13$. Note that the orientation of $R^{\text{in}}$ has to be chosen compatibly with the $D_{3h}$ symmetry of $R^{\text{out}}$.

Relaxing the above field configuration with exact $D_{3h}$ symmetry produces a solution that resembles the initial condition, but it appears that this solution is a saddle point, for all values of $m$. For small values of $m$, a symmetry breaking perturbation relaxes to a hollow polyhedron with tetrahedral symmetry, which is the form taken by the minimal energy $B = 12$ Skyrmion when $m = 0$. For larger values of $m$, in particular for $m$ of order 1, the saddle point solution is unstable to a deformation in which the Skyrmion in the centre moves down or up to merge with the bottom or top face of the triangular structure, filling a hole in the baryon density there. The energy is negligibly affected by this deformation, but the symmetry is reduced to $C_{3v}$. This $C_{3v}$-symmetric Skyrmion is shown in Fig. 1.6. Its energy is $E/12\pi^2 B = 1.288$. It can be verified that the structure of the inertia tensor is the same, whether the symmetry is $D_{3h}$ or $C_{3v}$.

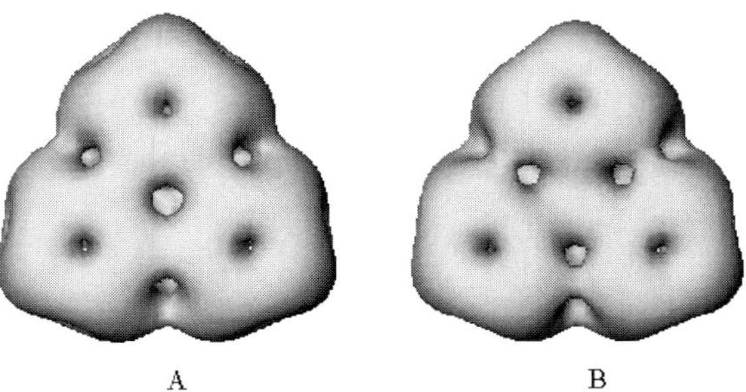



Fig. 1.6.   Top and bottom views of the $B = 12$ Skyrmion with $C_{3v}$ symmetry.

Battye and Sutcliffe found another $B = 12$ solution with $C_3$ symmetry, with energy $E/12\pi^2 B = 1.289$.[15] It is a general observation that rearrangements of clusters have only a tiny effect on the energy of a Skyrmion, so as $B$ increases one expects an increasingly large number of local minima with extremely close energies. Rearranged solutions are analogous to the rearrangements of the $\alpha$-particles which model excited states of nuclei. An example is the Skyrme model analogue of the three $\alpha$-particles in a chain configuration modelling the 7.65 MeV excited state of $^{12}$C.[34,35] This is obtained from three $B = 4$ cubes placed next to each other in a line, with the middle cube twisted relative to the other two by 90° around the axis of the chain. The relaxed solution is displayed in Fig. 1.7 and has energy $E/12\pi^2 B = 1.285$. This may be the lowest energy of the $B = 12$ solutions, but note that the energy difference 7.65 MeV is less than 0.1% of the total energy of a $^{12}$C nucleus, smaller than the numerical errors in the Skyrmion energies.

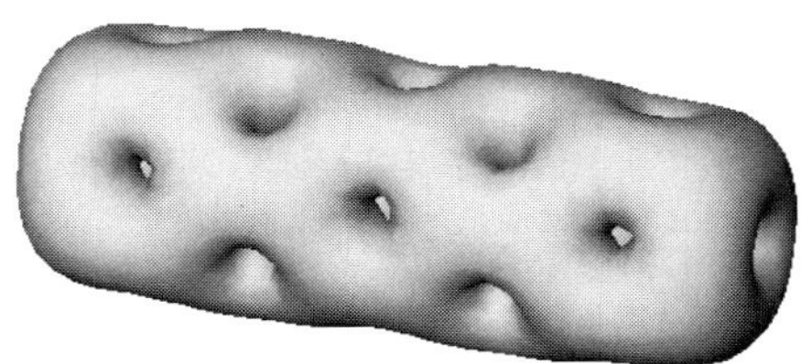

Fig. 1.7.   $B = 12$ Skyrmion formed from three cubes in a line, with the middle cube being rotated by 90° around the line of the cubes.

### 1.4.5.   $B = 16$

There is a tetrahedrally symmetric $B = 16$ solution which is an arrangement of four $B = 4$ cubes. It may be created using the double rational map ansatz as a starting point. There is a $T_d$-symmetric map $R^{\text{out}}$ of degree 12, and this can be combined

20          *R.A. Battye, N.S. Manton and P.M. Sutcliffe*

with the $O_h$-symmetric degree 4 map familiar from the $B = 4$ Skyrmion, giving $T_d$ symmetry overall. The maps are

$$R^{\text{out}} = \frac{a p_+^3 + b p_-^3}{p_+^2 p_-} \tag{1.4.20}$$

$$R^{\text{in}} = \frac{p_+}{p_-}, \tag{1.4.21}$$

where $p_\pm(z) = z^4 \pm 2\sqrt{3}\,i z^2 + 1$, $a = -0.53$ and $b = 0.78$. Letting the field $U$ relax, preserving the $T_d$ symmetry, results in the solution displayed in Fig. 1.8(a), in which $U = -\mathbf{1}$ at 16 points clustered into groups of four close to the centre of each cube. For $m = 1$ the energy of this solution is $E/12\pi^2 B = 1.288$.

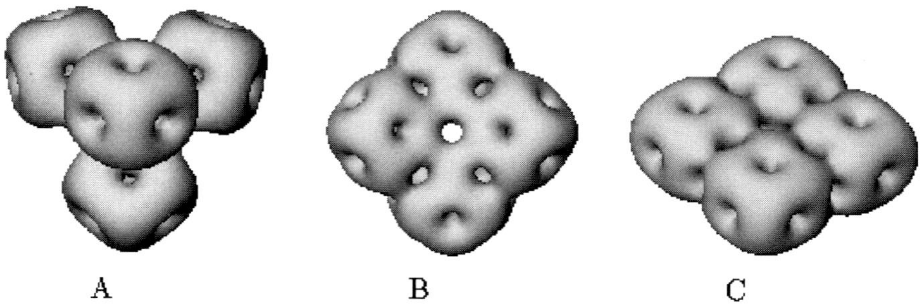

A             B             C

Fig. 1.8.   $B = 16$ Skyrmions composed of four cubes. (a) Tetrahedral arrangement; (b) bent square; (c) flat square.

This tetrahedral solution is only a saddle point. It is energetically more favourable for the two cubes on a pair of opposite edges of the tetrahedron to open out, leading to the $D_{2d}$-symmetric solution in Fig. 1.8(b), which has the slightly lower energy $E/12\pi^2 B = 1.284$. An $\alpha$-particle molecule of similar shape has also been found, termed a "bent rhomb".[36]

A stable tetrahedral solution would be phenomenologically preferable, since the closed shell structure of $^{16}$O is known to be compatible with clustering into a tetrahedral arrangement of four $\alpha$-particles. Moreover, the $^{16}$O ground state and the excited states at 6.1 MeV and 10.4 MeV, with spin/parity $0^+$, $3^-$ and $4^+$, and some higher states, look convincingly like a rotational band for a tetrahedral intrinsic structure.[37,38]

Other low energy solutions are also known. For example, a solution in which four $B = 4$ cubes all have the same orientation, and are connected together to form a flat square (Fig. 1.8(c)), has energy $E/12\pi^2 B = 1.293$.

### 1.4.6.  $B = 32$

Even for relatively small values of $m$, the $B = 32$ Skyrmion is cubic, and has lower energy than the minimal energy, hollow polyhedral structure.[14] The solution may be thought of as eight $B = 4$ cubic Skyrmions placed on the vertices of a cube, each with the same spatial and isospin orientations, and it may also be created by cutting out a cubic $B = 32$ chunk from the infinite, triply-periodic Skyrme crystal.[39]

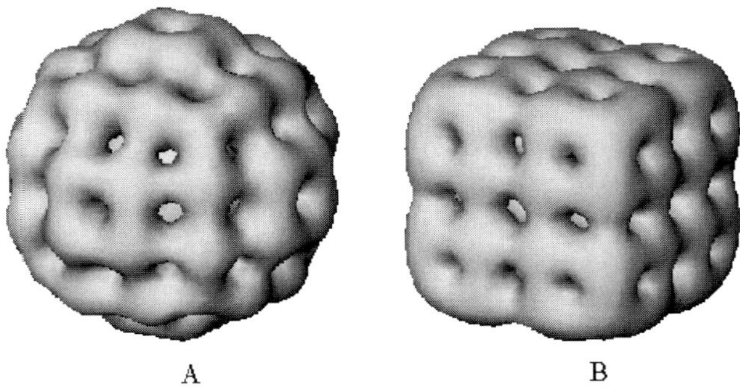


Fig. 1.9.  (a) Initial condition of the cubic $B = 4$ Skyrmion inside a cubic $B = 28$ configuration; (b) the final relaxed $B = 32$ Skyrmion, which is a chunk of the Skyrme crystal.

Alternatively, it may be obtained beginning with the double rational map ansatz. One places a $B = 4$ cube inside a $B = 28$ configuration with cubic symmetry using the maps

$$R^{\text{out}} = \frac{p_+(ap_+^6 + bp_+^3 p_-^3 - p_-^6)}{p_-(p_+^6 - bp_+^3 p_-^3 - ap_-^6)} \tag{1.4.22}$$

$$R^{\text{in}} = \frac{p_+}{p_-}, \tag{1.4.23}$$

where $a = 0.33$ and $b = 1.64$, and $p_\pm(z)$ are as before. This is displayed in Fig. 1.9(a). Numerical relaxation yields the solution in Fig. 1.9(b), which is the $B = 32$ Skyrmion for $m = 1$, with energy $E/12\pi^2 B = 1.274$. Note that slicing the $B = 32$ Skyrmion in half produces the square $B = 16$ solution of Fig. 1.8(c).

Further solutions which look like clusters of several $B = 4$ cubes have been found, for $B = 16, 20, 24$ and $28$. They are less symmetric, and not necessarily stable. It would be interesting to find stable solutions which have the same shapes and symmetries as those suggested by the $\alpha$-particle model and by many-body models with tensor-correlated nucleons, that is, a double triangular pyramid for $B = 20$, and a double tetrahedron structure (with a shared edge) for $B = 24$.[40]

## 1.5. Quantization

The quantization of Skyrmions has been a vital issue from the beginning, because Skyrmions are supposed to model physical nucleons (protons and neutrons) and nuclei, and a nucleon is a spin half fermion. One quantizes a Skyrmion as a fermion by lifting the classical field configuration space to its simply connected covering space. In the $SU(2)$ Skyrme model, this is a double cover for any value of $B$. Because of the formal connection between the Skyrme model and QCD, states should be multiplied by a factor of $-1$ when acted upon by any operation corresponding to a circuit around a non-contractible loop in the configuration space.[41] Equivalently, the wavefunction has opposite signs on the two points of the covering space that cover one point in the configuration space. A $2\pi$ rotation of a $B = 1$ Skyrmion is a non-contractible loop, which allows the Skyrmion to be quantized as a spin half fermion.[42] Finkelstein and Rubinstein showed that the exchange of two $B = 1$ Skyrmions is a loop which is homotopic to a $2\pi$ rotation of one of the $B = 1$ Skyrmions, in agreement with the spin-statistics result.[8] More generally, a $2\pi$ rotation and a $2\pi$ isorotation of a Skyrmion of baryon number $B$ are both non-contractible loops if $B$ is odd and contractible if $B$ is even.[43] The spin and isospin are therefore half-integral for odd $B$ and integral for even $B$.

A practical, approximate quantum theory of Skyrmions is achieved by a rigid-body quantization of the spin and isospin rotations. This can now be done for Skyrmions up to baryon number 12. We shall summarize the considerable recent progress that has been made using this finite-dimensional truncation of the theory. Quantized translational motion gives a Skyrmion a non-zero momentum, but this will not be discussed further. Quantized vibrational modes will be mentioned briefly at some points.

The kinetic energy of a rigidly rotating Skyrmion is of the form

$$T = \frac{1}{2}a_i U_{ij} a_j - a_i W_{ij} b_j + \frac{1}{2} b_i V_{ij} b_j \,, \tag{1.5.24}$$

where $b_i$ and $a_i$ are the angular velocities in space and isospace respectively, and $U_{ij}$, $V_{ij}$ and $W_{ij}$ are inertia tensors.[6,44] The inertia tensors are determined from the kinetic terms of the Skyrme Lagrangian to be

$$U_{ij} = -\int \mathrm{Tr}\left(T_i T_j + \frac{1}{4}[R_k, T_i][R_k, T_j]\right) d^3x \,, \tag{1.5.25}$$

$$W_{ij} = \int \epsilon_{jlm} x_l \, \mathrm{Tr}\left(T_i R_m + \frac{1}{4}[R_k, T_i][R_k, R_m]\right) d^3x \,, \tag{1.5.26}$$

$$V_{ij} = -\int \epsilon_{ilm} \epsilon_{jnp} x_l x_n \, \mathrm{Tr}\left(R_m R_p + \frac{1}{4}[R_k, R_m][R_k, R_p]\right) d^3x \,, \tag{1.5.27}$$

where $R_k = (\partial_k U)U^{-1}$ is the $\mathfrak{su}(2)$ current that appears in the Skyrme energy function, and

$$T_i = \frac{i}{2}[\tau_i, U]U^{-1} \tag{1.5.28}$$

is also an $\mathfrak{su}(2)$ current.

The momenta conjugate to $b_i$ and $a_i$ are the body-fixed spin and isospin, $L_i$ and $K_i$. The quantum Hamiltonian $H$ is obtained by re-expressing $T$ in terms of these quantities, which are then treated as operators with standard angular momentum commutation relations. $H$ is the Hamiltonian of coupled rigid bodies in space and isospace.

Continuous and discrete symmetries of the classical Skyrmion solutions give rise to further Finkelstein–Rubinstein(FR) constraints on quantum states $|\Psi\rangle$. These constraints are of the form

$$e^{i\theta_2 \mathbf{n}_2 \cdot \mathbf{L}} e^{i\theta_1 \mathbf{n}_1 \cdot \mathbf{K}} |\Psi\rangle = \chi_{\mathrm{FR}} |\Psi\rangle \,, \tag{1.5.29}$$

where $\mathbf{n}_1$, $\mathbf{n}_2$ and $\theta_1$, $\theta_2$ are, respectively, the axes and angles defining the rotations in isospace and space associated with a particular symmetry, and $\chi_{\mathrm{FR}} = \pm 1$. Each symmetry gives rise to a loop in configuration space, by simultaneously letting the isorotation angle increase from 0 to $\theta_1$ and the rotation angle increase from 0 to $\theta_2$, and

$$\chi_{FR} = \begin{cases} +1 \text{ if the loop is contractible,} \\ -1 \text{ if the loop is non-contractible.} \end{cases} \tag{1.5.30}$$

The FR signs $\chi_{\mathrm{FR}}$ define a 1-dimensional representation of the symmetry group of the Skyrmion. Krusch has found the following convenient way to calculate them for any Skyrmion that has the same symmetries as an approximate Skyrmion constructed using the rational map ansatz, with rational map $R(z)$.[9,45] The method exploits the known topology of the space of rational maps.[46]

The rational map $R(z)$ has the above symmetry if

$$R(z) = M_1(R(M_2(z))) \,, \tag{1.5.31}$$

where $M_1$ is the $SU(2)$ Möbius transformation corresponding to the isorotation by angle $\theta_1$ around $\mathbf{n}_1$, and $M_2$ is the Möbius transformation corresponding to the rotation by $\theta_2$ around $\mathbf{n}_2$. For non-zero $\theta_2$, $M_2$ only leaves the antipodal points

$$z_{\mathbf{n}_2} = \frac{(\mathbf{n}_2)_1 + i(\mathbf{n}_2)_2}{1 + (\mathbf{n}_2)_3} \quad \text{and} \quad z_{-\mathbf{n}_2} = -\frac{(\mathbf{n}_2)_1 + i(\mathbf{n}_2)_2}{1 - (\mathbf{n}_2)_3} \tag{1.5.32}$$

fixed. Similarly, $M_1$ only leaves the antipodal target space points $R_{\pm\mathbf{n}_1}$ fixed, where $R_{\pm\mathbf{n}_1}$ are defined similarly. The symmetry (1.5.31) implies that $R(z_{-\mathbf{n}_2}) = R_{\mathbf{n}_1}$ or $R_{-\mathbf{n}_1}$. One should fix the relative orientations, by reversing the signs of $\mathbf{n}_1$ and $\theta_1$ if necessary, so that

$$R(z_{-\mathbf{n}_2}) = R_{\mathbf{n}_1} \,. \tag{1.5.33}$$

Then, in terms of $\theta_1$ and $\theta_2$, Krusch's formula for $\chi_{FR}$ is

$$\chi_{FR} = (-1)^{\mathcal{N}} \,, \quad \text{where } \mathcal{N} = \frac{B}{2\pi}(B\theta_2 - \theta_1) \,. \tag{1.5.34}$$

The space of states $|\Psi\rangle$ has a basis given by the products $|J, L_3\rangle \otimes |I, K_3\rangle$, the tensor products of states for a rigid body in space and a rigid body in isospace. $J$

and $I$ are the total spin and isospin quantum numbers, $L_3$ and $K_3$ the projections on to the third body-fixed axes, and the space projection labels (which are the physical third components of spin and isospin, $J_3$ and $I_3$) are suppressed. The FR constraints only allow a subspace of these states as physical states.

A parity operator is introduced by considering a Skyrmion's reflection symmetries. Generally, the parity operation in the Skyrme model is an inversion in space and isospace, $\mathcal{P} : U(\mathbf{x}) \to U^\dagger(-\mathbf{x})$. One cannot directly calculate its eigenvalue by acting on a rigid-body state $|\Psi\rangle$. However, if the Skyrmion possesses some reflection symmetry (in space and isospace), then the above parity operation can be obtained from this by acting with a further rotation operator (in space and isospace). The eigenvalue of this latter operator, acting on a physical state, is taken to be the parity $P$ of the state. Quantum states are therefore labelled by the usual quantum numbers: spin/parity $J^P$, and isospin $I$. Note that we attach the parity label to the spin quantum number, as conventionally done, despite the fact that it is associated with a combination of rotations in space and isospace.

For the $B = 1$ Skyrmion, this quantization was carried out by Adkins, Nappi and Witten,[5] who showed that the lowest energy states have spin/parity $J^P = \frac{1}{2}^+$, and may be identified with the proton/neutron isospin doublet. The next lowest states are identified with the $J^P = \frac{3}{2}^+$ delta resonances, with isospin $\frac{3}{2}$.

The Skyrmions with baryon numbers $B = 2$, 3 and 4 have the right properties to model the deuteron $^2\mathrm{H}$, the isospin doublet $^3\mathrm{H}/^3\mathrm{He}$, and the $\alpha$-particle $^4\mathrm{He}$.[6,47–50] In each case, the rigid-body quantization is constrained by the symmetries of the classical solution. The resulting lowest energy states for $B = 2$, 3 and 4 have spin/parity, respectively, $J^P = 1^+, \frac{1}{2}^+$ and $0^+$, with isospin zero for $B = 2$ and 4, and isospin half for $B = 3$, agreeing with the ground states of the above nuclei.

Irwin found the allowed states of the $B = 6$ Skyrmion,[51] finding a ground state of spin/parity $1^+$ and isospin zero, modelling the nucleus $^6\mathrm{Li}$. Irwin also determined some allowed isospin excited states for $B = 4$ and 6. This was extended by Krusch to a much larger set of Skyrmions.[9,45] However, the inertia tensors were not computed, so the energy spectra were not determined. Some quantitative energy spectra of the $B = 4$, 6 and 8 Skyrmions have been calculated using approximate Skyrmion solutions and their inertia tensors.[16,44] The inertia tensors have the right symmetries (or slightly too much symmetry). The results were encouraging, in that the allowed spin and isospin states match experimental data quite well. For example, one could see the $0^+$, $2^+$ and $4^+$ rotational band of states of $^8\mathrm{Be}$.

More recently the $B = 4$, 6, 8, 10 and 12 Skyrmions have all been calculated afresh, for several non-zero values of $m$.[7] For each of these Skyrmions, all the FR-allowed quantum states have been determined, working up to spin and isospin values just beyond what is experimentally accessible. We summarize some of these recent results below. In most of the cases we do not explain in detail the analysis of the symmetries and FR constraints, but refer the reader to earlier papers. In the cases of $B = 10$ and $B = 12$, however, we give some explanation of these calculations.

Odd baryon numbers have caused more difficulty. Rigid-body quantization of the dodecahedral $B = 7$ Skyrmion leads to a lowest energy state with spin $J^P = \frac{7}{2}^-$ and isospin half,[9,51] disagreeing with the experimental value $J^P = \frac{3}{2}^-$ for the ground state of the isospin doublet $^7$Li/$^7$Be. The only encouragement here is that experimentally there are $\frac{7}{2}^-$ states with relatively low energy. The dodecahedral Skyrmion appears too symmetric to model the ground state and it would be preferable if a less symmetric solution existed, with a larger classical energy, which could be quantized with a lower spin. There is some progress in this direction.[52] Quantum states of the $B = 5$ Skyrmion also differ from those of $^5$He/$^5$Li, but these nuclei are highly unstable.

### 1.5.1. $B = 4$

The $B = 4$ Skyrmion has $O_h$ symmetry, one of the largest point symmetry groups. This leads to particularly restrictive FR constraints on the space of allowed states. The $O_h$ symmetry implies that the inertia tensors are diagonal with $U_{11} = U_{22}$ and $U_{33}$ different, $V_{ij}$ proportional to the identity matrix and $W_{ij} = 0$. The quantum collective coordinate Hamiltonian is therefore the sum of a spherical top in space and a symmetric top in isospace,

$$H = \frac{1}{2V_{11}}\mathbf{J}^2 + \frac{1}{2U_{11}}\mathbf{I}^2 + \left(\frac{1}{2U_{33}} - \frac{1}{2U_{11}}\right) K_3^2 . \tag{1.5.35}$$

For a derivation of its quantum states we refer the reader to Ref. 44.

The lowest state is a $J^P = 0^+$ state with isospin 0, agreeing with the quantum numbers of the $\alpha$-particle in its ground state. The first excited state with isospin 0 is a $4^+$ state, which has not been experimentally observed, probably because of its high energy. It should be regarded as a success of the Skyrme model that because of the $O_h$ symmetry there is no rotational $2^+$ state. The lowest state with isospin 1 is a $2^-$ state, which matches the observed isotriplet of nuclei including the $^4$H and $^4$Li ground states.

### 1.5.2. $B = 6$

The $B = 6$ Skyrmion has $D_{4d}$ symmetry. The quantum Hamiltonian is that of a system of coupled symmetric tops:

$$H = \frac{1}{2V_{11}}\mathbf{J}^2 + \frac{1}{2U_{11}}\mathbf{I}^2 + \left(\frac{U_{33}}{2\Delta_{33}} - \frac{1}{2V_{11}}\right) L_3^2 + \left(\frac{V_{33}}{2\Delta_{33}} - \frac{1}{2U_{11}}\right) K_3^2 + \frac{W_{33}}{\Delta_{33}} L_3 K_3 , \tag{1.5.36}$$

where $\Delta_{33} = U_{33}V_{33} - W_{33}^2$. Its allowed quantum states, discussed in Ref. 44, are listed in Table 1.3. Recall that the notation is $|J, L_3\rangle \otimes |I, K_3\rangle$.

This qualitatively reproduces the experimental spectrum of $^6$Li and its isobars, which is shown in Fig. 1.10. There are rather few complete isospin multiplets here.

Table 1.3.    States of the quantized $B = 6$ Skyrmion.

| $I$ | $J^P$ | Quantum State |
|---|---|---|
| 0 | $1^+$ | $\lvert 1,0\rangle \otimes \lvert 0,0\rangle$ |
| | $3^+$ | $\lvert 3,0\rangle \otimes \lvert 0,0\rangle$ |
| | $4^-$ | $(\lvert 4,4\rangle - \lvert 4,-4\rangle) \otimes \lvert 0,0\rangle$ |
| | $5^+$ | $\lvert 5,0\rangle \otimes \lvert 0,0\rangle$ |
| 1 | $0^+$ | $\lvert 0,0\rangle \otimes \lvert 1,0\rangle$ |
| | $2^+$ | $\lvert 2,0\rangle \otimes \lvert 1,0\rangle$ |
| | | $\lvert 2,2\rangle \otimes \lvert 1,1\rangle + \lvert 2,-2\rangle \otimes \lvert 1,-1\rangle$ |
| | $2^-$ | $\lvert 2,2\rangle \otimes \lvert 1,-1\rangle + \lvert 2,-2\rangle \otimes \lvert 1,1\rangle$ |
| | $3^+$ | $\lvert 3,2\rangle \otimes \lvert 1,1\rangle - \lvert 3,-2\rangle \otimes \lvert 1,-1\rangle$ |
| | $3^-$ | $\lvert 3,2\rangle \otimes \lvert 1,-1\rangle - \lvert 3,-2\rangle \otimes \lvert 1,1\rangle$ |
| | $4^+$ | $\lvert 4,0\rangle \otimes \lvert 1,0\rangle$ |
| | | $\lvert 4,2\rangle \otimes \lvert 1,1\rangle + \lvert 4,-2\rangle \otimes \lvert 1,-1\rangle$ |
| | $4^-$ | $\lvert 4,2\rangle \otimes \lvert 1,-1\rangle + \lvert 4,-2\rangle \otimes \lvert 1,1\rangle$ |
| | | $(\lvert 4,4\rangle + \lvert 4,-4\rangle) \otimes \lvert 1,0\rangle$ |
| 2 | $0^-$ | $\lvert 0,0\rangle \otimes (\lvert 2,2\rangle - \lvert 2,-2\rangle)$ |
| | $1^+$ | $\lvert 1,0\rangle \otimes \lvert 2,0\rangle$ |
| | $1^-$ | $\lvert 1,0\rangle \otimes (\lvert 2,2\rangle + \lvert 2,-2\rangle)$ |
| | $2^+$ | $\lvert 2,2\rangle \otimes \lvert 2,1\rangle - \lvert 2,-2\rangle \otimes \lvert 2,-1\rangle$ |
| | $2^-$ | $\lvert 2,0\rangle \otimes (\lvert 2,2\rangle - \lvert 2,-2\rangle)$ |
| | | $\lvert 2,2\rangle \otimes \lvert 2,-1\rangle - \lvert 2,-2\rangle \otimes \lvert 2,1\rangle$ |

Beyond the low-lying $J^P = 1^+$ and $3^+$ states, further isospin zero states of $^6$Li with $J^P = 4^-$ and $5^+$ are predicted. The lowest isospin 1 states have $J^P = 0^+$ and $2^+$, matching those observed in $^6$He, $^6$Li and $^6$Be. The lowest isospin 2 state, matching the ground state of $^6$H, is predicted to have $J^P = 0^-$.

### 1.5.3.  $B = 8$

When $m = 1$, the stable $B = 8$ Skyrmion is $D_{4h}$-symmetric, and resembles two touching $B = 4$ cubes, matching the known physics that $^8$Be is an almost bound configuration of two $\alpha$-particles (see Fig. 1.4C). The quantum Hamiltonian is the sum of a symmetric top in space and an asymmetric top in isospace[7]:

$$H = \frac{1}{2V_{11}}\mathbf{J}^2 + \left(\frac{1}{2V_{33}} - \frac{1}{2V_{11}}\right) L_3^2 + \frac{1}{2U_{11}}K_1^2 + \frac{1}{2U_{22}}K_2^2 + \frac{1}{2U_{33}}K_3^2 . \qquad (1.5.37)$$

The numbers of independent FR-allowed energy eigenstates, $n$, for a range of $I$ and $J^P$ values, are listed in Table 1.4. For comparison, Fig. 1.11 is an energy level diagram for nuclei with baryon number 8. The Skyrme model predictions for positive parity states agree well with experiment. However, of particular interest is the prediction of an additional isospin triplet of $J^P = 0^-$ states, and further negative parity states, which if established experimentally could include new ground states of the $^8$Li and $^8$B nuclei. Low-lying spin 0, negative parity states could be difficult to observe, as experienced in the search for the bottomonium and charmonium ground state mesons $\eta_b$ and $\eta_c$.[54,55]

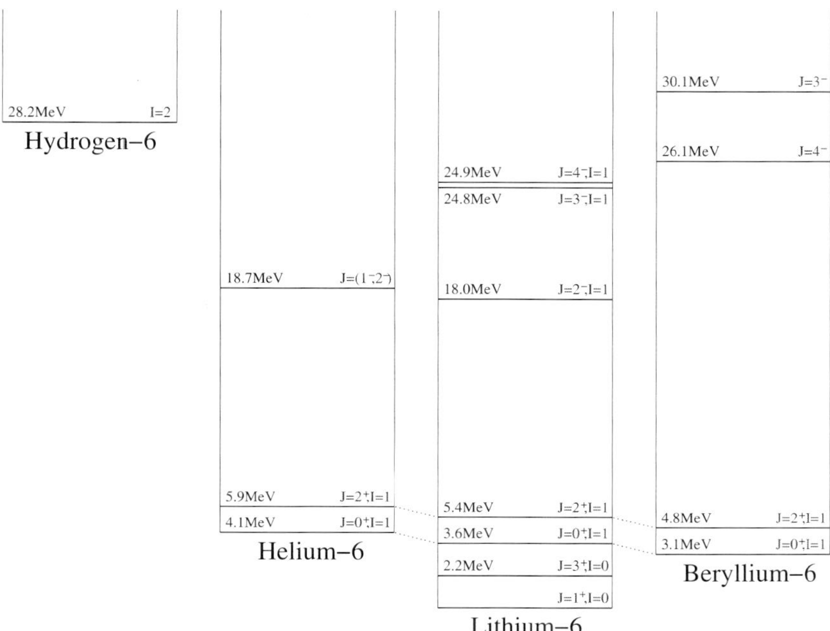


Fig. 1.10. Energy level diagram for nuclei with baryon number 6. Here, and similarly in later figures, individual isobars are shifted vertically for clarity, and mass splittings between nuclei are adjusted to eliminate the proton/neutron mass difference and remove Coulomb effects, as described in Ref. 53.

### 1.5.4. $B = 10$

As described earlier, the $B = 10$ Skyrmion has $D_{2h}$ symmetry[15] and it can be thought of as a pair of $B = 4$ cubes with two $B = 1$ Skyrmions between them. This Skyrmion was recently quantized for the first time, using the rational map ansatz to determine its FR constraints.[7] We give some details of this here. Using the rational map (1.4.16), which shares the same symmetry as the exact solution, the $D_2$ rotation group is realized as

$$R(-z) = R(z), \quad R(1/z) = 1/R(z). \tag{1.5.38}$$

The first symmetry involves no isorotation, but the second one combines the spatial rotation with a 180° rotation about the 1-axis in isospace. The integers $\mathcal{N}$, determined using (1.5.34), are therefore 50 and 45 respectively, so the signs $\chi_{\mathrm{FR}}$ are 1 and $-1$, and generate one of the non-trivial 1-dimensional representations of $D_2$. The FR constraints are

$$e^{i\pi L_3}|\Psi\rangle = |\Psi\rangle, \quad e^{i\pi L_1}e^{i\pi K_1}|\Psi\rangle = -|\Psi\rangle. \tag{1.5.39}$$

A rational map is invariant under parity if it satisfies

$$R(-1/\bar{z}) = -1/\overline{R(z)}. \tag{1.5.40}$$

R.A. Battye, N.S. Manton and P.M. Sutcliffe

Table 1.4. States of the quantized $B = 8$ Skyrmion.

| $I$ | $J^P$ | $n$ |
|---|---|---|
| 0 | $0^+$ | 1 |
|   | $2^+$ | 1 |
|   | $4^+$ | 2 |
| 1 | $0^-$ | 1 |
|   | $2^+$ | 1 |
|   | $2^-$ | 2 |
|   | $3^+$ | 1 |
|   | $3^-$ | 1 |
|   | $4^+$ | 1 |
|   | $4^-$ | 3 |
| 2 | $0^+$ | 2 |
|   | $0^-$ | 1 |
|   | $2^+$ | 3 |
|   | $2^-$ | 2 |

The rational map (1.4.16) does not have this symmetry, but has the closely related reflection symmetry

$$R(-1/\bar{z}) = 1/\overline{R(z)}\,. \tag{1.5.41}$$

The parity operator in this case is therefore equivalent to a single rotation in isospace, given by $\mathcal{P} = e^{i\pi K_3}$, whose eigenvalue determines the parity $P$ of a quantum state.

The symmetries of the $B = 10$ Skyrmion, as seen from its rational map, imply that the inertia tensors $U_{ij}$ and $V_{ij}$ are diagonal, and the only non-zero component of the mixed inertia tensor $W_{ij}$ is $W_{33}$. The quantum Hamiltonian is that of a system of coupled asymmetric tops:

$$H = \frac{1}{2V_{11}}L_1^2 + \frac{1}{2V_{22}}L_2^2 + \frac{U_{33}}{2\Delta_{33}}L_3^2 + \frac{1}{2U_{11}}K_1^2 + \frac{1}{2U_{22}}K_2^2 + \frac{V_{33}}{2\Delta_{33}}K_3^2 + \frac{W_{33}}{\Delta_{33}}L_3K_3\,, \tag{1.5.42}$$

where $\Delta_{33} = U_{33}V_{33} - W_{33}^2$ as before.

In Table 1.5 we list the number of independent FR-allowed states, $n$, for different combinations of spin and isospin, up to isospin 3. The calculation of energy levels requires a matrix diagonalization, separately for each combination of $J^P$ and $I$.

For isospin 0, all states have positive parity. The lowest allowed state has $J^P = 1^+$, and there are various excited states including one $2^+$ state and two $3^+$ states. For isospin 1 just about every spin/parity pairing is allowed. Only $J^P = 1^+$ is forbidden. For higher isospin, no $J^P$ combination is forbidden.

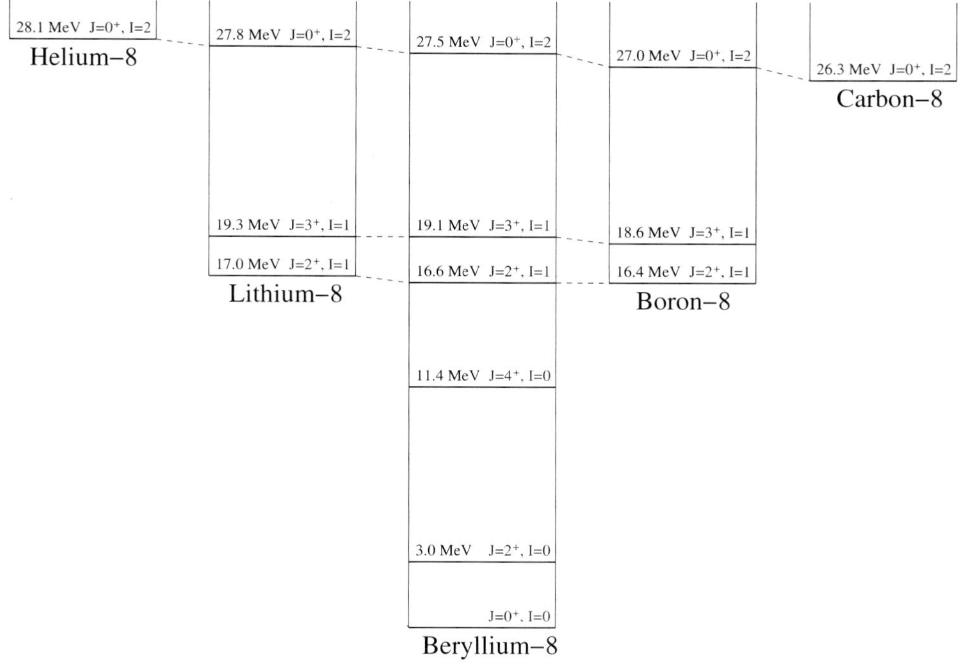


Fig. 1.11. Energy level diagram for nuclei with baryon number 8.[56]

The experimental energy spectrum for $B = 10$ nuclei is shown in Fig. 1.12. The physical ground state of $^{10}$B has $J^P = 3^+$ and isospin zero, and its first excited state has $J^P = 1^+$. We incorrectly predict the $1^+$ state as the ground state, and the $3^+$ states as excited states. However, this problem arises in many models of $^{10}$B, for example in models involving nucleon-nucleon potentials in chiral perturbation theory.[57] We also predict that the $2^+$ state lies below the lowest $3^+$ state, although experimentally it is higher.

We predict an isospin 1 triplet of $0^+$ states, which match the ground states of $^{10}$Be and $^{10}$C and an excited state of $^{10}$B. We also predict two $2^+$ states for these nuclei, whereas experimentally three are seen. In agreement with the model, no isospin 1 states with $J^P = 1^+$ are observed. Isospin 2 states are predicted, matching the incomplete quintet of observed states, including the $^{10}$Li and $^{10}$N ground states, whose spins are not certain and apparently not the same. The lowest isospin 3 state matches the ground state of $^{10}$He.

Apart from missing the spin $3^+$ ground state of $^{10}$B, the Skyrme model does quite well in the $B = 10$ sector. This is probably because the shape of the classical Skyrmion and its symmetries are what is expected from the cluster model picture, with two $\alpha$-particles and two additional nucleons between them.[10] This picture has previously been successful in modelling $^{10}$Be and $^{10}$C at least.

Table 1.5. States of the quantized $B = 10$ Skyrmion.

| $I$ | $J^P$ | $n$ |
|---|---|---|
| 0 | $1^+$ | 1 |
|   | $2^+$ | 1 |
|   | $3^+$ | 2 |
|   | $4^+$ | 2 |
| 1 | $0^+$ | 1 |
|   | $0^-$ | 1 |
|   | $1^-$ | 1 |
|   | $2^+$ | 2 |
|   | $2^-$ | 3 |
|   | $3^+$ | 1 |
|   | $3^-$ | 3 |
|   | $4^+$ | 3 |
|   | $4^-$ | 5 |
| 2 | $0^+$ | 1 |
|   | $0^-$ | 1 |
|   | $1^+$ | 2 |
|   | $1^-$ | 1 |
|   | $2^+$ | 4 |
|   | $2^-$ | 3 |
|   | $3^+$ | 5 |
|   | $3^-$ | 3 |
| 3 | $0^+$ | 2 |
|   | $0^-$ | 2 |
|   | $1^+$ | 1 |
|   | $1^-$ | 2 |
|   | $2^+$ | 5 |
|   | $2^-$ | 6 |

### 1.5.5. $B = 12$

We described earlier the triangular $B = 12$ Skyrmion with $D_{3h}$ symmetry, and its approximation using the double rational map ansatz. The symmetry generators are a combined $120°$ rotation and $120°$ isorotation, and a combined $180°$ rotation and $180°$ isorotation about an orthogonal pair of axes. As the baryon number is a multiple of four, the FR signs are all $+1$, and the FR constraints are

$$e^{i\frac{2\pi}{3}L_3}e^{i\frac{2\pi}{3}K_3}|\Psi\rangle = |\Psi\rangle, \quad e^{i\pi L_1}e^{i\pi K_1}|\Psi\rangle = |\Psi\rangle. \tag{1.5.43}$$

The rational maps satisfy the reflection symmetry

$$R(1/\bar{z}) = 1/\overline{R(z)}, \tag{1.5.44}$$

which differs from the parity operation by a pair of minus signs, so the parity operator is equivalent to $\mathcal{P} = e^{i\pi L_3}e^{i\pi K_3}$.

The $D_{3h}$ symmetry implies that the inertia tensors are diagonal, with $U_{11} = U_{22}$, $V_{11} = V_{22}$ and $W_{11} = W_{22}$, so the quantum Hamiltonian is that of a system of

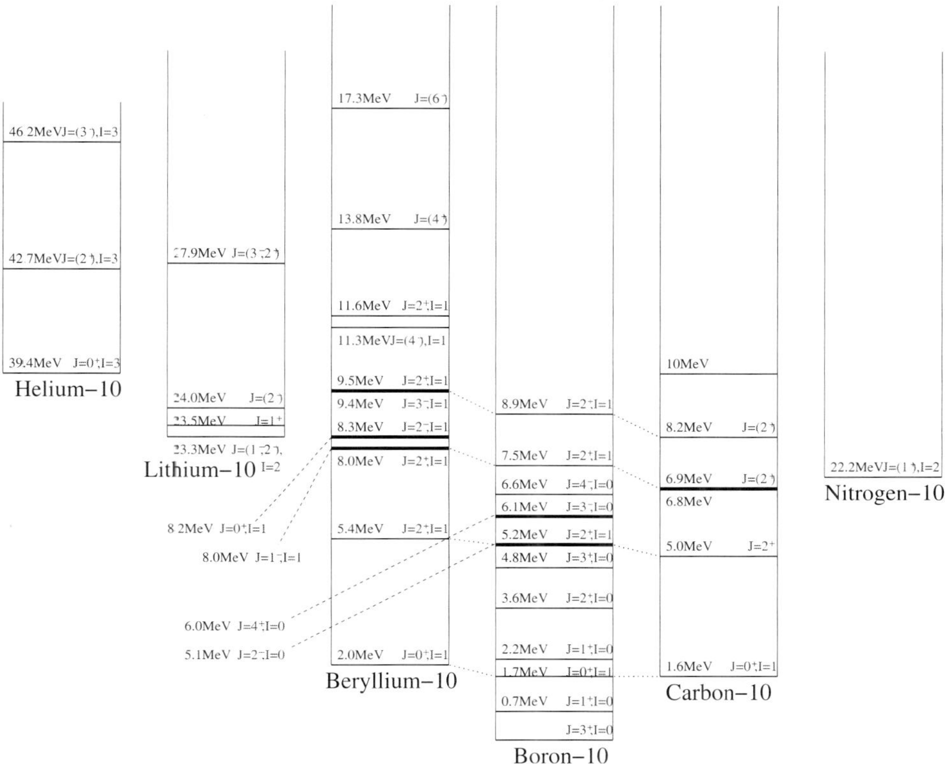



Fig. 1.12. Energy level diagram for nuclei with baryon number 10.[56]

coupled symmetric tops[7]:

$$H = \left(\frac{U_{11} - W_{11}}{2\Delta_{11}}\right)\mathbf{J}^2 + \left(\frac{V_{11} - W_{11}}{2\Delta_{11}}\right)\mathbf{I}^2 + \left(\frac{W_{11}}{2\Delta_{11}}\right)\mathbf{M}^2$$

$$+ \left(\frac{U_{33}}{2\Delta_{33}} - \frac{U_{11}}{2\Delta_{11}}\right)L_3^2 + \left(\frac{V_{33}}{2\Delta_{33}} - \frac{V_{11}}{2\Delta_{11}}\right)K_3^2 + \left(\frac{W_{33}}{\Delta_{33}} - \frac{W_{11}}{\Delta_{11}}\right)L_3 K_3 , \quad (1.5.45)$$

where $\mathbf{M} = \mathbf{L} - \mathbf{K}$, $\Delta_{33}$ is as before, and $\Delta_{11} = U_{11}V_{11} - W_{11}^2$.

The states that are allowed by the FR constraints are shown in Table 1.6. Each of the allowed states with isospin 0 is also an eigenstate of the Hamiltonian, with an energy that is easily determined. These isospin 0 states also result from the rigid-body quantization of an equilateral triangle with $\alpha$-particles at its vertices, and are not a prediction characteristic of the Skyrme model itself. The states fall into rotational bands labelled by $|L_3| = 0, 3, 6, \dots$. These fit the $^{12}$C data quite well, provided, as in Ref. 58, we reassign the lowest experimental $J^P = 2^-$ state of $^{12}$C as a $4^-$ state. Again as in Ref. 58, we predict a relatively low-energy $6^+$

state with $|L_3| = 6$, which has not yet been seen experimentally. The experimental spectrum for $B = 12$ nuclei is shown in Fig. 1.13.

To find the isospin 1 states that are FR-allowed and also eigenstates of the Hamiltonian requires a matrix diagonalization.[7] Isospin 2 states require a similar treatment. Because of off-diagonal elements in these matrices, the eigenstates mix the $|L_3|$ and $|K_3|$ quantum numbers, but this mixing is small, so states can be labelled by their dominant $|L_3|$ and $|K_3|$ values. Table 1.6 lists these dominant quantum numbers. We predict two $J^P = 1^+$, isospin 1 triplets. One such isotriplet is observed, and includes the ground states of $^{12}$B and $^{12}$N. We also predict a $2^+$ and a $2^-$ isotriplet. Both of these are seen experimentally, but in the opposite energy order. The observed, relatively high-lying $0^+$ isotriplet is not explained. An (incomplete) $J^P = 0^+$, isospin 2 quintet is observed experimentally, which includes the ground states of $^{12}$Be and $^{12}$O. We predict such an isoquintet, but at a higher energy than a $J^P = 1^-$ isoquintet.

The spin and isospin moments of inertia of the $B = 8$ Skyrmion can be estimated by treating it as a "double cube" of $B = 4$ Skyrmions, and this is quite accurate.[44] Similarly, one can treat the $B = 12$ Skyrmion as a triangle of cubes, and estimate its inertia tensors in terms of those of the $B = 4$ constituents, using the parallel axis theorem.[7] It is found that the inertia tensors are simpler than the exact ones, since the tensor $W_{ij}$ vanishes, and it is easier to determine the eigenstates of the quantum Hamiltonian and the energy spectrum. This provides a check on the more difficult exact calculations, and some physical insight. With $W_{ij}$ vanishing, the quantum Hamiltonian simplifies to the sum of a symmetric top in space and a symmetric top in isospace:

$$H = \frac{1}{2V_{11}}\mathbf{J}^2 + \frac{1}{2U_{11}}\mathbf{I}^2 + \left(\frac{1}{2V_{33}} - \frac{1}{2V_{11}}\right)L_3^2 + \left(\frac{1}{2U_{33}} - \frac{1}{2U_{11}}\right)K_3^2. \qquad (1.5.46)$$

$|L_3|$ and $|K_3|$ become good quantum numbers.

$^{12}$C has an excited $0^+$ state at 7.65 MeV, the Hoyle state. Unfortunately our method of rigid-body quantization prohibits two independent spin 0, isospin 0 states. We mentioned earlier that the lowest-lying quantum state of an alternative $B = 12$ solution, such as the solution with three $B = 4$ Skyrmions in a linear chain, could be interpreted as this excited $0^+$ state.

## 1.6. Calibration and Energy Levels

The free parameters of the Skyrme model are $F_\pi$, $e$, and $m_\pi$. Of these, two set an energy scale and a length scale, which have been scaled out in eq. (2.2) and there is one remaining dimensionless combination, $m$, which appears in (2.2). It is convenient to work with the combinations $F_\pi/4e$ and $2/eF_\pi$. $F_\pi/4e$ has dimensions of energy and is numerically a few MeV. The dimensionless classical Skyrmion energy $E$ is turned into a physical energy by multiplying by $F_\pi/4e$. $2/eF_\pi$ is a length scale, and numerically about 1 fm. (More precisely it is an inverse energy scale which is converted

Table 1.6. Quantum states of the $B = 12$ Skyrmion. To each state there correspond dominant values of $|L_3|$ and $|K_3|$.

| $I$ | $J^P$ | $|L_3|$ | $|K_3|$ |
|---|---|---|---|
| 0 | $0^+$ | 0 | 0 |
| | $2^+$ | 0 | 0 |
| | $3^-$ | 3 | 0 |
| | $4^-$ | 3 | 0 |
| | $4^+$ | 0 | 0 |
| | $5^-$ | 3 | 0 |
| | $6^-$ | 3 | 0 |
| | $6^+$ | 0 | 0 |
| | | 6 | 0 |
| 1 | $1^+$ | 1 | 1 |
| | | 0 | 0 |
| | $2^-$ | 2 | 1 |
| | $2^+$ | 1 | 1 |
| | $3^-$ | 3 | 0 |
| | | 2 | 1 |
| | $3^+$ | 1 | 1 |
| | | 0 | 0 |
| | $4^-$ | 4 | 1 |
| | | 3 | 0 |
| | | 2 | 1 |
| | $4^+$ | 1 | 1 |
| 2 | $0^+$ | 0 | 0 |
| | $1^-$ | 1 | 2 |
| | $1^+$ | 1 | 1 |
| | $2^-$ | 1 | 2 |
| | | 2 | 1 |
| | $2^+$ | 2 | 2 |
| | | 1 | 1 |
| | | 0 | 0 |

to a length scale through the conversion factor $\hbar c = 197.3$ MeV fm.) A further combination is $m_\pi = m(2/eF_\pi)^{-1}$, which is the tree level pion mass in the Skyrme Lagrangian. Also important is the derived quantity $e^3 F_\pi = (F_\pi/4e)^{-1}(2/eF_\pi)^{-2}$. This is the quantum energy scale, which determines the physical contributions of the quantized spin and isospin to the total Skyrmion energy.

$F_\pi$ is supposed to be the "pion decay constant", but in most work on Skyrmions, it has not been given its physical value. Nevertheless, one normally uses the physical pion mass $m_\pi = 138$ MeV to determine $m$ from the length scale $2/eF_\pi$. However, one might do something different, as the tree level pion mass is not necessarily the physical value.

Adkins, Nappi and Witten originally calibrated the Skyrme model by matching the masses of the nucleons and deltas to the quantized, rigidly rotating $B = 1$ Skyrmion.[5] The classical $B = 1$ Skyrmion then has energy about 860 MeV, and

     *R.A. Battye, N.S. Manton and P.M. Sutcliffe*

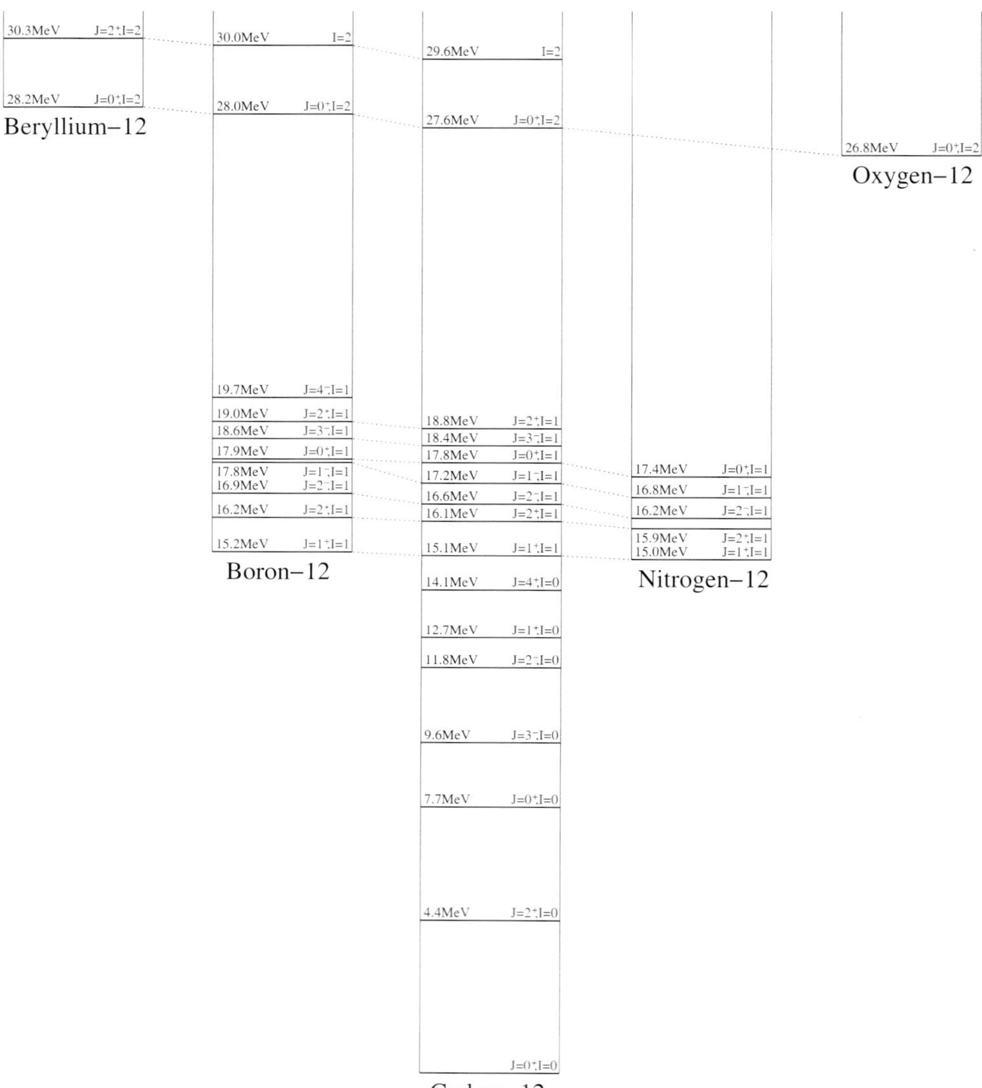


Fig. 1.13.   Energy level diagram for nuclei with baryon number 12.   Mass splittings between nuclei are adjusted to eliminate the proton/neutron mass difference and remove Coulomb effects, as described in Ref. 59.

the physical particles have additional spin energy. But this was done in the approximation of zero pion mass. An improved calibration along the same lines was performed by Adkins and Nappi, using the physical pion mass.[12]  However, the delta is a broad resonance about 300 MeV above the nucleon ground state, and it strongly radiates pions. Related to this, it has been observed that the spin of the delta strongly deforms the $B = 1$ Skyrmion,[13,60] and if this is taken into account, it has a rather complicated effect on the calibration of the model. Moreover, it has been found that the Adkins and Nappi parameters lead to higher baryon number Skyrmions that are too tightly bound and too small to model the corresponding nuclei.

To create larger, less tightly bound nuclei, Leese et al.[48] in their Skyrme model analysis of the deuteron state, and also Walhout,[50] analysing the $\alpha$-particle, took some account of selected low-frequency vibrational modes. This works quite well. However, for larger baryon numbers, it becomes more difficult to make any allowance for vibrational motion of Skyrmions, so subsequent work has considered only the quantized rigid-body motion. In any case, the quantized harmonic oscillator approximation to Skyrmion vibrations is quite poor. Many vibrational modes lead to Skyrmions breaking into smaller clusters, and here one would wish to model the multi-dimensional potential energy as tending to a constant at large amplitude, and not rising quadratically.

It may be adequate to ignore explicit vibrational motion when modelling larger nuclei, but to take some account of the ground state vibrational motion by a "renormalization" of the Skyrme parameters. What this means in practice is that to model nuclei in the Skyrme model, the length scale should be about twice as large as that proposed by Adkins and Nappi. Also, one may need to let these parameters vary as the baryon number changes. This slightly reduces the predictive power of the Skyrme model.

In Ref. 16, a recalibration of the Skyrme model around the properties of the $^6$Li nucleus was performed. The $B = 6$ Skyrmion is well-known, and can be approximated by the rational map ansatz, which is useful when estimating the energy and size, and calculating the quantum states. The rotational motion of the Skyrmion is quite non-relativistic, not leading to strong pion radiation, nor to significant Skyrmion deformation. This is because the ground state of spin 1 and the first excited state of spin 3 are separated by just a few MeV, whereas the mass of the nucleus is approximately 5600 MeV. Because the isospin is zero, the electric charge density (within the Skyrme model) is half the baryon density, so it is straightforward to estimate the charge radius of the quantized Skyrmion.

In detail, $F_\pi/4e$ is determined by requiring the $B = 6$ Skyrmion to have mass 5600 MeV ($^6$Li has mass 5601 MeV, of which it is estimated that 1 MeV comes from the spin energy in the ground state), and $2/eF_\pi$ is determined from the $^6$Li charge radius of 2.6 fm. $m$ is determined from the physical pion mass, but because of the significant change of length scale, it is about double the traditional value, and is now $m = 1.125$. This value of $m$ is in the range where the solutions described in Section 1.4, constructed from $B = 4$ cubes, are favoured.

Recently, we have carried out detailed calculations of the Skyrmion solutions for all the even values of $B$ up to $B = 12$, finding their masses and moments of inertia, for a range of values of $m$ between 0.5 and 1.5.[7] We have also calibrated the Skyrme model parameters separately for each of these values of $B$, using the known masses and charge radii of the isospin zero nuclei in each case ($^4$He, $^6$Li, $^8$Be, $^{10}$B and $^{12}$C). As expected, $F_\pi/4e$ is approximately constant at 6 MeV, reflecting the fact that the nuclear masses are almost exactly proportional to $B$, and the Skyrmion masses are too. The length scale $2/eF_\pi$ is more variable, because Skyrmions increase in size steadily with $B$, but the physical charge radii between $^6$Li and $^{12}$C are almost constant, and slightly decreasing. Consequently, $m$ is varying, but is still around the value 1.

With the parameters fixed, the energy spectrum of isospin and spin excitations can be regarded as a quantitative prediction of the Skyrme model. For the details, the reader in referred to Ref. 7. The isospin splittings are about 10 MeV between $I = 0$ and $I = 1$ states, increasing to about 60 MeV between $I = 0$ and $I = 3$ states. This is satisfactory, although the high isospin states are rather too high in energy. The spin splittings vary from a few MeV between $J = 0$ and $J = 1$ states, up to 10-40 MeV between $J = 0$ and $J = 4$ states. The varying length scale generally has a favourable effect. For example the large size of $^6$Li is consistent with the relatively small spin splittings. Better quantitatively is the fit to the rotational bands of states of $^8$Be and $^{12}$C. The splittings between the lowest $0^+$, $2^+$ and $4^+$ states come out well, being slightly greater for $^{12}$C than for $^8$Be, consistent with the smaller length scale for $^{12}$C.

## 1.7. Conclusion

We have reviewed Skyrme's original model and its application to modelling nuclei. Work over the last twenty years has led to a much greater understanding of the classical Skyrmion solutions for quite a large range of baryon numbers. A key discovery is that for the physical value of the pion mass the solutions have a much more compact structure than for massless pions. These compact structures have many similarities to $\alpha$-particle and other cluster models of nuclei.

We have reviewed the rigid body quantization of these Skyrmions. The quantization gives a uniform understanding of spin and isospin excitations. The most recent calculations give reasonable energy spectra for nuclei with baryon numbers 8, 10 and 12.[7]

Some details of the spectra are not satisfactory and it would be useful to explore whether variants of the Skyrme model offer improvements. It would be very helpful if the Skyrme model could be better related to chiral effective field theory.

## Acknowledgements

This review is partly based on papers written jointly with O.M. Manko and S.W. Wood. We would like to thank them for their contribution to the research described here.

# References


1. T.H.R. Skyrme, A nonlinear field theory, *Proc. R. Soc. A* **260** (1961) 127.
2. T.H.R. Skyrme, A unified field theory of mesons and baryons, *Nucl. Phys.* **31** (1962) 556.
3. M.J. Esteban, A direct variational approach to Skyrme's model for meson fields, *Commun. Math. Phys.* **105** (1986) 571.
4. N.S. Manton, B.J. Schroers and M.A. Singer, The interaction energy of well-separated Skyrme solitons, *Commun. Math. Phys.* **245** (2004) 123.
5. G.S. Adkins, C.R. Nappi and E. Witten, Static properties of nucleons in the Skyrme model, *Nucl. Phys. B* **228** (1983) 552.
6. E. Braaten and L. Carson, Deuteron as a toroidal Skyrmion, *Phys. Rev. D* **38** (1998) 3525.
7. R.A. Battye, N.S. Manton, P.M. Sutcliffe and S.W. Wood, in preparation.
8. D. Finkelstein and J. Rubinstein, Connection between spin, statistics and kinks, *J. Math. Phys.* **9** (1968) 1762.
9. S. Krusch, Homotopy of rational maps and the quantization of Skyrmions, *Ann. Phys.* **304** (2003) 103.
10. W. von Oertzen, M. Freer and Y. Kanada-En'yo, Nuclear clusters and nuclear molecules, *Phys. Rep.* **432** (2006) 43.
11. N. Manton and P. Sutcliffe, *Topological Solitons* (Chapter 9), Cambridge University Press, Cambridge, 2004.
12. G.S. Adkins and C.R. Nappi, The Skyrme model with pion masses, *Nucl. Phys. B* **233** (1984) 109.
13. R.A. Battye, S. Krusch and P.M. Sutcliffe, Spinning Skyrmions and the Skyrme parameters, *Phys. Lett. B* **626** (2005) 120.
14. R.A. Battye and P.M. Sutcliffe, Skyrmions and the pion mass, *Nucl. Phys. B* **705** (2005) 384.
15. R.A. Battye and P.M. Sutcliffe, Skyrmions with massive pions, *Phys. Rev. C* **73** (2006) 055205.
16. N.S. Manton and S.W. Wood, Reparametrising the Skyrme model using the lithium-6 nucleus, *Phys. Rev. D* **74** (2006) 125017.
17. R.A. Battye and P.M. Sutcliffe, Symmetric Skyrmions, *Phys. Rev. Lett.* **79** (1997) 363.
18. R.A. Battye and P.M. Sutcliffe, Solitonic fullerene structures in light atomic nuclei, *Phys. Rev. Lett.* **86** (2001) 3989.
19. R.A. Battye and P.M. Sutcliffe, Skyrmions, fullerenes and rational maps, *Rev. Math. Phys.* **14** (2002) 29.
20. J.L. Forest, V.R. Pandharipande, S.C. Pieper, R.B. Wiringa, R. Schiavilla and A. Arriaga, Femtometer toroidal structures in nuclei, *Phys. Rev. C* **54** (1996) 646.
21. E. Piasetzky et al., Evidence for strong dominance of proton-neutron correlations in nuclei, *Phys. Rev. Lett.* **97** (2006) 162504.
22. R.A. Battye, N.S. Manton and P.M. Sutcliffe, Skyrmions and the $\alpha$-particle model of nuclei, *Proc. R. Soc. A* **463** (2007) 261.
23. R.A. Battye and P.M. Sutcliffe, A Skyrme lattice with hexagonal symmetry, *Phys. Lett. B* **416** (1998) 385.
24. N.J. Hitchin, N.S. Manton and M.K. Murray, Symmetric monopoles, *Nonlinearity* **8** (1995) 661.
25. C.J. Houghton and P.M. Sutcliffe, Octahedral and dodecahedral monopoles, *Nonlinearity* **9** (1996) 385.

26. C.J. Houghton and P.M. Sutcliffe, Tetrahedral and cubic monopoles, *Commun. Math. Phys.* **180** (1996) 343.

27. S. Jarvis, A rational map for Euclidean monopoles via radial scattering, *J. reine angew. Math.* **524** (2000) 17.

28. C.J. Houghton, N.S. Manton and P.M. Sutcliffe, Rational maps, monopoles and Skyrmions, *Nucl. Phys. B* **510** (1998) 507.

29. N.S. Manton and B.M.A.G. Piette, Understanding Skyrmions using rational maps, *Prog. Math.* **201** (2001) 469.

30. J.M. Blatt and V.F. Weisskopf, *Theoretical Nuclear Physics* (p. 292), Wiley, New York, 1952.

31. D.M. Brink, H. Friedrich, A. Weiguny and C.W. Wong, Investigation of the alpha-particle model for light nuclei, *Phys. Lett. B* **33** (1970) 143.

32. A.H. Wuosmaa, R.R. Betts, M. Freer and B.R. Fulton, Recent advances in the study of nuclear clusters, *Ann. Rev. Nucl. Part. Sci.* **45** (1995) 89.

33. N.S. Manton, Skyrmions and their pion multipole moments, *Acta Phys. Pol. B* **25** (1994) 1757.

34. H. Morinaga, Interpretation of some of the excited states of 4n self-conjugate nuclei, *Phys. Rev.* **101** (1956) 254.

35. H. Friedrich, L. Satpathy and A. Weiguny, Why is there no rotational band based on the 7.65 MeV $0^+$ state in $^{12}$C? *Phys. Lett. B* **36** (1971) 189.

36. W. Bauhoff, H. Schultheis and R. Schultheis, Alpha cluster model and the spectrum of $^{16}$O, *Phys. Rev. C* **29** (1984) 1046.

37. D.M. Dennison, Energy levels of the $^{16}$O nucleus, *Phys. Rev.* **96** (1954) 378.

38. D. Robson, Evidence for the tetrahedral nature of $^{16}$O, *Phys. Rev. Lett.* **42** (1979) 876.

39. W.K. Baskerville, Making nuclei out of the Skyrme crystal, *Nucl. Phys. A* **596** (1996) 611.

40. T. Neff and H. Feldmeier, Short-ranged radial and tensor correlations in nuclear many-body systems, *Proc. Int. Workshop XXXI on Nuclear Structure and Dynamics, (Hirschegg, Austria, 2003).* arXiv:nucl-th/0303007.

41. D.S. Freed, Pions and generalized cohomology, *J. Differential Geom.* **80** (2008) 45.

42. J.G. Williams, Topological analysis of a nonlinear field theory, *J. Math. Phys.* **11** (1970) 2611.

43. D. Giulini, On the possibility of spinorial quantization in the Skyrme model, *Mod. Phys. Lett. A* **8** (1993) 1917.

44. O.V. Manko, N.S. Manton and S.W. Wood, Light nuclei as quantized Skyrmions, *Phys. Rev. C* **76** (2007) 055203.

45. S. Krusch, Finkelstein–Rubinstein constraints for the Skyrme model with pion masses, *Proc. R. Soc. A* **462** (2006) 2001.

46. G. Segal, The topology of the space of rational maps, *Acta Math.* **143** (1979) 39.

47. V.B. Kopeliovich, Quantization of the axially-symmetric system's rotations in the Skyrme model (in Russian), *Yad. Fiz.* **47** (1988) 1495.

48. R.A. Leese, N.S. Manton and B.J. Schroers, Attractive channel Skyrmions and the deuteron, *Nucl. Phys. B* **442** (1995) 228.

49. L. Carson, $B = 3$ nuclei as quantized multi-Skyrmions, *Phys. Rev. Lett.* **66** (1991) 1406.

50. T.S. Walhout, Quantizing the four-baryon Skyrmion, *Nucl. Phys. A* **547** (1992) 423.

51. P. Irwin, Zero mode quantization of multi-Skyrmions, *Phys. Rev. D* **61** (2000) 114024.

52. O.V. Manko and N.S. Manton, On the spin of the $B = 7$ Skyrmion, *J. Phys. A* **40** (2007) 3683.

53. D.R. Tilley et al., Energy levels of light nuclei A=5, 6, 7, *Nucl. Phys. A* **708** (2002) 3.
54. B. Aubert et al., Observation of the bottomonium ground state in the decay $\Upsilon(3S)$ $\to \gamma\eta_b$, *Phys. Rev. Lett.* **101** (2008) 071801.
55. R. Partridge et al., Observation of an $\eta_c$ candidate state with mass $2978 \pm 9$ MeV, *Phys. Rev. Lett.* **45** (1980) 1150.
56. D.R. Tilley et al., Energy levels of light nuclei A=8, 9, 10, *Nucl. Phys. A* **745** (2004) 155.
57. P. Navrátil and E. Caurier, Nuclear structure with accurate chiral perturbation theory nucleon-nucleon potential: Application to $^6$Li and $^{10}$B, *Phys. Rev. C* **69** (2004) 014311.
58. R. Álvarez-Rodríguez, E. Garrido, A.S. Jensen, D.V. Fedorov and H.O.U. Fynbo, Structure of low-lying 12C-resonances, *Eur. Phys. J. A* **31** (2007) 303.
59. F. Ajzenberg-Selove, Energy levels of light nuclei A=11-12, *Nucl. Phys. A* **506** (1990) 1.
60. C.J. Houghton and S. Magee, A zero-mode quantization of the Skyrmion, *Phys. Lett. B* **632** (2006) 593.

# Chapter 2

# Electromagnetic Form Factors of the Nucleon
# in Chiral Soliton Models

Gottfried Holzwarth

*Fachbereich Physik, Universität Siegen,*
*D-57068 Siegen, Germany*

The ratio of electric to magnetic proton form factors $G_{\mathrm{E}}^{\mathrm{p}}/G_{\mathrm{M}}^{\mathrm{p}}$ as measured in polarization transfer experiments shows a characteristic linear decrease with increasing momentum transfer $Q^2 (< 10\ (\mathrm{GeV/c})^2)$. We present a simple argument how such a decrease arises naturally in chiral soliton models. For a detailed comparison of model results with experimentally determined form factors it is necessary to employ a boost from the soliton rest frame to the Breit frame. To enforce asymptotic counting rules for form factors, the model must be supplemented by suitably chosen interpolating powers $n$ in the boost prescription. Within the minimal $\pi$-$\varrho$-$\omega$ soliton model, with the same $n$ for both, electric and magnetic form factors, it is possible to obtain a very satisfactory fit to all available proton data for the magnetic form factor and to the recent polarization results for the ratio $G_{\mathrm{E}}^{\mathrm{p}}/G_{\mathrm{M}}^{\mathrm{p}}$. At the same time the small and very sensitive neutron electric form factor is reasonably well reproduced. The results show a systematic discrepancy with presently available data for the neutron magnetic form factor $G_{\mathrm{M}}^{\mathrm{n}}$ for $Q^2 > 1\ (\mathrm{GeV/c})^2$. We additionally comment on the possibility to extract information about the form factors in the time-like region and on two-photon exchange contributions to unpolarized elastic scattering which specifically arise in soliton models.


## Contents


2.1  Introduction . . . . . . . . . . . . . . . . . . . . . . . . . . . . . . . . . . . .  41
2.2  Characteristic Feature of the Electric Proton Form Factor . . . . . . . . . . . . . . . .  43
2.3  Chiral $\pi$-$\rho$-$\omega$—Meson Model . . . . . . . . . . . . . . . . . . . . . . . . . . . .  44
2.4  Boost to the Breit Frame . . . . . . . . . . . . . . . . . . . . . . . . . . . . . . . .  46
2.5  Results . . . . . . . . . . . . . . . . . . . . . . . . . . . . . . . . . . . . . . . . .  47
2.6  Extension to Time-Like $Q^2$ . . . . . . . . . . . . . . . . . . . . . . . . . . . . . . .  51
2.7  Two-Photon Amplitudes in Soliton Models . . . . . . . . . . . . . . . . . . . . . . . .  52
References . . . . . . . . . . . . . . . . . . . . . . . . . . . . . . . . . . . . . . . . . . .  54


## 2.1. Introduction

Baryons are spatially extended objects. Soliton models provide spatial profiles for baryons already in leading classical approximation from the underlying effective action. Therefore all types of form factors may readily be extracted from soliton models. Specifically, the wealth of experimental data for electromagnetic nucleon form factors pose a severe challenge for chiral soliton models.

Electron-nucleon scattering experiments which measure ratios of polarization variables have confirmed that with increasing momentum transfer $Q^2 = -q_\mu q^\mu$ the proton electric form factor $G_E^p(Q^2)$ decreases significantly faster than the proton magnetic form factor $G_M^p(Q^2)$. This characteristic feature of the electric proton form factor arises naturally in chiral soliton models of the nucleon and has been predicted previously from such models.[1] In the following section we give a very simple and transparent argument for the origin of this result.

We then present a detailed comparison of presently available experimental data with results from the soliton solution of the minimal $\pi$-$\rho$-$\omega$−meson model. In Section 1.3 we simply state the relevant classical action for the meson fields without derivation or comment. It has been discussed extensively in the literature to which we refer. Similarly, we do not repeat here the derivation of the detailed expressions for the form factors. We state them explicitly only for the simple purely pionic Skyrme model, and indicate the modifications brought about by including dynamical vector mesons.

Form factors in soliton models are obtained in the rest frame of the soliton. A severe source of uncertainty lies in the fact that comparison with experimental data requires a boost to the Breit frame. This difficulty applies to all kinds of models for extended objects with internal structure. Ambiguities due to differences in boost prescriptions become increasingly significant for $Q^2$ around and above $(2M)^2$ (with nucleon mass $M$). In order to enforce superconvergence for $Q^2 \to \infty$, we use in the following a boost prescription with the same interpolating power $n = 2$ for both, electric and magnetic form factors.

In Section 1.5 we then show that within this rather restricted framework it is possible to obtain a satisfactory fit to the presently available data for the electromagnetic proton form factors over more than three orders of magnitude of momentum transfer $Q^2$. This can be achieved with the relevant parameters of the effective action at (or close to) their empirical values. The electric neutron form factor is a small difference between two larger quantities. So it is remarkable that the observed $Q^2$-dependence is also essentially reproduced. The absolute size is closely linked to the effective $\pi$-$\omega$ and $\gamma$-$\omega$ coupling strengths, and it is sensitive to the number of flavors considered. So it is not difficult to bring also this delicate quantity close to the corresponding data. Altogether, this fit then results in a prediction for the magnetic neutron form factor $G_M^n(Q^2)$. It turns out that for $Q^2 > 1(\mathrm{GeV}/c)^2$ where new data are still lacking, the calculated result for $G_M^n(Q^2)$ rises above the magnetic proton form factor. This is in conflict with existing older data.

Prospects to obtain results from soliton models for form factors in the time-like region are briefly discussed in Section 1.6.

Finally, leading contributions to the $2\gamma$-exchange amplitudes in soliton models are outlined, which may help to reduce the discrepancies between form factors extracted via Rosenbluth separation from unpolarized elastic electron-nucleon scattering and those obtained from ratios of polarization observables.

## 2.2. Characteristic Feature of the Electric Proton Form Factor

Chiral soliton models for the nucleon naturally account for a characteristic decrease of the ratio $G_{\mathrm{E}}^{\mathrm{p}}/(G_{\mathrm{M}}^{\mathrm{p}}/\mu_p)$ with increasing $Q^2$. The reason for this behaviour basically originates in the fact that in soliton models the isospin for baryons is generated by rotating the soliton in isospace. The hedgehog structure of the soliton couples the isorotation to a spatial rotation. Therefore, in the rest frame of the soliton, the isovector ($I = 1$) form factors measure the (rotational) inertia density $B_1(r)$, as compared to the isoscalar baryon density $B_0(r)$ for the isoscalar ($I = 0$) form factors. This becomes evident from the explicit form of the isoscalar and isovector form factors in the simple purely pionic soliton model:[2]

$$G_{\mathrm{E}}^0(k^2) = \frac{1}{2} \int d^3r\, j_0(kr) B_0(r) \tag{2.2.1}$$

$$G_{\mathrm{M}}^0(k^2)/\mu_0 = \frac{3}{r_B^2} \int d^3r\, r^2\, \frac{j_1(kr)}{kr} B_0(r) \tag{2.2.2}$$

$$G_{\mathrm{E}}^1(k^2) = \frac{1}{2} \int d^3r\, j_0(kr) B_1(r) \tag{2.2.3}$$

$$G_{\mathrm{M}}^1(k^2)/\mu_1 = 3 \int d^3r\, \frac{j_1(kr)}{kr} B_1(r), \tag{2.2.4}$$

(with mean square isoscalar baryon radius $r_B^2$, isoscalar and isovector magnetic moments $\mu_0, \mu_1$, and normalization $\int B_0(r)d^3r = \int B_1(r)d^3r = 1$).

Evidently, if the inertia density were obtained from rigid rotation of the baryon density $B_1(r) = (r^2/r_B^2)B_0(r)$, the normalized isoscalar and isovector magnetic form factors would satisfy the scaling relation

$$G_{\mathrm{M}}^1(k^2)/\mu_1 = G_{\mathrm{M}}^0(k^2)/\mu_0, \tag{2.2.5}$$

while for the electric form factors the same argument leads to

$$G_{\mathrm{E}}^1(k^2) = -\frac{1}{r_B^2} \left(\frac{\partial}{\partial \vec{k}}\right)^2 G_{\mathrm{E}}^0(k^2). \tag{2.2.6}$$

For a Gaussian baryon density $B_0(r) \propto \exp(-(3r^2)/(2r_B^2))$ the 'scaling' property (2.2.5) includes also the isoscalar electric form factor

$$G_{\mathrm{M}}^1(k^2)/\mu_1 = G_{\mathrm{M}}^0(k^2)/\mu_0 = 2G_{\mathrm{E}}^0(k^2), \tag{2.2.7}$$

and Eq. (2.2.6) then leads to

$$G_{\mathrm{E}}^1(k^2) = \left(1 - \frac{1}{9}k^2 r_B^2\right) G_{\mathrm{E}}^0(k^2). \tag{2.2.8}$$

Therefore, for proton form factors

$$G_{\mathrm{E,M}}^{\mathrm{p}} = G_{\mathrm{E,M}}^0 + G_{\mathrm{E,M}}^1, \tag{2.2.9}$$

the ratio $G_\mathrm{E}^\mathrm{p}/(G_\mathrm{M}^\mathrm{p}/\mu_p)$ resulting from Eqs. (2.2.5), (2.2.7) and (2.2.8), is

$$R(k^2) = G_\mathrm{E}^\mathrm{p}(k^2)/(G_\mathrm{M}^\mathrm{p}(k^2)/\mu_p) = \left(1 - \frac{1}{18}k^2 r_B^2\right). \qquad (2.2.10)$$

With $r_B^2 \approx 2.3 \ (\mathrm{GeV/c})^{-2} \approx (0.3 \ \mathrm{fm})^2$, this simple consideration provides an excellent fit (see Fig.2.3) through the polarization data for $R(k^2)$. Of course, in typical soliton models $B_1(r)$ is not exactly proportional to $r^2 B_0(r)$ and the baryon density is not really Gaussian (cf. Fig. 2.1). Furthermore, to compare with experimentally extracted form factors, the $k^2$-dependence of the form factors in the soliton rest frame must be subject to the Lorentz boost from the rest frame to the Breit frame (which compensates for the fact that typical baryon radii obtained in soliton models are near 0.4-0.5 fm).

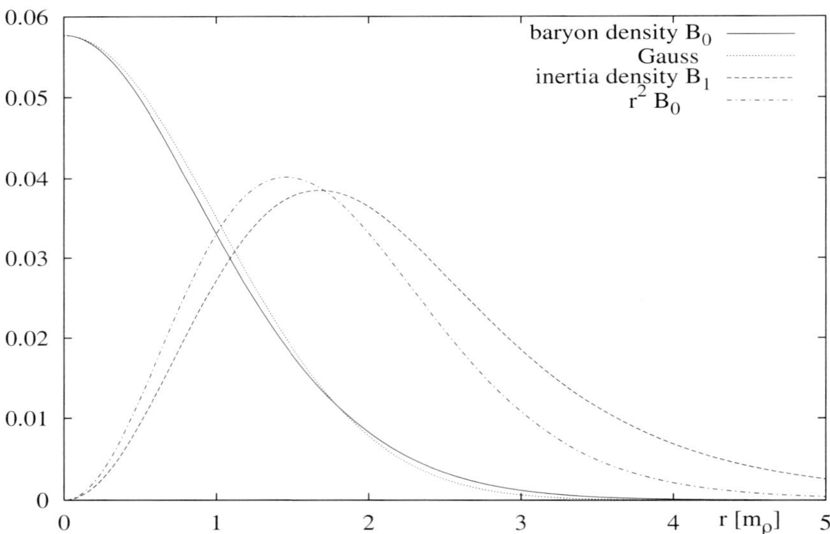


Fig. 2.1. Comparison between the topological baryon density $B_0$ and a Gaussian, and between the inertia density $B_1$ and $r^2 B_0$, for the standard pionic Skyrme model (2.3.11)–(2.3.13) with $e = 4.25$.

But still, we may conclude from these simple considerations that a strong decrease of the ratio (2.2.10) from $R = 1$ towards an eventual zero near $k^2 \sim 10 \ (\mathrm{GeV/c})^2$ appears as a natural and characteristic feature of proton electromagnetic form factors in chiral soliton models.

## 2.3. Chiral $\pi$-$\rho$-$\omega$−Meson Model

After the above rather general remarks we consider a specific realistic model which includes also vector mesons. They are known to play an essential role in the coupling of baryons to the electromagnetic field and different possibilities for their explicit inclusion in a chirally invariant effective meson theory have been suggested.[3] We adopt the pionic Skyrme model for the chiral SU(2)-field $U$

$$\mathcal{L}^{(\pi)} = \mathcal{L}^{(2)} + \mathcal{L}^{(4)}, \tag{2.3.11}$$

$$\mathcal{L}^{(2)} = \frac{f_\pi^2}{4} \int \left(-tr L_\mu L^\mu + m_\pi^2 tr(U + U^\dagger - 2)\right) d^3x, \tag{2.3.12}$$

$$\mathcal{L}^{(4)} = \frac{1}{32e^2} \int tr[L_\mu, L_\nu]^2 d^3x, \tag{2.3.13}$$

($L_\mu$ denotes the chiral gradients $L_\mu = U^\dagger \partial_\mu U$, the pion decay constant is $f_\pi$=93 MeV, and the pion mass $m_\pi$=138 MeV). Without explicit vector mesons the Skyrme parameter $e$ is well established near $e$=4.25. Minimal coupling to the photon field is obtained through the local gauge transformation $U \to \exp(i\epsilon \hat{Q})\, U \exp(-i\epsilon \hat{Q})$ with the charge operator $\hat{Q} = (1/3 + \tau_3)/2$. The isoscalar part of the coupling arises from gauging the standard Wess-Zumino term in the SU(3)-extended version of the model.

Vector mesons may be explicitly included as dynamical gauge bosons. In the minimal version the axial vector mesons are eliminated in chirally invariant way.[4–6] This leaves two gauge coupling constants $g_\rho, g_\omega$ for $\rho$- and $\omega$-mesons,

$$\mathcal{L} = \mathcal{L}^{(\pi)} + \mathcal{L}^{(\rho)} + \mathcal{L}^{(\omega)} \tag{2.3.14}$$

$$\mathcal{L}^{(\rho)} = \int \left(-\frac{1}{8} tr \rho_{\mu\nu} \rho^{\mu\nu} + \frac{m_\rho^2}{4} tr(\rho_\mu - \frac{i}{2g_\rho}(l_\mu - r_\mu))^2\right) d^3x, \tag{2.3.15}$$

$$\mathcal{L}^{(\omega)} = \int \left(-\frac{1}{4}\omega_{\mu\nu}\omega^{\mu\nu} + \frac{m_\omega^2}{2}\omega_\mu\omega^\mu + 3g_\omega\omega_\mu B^\mu\right) d^3x, \tag{2.3.16}$$

with the topological baryon current $B_\mu = 1/(24\pi^2)\epsilon_{\mu\nu\rho\sigma} tr\,(L^\nu L^\rho L^\sigma)$, and $l_\mu = \xi^\dagger \partial_\mu \xi$, $r_\mu = \partial_\mu \xi \xi^\dagger$, where $\xi^2 = U$.

The contributions of the vector mesons to the electromagnetic currents arise from the local gauge transformations

$$\rho^\mu \to e^{i\epsilon \hat{Q}_V} \rho^\mu e^{-i\epsilon \hat{Q}_V} + \frac{\hat{Q}_V}{g_\rho}\partial^\mu \epsilon, \qquad \omega^\mu \to \omega^\mu + \frac{\hat{Q}_0}{g_0}\partial^\mu \epsilon \tag{2.3.17}$$

(with $\hat{Q}_0 = 1/6$ , $\hat{Q}_V = \tau_3/2$). The resulting form factors are expressed in terms of three static and three rotationally induced profile functions which characterize the rotating $\pi$-$\rho$-$\omega$−hedgehog soliton with baryon number $B = 1$.

Because the Skyrme term $\mathcal{L}^{(4)}$ at least partly accounts for static $\rho$-meson effects its strength should be reduced in the presence of dynamical $\rho$-mesons, as compared to the plain Skyrme model. The coupling constant $g_\rho$ can be fixed by the KSRF relation $g_\rho = m_\rho/(2\sqrt{2}f_\pi)$, but small deviations from this value are tolerable. The $\omega$-mesons introduce two gauge coupling constants, $g_\omega$ to the baryon current in $\mathcal{L}^{(\rho)}$, and $g_0$ for the isoscalar part of the charge operator. Within the $SU(2)$ scheme we can in principle allow $g_0$ to differ from $g_\omega$ and thus exploit the freedom in the electromagnetic coupling of the isoscalar $\omega$-mesons.

The general structure of the form factors as given in Eqs. (2.2.1–2.2.4) for the purely pionic model remains almost unchanged in the $\pi$-$\rho$-$\omega$−model. In the isoscalar form factors the topological baryon density $B_0(r)$ is replaced by the total isoscalar charge density. After insertion of the equation of motion for the $\omega$-mesons we have to replace in Eqs. (2.2.1) and (2.2.2)

$$B_0(r) \Longrightarrow \left(1 + \frac{g_\omega}{g_0}\left(\frac{m_\omega^2}{k^2 + m_\omega^2} - 1\right)\right) B_0(r). \tag{2.3.18}$$

This shows explicitly how the $\omega$-meson pole is introduced into the isoscalar form factors. For the isovector electric $G_{\rm E}^1(k^2)$ in Eq. (2.2.3) the function $B_1(r)$ again is given by the rotational inertia density, which now, however, receives also contributions from the rotationally induced $\rho$ and $\omega$ components. In the isovector magnetic $G_{\rm M}^1(k^2)$ in Eq. (2.2.4) the function which replaces $B_1(r)$ includes also contributions from the static $\rho$ and $\omega$ profiles and no longer coincides with the rotational inertia density. The detailed expressions of the form factors which we use here in the minimal $\pi$-$\rho$-$\omega$−model (making use of the KSRF relation for $g_\rho$) are given in Ref. 6.

### 2.4. Boost to the Breit Frame

For all dynamical models of spatially extended clusters it is difficult to relate the non-relativistic form factors evaluated in the rest frame of the cluster to the relativistic $Q^2$-dependence in the Breit frame where the cluster moves with velocity $v$ relative to the rest frame. For the associated Lorentz-boost factor $\gamma$ we have

$$\gamma^2 = (1 - v^2)^{-1} = 1 + \frac{Q^2}{(2M)^2}, \tag{2.4.19}$$

where $M$ is the rest mass of the cluster. For elastic scattering of clusters composed of $\nu$ constituents dimensional scaling arguments[7] require that the leading power in the asymptotic behaviour of relativistic form factors is $\sim Q^{2-2\nu}$. Boost prescriptions of the general form

$$G_{\rm M}^{\rm Breit}(Q^2) = \gamma^{-2n_{\rm M}} \, G_{\rm M}^{\rm rest}(k^2), \qquad G_{\rm E}^{\rm Breit}(Q^2) = \gamma^{-2n_{\rm E}} \, G_{\rm E}^{\rm rest}(k^2) \tag{2.4.20}$$

with

$$k^2 = \gamma^{-2} \, Q^2 \tag{2.4.21}$$

have been suggested with various values for the interpolating powers $n_{\rm M}, n_{\rm E},$[8,9] where $M$ takes the role of an effective mass.

This boost prescription has the appreciated feature that a low-$k^2$ region in the rest frame $(0 < k^2 < 1 \ ({\rm GeV/c})^2$, say), where we trust the physical content of the rest frame form factors, appears as an appreciably extended $Q^2$-regime in the Breit frame. So, through the boost (2.4.21) from rest frame to Breit frame, the region

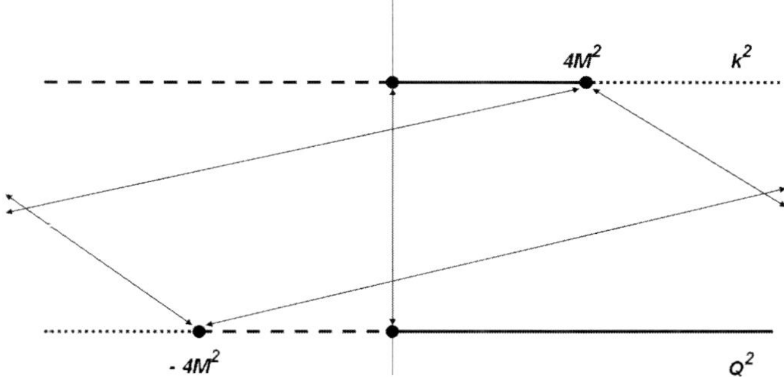


Fig. 2.2.   The boost (2.4.21) maps the dashed, solid, and dotted intervals of rest frame $k^2$ onto the dashed, solid, and dotted intervals of the Breit frame momentum transfer $Q^2$.

of validity of soliton form factors for spatial $Q^2$ is extended. Evidently, the boost in Eq. (2.4.21) maps $G^{\mathrm{rest}}(k^2 \to 4M^2) \;\to\; G^{\mathrm{Breit}}(Q^2 \to \infty)$. But, even though $G^{\mathrm{rest}}(4M^2)$ may be very small, it generally does not vanish exactly. So, unless $n_{\mathrm{M}}, n_{\mathrm{E}} \geq 2$, this shows up, of course, very drastically in the asymptotic behaviour, if the resulting form factors are divided by the standard dipole

$$G_{\mathrm{D}}(Q^2) = 1/(1 + Q^2/0.71)^2, \tag{2.4.22}$$

which is the common way to present the nucleon form factors and accounts for the proper asymptotic $Q^{2-2\nu}$ behaviour of an $\nu = 3$ quark cluster. So it is vital for a comparison with experimentally determined form factors for $Q^2 \gg M^2$ to employ a boost prescription which preserves at least the 'superconvergence' property $Q^2 G(Q^2) \to 0$ for $Q^2 \to \infty$. In accordance with an early suggestion by Mitra and Kumari[10,11] we use $n_{\mathrm{M}} = n_{\mathrm{E}} = 2$. In any case, the high-$Q^2$ behaviour is not a profound consequence of the model but simply reflects the boost prescription. There is no reason anyway, why low-energy effective models should provide any profound answer for the high-$Q^2$ limit. Note that the position of an eventual zero in $G_{\mathrm{E}}^{\mathrm{Breit}}(Q^2)$ is not affected by the choice of the interpolating power $n_{\mathrm{E}}$, and the ratio $G_{\mathrm{E}}/G_{\mathrm{M}}$ is independent of the interpolating power, as long as $n_{\mathrm{M}} = n_{\mathrm{E}}$.

## 2.5.  Results

To demonstrate the amount of agreement with experimental data that can be achieved within the framework of such models we present in Fig. 2.3 typical results from the $\pi$-$\rho$-$\omega$−model with essential parameters of the model fixed at their physical values: the pion decay constant $f_\pi = 93$ MeV, the pion mass $m_\pi = 138$ MeV, $\rho$-mass $m_\rho = 770$ MeV, $\omega$-mass $m_\omega = 783$ MeV, and the $\pi$-$\rho$-coupling constant at its physical KSRF-value $g_\rho = 2.9$. As variable parameters remain the $\pi$-$\omega$ coupling constant $g_\omega$, and the $\omega$-photon coupling constant $g_0$. Due to the presence of dynamical $\rho$-mesons the strength $1/e^2$ of the fourth-order Skyrme term $\mathcal{L}^{(4)}$

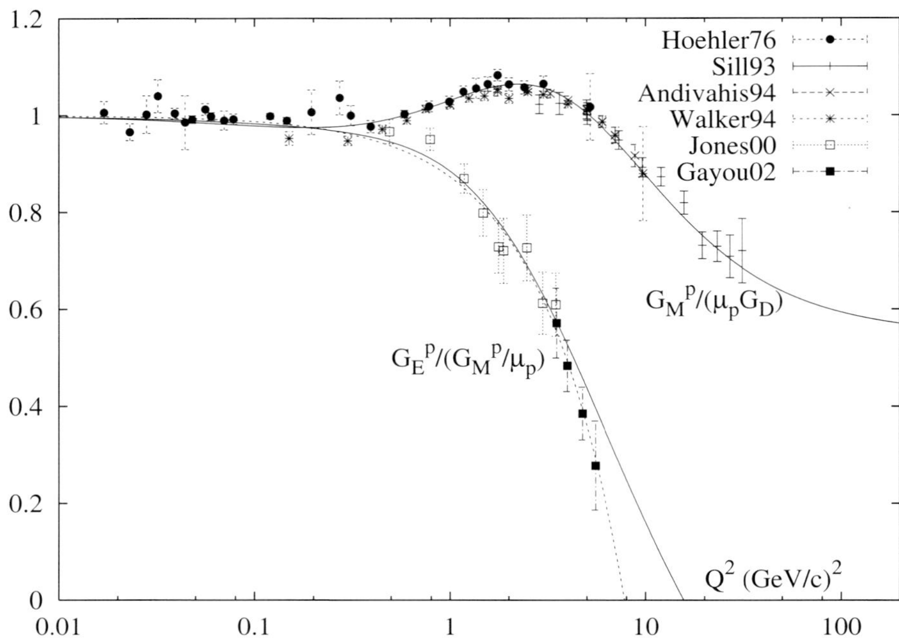



Fig. 2.3. Magnetic and electric proton form factors $G_M^p/(\mu_p G_D)$ and $G_E^p/(G_M^p/\mu_p)$ for the $\pi$-$\rho$-$\omega$-model with the set of parameters given in the text. The dotted line shows the result of Eq. (2.2.10) with $r_B = 0.3$ fm. The abscissa shows $Q^2(\mathrm{GeV/c})^2$ on logarithmic scale. The experimental data are from Refs. 14–19.

should be reduced as compared to its standard value; it may even be omitted altogether. In addition to these three coupling constants, the high-$Q^2$ behaviour of the form factors is, of course, very sensitive to the effective kinematical mass $M$ which appears in the Lorentz-boost (2.4.19).

Altogether, while the general features are generic to the soliton model, we use in the following these four parameters $g_\omega = 1.4$, $g_\omega/g_0 = 0.75$, $e = 7.5$, and $M = 1.23$ GeV, for the fine-tuning of the proton form factors as shown in Fig. 2.3. Of course, these four parameters are not independent. Changes in the calculated form factors due to variations in one of these parameters may be compensated by suitable variations in the others for comparable quality of the fits. (For example, the agreement shown in Fig. 2.3 could also be obtained in a three-parameter fit without Skyrme term (i.e. $1/e^2 = 0$) with $g_\omega = 2.4$, $g_\omega/g_0 = 0.7$, and $M = 1.16$ GeV).

The absolute size of the neutron electric form factor $G_E^n$ is closely related to the choice of $g_\omega/g_0$. For the chosen set of parameters the maximum of $G_E^n$ exceeds the Galster parametrization by a factor of about 1.3 (cf. Fig. (2.4)). Correspondingly, the calculated values for the electric neutron square radii exceed the experimental

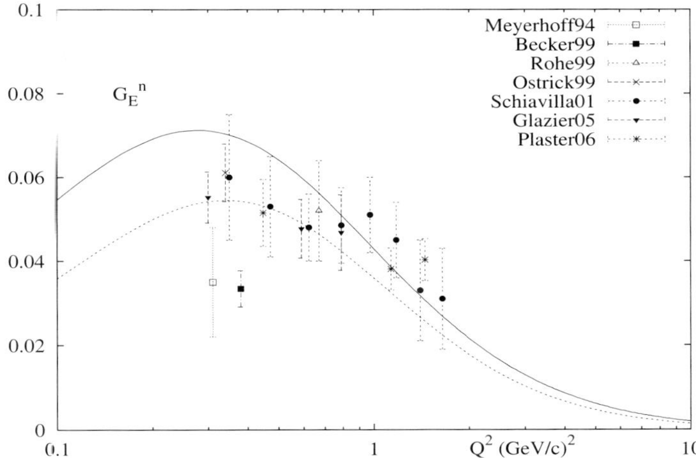

Fig. 2.4. The neutron electric form factor $G_E^n$ as obtained in the $\pi$-$\rho$-$\omega$−model with the set of parameters given in the text. The dotted line is the standard Galster parametrization $G_E^n = -\mu_n \tau/(1 + 5.6\tau) \cdot G_D$ with $\tau = Q^2/(4M_n^2)$. Experimental results for $G_E^n$ are mainly from more recent polarization data.[20−26]

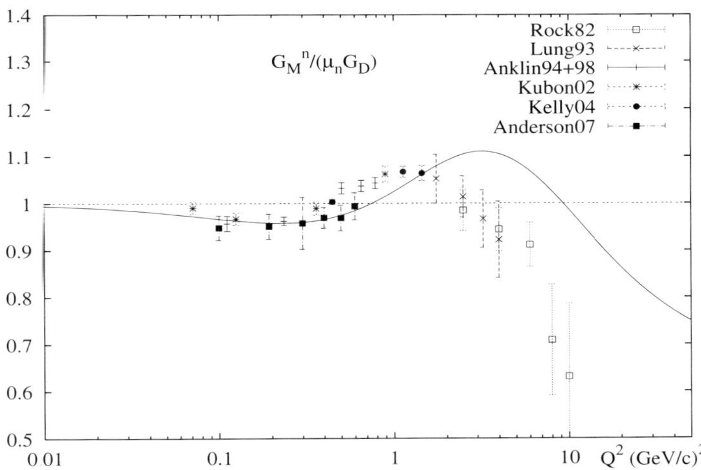

Fig. 2.5. The magnetic neutron form factor (normalized to the standard dipole) $G_M^n/(\mu_n G_D)$ in the same model. Here the data are from Refs. 27–33.

value by about a factor of 2, and we found it difficult to lower them, for reasonable parametrizations within the $SU(2)$ framework. But otherwise the shape of $G_E^n$ follows the Galster parametrization rather well, with the maximum slightly shifted to lower $Q^2$. In the $SU(3)$-embedding of the Skyrme model the mixing coefficients for isoscalar, isovector, and kaonic contributions to the electromagnetic form factors cause a sizable reduction of the electric neutron form factor as compared to the $SU(2)$ scheme. The relevant coefficients are listed in Ref. 12 for the case of exact flavor symmetry; when symmetry breaking is included, their numerical values reduce the square radius $\langle r^2 \rangle_{\mathrm{E}}^n$ by a factor of about one-half as compared to $SU(2)$, while the results for the proton remain almost unaffected.[12,13] This cures the discrepancy for $G_{\mathrm{E}}^n$ in Fig. (2.4) and for $\langle r^2 \rangle_{\mathrm{E}}^n$ shown in Table 1. However, we are not aware of calculations of electromagnetic form factors for $Q^2 > 1(\mathrm{GeV/c})^2$ in the $SU(3)$-embedded Skyrme and vector meson model.

In Fig.2.5 we also present the resulting magnetic neutron form factor $G_{\mathrm{M}}^n$, normalized to the standard dipole $G_{\mathrm{D}}$. For $Q^2 \leq 1(\mathrm{GeV/c})^2$ the model result is in perfect agreement with the latest data[32](as quoted in[26]),.[33] For $Q^2 > 1(\mathrm{GeV/c})^2$, however, the model prediction deviates substantially from the available older data.[27,28] The ratio of the normalized proton and neutron magnetic form factors $G_{\mathrm{M}}^n \mu_p / (G_{\mathrm{M}}^p \mu_n)$ is independent of the choice of the interpolating power $n_{\mathrm{M}}$ in the boost prescription. Therefore it would be desirable to compare directly with data for this ratio. Experimentally it is accessible from quasielastic scattering on deuterium with final state protons and neutrons detected. The generic scaling relation (2.2.5) predicts this ratio to be equal to one, $G_{\mathrm{M}}^n \mu_p / (G_{\mathrm{M}}^p \mu_n) = 1$, so deviations from this value indicate, how the function $B_1(r)$ which appears in $G_{\mathrm{M}}^1(k^2)/\mu_1$ differs from $r^2 B_0(r)$ in the specific model considered. Both, the Skyrme model and the $\pi$-$\rho$-$\omega$−model considered here, consistently predict this ratio to increase above 1 by up to 15% for $1 < Q^2(\mathrm{GeV/c})^2 < 10$. However, also in this case an $SU(3)$ embedding may change this prediction appreciably. The presently available data do not show such an increase for this ratio, in fact they indicate the opposite tendency. This conflict was already noticed in Refs. 1 and 34. Preliminary data from CLAS[35] apparently are compatible with $G_{\mathrm{M}}^n / (\mu_n G_{\mathrm{D}}) = 1$ in the region $1 < Q^2(\mathrm{GeV/c})^2 < 4.5$.

Table 2.1.  Nucleon quadratic radii and magnetic moments as obtained from the chiral $\pi$-$\rho$-$\omega$−model, for the parameters given in the text. The experimental values are from Ref. 36.

| | $\langle r^2 \rangle_{\mathrm{E}}^{\mathrm{P}}$ | $\langle r^2 \rangle_{\mathrm{M}}^{\mathrm{P}}$ | $\langle r^2 \rangle_{\mathrm{E}}^{n}$ | $\langle r^2 \rangle_{\mathrm{M}}^{n}$ | $\mu_p$ | $\mu_n$ |
|---|---|---|---|---|---|---|
| Model | 0.74 | 0.72 | −0.24 | 0.76 | 1.82 | −1.40 |
| Exp. | 0.77 | 0.74 | −0.114 | 0.77 | 2.79 | −1.91 |

In Table 2.1 we list quadratic radii and magnetic moments as they arise from the fit given above. Notoriously low are the magnetic moments. This fact is common to chiral soliton models and well known. Quantum corrections will partly be helpful in this respect (see Ref. 37), as they certainly are for the absolute value of the nucleon mass.

Of course, such models can be further extended. Addition of higher-order terms in the skyrmion lagrangian, explicit inclusion of axial vector mesons, non-minimal photon-coupling terms, provide more flexibility through additional parameters. Our point here, however, is to demonstrate that a minimal version as described above is capable of providing the characteristic features for both proton form factors and for the electric neutron form factor in remarkable detail. In fact, the unexpected decrease of $G_E^p$ was predicted by these models, and it will be interesting to compare with new data for $G_M^n$ concerning the conflict indicated in Fig. 2.5.

## 2.6. Extension to Time-Like $Q^2$

In the soliton rest frame the extension to time-like $k^2$ amounts to finding the spectral functions $\Gamma(\nu^2)$ as Laplace transforms of the relevant densities $B(r)$, e.g. for the isoscalar electric case

$$rB_0(r) = \frac{1}{\pi^2} \int_{\nu_0^2}^{\infty} e^{-\nu r} \nu \Gamma_0(\nu^2) d\nu, \qquad (2.6.23)$$

and similarly for other cases. In soliton models the densities are obtained numerically on a spatial grid, therefore the spectral functions cannot be determined uniquely. Results will always depend on the choice of constraints which have to be imposed on possible solutions. But with reasonable choices it seems possible to stabilize the spectral functions in the regime from the 2- or 3-pion threshold to about two $\rho$-meson masses and distinguish continuous and discrete structures in this regime.[1]

We note (cf. Fig. (2.2)) that the transformation to the Breit frame (2.4.21) formally maps the rest frame form factors $G^{\mathrm{rest}}(k^2)$ for the whole time-like regime $-\infty < k^2 < 0$ onto the Breit-frame form factors $G^{\mathrm{Breit}}(Q^2)$ in the unphysical time-like regime up to the nucleon-antinucleon threshold $-4M^2 < Q^2 < 0$. On the other hand, the physical time-like regime $-\infty < Q^2 < -4M^2$ in the Breit frame is obtained as the image of the spacelike regime $4M^2 < k^2 < \infty$ of form factors in the rest frame. So the (real parts) of the Breit-frame form factors for time-like $Q^2$ beyond the nucleon-antinucleon threshold are formally fixed through Eq. (2.4.20). However, apart from the probably very limited validity of the boost prescription (2.4.20), we do not expect that the form factors in the soliton rest frame for $k^2 > 4M^2$ contain sufficiently reliable physical information. Specifically, oscillations which the rest frame form factors may show for $k^2 \to \infty$, are sqeezed by the transformation (2.4.21) into the vicinity of the physical threshold $Q^2 < -4M^2$. With $G^{\mathrm{rest}}(k^2) \to 0$ for $k^2 \to \infty$, the Breit-frame form factors are undetermined at threshold $Q^2 \to -4M^2$.

Attempts to obtain form factors for time-like $Q^2$ from soliton-antisoliton configurations in the baryon number $B = 0$ sector face the difficulty that in this sector the only stable classical configuration is the vacuum. So, any result will reflect the arbitrariness in the construction of nontrivial configurations.

Altogether we conclude, that presently we see no reliable way for extracting profound information about electromagnetic form factors in the physical time-like regime from soliton models.

## 2.7. Two-Photon Amplitudes in Soliton Models

The discrepancies between form factors extracted through the Rosenbluth separation from unpolarized elastic scattering data[38] and ratios directly obtained from polarization transfer measurements[18,19] have lead to the difficult situation that two distinct methods to experimentally determine fundamental nucleon properties yield inconsistent results.[39] As a possible remedy, the theoretical focus has shifted to two-photon amplitudes which enter the unpolarized cross section and polarization variables in different ways. Two-photon exchange diagrams involve the full response of the nucleon to doubly virtual Compton scattering and therefore rely heavily on specific nucleon models. Simple box diagrams which iterate the single-photon exchange, require virtual intermediate nucleons and resonances with unknown off-shell form factors. They have been analysed with various assumptions for the intermediate states and have been found helpful for a partial reduction of the discrepancies.[40,41]

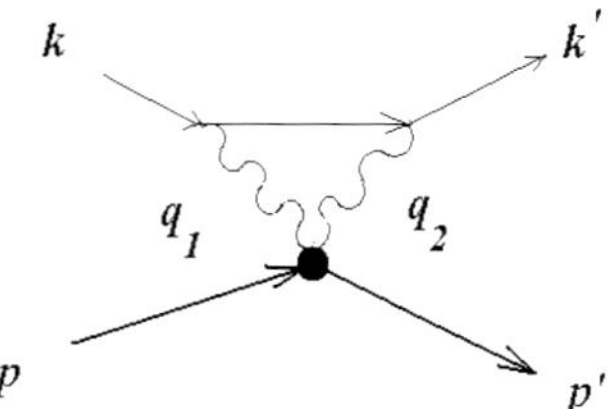

Fig. 2.6. Electron-nucleon scattering $2\gamma$-exchange amplitude with local $2\gamma$-soliton vertex with momentum transfer $q = q_1 - q_2 = k - k' = p' - p$.

It is interesting to note that, in addition to box diagrams, soliton models contain $2\gamma$-exchange contributions where the two virtual photons interact with the pion cloud of the baryon at *local* two-photon vertices. Products of covariant derivatives

$$D_\mu U = \partial_\mu U + i[\hat{Q}, U]A_\mu \tag{2.7.24}$$

which appear in all terms of the derivative expansion after gauging the chiral fields with the electric charge $\hat{Q}$, naturally produce these local two-photon couplings. The simplest ones originate from the quadratic nonlinear $\sigma$-term and from the gauged Wess-Zumino anomalous action

$$\mathcal{L}_{nl\sigma}^{(2\gamma)} = -\frac{f_\pi^2}{4} A_\mu A^\mu 2\mathrm{tr}(\hat{Q}U\hat{Q}U^\dagger - \hat{Q}^2), \tag{2.7.25}$$

$$\mathcal{L}_{WZ}^{(2\gamma)} = i\frac{N_c}{48\pi^2}\varepsilon^{\mu\nu\varrho\sigma}(\partial_\mu A_\nu)A_\varrho \mathrm{tr}\left(\hat{Q}\partial_\sigma U\hat{Q}U^\dagger - \hat{Q}U\hat{Q}\partial_\sigma U^\dagger + 2\hat{Q}^2(U^\dagger\partial_\sigma U - U\partial_\sigma U^\dagger)\right). \tag{2.7.26}$$

After quantization of the collective coordinates the matrix elements of these $2\gamma$-vertices sandwiched between incoming and outgoing nucleon states are obtained, without additional parameters, with form factors fixed through the soliton profiles. Then the interference terms with the single-photon-exchange amplitudes for the unpolarized elastic cross section can be evaluated. It turns out that the contribution from $\mathcal{L}_{nl\sigma}^{(2\gamma)}$ interferes only with the electric part of the 1-photon-exchange Born term and vanishes after spin averaging. On the other hand, the scattering amplitude following from $\mathcal{L}_{WZ}^{(2\gamma)}$ interferes only with the magnetic part of the Born amplitude, so that apart from kinematical factors the unpolarized elastic electron-nucleon cross section has the general structure

$$\frac{d\sigma}{d\Omega} \propto \left( G_{\mathrm{M}}^2(Q^2) + \frac{\epsilon}{\tau}G_{\mathrm{E}}^2(Q^2) + \nu(1-\epsilon)G_{\mathrm{M}}(Q^2)F_{WZ}^{(2\gamma)}(Q^2) \right) \tag{2.7.27}$$

with Lorentz invariants $\tau = Q^2/(4M^2)$, and

$$\nu = \frac{1}{4}(k+k')\cdot(p+p') = \sqrt{\tau(1+\tau)\frac{1+\epsilon}{1-\epsilon}} \, . \tag{2.7.28}$$

The form factor $F_{WZ}^{(2\gamma)}$ is of the order of the electromagnetic coupling constant $\alpha$, and involves a loop integral and Fourier transforms of soliton profiles. Due to its origin from the Wess-Zumino action, it is parameter free. The possibility to obtain parameter free information about the influence of two-photon exchange contributions, makes this scheme very attractive. However, it should be mentioned that the infinite part of the loop integral requires a counterterm which has to be fixed by other experimental input. This program has been performed in Ref. 42. The corrections obtained have been found to reduce the observed discrepancies, with an absolute size, however, which by itself is also not sufficient to resolve the problem. It has to be supplemented by iterated single-photon exchange.

The $\epsilon$-dependence through $\sqrt{(1+\epsilon)/(1-\epsilon)}$ as contained in $\nu$ is a general symmetry and consistency requirement for the two-photon interaction.[43] There is, however, experimental evidence that within the present error limits the unpolarized elastic cross section is consistent with a linear $\epsilon$-dependence.[44,45] This still allows to extract via Rosenbluth separation, effective electric and magnetic form factors which then comprise also the sum of all relevant $2\gamma$-contributions. Their ratios may differ appreciably from ratios of the single-photon-exchange form factors $G_{\mathrm{E}}^{\mathrm{p}}/G_{\mathrm{M}}^{\mathrm{p}}$ as extracted from polarization transfer data, which are believed to remain mostly unaffected by $2\gamma$-contributions.[41] Although at present the situation is not yet fully understood, there is strong evidence that $2\gamma$-exchange effects may in fact account for most of the observed differences,[46] and electromagnetic form factors remain the challenging testing ground for models of the nucleon.

The fact that the unexpected results of the polarization transfer experiments follow as generic consequence from soliton models; that within a minimal specific model form factors can be reproduced in detail; and that, in addition to the usual box diagrams, standard gauging provides a new class of radiative corrections with local $2\gamma$-nucleon coupling; all of this once again underlines the strength of the soliton approach to baryons.

## Acknowledgements

The author is very much indebted to H. Walliser and H. Weigel for numerous discussions.

## References


1. G. Holzwarth, *Z. Phys. A* **356** (1996) 339; Proc. 6th Int. Symp. Meson-Nucleon Phys., $\pi N$ Newsletter **10** (1995) 103.
2. E. Braaten, S.M. Tse and C. Willcox, *Phys. Rev. D* **34** (1986) 1482; *Phys. Rev. Lett.* **56** (1986) 2008.
3. O. Kaymakcalan and J. Schechter, *Phys. Rev. D* **31** (1985) 1109;
   M. Bando, T. Kugo, S Uehara, K. Yamawaki and T. Yanagida, *Phys. Rev. Lett.* **54** (1985) 1215.
4. U.G. Meissner, N. Kaiser and W. Weise, *Nucl. Phys. A* **466** (1987) 685;
   U.G. Meissner, *Phys. Reports* **161** (1988) 213.
5. B. Schwesinger, H. Weigel, G. Holzwarth and A. Hayashi, *Phys. Reports* **173** (1989) 173.
6. F. Meier, in *Baryons as Skyrme solitons*, ed. G. Holzwarth (World Scientific, Singapore 1993), 159.
7. V.A. Matveev, R.M. Muradyan and A.N. Tavkhelidze, *Lett. Nuovo Cim.* **7** (1973) 719.
8. A.L. Licht and A. Pagnamenta, *Phys. Rev. D* **2** (1976) 1150; and 1156.
9. X. Ji, *Phys. Lett. B* **254** (1991) 456.
10. A. N. Mitra and I. Kumari, *Phys. Rev. D* **15** (1977) 261.
11. J.J. Kelly, *Phys. Rev. C* **66** (2002) 065203.
12. H. Weigel, *Chiral Soliton Models for Baryons*, Lect. Notes Phys. **743** (Springer, Berlin Heidelberg 2008), p.118.
13. H. Walliser, private communication.
14. G. Höhler et al., *Nucl. Phys. B* **114** (1976) 505.
15. A.F. Sill et al., *Phys. Rev. D* **48** (1993) 29.
16. L. Andivahis et al., *Phys. Rev. D* **50** (1994) 5491.
17. R.C. Walker et al., *Phys. Rev. D* **49** (1994) 5671.
18. M.K. Jones et al., *Phys. Rev. Lett.* **84** (2000) 1398.
19. O. Gayou et al., *Phys. Rev. Lett.* **88** (2002) 092301.
20. M. Ostrick et al., *Phys. Rev. Lett.* **83** (1999) 276.
21. M. Meyerhoff et al., *Phys. Lett. B* **327** (1994) 201.
22. J. Becker et al., *Eur. Phys. J. A* **6** (1999) 329.
23. D. Rohe et al., *Phys. Rev. Lett.* **83** (1999) 4257.
24. R. Schiavilla and I. Sick, *Phys. Rev. C* **64** (2001) 041002-1.
25. R. Glazier et al., *Eur. Phys. J. A* **24** (2005) 101.
26. B. Plaster et al. [E93-038 Collaboration], *Phys. Rev. C* 73 (2006) 025205;
    R. Madey et al., *Phys. Rev. Lett.* **91** (2003) 122002.
27. S. Rock et al., *Phys. Rev. Lett.* **49** (1982) 1139.
28. A. Lung et al., *Phys. Rev. Lett.* **70** (1993) 718.
29. H. Anklin et al., *Phys. Lett. B* **336** (1994) 313.

30. H. Anklin et al., *Phys. Lett. B* **428** (1998) 248.

31. G. Kubon, H. Anklin et al., *Phys. Lett. B* **524** (2002) 26.

32. J.J. Kelly, *Phys. Rev. C* **70** (2004) 068202.

33. B. Anderson et al. [E95-001 Collaboration], *Phys. Rev. C* **75** (2007) 034003;
    W. Xu et al., *Phys. Rev. Lett.* **85** (2000) 2900; *Phys. Rev. C* **67** (2003) 012201.

34. G. Holzwarth, [arXiv:hep-ph/0201138].

35. W.K. Brooks and J.D. Lachniet [CLAS E94-017 Collaboration], *Nucl. Phys. A* **755** (2005) 261.

36. G. Simon et al., *Z. Naturforsch.* **35A** (1980) 1.

37. F. Meier and H. Walliser, *Phys. Reports* **289** (1997) 383.

38. M.E. Christy et al. [E94110 Collaboration], *Phys. Rev. C* **70** (2004) 015206;
    I.A. Qattan et al., *Phys. Rev. Lett.* **94** (2005) 142301.

39. J. Arrington, *Phys. Rev. C* **68** (2003) 034325; *Phys. Rev. C* **69** (2004) 022201; *Phys. Rev. C* **71** (2005) 015202;
    H. Gao, *Int. J. Mod. Phys. A* **20** (2005) 1595.

40. P.A.M. Guichon and M. Vanderhaeghen, *Phys. Rev. Lett.* **91** (2003) 142303;
    P.G. Blunden, W. Melnitchouk and J.A. Tjon, *Phys. Rev. Lett.* **91** (2003) 142304;
    *Phys. Rev. C* **72** (2005) 034612;
    Y.C. Chen et al., *Phys. Rev. Lett.* **93** (2004) 122301;
    A.V. Afanasev et al., *Phys. Rev. D* **72** (2005) 013008;
    S. Kondratyuk et al., *Phys. Rev. Lett.* **95** (2005) 172503;
    P. Jain et al., arXiv:hep-ph/0606149;
    D. Borisyuk and A. Kobushkin, *Phys. Rev. C* **74** (2006) 065203;
    C.F. Perdrisat, V. Punjabi and M. Vanderhaeghen, *Prog. Part. Nucl. Phys.* **59** (2007) 694.

41. C.E. Carlson and M. Vanderhaeghen, *Annu. Rev. Nucl. Part. Sci.* **57** (2007) 171.

42. M. Kuhn and H. Weigel, arXiv:0804.3334 [nucl-th], to be publ. in *Eur. Phys. J. A*.

43. M.P. Rekalc and E. Tomasi-Gustafsson, *Eur. Phys. J. A* **22** (2004) 331; *Nucl. Phys. A* **742** (2004) 322.

44. E. Tomasi-Gustafsson and G.I. Gakh, *Phys. Rev. C* **72** (2005) 015209;
    V. Tvaskis et al., *Phys. Rev. C* **73** (2006) 025206.

45. Y.C. Chen, C.W. Kao and S.N. Yang, *Phys. Lett. B* **652** (2007) 269.

46. J. Arrington, W. Melnitchouk and J.A. Tjon, *Phys. Rev. C* **76** (2007) 035205.

# Chapter 3

# Exotic Baryon Resonances in the Skyrme Model

Dmitri Diakonov and Victor Petrov

*Petersburg Nuclear Physics Institute,*
*Gatchina, 188300, St. Petersburg, Russia*
*diakonov@thd.pnpi.spb.ru,*
*victorp@thd.pnpi.spb.ru*

We outline how one can understand the Skyrme model from the modern perspective. We review the quantization of the $SU(3)$ rotations of the Skyrmion, leading to the exotic baryons that cannot be made of three quarks. It is shown that in the limit of large number of colours the lowest-mass exotic baryons can be studied from the kaon-Skyrmion scattering amplitudes, an approach known after Callan and Klebanov. We follow this approach and find, both analytically and numerically, a strong $\Theta^+$ resonance in the scattering amplitude that is traced to the rotational mode. The Skyrme model *does* predict an exotic resonance $\Theta^+$ but grossly overestimates the width. To understand better the factors affecting the width, it is computed by several different methods giving, however, identical results. In particular, we show that insofar as the width is small, it can be found from the transition axial constant. The physics leading to a narrow $\Theta^+$ resonance is briefly reviewed and affirmed.



## Contents



3.1   How to Understand the Skyrme Model . . . . . . . . . . . . . . . . . . . . . . 58
3.2   Rotational States of the $SU(3)$ Skyrmion . . . . . . . . . . . . . . . . . . 61
3.3   Rotational Multiplets at Arbitrary $N_c$ . . . . . . . . . . . . . . . . . . . . 64
3.4   Rotational Wave Functions . . . . . . . . . . . . . . . . . . . . . . . . . . 69
3.5   Kaons Scattering off the Skyrmion . . . . . . . . . . . . . . . . . . . . . . 71
3.6   Physics of the Narrow $\Theta^+$ Width . . . . . . . . . . . . . . . . . . . . . 76
3.7   Getting a Narrow $\Theta^+$ in the Skyrme Model . . . . . . . . . . . . . . . 79
      3.7.1   Vanishing $m_\Theta - m_N$, vanishing $\Gamma_\Theta$ . . . . . . . . . . . . . 79
      3.7.2   Finite $m_\Theta - m_N$, vanishing $\Gamma_\Theta$ . . . . . . . . . . . . . . . 82
3.8   Goldberger–Treiman Relation and the $\Theta^+$ Width . . . . . . . . . . . . 83
3.9   Finite-$N_c$ Effects in the $\Theta^+$ Width . . . . . . . . . . . . . . . . . . . 85
3.10 Conclusions . . . . . . . . . . . . . . . . . . . . . . . . . . . . . . . . . . 87
References . . . . . . . . . . . . . . . . . . . . . . . . . . . . . . . . . . . . . . 89

## 3.1. How to Understand the Skyrme Model

It is astounding that Skyrme had suggested his model[1] as early as in 1961 before it has been generally accepted that pions are (pseudo) Goldstone bosons associated with the spontaneous breaking of chiral symmetry, and of course long before Quantum Chromodynamics (QCD) has been put forward as the microscopic theory of strong interactions.

The revival of the Skyrme idea in 1983 is due to Witten[2] who explained the *raison d'être* of the Skyrme model from the viewpoint of QCD. In the chiral limit when the light quark masses $m_u, m_d, m_s$ tend to zero, such that the octet of the pseudoscalar mesons $\pi, K, \eta$ become nearly massless (pseudo) Goldstone bosons, they are the lightest degrees of freedom of QCD. The effective chiral Lagrangian (E$\chi$L) for pseudoscalar mesons, understood as an infinite expansion in the derivatives of the pseudoscalar (or chiral) fields, encodes, in principle, full information about QCD. The famous two-term Skyrme Lagrangian can be understood as a low-energy truncation of this infinite series. Witten has added an important four-derivative Wess–Zumino term[3] to the original Skyrme Lagrangian and pointed out that the overall coefficient in front of the E$\chi$L is proportional to the number of quark colours $N_c$.

Probably most important, Witten has shown that Skyrme's original idea of getting the nucleon as a soliton of the E$\chi$L is justified in the limit of large $N_c$ (since quantum corrections to a classical saddle point die out as $1/N_c$) and that the 'Skyrmion' gets correct quantum numbers upon quantization of its rotations in ordinary and flavour spaces. Namely, if one restricts oneself to two light flavours $u, d$, the lowest rotational states of a Skyrmion are the nucleon with spin $J = \frac{1}{2}$ and isospin $T = \frac{1}{2}$ and the $\Delta$ resonance with $J = \frac{3}{2}$ and $T = \frac{3}{2}$. For three light flavours $u, d, s$ the lowest rotational state is the $SU(3)$ octet with spin $\frac{1}{2}$ and the next is the decuplet with spin $\frac{3}{2}$, in full accordance with reality. The statement appeared in Witten's 'note added in proof' without a derivation but a number of authors[4] have derived the result (it is reproduced in Section 2). Almost all of those authors noticed that formally the next rotational excitation of the Skyrmion is an exotic baryon *anti*decuplet, again with spin $\frac{1}{2}$, however few took it seriously. It was only after the publication of Ref.[5] where it was predicted that the lightest member of the antidecuplet, the $\Theta^+$ baryon, must be light and narrow, that a considerable experimental and theoretical interest in the exotic baryons has been aroused.

Soon after Witten's work it has been realized that it is possible to bring the Skyrme model and the Skyrmion even closer to QCD and to the more customary language of constituent quarks. It has been first noticed[6–8] that a simple chiral-invariant Lagrangian for massive (constituent) quarks $Q$ interacting with the octet chiral field $\pi^A (A = 1, ..., 8)$,

$$\mathcal{L} = \bar{Q} \left( \partial\!\!\!/ - M e^{\frac{i\pi^A \lambda^A \gamma_5}{F_\pi}} \right) Q, \qquad \pi^A = \pi, K, \eta, \tag{3.1.1}$$

induces, via a quark loop in the external pseudoscalar fields (see Fig. 3.1), the E$\chi$L whose lowest-derivative terms coincide with the Skyrme Lagrangian, including automatically the Wess–Zumino term, with the correct coefficient!

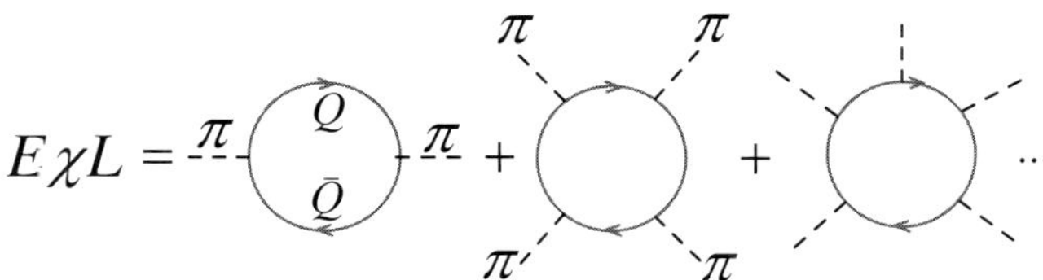


Fig. 3.1.   The effective chiral lagrangian (E$\chi$L) is the quark loop in the external chiral field, or the determinant of the Dirac operator (3.1.1). Its real part is the kinetic energy term for pions, the Skyrme term and, generally, an infinite series in derivatives of the chiral field. Its imaginary part is the Wess–Zumino term, plus also an infinite series in derivatives.[6,7,9]

A step in the same direction, namely in bringing the Skyrme model closer to the language of quarks, has been made in the chiral bag model by Brown, Rho and collaborators,[10] for a review see Ref. 11.

In fact, Eq. (3.1.1) can be derived in the instanton liquid model for the spontaneous chiral symmetry breaking[8] where a dynamical momentum-dependent quark mass $M(p)$ is generated as an originally massless quark propagates through the random ensemble of instantons and anti-instantons, each time flipping its helicity. The low-energy quark Lagrangian (3.1.1) is generally speaking nonlocal which provides a natural ultraviolet cutoff. At low momenta, however, one can treat the dynamical mass as a constant $M(0) \approx 350\,\mathrm{MeV}$.[8]

It is implied that all gluon degrees of freedom, perturbative and not, are integrated out when one comes to the effective low-energy quark Lagrangian of the type given by Eq. (3.1.1). Important, one does not need to add explicitly, say, the kinetic energy term for pions to Eq. (3.1.1) (as several authors have originally suggested[12–14]) since the pion is a $Q\bar{Q}$ state itself and it propagates through quark loops, as exhibited in the first graph in Fig. 3.1.

Understanding the quark origin of the E$\chi$L it becomes possible to formulate what is the Skyrmion in terms of quarks and demystify the famous prescription of the Skyrme model that a chiral soliton with a topological (or winding) number equal to unity, is in fact a fermion.

To that end, one looks for a trial chiral field capable of binding constituent quarks. Let there be such a field $\pi(\mathbf{x})$ that creates a bound-state level for "valence" quarks, $E_{\mathrm{val}}$. Actually, one can put $N_c$ quarks at that level in the antisymmetric colour state, as the chiral field is colour-blind. The energy penalty for creating the trial field is given by the same Lagrangian (3.1.1). It is the aggregate energy of the negative-energy Dirac sea of quarks distorted by the trial field, $E_{\mathrm{sea}}$; it should be also multiplied by $N_c$ since all negative-energy levels should be occupied and they are $N_c$-fold degenerate in colour. Therefore, the full energy of a state with baryon number unity and made of $N_c$ quarks, is a sum of two functionals,[9,15]

$$\mathcal{M}_N = N_c \left( E_{\mathrm{val}}[\pi(\mathbf{x})] + E_{\mathrm{sea}}[\pi(\mathbf{x})] \right). \tag{3.1.2}$$

Schematically it is shown in Fig. 3.2. The self-consistent (or mean) pion field binding quarks is the one minimizing the nucleon mass. Quantum fluctuations about it are suppressed insofar as $N_c$ is large. The condition that the winding number of the trial field is unity needs to be imposed to get a deeply bound state, that is to guarantee that the baryon number is unity.[9] The Skyrmion is, thus, nothing but the **mean chiral field binding quarks in a baryon**.

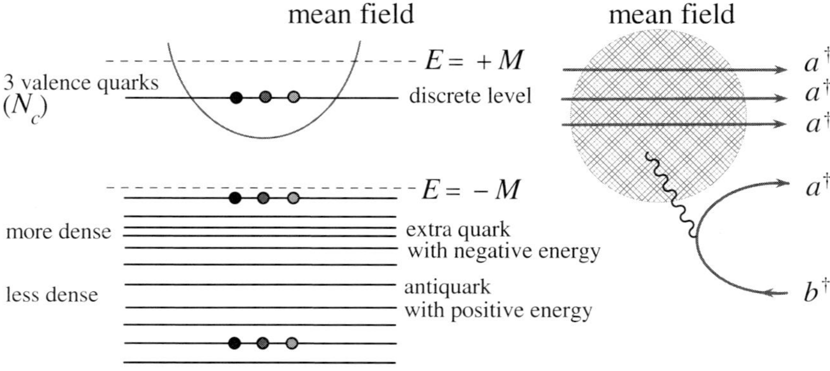


Fig. 3.2. Left: If the trial pion field is large enough (shown schematically by the solid curve), there is a discrete bound-state level for $N_c$ 'valence' quarks, $E_{\mathrm{val}}$. One has also to fill in the negative-energy Dirac sea of quarks (in the absence of the trial pion field it corresponds to the vacuum). The spectrum of the negative-energy levels is shifted in the trial pion field, its aggregate energy, as compared to the free case, being $E_{\mathrm{sea}}$. Right: Equivalent view of baryons, where the polarized Dirac sea is presented as $Q\bar{Q}$ pairs.

This model of baryons, called the *Chiral Quark Soliton Model* or the *Relativistic Mean Field Approximation*, apparently interpolates between the nonrelativistic constituent quark model and the Skyrme model, making sense and showing the limitations of both. Indeed, if the mean field happens to be small, the valence level is shallow, the Dirac sea is weakly distorted, and there are few antiquarks. In this case the model reproduces the well-known spin and space quark wave functions of the nonrelativistic models for baryons.[16] If, on the contrary, the mean field happens to be very broad, the valence level dives towards the negative-energy Dirac sea, and there are $\mathcal{O}(N_c)$ (that is many!) additional $Q\bar{Q}$ pairs in a baryon, whose energy can be approximated by the two- and four-derivative Skyrme Lagrangian. The realistic picture is somewhere in between the two extremes.

Decoding the Skyrme model in terms of quarks allows one to answer important questions that cannot even be asked in the Skyrme model. For example, one can find out parton distributions in nucleons, satisfying all general sum rules and positivity constraints,[17] the light-cone distribution amplitudes[18] or, *e.g.* the wave functions of the 5-quark components in nucleons.[16] For reviews of the model see Refs. 19 and 20.

To summarize this introduction: The original Skyrme's idea is well founded from the modern QCD viewpoint. There is no mystics in the identification of the pion field winding number with the baryon number, and in the Skyrmion being a fermion (at odd $N_c$). The chiral soliton field, the Skyrmion, is nothing but the self-consistent mean field binding $N_c$ valence quarks and distorting the Dirac sea such that additional $Q\bar{Q}$ pairs are necessary present in a baryon.

At the same time, one cannot expect a fully quantitative description of reality in the concrete two-terms Skyrme's original model as an infinite series in the derivatives in the E$\chi$L is truncated: it is similar to replacing $e^{-x}$ by $1-x$. What is even worse, there are no explicit valence quarks in the Skyrme model as they cannot be separated from the sea.

In what follows, we shall nevertheless mainly deal with the concrete model by Skyrme (supplemented by the Wess–Zumino term) in order to study certain qualitative features of the exotic baryon resonances, *i.e.* those that by quantum numbers cannot be composed of three quarks only but need additional $Q\bar{Q}$ pairs.

## 3.2. Rotational States of the $SU(3)$ Skyrmion

The results of this section are general in the sense that they are independent on whether one takes literally the Skyrme model or a more sophisticated chiral quark model.

The standard choice of the saddle point field is the "upper-left corner hedgehog" Ansatz:

$$U_0(\mathbf{x}) \equiv e^{i\pi_0^A(\mathbf{x})\lambda^A} = \begin{pmatrix} e^{i(\boldsymbol{n}\cdot\boldsymbol{\tau})P(r)} & & 0 \\ & & 0 \\ 0 & 0 & 1 \end{pmatrix}, \qquad \mathbf{n} = \frac{\mathbf{x}}{r}, \qquad (3.2.3)$$

where $\lambda^A$ are eight Gell-Mann matrices, $\tau^i$ are three Pauli matrices, and $P(r)$ is a spherically symmetric function called the profile of the Skyrmion. In the chiral limit $m_u = m_d = m_s = 0$ any $SU(3)$ rotation of the saddle point field, $RU_0R^\dagger$, $R \in SU(3)$, is also a saddle point. We consider a slowly rotating Ansatz,

$$U(\mathbf{x}, t) = R(t)U_0(\mathbf{x})R^\dagger(t) \qquad (3.2.4)$$

and plug it into the E$\chi$L. The degeneracy of the saddle point in the flavour rotations means that the action will not depend on $R$ itself but only on the time derivatives $\dot{R}$. We do not consider the rotation angles as small but rather expand the action in angular velocities. In fact, one has to distinguish between the 'right' ($\Omega_A$) and 'left' ($\omega_A$) angular velocities defined as

$$\Omega_A = -i\text{Tr}(R^\dagger \dot{R}\lambda^A), \qquad \omega_A = -i\text{Tr}(\dot{R}R^\dagger\lambda^A), \qquad \Omega^2 = \omega^2 = 2\text{Tr}\dot{R}^\dagger\dot{R}. \qquad (3.2.5)$$

Given the Ansatz (3.2.3) one expects on symmetry grounds the following Lagrangian for slow rotations:

$$\mathcal{L}_{\text{rot}} = \frac{I_1}{2}\left(\Omega_1^2 + \Omega_2^2 + \Omega_3^2\right) + \frac{I_2}{2}\left(\Omega_4^2 + \Omega_5^2 + \Omega_6^2 + \Omega_7^2\right) - \frac{N_c B}{2\sqrt{3}}\Omega_8 \tag{3.2.6}$$

where $I_{1,2}$ are the two soliton moments of inertia that are functionals of the profile function $P(r)$. Rotation along the 8th axis in flavour space, $R = \exp(i\alpha_8\lambda^8)$, commutes with the 'upper-left-corner' Ansatz, therefore there is no quadratic term in $\Omega_8$. However there is a Wess–Zumino term resulting in a term linear in $\Omega_8$ proportional to the baryon number $B$. In the chiral quark models this term arises from the extra bound-state levels for quarks.[21]

To quantize this rotational Lagrangian one uses the canonical quantization procedure. Namely, one introduces eight 'right' angular momenta $J_A$ canonically conjugate to 'right' angular velocities $\Omega_A$,

$$J_A = -\frac{\partial\mathcal{L}_{\text{rot}}}{\partial\Omega_A}, \tag{3.2.7}$$

and writes the rotational Hamiltonian as

$$\mathcal{H}_{\text{rot}} = \Omega_A J_A - \mathcal{L}_{\text{rot}} = \frac{J_1^2 + J_2^2 + J_3^2}{2I_1} + \frac{J_4^2 + J_5^2 + J_6^2 + J_7^2}{2I_2} \tag{3.2.8}$$

with the additional quantization prescription following from Eq. (3.2.7),

$$J_8 = \frac{N_c B}{2\sqrt{3}}. \tag{3.2.9}$$

The quantization amounts to replacing classical angular momenta $J_A$ by $SU(3)$ generators satisfying the $su(3)$ algebra: $[J_A J_B] = if_{ABC}J_C$ where $f_{ABC}$ are the $su(3)$ structure constants. These generators act on the matrix $R$ on the right, $\exp(i\alpha^A J_A))R\exp(i(-\alpha_A J_A)) = R\exp(-i\alpha^A\lambda^A/2)$. For the first three generators $(A = 1, 2, 3)$ this is equivalent, thanks to the hedgehog Ansatz (3.2.3), to rotating the space axes $x, y, z$. Therefore, $J_{1,2,3}$ are in fact spin generators.

One can also introduce 'left' angular momenta $T_A$ canonically conjugate to the 'left' angular velocities $\omega_A$; they satisfy the same $su(3)$ algebra, $[T_A T_B] = if_{ABC}T_C$, whereas $[T_A J_B] = 0$. These generators act on the matrix $R$ on the left, $\exp(i\alpha^A T_A))R\exp(i(-\alpha_A T_A)) = \exp(i\alpha^A\lambda^A/2)R$, and hence have the meaning of $SU(3)$ flavour generators. The quadratic Casimir operator can be written using either 'left' or 'right' generators as

$$J_A J_A = T_A T_A = C_2(p, q) = \frac{1}{3}\left[p^2 + q^2 + pq + 3(p + q)\right] \tag{3.2.10}$$

where $C_2(p, q)$ is the eigenvalue of the quadratic Casimir operator for an irreducible representation $r$ of $SU(3)$, labeled by two integers $(p, q)$. The rotational wave functions of chiral soliton are thus finite $SU(3)$ rotation matrices $D^r_{T,T_3,Y;J,J_3,Y'}(R)$ characterized by the eigenvalues of the commuting generators. For the $SU(2)$ group they are called Wigner finite-rotation matrices and depend on 3 Euler angles; in

$SU(3)$ there are 8 'Euler' angles. The general rotational functions (with important sign subtleties) are given in the Appendix of Ref. 21, and practically useful examples are given explicitly in Ref. 16.

One can visualize the rotational wave functions as a product of two same $SU(3)$ weight diagrams: one for the eigenvalues of the flavour ('left') generators, and the other for the eigenvalues of 'right' generators including the spin. Important, the quantization condition (3.2.9) means that not all $SU(3)$ representations can be viewed as rotational states of a Skyrmion. Taking baryon number $B=1$ and $N_c = 3$ and recalling that $J_8 = Y'\sqrt{3}/2$ where $Y$ is the hypercharge, the condition (3.2.9) means that only those multiplets are rotational states that contain particles with $Y' = 1$ or, more generally,

$$Y' = \frac{N_c}{3}. \tag{3.2.11}$$

The lowest $SU(3)$ multiplets meeting this condition are the octet, the decuplet and the antidecuplet, see Fig. 3.3.

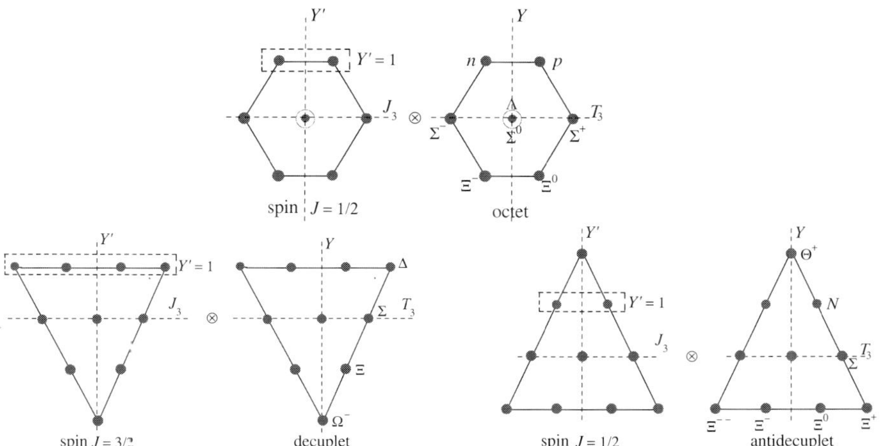

Fig. 3.3. The lowest rotational states of a Skyrmion, satisfying the condition $Y' = 1$: $\left(\mathbf{8}, \frac{1}{2}\right)$, $\left(\mathbf{10}, \frac{3}{2}\right)$, $\left(\overline{\mathbf{10}}, \frac{1}{2}\right)$. The number of states with $Y' = 1$, if one equates it to $2J + 1$, determines the spin $J$ of the particles in the multiplet.

It is remarkable that the lowest rotational states of the Skyrmion are exactly those observed in nature. The third is the antidecuplet with spin $\frac{1}{2}$. In the three vertices of the big triangle shown in Fig. 3.3, bottom right, there are baryons that are explicitly exotic, in the sense that they cannot be composed of three quarks but need an additional quark-antiquark pair. For example, the $\Theta^+$ baryon at the top of the triangle can be composed minimally of $uudd\bar{s}$ quarks, *i.e.* it is a *pentaquark*. Seven baryons that are not in the vertices of the antidecuplet are cryptoexotic, in the sense that their quantum numbers can be, in principle, arranged from three quarks, however their expected properties are quite different from those of the similar members of a baryon octet.

It should be remembered, however, that strictly speaking the whole Skyrmion approach to baryons is justified in the limit of large $N_c$. Whether $N_c = 3$ is "large enough" is a question to which there is no unique answer: it depends on how large are the $1/N_c$ corrections to a particular physical quantity. Therefore, one has to be able to write equations with $N_c$ being a free parameter. In particular, at arbitrary $N_c$ one has to construct explicitly $SU(3)$ flavor multiplets that generalize the lightest baryon multiplets $\left(\mathbf{8}, \frac{1}{2}\right)$, $\left(\mathbf{10}, \frac{3}{2}\right)$, $\left(\overline{\mathbf{10}}, \frac{1}{2}\right)$, *etc.*, to arbitrary $N_c$. We do it in the next section following Ref. 22 that generalizes previous work on this subject.[23,24]

### 3.3. Rotational Multiplets at Arbitrary $N_c$

We remind the reader that a generic $SU(3)$ multiplet or irreducible representation is uniquely determined by two non-negative integers $(p, q)$ having the meaning of upper (lower) components of the irreducible $SU(3)$ tensor $T^{\{f_1 \ldots f_p\}}_{\{g_1 \ldots g_q\}}$ symmetrized both in upper and lower indices and with a contraction with any $\delta^{g_n}_{f_m}$ being zero. Schematically, $q$ is the number of boxes in the lower line of the Young tableau depicting an $SU(3)$ representation and $p$ is the number of extra boxes in its upper line, see Fig. 3.4.

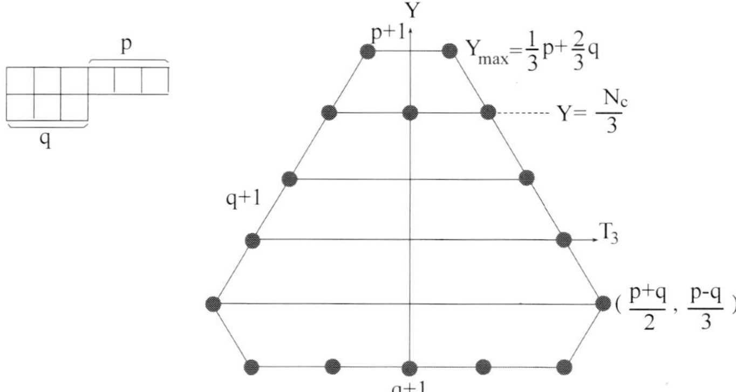

Fig. 3.4. A generic $SU(3)$ multiplet is, on the one hand, defined by the Young tableau and on the other hand can be characterized by quantum numbers $(T_3, Y)$ of its members filling a hexagon in the $(T_3, Y)$ axes (the weight diagram).

The dimension of a representation or the number of particles in the multiplet is

$$\mathrm{Dim}(p, q) = (p + 1)(q + 1)\left(1 + \frac{p + q}{2}\right). \qquad (3.3.12)$$

On the weight $(T_3, Y)$ diagram where $T_3$ is the third projection of the isospin and $Y$ is the hypercharge, a generic $SU(3)$ representation is depicted by a hexagon, whose upper horizontal side contains $p + 1$ 'dots' or particles, the adjacent sides contain $q + 1$ particles, with alternating $p + 1$ and $q + 1$ particles in the rest sides, the corners included — see Fig. 3.4. If either $p$ or $q$ is zero, the hexagon reduces to a triangle. Particles on the upper (horizontal) side of the hexagon have the hypercharge

$$Y_{\max} = \frac{1}{3}p + \frac{2}{3}q \qquad (3.3.13)$$

being the maximal possible hypercharge of a multiplet with given $(p, q)$. Given that

$$\sum_{A=1}^{3} J_A^2 = J(J+1), \qquad \sum_{A=1}^{8} J_A^2 = C_2(p, q), \qquad J_8^2 = \frac{N_c^2}{12}, \qquad (3.3.14)$$

one gets from Eq. (3.2.8) the rotational energy of baryons with given spin $J$ and belonging to representation $(p, q)$:

$$\mathcal{E}_{\mathrm{rot}}(p, q, J) = \frac{C_2(p, q) - J(J+1) - \frac{N_c^2}{12}}{2I_2} + \frac{J(J+1)}{2I_1}. \qquad (3.3.15)$$

Only those multiplets are realized as rotational excitations that have members with hypercharge $Y = \frac{N_c}{3}$; if the number of particles with this hypercharge is $n$ the spin $J$ of the multiplet is such that $2J + 1 = n$. It is easily seen that the number of particles with a given $Y$ is $\frac{4}{3}p + \frac{2}{3}q + 1 - Y$ and hence the spin of the allowed multiplet is

$$J = \frac{1}{6}(4p + 2q - N_c). \qquad (3.3.16)$$

A common mass $\mathcal{M}_0$ must be added to Eq. (3.3.15) to get the mass of a particular multiplet. Throughout this section we disregard the splittings inside multiplets as due to nonzero current strange quark mass.

The condition that a horizontal line $Y = \frac{N_c}{3}$ must be inside the weight diagram for the allowed multiplet leads to the requirement

$$\frac{N_c}{3} \leq Y_{\max} \qquad \text{or} \quad p + 2q \geq N_c \qquad (3.3.17)$$

showing that at large $N_c$ multiplets must have a high dimension!

We introduce a non-negative number which we name "exoticness" $X$ of a multiplet defined as[22]

$$Y_{\max} = \frac{1}{3}p + \frac{2}{3}q \equiv \frac{N_c}{3} + X, \qquad X \geq 0. \qquad (3.3.18)$$

Combining Eqs. (3.3.16) and (3.3.18) we express $(p, q)$ through $(J, X)$:

$$p = 2J - X,$$
$$q = \frac{1}{2}N_c + 2X - J. \qquad (3.3.19)$$

The total number of boxes in Young tableau is $2q + p = N_c + 3X$. Since we are dealing with unity baryon number states, the number of quarks in the multiplets we discuss is $N_c$, *plus* some number of quark-antiquark pairs. In the Young tableau, quarks are presented by single boxes and antiquarks by double boxes. It explains

Fig. 3.5.  Rotational excitations form a sequence of bands.

the name "exoticness": $X$ gives the minimal number of additional quark-antiquark pairs one needs to add on top of the usual $N_c$ quarks to compose a multiplet.

Putting $(p, q)$ from Eq. (3.3.19) into Eq. (3.3.15) we obtain the rotational energy of a soliton as function of the spin and exoticness of the multiplet:

$$\mathcal{E}_{\rm rot}(J, X) = \frac{X^2 + X(\frac{N_c}{2} + 1 - J) + \frac{N_c}{2}}{2I_2} + \frac{J(J+1)}{2I_1}. \tag{3.3.20}$$

We see that for given $J \le \frac{N_c}{2} + 1$ the multiplet mass is a monotonically growing function of $X$: the minimal-mass multiplet has $X = 0$. Masses of multiplets with increasing exoticness are:

$$\mathcal{M}_{X=0}(J) = \mathcal{M}'_0 + \frac{J(J+1)}{2I_1}, \qquad \text{where} \quad \mathcal{M}'_0 \equiv \mathcal{M}_0 + \frac{N_c}{4I_2}, \tag{3.3.21}$$

$$\mathcal{M}_{X=1}(J) = \mathcal{M}'_0 + \frac{J(J+1)}{2I_1} + 1 \cdot \frac{\frac{N_c}{2} + 2 - J}{2I_2}, \tag{3.3.22}$$

$$\mathcal{M}_{X=2}(J) = \mathcal{M}'_0 + \frac{J(J+1)}{2I_1} + 2 \cdot \frac{\frac{N_c}{2} + 2 - J}{2I_2} + \frac{1}{I_2}, \qquad \text{etc.} \tag{3.3.23}$$

At this point it should be recalled that both moments of inertia $I_{1,2} = O(N_c)$, as is $\mathcal{M}_0$. We see from Eqs. (3.3.21)–(3.3.23) that multiplets fall into a sequence of rotational bands each labeled by its exoticness with small $O(1/N_c)$ splittings inside the bands. The separation between bands with different exoticness is $O(1)$. The corresponding masses are schematically shown in Fig. 3.5.

The lowest band is non-exotic ($X=0$); the multiplets are determined by $(p, q) = \left(2J, \frac{N_c}{2} - J\right)$, and their dimension is $\text{Dim} = (2J + 1)(N_c + 2 - 2J)(N_c + 4 + 2J)/8$ which in the particular (but interesting) case of $N_c = 3$ becomes 8 for spin one half and 10 for spin 3/2. These are the correct lowest multiplets in real world, and the above multiplets are their generalization to arbitrary values of $N_c$. To make baryons fermions one needs to consider only odd $N_c$.

Recalling that $u, d, s$ quarks' hypercharges are $1/3$, $1/3$ and $-2/3$, respectively, one observes that all baryons of the non-exotic $X = 0$ band can be made of $N_c$ quarks. The upper side of their weight diagrams (see Fig. 3.6) is composed of $u, d$ quarks only; in the lower lines one consequently replaces $u, d$ quarks by the $s$ one. This is how the real-world $\left(\mathbf{8}, \frac{1}{2}\right)$ and $\left(\mathbf{10}, \frac{3}{2}\right)$ multiplets are arranged and this property is preserved in their higher-$N_c$ generalizations. The construction coincides with that of Ref. 23. At high $N_c$ there are further multiplets with spin $5/2$ and so on. The maximal possible spin at given $N_c$ is $J_{\max} = \frac{N_c}{2}$: if one attempts higher spin, $q$ becomes negative.

The rotational bands for $X = 1$ multiplets are shown in Fig. 3.7, left and middle graphs. The upper side of the weight diagram is exactly one unit higher than the line $Y = \frac{N_c}{3}$ which is non-exotic, in the sense that its quantum numbers can be, in principle, achieved from exactly $N_c$ quarks. However, particles corresponding to the upper side of the weight diagram cannot be composed of $N_c$ quarks but require at least one additional $\bar{s}$ quark and hence *one additional quark-antiquark pair* on top of $N_c$ quarks.

The multiplet shown in Fig. 3.7, left, has only one particle with $Y = \frac{N_c}{3} + 1$. It is an isosinglet with spin $J = \frac{1}{2}$, and in the quark language is built of $(N_c + 1)/2$ *ud* pairs and one $\bar{s}$ quark. It is the generalization of the $\Theta^+$ baryon to arbitrary odd $N_c$. As seen from Eqs. (3.3.12) and (3.3.19), the multiplet to which the "$\Theta^+$" belongs is characterized by $(p, q) = (0, (N_c + 3)/2)$, its dimension is $(N_c + 5)(N_c + 7)/8$ becoming the $\left(\overline{\mathbf{10}}, \frac{1}{2}\right)$ at $N_c = 3$. Its splitting with the $N_c$ generalization of the non-exotic $\left(\mathbf{8}, \frac{1}{2}\right)$ multiplet follows from Eq. (3.3.22):

$$\mathcal{M}_{\overline{10},\frac{1}{2}} - \mathcal{M}_{8,\frac{1}{2}} = \frac{N_c + 3}{4 I_2}, \tag{3.3.24}$$

a result first found in Ref. 24. Here and in what follows we denote baryon multiplets by their dimension at $N_c = 3$ although at $N_c > 3$ their dimension is higher, as given by Eq. (3.3.12).

The second rotational state of the $X = 1$ sequence has $J = \frac{3}{2}$; it has $(p, q) = (2, (N_c + 1)/2)$ and dimension $3(N_c + 3)(N_c + 9)/8$ reducing to the multiplet $\left(\mathbf{27}, \frac{3}{2}\right)$ at $N_c = 3$, see Fig. 3.7, middle. In fact there are two physically distinct multiplets there. Indeed, the weights in the middle of the second line from top on the weight

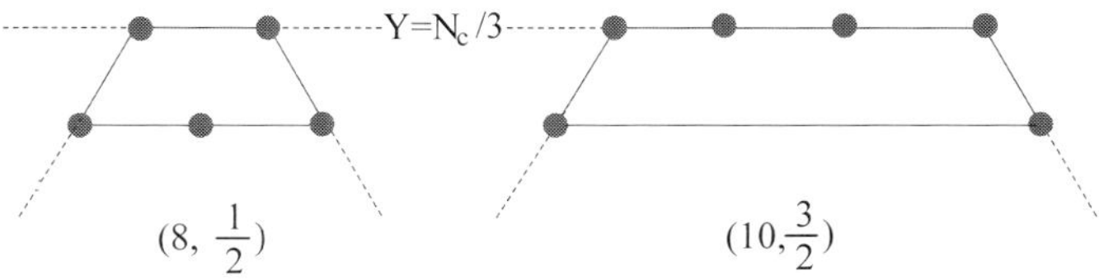


Fig. 3.6.   Non-exotic ($X = 0$) multiplets that can be composed of $N_c$ quarks.

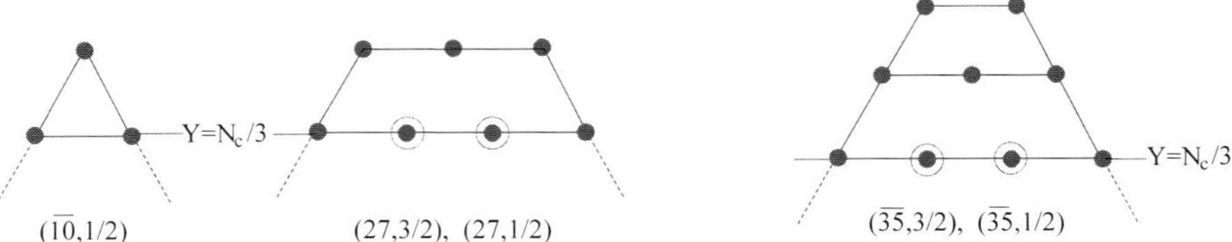



Fig. 3.7. Exotic ($X\!=\!1$) multiplets (left and middle graphs) that can be composed of $N_c$ quarks and one extra $Q\bar{Q}$ pair. An example of an $X = 2$ multiplet that can be composed with two additional $Q\bar{Q}$ pairs, is shown on the right.

diagram with $Y = \frac{N_c}{3}$ are twice degenerate, corresponding to spin $3/2$ and $1/2$. Therefore, there is another $3(N_c + 3)(N_c + 9)/8$-plet with unit exoticness, but with spin $1/2$. At $N_c\!=\!3$ it reduces to $\left(\mathbf{27}, \frac{1}{2}\right)$. The splittings with non-exotic multiplets are

$$\mathcal{M}_{27,\frac{3}{2}} - \mathcal{M}_{10,\frac{3}{2}} = \frac{N_c + 1}{4I_2}, \tag{3.3.25}$$

$$\mathcal{M}_{27,\frac{1}{2}} - \mathcal{M}_{8,\frac{1}{2}} = \frac{N_c + 7}{4I_2}. \tag{3.3.26}$$

The $X\!=\!1$ band continues to the maximal spin $J_{\max} = (N_c + 4)/2$ where $q$ becomes zero.

The $X = 2$ rotational band (see Fig. 3.7, right) starts from two states with spin $3/2$ and $1/2$ both belonging to the $SU(3)$ representation $(p, q, \text{Dim}) = (1, (N_c + 5)/2, (N_c + 7)(N_c + 11)/4)$. It reduces to the $\overline{\mathbf{35}}$ multiplet at $N_c = 3$. Their splittings with non-exotic multiplets are

$$\mathcal{M}_{\overline{35},\frac{3}{2}} - \mathcal{M}_{10,\frac{3}{2}} = \frac{N_c + 3}{2I_2}, \tag{3.3.27}$$

$$\mathcal{M}_{\overline{35},\frac{1}{2}} - \mathcal{M}_{8,\frac{1}{2}} = \frac{N_c + 6}{2I_2}. \tag{3.3.28}$$

The maximal spin of the $X\!=\!2$ rotational band is $J_{\max} = (N_c + 8)/2$.

The upper side in the weight diagram in Fig. 3.7, right, for the $X\!=\!2$ sequence has hypercharge $Y_{\max} = \frac{N_c}{3} + 2$. Therefore, one needs *two* $\bar{s}$ quarks to get that hypercharge and hence the multiplets can be minimally constructed of $N_c$ quarks plus *two additional quark-antiquark pairs*.

Disregarding the rotation along the 1,2,3 axes (for example taking only the lowest $J$ state from each band) we observe from Eq. (3.3.20) that at large $N_c$ the spectrum is equidistant in exoticness,

$$\mathcal{E}_{\text{rot}}(X) = \frac{N_c(X + 1)}{4I_2}, \tag{3.3.29}$$

with the spacing $\frac{N_c}{4I_2} = O(1)$. It is consistent with the fact explained in the next section, that at large $N_c$ the rotation corresponding to the excitations of exoticness is actually a small-angle precession equivalent to small oscillations whose quantization leads to an equidistant spectrum. We stress that there is no deformation of the Skyrmion by rotation until $X$ becomes of the order of $N_c$.[22] Eq. (3.3.29) means that each time we add a quark-antiquark pair it costs at large $N_c$ the same

$$\text{energy of a Q}\bar{\text{Q}}\text{ pair} = \omega_{\text{rot}} = \frac{N_c}{4I_2} = O(N_c^0). \tag{3.3.30}$$

Naively one may think that this quantity should be approximately twice the constituent quark mass $M \approx 350\,\text{MeV}$. Actually, it can be much less than that. For example, an inspection of $I_2$ in the Chiral Quark Soliton Model[9,21] shows that the pair energy is strictly less than $2M$; in fact $1/I_2$ tends to zero in the limit when the baryon size blows up.

In physical terms, the energy cost of adding a $Q\bar{Q}$ pair can be small if the pair is added in the form of a Goldstone boson. The energy penalty for making, say, the $\Theta^+$ baryon from a nucleon would be exactly zero in the chiral limit and were baryons infinitely large. In reality, one has to create a pseudo-Goldstone K-meson and to confine it inside the baryon of the size $\geq 1/M$. It costs roughly

$$m(\Theta) - m(N) \approx \sqrt{m_K^2 + \mathbf{p}^2} \leq \sqrt{495^2 + 350^2} = 606\,\text{MeV}. \tag{3.3.31}$$

Therefore, one should expect the exotic $\Theta^+$ around 1540 MeV where indeed it has been detected in a number of experiments!

## 3.4. Rotational Wave Functions

It is helpful to realize how do the rotational wave functions $\Psi(R)$ look like for various known (and unknown) baryons. To that end, one needs a concrete parameterization of the $SU(3)$ rotation matrix $R$ by 8 'Euler' angles: the wave functions are in fact functions of those angles.

In general, the parameter space of an $SU(N)$ group is a direct product of odd-dimensional spheres, $S^3 \times S^5 \times \ldots \times S^{2N-1}$. For $SU(3)$, it is a product of the spheres $S^3 \times S^5$. A general $SU(3)$ matrix $R$ can be written as $R = S_3 R_2$ where $R_2$ is a general $SU(2)$ matrix with three parameters, put in the upper-left corner, and $S_3$ is an $SU(3)$ matrix of a special type with five parameters, see Appendix A in Ref. 16.

To be specific, let us consider the rotational wave function corresponding to the exotic $\Theta^+$ baryon. For general $N_c$ its (complex conjugate) wave function is given by[16]

$$\Theta_k(R)^* = \left(R_3^3\right)^{N_c-1} R_k^3 \tag{3.4.32}$$

where $k = 1, 2$ is the spin projection and $R_k^3$ is the $k^{\text{th}}$ matrix element in the $3^{\text{d}}$ row of the $3 \times 3$ matrix $R$. Using the concrete parameterization of Ref. 16, Eq. (3.4.32) becomes

$$\Theta(R)^* \sim (\cos\theta \, \cos\phi)^{N_c}, \tag{3.4.33}$$

where $\theta, \phi \in (0, \frac{\pi}{2})$ are certain angles parameterizing the $S^5$ sphere; $\theta = \phi = 0$ corresponds to the North pole of that sphere. We see that although for $N_c = 3$ the typical angles in the wave function are large such that it is spread over both $S^3$ and $S^5$ globes, at $N_c \to \infty$ the wave function is concentrated near the North pole of $S^5$ since

$$\theta \sim \phi \sim \sqrt{\frac{2}{N_c}} \xrightarrow{N_c \to \infty} 0. \tag{3.4.34}$$

This is illustrated in Fig. 3.8.

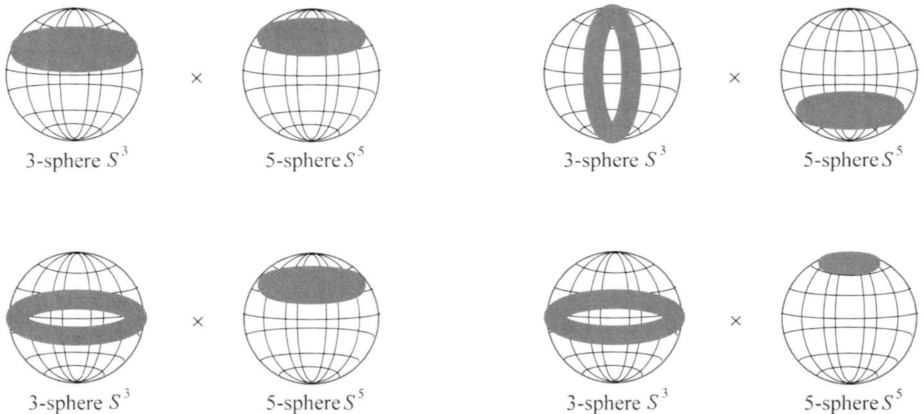

Fig. 3.8. A schematic view of the rotational wave functions of several baryons. The shaded areas indicate where the wave functions of the eight 'Euler' angles that parameterize the $S^3, S^5$ spheres, are large. Top left: proton, spin up; top right: $\Omega^-$, spin down-down; bottom left: $\Theta^+$, spin up; bottom, right: $\Theta^+$, spin up at $N_c = 37$.

Let us show that the limit $N_c \to \infty$ corresponds to the weak kaon field in the $\Theta^+$ baryon. To that end we use an alternative parameterization for the meson field fluctuations about the Skyrmion, suggested by Callan and Klebanov[25]:

$$U = \sqrt{U_0}\, U_K \sqrt{U_0}, \tag{3.4.35}$$

where $U_K$ is the meson $SU(3)$ unitary matrix which, for small meson fluctuations $\phi^A$ about the saddle-point Skyrmion field $U_0$ (3.2.3), is

$$U_K = \mathbf{1}_3 + i\phi^A \lambda^A, \qquad A = 1...8, \tag{3.4.36}$$

$$\pi^\pm = \frac{\phi^1 \pm i\phi^2}{\sqrt{2}}, \quad \pi^0 = \phi^3, \quad K^\pm = \frac{\phi^4 \pm i\phi^5}{\sqrt{2}}, \quad K^0, \overline{K^0} = \frac{\phi^6 \pm i\phi^7}{\sqrt{2}}, \quad \eta = \phi^8.$$

One can compare Eq. (3.4.35) with the rotational Ansatz, $U = R U_0 R^\dagger$, and find the meson fields in baryons corresponding to rotations. In particular, for rotations "near the North pole" *i.e.* at small angles $\theta, \phi$, one finds the kaon field[16]

$$\left. \begin{array}{l} K^+ = -\sqrt{2} \sin \frac{P(r)}{2} [\theta\, n_z + \phi\, (n_x - i n_y)] \\ K^0 = -\sqrt{2} \sin \frac{P(r)}{2} [\theta\, (n_x + i n_y) - \phi\, n_z] \end{array} \right\} = -\sqrt{2} \sin \frac{P(r)}{2} (\boldsymbol{n} \cdot \boldsymbol{\tau}) \begin{pmatrix} \theta \\ \phi \end{pmatrix}, \tag{3.4.37}$$

meaning that at large $N_c$ the amplitude of the kaon fluctuations in the prototype "$\Theta$" is vanishing as $\sim 1/\sqrt{N_c}$. Therefore, at large $N_c$ the rotation is in fact a small-angle precession about the North pole, that can be studied as a small kaon field fluctuation about the Skyrmion in a given particular model for the E$\chi$L.[26,27] It should be kept in mind, however, that in reality at $N_c{=}3$ the rotations by large angles $\theta$, $\phi$ are not suppressed. It means that in the real world the kaon field in the $\Theta^+$ is generally not small.

## 3.5. Kaons Scattering off the Skyrmion

As explained in the previous section, at large $N_c$ the kaon field in the exotic baryon $\Theta^+$ is weak, hence the resonance should manifest itself in the linear order in the kaon field perturbing the nucleon which, again at large $N_c$, can be represented by a Skyrmion.[25,27] In this section we look for the $\Theta^+$ by studying small kaon field fluctuations about the Skyrmion taking as a model for the E$\chi$L the Skyrme Lagrangian:

$$S = S_{\text{kin}} + S_{\text{Sk}} + S_{\text{WZ}} + S_m, \tag{3.5.38}$$

$$S_{\text{kin}} = \frac{F_\pi^2}{4} \int d^4x \, \text{Tr} L_\mu L_\mu, \qquad L_\mu := iU^\dagger \partial_\mu U, \tag{3.5.39}$$

$$S_{\text{Sk}} = -\frac{1}{32e^2} \int d^4x \, \text{Tr}[L_\mu L_\nu]^2, \tag{3.5.40}$$

$$S_{\text{WZ}} = \frac{N_c}{24\pi^2} \int d^4x \int_0^1 ds \, \epsilon^{\alpha\beta\gamma\delta} \, \text{Tr}\left(e^{-is\Pi}\partial_\alpha e^{is\Pi}\right)\binom{}{\beta}\binom{}{\gamma}\binom{}{\delta}, \quad e^{i\Pi} = U, \tag{3.5.41}$$

$$S_m = \int d^4x \, \frac{m_K^2 F_\pi^2}{2} \text{Tr}\left[(U + U^\dagger - 2\cdot 1_3)\,\text{diag}\left(\frac{m_u}{m_s}, \frac{m_d}{m_s}, 1\right)\right]. \tag{3.5.42}$$

We have written the Wess–Zumino term (3.5.41) in the explicit form suggested in Ref. 6. In the last, symmetry breaking term, we shall put $m_{u,d} = 0$.

Following the general approach of Callan and Klebanov[25] revived by Klebanov *et al.*[27] in the pentaquark era, we use the parameterization of $U(\mathbf{x}, t)$ in the form of Eq. (3.4.36) where we take the small kaon fluctuation in the form hinted by Eq. (3.4.37):

$$K^\alpha(\mathbf{x}, t) = (\boldsymbol{n} \cdot \boldsymbol{\tau})^\alpha_\beta \, \zeta^\beta \, \eta(r) \, e^{-i\omega t} \tag{3.5.43}$$

where $\zeta^\beta$ is a constant spinor. It corresponds to the $p$-wave kaon field.

Expanding the action (3.5.38) in the kaon field up to the second order one obtains[25,27] (we measure $r, t$ in conventional units of $1/(2F_\pi e) = \mathcal{O}(N_c^0)$)

$$S = S_0 + S_2, \tag{3.5.44}$$

$$S_0 = \frac{2\pi F_\pi}{e} \int dr\, r^2 \left[ \frac{d(r)}{2}\, (1 + 2s(r)) + s(r) \left( 1 + \frac{s(r)}{2} \right) \right], \tag{3.5.45}$$

$$S_2 = \frac{4\pi F_\pi}{e}\, \zeta^\dagger \zeta \int dr\, r^2\, \eta(r) \left\{ \omega^2 A(r) - 2\omega\gamma B(r) \right. \tag{3.5.46}$$

$$\left. + \left[ C(r) \frac{d^2}{dr^2} + D(r) \frac{d}{dr} - V(r) \right] \right\} \eta(r)$$

where one introduces short-hand notations:

$$b(r) := P''(r) \sin P(r) + P'^{\,2}(r)\, \cos P(r), \quad c(r) := \sin^2 \frac{P(r)}{2}, \quad d(r) := P'^{\,2}(r),$$

$$h(r) := \sin(2P(r))P'(r), \quad s(r) := \frac{\sin^2 P(r)}{r^2},$$

$$A(r) := 1 + 2s(r) + d(r), \quad B(r) := -N_c e^2 \frac{P'(r) \sin^2 P(r)}{2\pi^2 r^2},$$

$$C(r) := 1 + 2s(r), \quad D(r) := \frac{2}{r}\, [1 + h(r)],$$

$$V(r) := -\frac{1}{4}\, [d(r) + 2s(r)] - 2s(r)\, [s(r) + 2d(r)] + 2\frac{1 + d(r) + s(r)}{r^2}\, [1 - c(r)]^2$$

$$+ \frac{6}{r^2} \left[ s(r)\, (1 - c(r))^2 - b(r)(1 - c(r)) + \tfrac{1}{2} r^2 d(r)s(r) \right] + \mu_K^2. \tag{3.5.47}$$

Here $\mu_K$ is the dimensionless kaon mass, $\mu_K = m_K/(2F_\pi e)$. The term linear in $\omega$ in Eq. (3.5.46) arises from the Wess–Zumino term (3.5.41); the function $B(r)$ is the baryons number density in the Skyrme model. The coefficient $\gamma$ in front of it is unity in the chiral limit but in general is not universal. In what follows it is useful to analyze the results as one varies $\gamma$ from 0 to 1.

Varying $S_0$ with respect to $P(r)$ one finds the standard Skyrmion profile with $P(0) = \pi$ and $P(r) \overset{r \to \infty}{\longrightarrow} r_0^2/r^2$. Varying $S_2$ with respect to the kaon field profile $\eta(r)$ one obtains a Schrödinger-type equation

$$\left\{ \omega^2 A(r) - 2\omega\gamma B(r) + \left[ C(r) \frac{d^2}{dr^2} + D(r) \frac{d}{dr} - V(r) \right] \right\} \eta(r) = 0 \tag{3.5.48}$$

where the profile $P(r)$ found from the minimization of $S_0$ has to be substituted. In the chiral limit ($m_K \to 0$) the equations for $P(r)$, $\eta(r)$ are equivalent to the conservation of the axial current, $\partial_\mu j_{\mu 5}^A = 0$ since it is the equation of motion for the Skyrme model.

If $m_K = 0$, the $SU(3)$ symmetry is exact, and a small and slow rotation in the strange direction must be a zero mode of Eq. (3.5.48). Indeed, one can easily check that

$$\eta_{\text{rot}}(r) = \sin \frac{P(r)}{2} \tag{3.5.49}$$

is a zero mode of the square brackets in Eq. (3.5.48) and hence a zero mode of the full equation with $\omega = 0$. If in addition the Wess–Zumino coefficient $\gamma$ is set to zero, this mode is twice degenerate. These states are the large-$N_c$ prototypes of $\Lambda$ (strangeness $S = -1$) and $\Theta^+$ ($S = +1$).[25,27] At $\gamma > 0$ the two states split: $\Lambda$ remains a pole of the scattering amplitude at $\omega = 0$, and $\Theta^+$ moves into the lower semi-plane of the complex $\omega$ plane. If $m_K > 0$ the pole corresponding to the $\Lambda$ moves to $\omega < 0$ remaining on the real axis, whereas the $\Theta^+$ pole remains in the lower semi-plane with Re $\omega > 0$ and Im $\omega < 0$. Both poles are singularities of the same analytical function *i.e.* the scattering amplitude, see below. It is amusing that $\Lambda$ "knows" about $\Theta^+$ and its width through analyticity.

In what follows we shall carefully study the solutions of Eq. (3.5.48) and in particular the trajectory of the $\Theta^+$ pole, by combining numerical and analytical calculations. In numerics, we use the conventional choice of the constants in the Skyrme model: $F_\pi = 64.5$ MeV (*vs.* 93 MeV experimentally) and $e = 5.45$. These values fit the nucleon mass $m_N = 940$ MeV (with the account for its rotational energy) and the mass splitting between the nucleon and the $\Delta$-resonance.[28] These were the values used also by Klebanov *et al.*[27] who solved numerically Eq. (3.5.48) and found the phase shifts $\delta(\omega)$ defined from the large-$r$ asymptotics of the solutions of Eq. (3.5.48) regular at the origin,

$$\eta_{\mathrm{as}}(r) = \frac{kr + i}{r^2}e^{ikr + i\delta(\omega)} + \frac{kr - i}{r^2}e^{-ikr - i\delta(\omega)}, \qquad (k = \sqrt{\omega^2 - m_K^2}), \qquad (3.5.50)$$

being a superposition of the incoming and outgoing spherical waves. At $\gamma = 1$ and physical $m_K = 495$ MeV, the phase shift $\delta(\omega)$ has been found in Ref. 27 to be less than $45^o$ in the range of interest. This have lead the authors to the conclusion that $\Theta^+$ does not exist in the Skyrme model, at least in the large $N_c$ limit and small $m_K$. We reproduce their phase shifts with a high accuracy (as well as the phase shifts studied in Ref. 29 for another choice of the Skyrme model parameters) but come to the opposite conclusions.

In a situation when there is a resonance and a potential scattering together, the phase shift does not need to go through $90^\circ$ as it would be requested by the Breit–Wigner formula for an isolated resonance. A far better and precise way to determine whether there is a resonance, is to look not into the phase shifts but into the singularities of the scattering amplitude in the complex energy plane. A resonance is, by definition, a pole of the scattering amplitude in the lower semi-plane on the second Riemann sheet:

$$\sqrt{s}_{\mathrm{pole}} = m_{\mathrm{res}} - i\frac{\Gamma}{2} \qquad (3.5.51)$$

where $m_{\mathrm{res}}$ is the resonance mass and $\Gamma$ is its width.

The scattering amplitude $f(\omega)$ and the scattering matrix $S(\omega)$ (which in this case has only one element) are defined as

$$f(\omega) = \frac{1}{2ik}\left(e^{2i\delta(\omega)} - 1\right), \qquad S(\omega) = e^{2i\delta(\omega)}. \qquad (3.5.52)$$

A standard representation for the scattering amplitude is

$$f = \frac{1}{g(\omega) - ik}, \qquad g(\omega) = k \cot \delta(\omega). \tag{3.5.53}$$

This representation solves the unitarity condition for the $S$-matrix: $g$ is real on the real $\omega$ axis. The function $g(\omega)$ does not have cuts related to the $KN$ thresholds and $\omega^{2l} g(\omega)$ is Taylor-expandable at small $\omega$, therefore it is a useful concept.[30]

For $\omega$ in the lower complex semi-plane the first term in Eq. (3.5.50) becomes a rising exponent of $r$, and the second term becomes a falling exponent. Since the $S$ matrix is proportional to the ratio of the coefficient in front of $\exp(-ikr)$ to that in front of $\exp(ikr)$, the pole of the $S$ matrix and hence of the scattering amplitude corresponds to the situation where the wave function $\eta(r)$ regular at the origin, has no falling exponent at $r \to \infty$ but only a rising one. Physically, it corresponds to a resonance decay producing outgoing waves only.

For the conventional choice of the parameters we find the $\Theta^+$ pole position at

$$\sqrt{s}_{\text{pole}} = \begin{cases} (1115 - 145i) \text{ MeV} & \text{for } m_K = 0 \quad \text{(threshold at 940 MeV)} \\ (1449 - 44i) \text{ MeV} & \text{for } m_K = 495 \text{ MeV (threshold at 1435 MeV)} \end{cases} \tag{3.5.54}$$

We have recalculated here the pole position in $\omega$ to the relativistic-invariant $KN$ energy $s = m_N^2 + 2m_N\omega + m_K^2$. It is a perfectly normal resonance in the strong interactions standards with a width of 90 MeV. It would be by all means seen in a partial wave analysis (see Fig. 3.9, left panel) or just in the $KN$ total $T=0$ cross section which we calculate from the well-known equation $\sigma = 4\pi(2j + 1)|f|^2 = \frac{4\pi}{k^2}(2j + 1)\sin^2 \delta$ (Fig. 3.9, right panel).

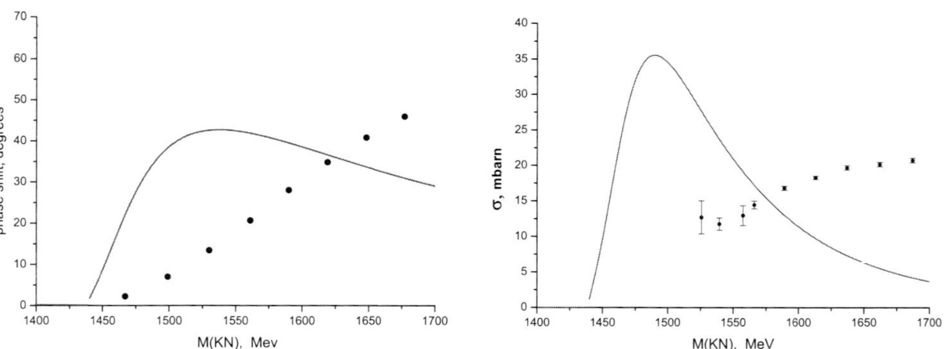

Fig. 3.9.   Left: the $T=0, L=1$ $KN$ scattering phase as function of the $KN$ invariant mass in the Skyrme model in the large-$N_c$ limit (it coincides with the phase found in Ref. 27), compared to the result of the partial wave analysis[31] shown by dots. Right: the ensuing $KN$ cross section in this partial wave exhibits a strong resonance around 1500 MeV, whereas the experimental data[32] for the sum over all partial waves shows no signs of a resonance.

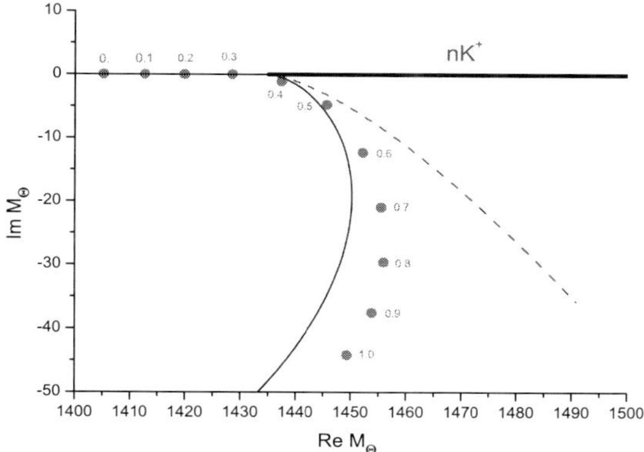


Fig. 3.10.   Trajectory of the pole in the $KN$ scattering amplitude for "realistic" parameters of the Skyrme model and physical $m_K = 495$ MeV at $\gamma = 0, 0.1, \ldots, 1.0$. The dashed and solid lines show the analytical calculation of the pole position in the first and second orders in $\gamma$, respectively.

At the maximum the cross section is as large as 35 mb, and it is a contribution of only one particular partial wave $P_{01}$! Needless to say, such a strong resonance is not observed. Varying the parameters of the Skyrme model[27,29] or modifying it[26] can make the exotic resonance narrower or broader but one cannot get rid of it. The reason is very general: poles in the scattering amplitude do not disappear as one varies the parameters but move in the complex plane.

One can check it in a very precise way by, say, varying artificially the coefficient in front of the Wess–Zumino term $\gamma$ from 0 to 1. At $\gamma = 0$ there is certainly an exotic bound state at $\omega = 0$ corresponding to the rotational zero mode (3.5.49). At $\gamma > 0$ the position of the pole of the $KN$ scattering amplitude moves into the complex $\omega$ plane such that

$$\text{Re } \omega_{\text{pole}} = a_1\gamma + a_3\gamma^3 + \ldots,$$
$$\text{Im } \omega_{\text{pole}} = b_2\gamma^2 + b_4\gamma^4 + \ldots \tag{3.5.55}$$

with analytically calculable coefficients in this Taylor expansion (we give explicitly the leading coefficients in Section 7). By comparing the numerical determination of the pole position with the analytical expressions we trace that the $\Theta^+$ pole (3.5.54) is a continuous deformation of the rotational mode, see Fig. 3.10.

Thus, the prediction of the Skyrme model is not that there is no $\Theta^+$ but just the opposite: **there must be a very strong resonance**, at least when the number of colours is taken to infinity. Since this prediction is of general nature and does not rely on the specifics of the Skyrme model, one must be worried why a strong exotic resonance is not observed experimentally!

The answer is that the large-$N_c$ logic in general and the concrete Skyrme model in particular grossly overestimate the resonance width (we explain it in the next sections). The resonance cannot disappear but in reality it becomes very narrow, and that is why it is so difficult to observe it.

One may object that the Skyrme model is a model anyway, and a modification of its parameters or a replacement by another chiral model can lead to an even larger width, say, of 600 MeV instead of 90 MeV obtained here from the "classical" Skyrme model. However, as we argue in Section 9, going from $N_c = \infty$ to the real world at $N_c = 3$ reduces the width by at least a factor of 5. Therefore, even a 600-MeV resonance at $N_c = \infty$ would become a normal 120-MeV resonance in the real world and would be observable.

Thus, the only way how a theoretically unavoidable resonance can escape observation is to become very narrow. We remark that the reanalysis of the old $KN$ scattering data[33] shows that there is room for the exotic resonance with a mass around 1530 MeV and width below 1 MeV.

## 3.6. Physics of the Narrow $\Theta^+$ Width

Quantum field theory says that baryons cannot be $3Q$ states only but necessarily have higher Fock components due to additional $Q\bar{Q}$ pairs; it is only a quantitative question how large are the $5Q$, $7Q$, ... components in ordinary baryons. Various baryon observables have varying sensitivity to the presence of higher Fock components. For example, the fraction of the nucleon momentum carried by antiquarks is, at low virtuality, less than 10%. However, the nucleon $\sigma$-term or nucleon spin are in fact dominated by antiquarks.[34,35] Both facts are in accord with a normalization of the $5Q$ component of the nucleon at the level of 30% from the $3Q$ component, meaning that 30% of the time nucleon is a pentaquark!

As to the exotic $\Theta^+$ and other members of the antidecuplet, their *lowest* Fock component is the $5Q$ one, nothing terrible. However it has dramatic consequences for the antidecuplet decay widths.

To evaluate the width of the $\Theta^+ \to K^+ n$ decay one has to compute the transition matrix element of the strange axial current, $< \Theta^+ |\bar{s}\gamma_\mu\gamma_5 u| n >$. There are two contributions to this matrix element: the "fall apart" process (Fig. 3.11(A)) and the "5-to-5" process where $\Theta^+$ decays into the $5Q$ component of the nucleon (Fig. 3.11(B)). One does not exist without the other: if there is a "fall apart" process it means that there is a non-zero coupling of quarks to pseudoscalar (and other) mesons, meaning that there is a transition term in the Hamiltonian between $3Q$ and $5Q$ states (Fig. 3.11(C)). Hence the eigenstates of the Hamiltonian must be a mixture of $3Q, 5Q, ...$ Fock components. Therefore, assuming there is process A, we have to admit that there is process B as well. Moreover, each of the amplitudes A and B are not Lorentz-invariant, only their sum is. Evaluating the "fall-apart" amplitude and forgetting about the "5-to-5" one makes no sense.

A convenient way to evaluate the sum of two graphs, A and B, in the chiral limit is to go to the infinite momentum frame (IMF) where only the process B survives,

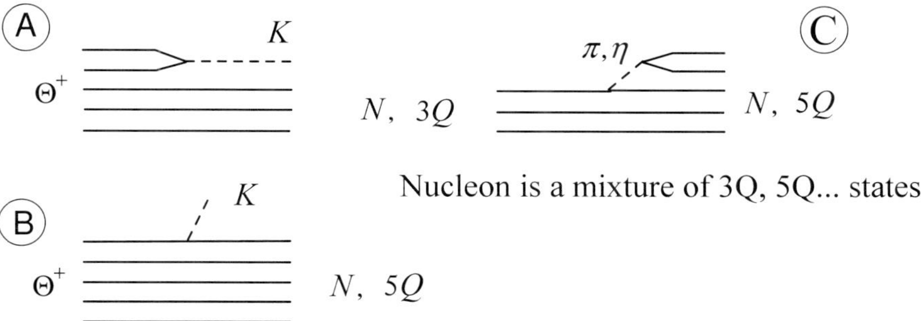


Fig. 3.11. "Fall-apart" (A) and "5-to-5" (B) contributions to the $\Theta^+ \to K^+ n$ decay.

as axial (and vector) currents with a finite momentum transfer do not create or annihilate quarks with infinite momenta. In the IMF the fall-apart process A is exactly zero in the chiral limit. The baryon matrix elements are thus non-zero only between Fock components with *equal number* of quarks and antiquarks.

The decays of ordinary (nonexotic) baryons are mainly due to the $3Q \to 3Q$ transitions with a small (30%) correction from $5Q \to 5Q$ transitions and even smaller corrections from higher Fock components, just because the $3Q$ components dominate. The nucleon axial constant is thus expected to be of the order of unity and indeed $g_A(N) = 1.27$.

However the $\Theta^+$ decay is dominated by the $5Q \to 5Q$ transition, and **the transition axial constant is suppressed to the extent the $5Q$ component in the nucleon is suppressed with respect to the $3Q$ one.**[36]

A quantitative estimate of this effect can be made in a relativistic model (since pair creation and annihilation is a relativistic effect) where it is possible to calculate both the $3Q$ and the $5Q$ wave functions of the nucleon and the $5Q$ wave function of the $\Theta^+$. We claim that in any such model of baryons the $\Theta^+$ will be narrow if the model tells that the $5Q$ component in the nucleon is suppressed with respect to the $3Q$ one. An example is provided by the Chiral Quark Soliton Model where, indeed, the $5Q \to 5Q$ axial constant has been estimated[16,37] as $g_A(\Theta \to KN) \approx 0.14$ yielding

$$\Gamma_\Theta \approx 2\,\mathrm{MeV}. \tag{3.6.56}$$

Apart from the suppression of general nature discussed above there is an additional suppression of $g_A(\Theta \to KN)$ due to $SU(3)$ group factors in the $\overline{10} \to 8$ transition. This estimate has been performed assuming the chiral limit ($m_K = 0$) and zero momentum transfer in the decay. In more realistic settings the width can only go down. A recent calculation with account for the decay via higher Fock components and also for $m_s$ corrections gives $\Gamma_\Theta = 0.7\,\mathrm{MeV}$.[38]

As stressed in Ref. 5 where the narrow $\Theta^+$ has been first predicted, in the imaginary nonrelativistic limit when ordinary baryons are made of three quarks only with no admixture of $Q\bar{Q}$ pairs the $\Theta^+$ width tends to zero strictly.

It may seem that by the same argument all members of the exotic multiplets $(\overline{\mathbf{10}}, \frac{1}{2})$, $(\mathbf{27}, \frac{3}{2})$ *etc.*, should be very narrow resonances but the above argument applies only to the transitions with an emission of one pseudoscalar meson. As a matter of fact it applies also to the $BBV$ transitions where $V$ is a vector meson that couples to baryons via the conserved vector current; such transitions are also expected to be strongly suppressed,[39] and the calculations[38] support it. However, the argument does not work for transitions with two or more pseudoscalar mesons emission. Therefore, if the phase volume allows for a decay of an exotic resonance to two or more mesons, the width does not need to be particularly narrow; it should be studied on case-to-case basis.

We now return to the Skyrme model and discuss why does it give a wide $\Theta^+$ in the large $N_c$ limit. As explained in Section 1, the Skyrme model is an idealization of nature: It implies that the chiral field is broad, the valence quarks are close to the negative-energy Dirac sea and cannot be separated from the sea, whereas the distortion of the sea is large. The number of $Q\bar{Q}$ pairs originating from the strongly deformed sea is $\mathcal{O}(1)$, times $N_c$. For example, it has been known for quite a while[40] that the fraction of nucleon spin carried by valence quarks is zero in the Skyrme model. Therefore, the Skyrme model implies the situation which is the opposite extreme from the nonrelativistic quarks where there are $N_c$ valence quarks and zero $Q\bar{Q}$ pairs. From the viewpoint of parton distributions, both limits are discussed in some detail in Ref. 17.

Therefore, the Skyrme model fails to accommodate the real-world physics explaining the narrow $\Theta^+$, in two essential points:

- At large $N_c$ justifying the study of the $\Theta^+$ resonance from the kaon-Skyrmion scattering both the nucleon and the $\Theta^+$ have an equal (and infinite) number of $Q\bar{Q}$ pairs; hence the $\Theta \to KN$ transition is not suppressed at all. [This is why we have obtained a large 90 MeV width in Section 5]

- Even if one takes a moderate $N_c = 3$, the Skyrme model implies that the $5Q$ component of the nucleon remains large, and there is no argument why the $\Theta^+$ width should be particularly small, although it must be less than in the infinite-$N_c$ limit.

Having this understanding in mind, in the next section we return to the Skyrme model to see if it is possible to play with its parameters in such a way that it would mimic to some extent the nonrelativistic limit. Then the "$\Theta^+$" of such a model should be narrow even if $N_c$ is large.

## 3.7. Getting a Narrow $\Theta^+$ in the Skyrme Model

Unfortunately, there are very few handles in the Skyrme model to play with. In fact, there are only three constants: $F_\pi$, $e$ and the coefficient in front of the Wess–Zumino term, $\gamma$. The last one is unity in the ideal case but is not universal if, for example, $m_K \neq 0$. A general statement is that $\gamma$ decreases as $m_K$ increases. The constant $F_\pi$ has to be taken 1.5 times less than its physical value to fit 940 MeV for the nucleon mass. The dimensionless coefficient $e$ is also rather arbitrary, it is not being fixed from the experimental $d$-wave pion scattering. Therefore, we feel free to modify these constants at will, in order to make a theoretical point. The models we are going to present are not realistic, of course. To get more realistic models, one has either to add vector mesons[26] or take the Chiral Quark Soliton Model, or do something else.

### 3.7.1. *Vanishing $m_\Theta - m_N$, vanishing $\Gamma_\Theta$*

We start with a simple exercise, making $\gamma$ a small number. To simplify the presentation we take the chiral limit $m_K = 0$ but give the final results for $m_K \neq 0$ at the end of this subsection.

At small $\gamma$, one can compute everything analytically. In particular, one can find the regular function $g(\omega)$ (3.5.53) in the range of interest $\omega = k \sim \gamma$. This is done by comparing the asymptotics of the wave function (3.5.50) in the range $kr \ll 1$ but $r \gg r_0$ where $r_0$ is the coefficient in the asymptotics of the profile function $P(r) \approx r_0^2/r^2$, with the asymptotics of $\eta(r)$ at $r \gg r_0$ being

$$\eta(r) = \frac{C_1}{r^2} + C_2\, r, \quad \text{from where} \quad g(\omega) = \frac{3C_2}{C_1 \omega^2}. \tag{3.7.57}$$

The coefficients $C_{1,2}$ are found from the following considerations. In the range of interest, $k \sim \gamma \ll 1$, the wave function $\eta(r)$ differs from the rotational wave function $\eta_{\rm rot}(r)$ (3.5.49), being the exact solution of Eq. (3.5.48) at $\omega = 0$, by terms of the order of $\gamma^2$. $\eta_{\rm rot}(r)$ falls off at large distances as $r_0^2/2r^2 + 0 \cdot r$. Therefore, $C_1 = r_0^2/2 + \mathcal{O}(\gamma^2)$ whereas $C_2 = \mathcal{O}(\gamma^2)$ and can be computed, in the leading order, as a matrix element (3.5.46) with $\eta(r)$ substituted by $\eta_{\rm rot}(r)$. We find

$$C_1 = \frac{r_0^2}{2} + \mathcal{O}(\gamma^2), \qquad C_2 = \frac{1}{6\pi r_0^2}\left(\gamma e^2 N_c\, \omega - \tilde{I}_2\, \omega^2\right) = \mathcal{O}(\gamma^2) \tag{3.7.58}$$

where the second moment of inertia $\tilde{I}_2$ arises here as

$$\tilde{I}_2 = 4\pi \int dr\, r^2 \eta_{\rm rot} A(r) \eta_{\rm rot}(r).$$

The physical moment of inertia (in MeV) is obtained from the dimensionless $\tilde{I}_2$ as $I_2 = \tilde{I}_2/(8e^3 F_\pi)$.

Technically, we obtain $C_2$ by the following trick: We integrate Eq. (3.5.48) multiplied to the left by $\eta_{\rm rot}(r)$ from zero to some $r \gg r_0$, and subtract the same integral with $\eta(r)$ and $\eta_{\rm rot}(r)$ interchanged. The terms with the first and second derivatives of $\eta, \eta_{\rm rot}$ become a full derivative that can be evaluated at the integration end point, while terms with no derivatives are fast convergent such that one can extend the integration range to infinity and also replace $\eta \to \eta_{\rm rot}$ in the leading order.

The poles of the scattering amplitude are found from the equation $g(\omega) = i\omega$. Using Eqs. (3.7.57) and (3.7.58) we get the real and imaginary parts of the pole position:

$$\Delta = \gamma\omega_{\rm rot} = \mathcal{O}(\gamma), \qquad \omega_{\rm rot} = \frac{N_c}{4I_2},$$

$$\Gamma = \frac{8\pi F_\pi^2 r_0^4}{\gamma N_c} \Delta^3 = \mathcal{O}(\gamma^2) \ll \Delta, \tag{3.7.59}$$

where $\Delta = m_\Theta - m_N$, $\Gamma$ and $r_0$ are in physical units. These are actually the first terms in the expansion (3.5.55). We have developed a perturbation theory in $\gamma$ and found analytically higher order terms as well. The first few orders follow the numerical determination of the pole position all the way up to $\gamma=1$, see Fig. 3.10. It makes us confident that even at finite width the $\Theta^+$ resonance is a continuous deformation of the rotational would-be zero mode.

Since in the limit of small $\gamma$ everything is analytically calculable, one can check various facts. In particular, we have computed the transition axial constant $g_A(\Theta \to KN)$ from the asymptotics (3.5.50) of the kaon scattering wave $\eta(r)$. It gives the needed (massless) kaon pole $1/(\omega^2-k^2)$; the axial constant is the residue of this pole, for more details see Section 8. The overall spinor coefficient $\zeta^\alpha$ in the kaon wave is fixed from the quantization condition requesting that a state with strangeness $S = +1$ and exoticness $X = 1$ is formed.[25] It gives

$$\langle\zeta_\alpha^\dagger\zeta^\alpha\rangle = \frac{8}{\gamma N_c}. \tag{3.7.60}$$

We obtain

$$g_A(\Theta \to KN) = 8\pi\frac{F_\pi^2 r_0^2}{\sqrt{2\gamma N_c}}.$$

Given the axial constant, the pseudoscalar coupling $G(\Theta \to KN)$ can be found from the Goldberger–Treiman relation

$$G(\Theta \to KN) = \frac{m_N \, g_A(\Theta \to KN)}{F_\pi}.$$

The pseudoscalar coupling and the decay phase volume determines the $\Theta^+$ width:

$$\Gamma = \frac{G^2(\Theta \to KN)\Delta^3}{4\pi m_N^2} = \frac{8\pi F_\pi^2 r_0^4}{\gamma N_c}\Delta^3.$$

Comparing it with the determination of $\Gamma$ from the pole position (3.7.59), we see that the two ways of determining the width coincide!

We can determine the width in a third way – from the *radiation* of the kaon field by the resonance.[41] According to Bohr's correspondence principle, the quantum width is determined as the inverse time during which a resonance looses, through classical radiation, the energy difference between the neighbour states:

$$\Gamma = \frac{W}{\Delta} \qquad (3.7.61)$$

where $W$ is the radiation intensity, *i.e.* the energy loss per unit time. Strictly speaking, Bohr's principle is semiclassical and applies only to the decays of the highly excited levels. In our case, however, we linearize in the kaon field at large $N_c$, therefore it is essentially a problem for a set of harmonic oscillators for which semiclassics is exact starting from the first excited level, here the $\Theta^+$.

To find the radiation intensity $W$, we look for the solution of Eq. (3.5.48) with only the outgoing wave in the asymptotics, $\eta(r) = c(kr + i)e^{ikr}/r^2$. The coefficient $c$ is found from matching it in the range $r_0 \ll r \ll 1/k$ with the solution regular at the origin. In the leading order in $\gamma$ it is the rotational mode $\eta_{\rm rot}(r) \approx r_0^2/2r^2$; it gives $c = -ir_0^2/2$. The radiation intensity corresponding to this outgoing wave is found as the flux through a distant surface of the Pointing vector

$$T_{0i} \overset{r\to\infty}{=} \frac{F_\pi^2}{2}\left(\partial_0 K_\alpha^\dagger \partial_i K^\alpha + \partial_i K_\alpha^\dagger \partial_0 K^\alpha\right) \overset{r\to\infty}{=} \frac{F_\pi^2 r_0^4 k^4}{4r^2}\zeta_\alpha^\dagger \zeta^\alpha.$$

Using the normalization (3.7.60) we find the radiation intensity

$$W = \int d^2 S_i\, T_{0i} = 4\pi \frac{F_\pi^2 r_0^4 k^4}{4}\frac{8}{\gamma N_c}$$

where in the massless kaon limit $k = \omega = \Delta$. Note that the $k^4$ dependence is typical for the dipole radiation; it is dipole as we look for kaon radiation in $p$-wave. [Another characteristic feature of the dipole radiation — the $\cos^2\theta$ angular dependance — is not seen here because we have in fact averaged over the $\Theta^+$ spin.] From Bohr's equation (3.7.61) we obtain the $\Theta^+$ width

$$\Gamma = \frac{W}{\Delta} = \frac{8\pi F_\pi^2 r_0^4}{N_c \gamma}\Delta^3$$

again coinciding with the determination of the width from the pole position, Eq. (3.7.59). One can also compute the width as the inverse time during which one unit of strangeness is lost through kaon radiation, with the same result. Yet another (a 5th!) way of computing $\Gamma$ — from the asymptotics of the classical profile function of the Skyrmion — will be presented in Section 9.

The derivation of $\Gamma$ and $\Delta$ can be repeated for $m_K \neq 0$ in which case we find

$$\Delta = \Delta_0\left(\frac{1}{2} + \sqrt{\frac{1}{4} + \frac{b\,m_K^2}{\Delta_0^2}}\right), \qquad \Gamma = \Gamma_0\frac{\Delta_0}{2\Delta - \Delta_0}\frac{(\Delta^2 - m_K^2)^{\frac{3}{2}}}{\Delta_0^3}, \qquad (3.7.62)$$

where $\Delta = m_\Theta - m_N$ when $m_K \neq 0$ while the subscript 0 refers to the case of $m_K = 0$. It is remarkable that the imaginary part of the pole position $\Gamma$ apparently "knows" — through unitarity — about the decay phase volume clearly visible in Eq. (3.7.62). The numerical coefficient $b$ is defined as $b = 4\pi \int dr\, r^2 \eta_{\rm rot}^2(r)/ \tilde{I}_2 \approx 0.705$.

To conclude this subsection: in the case when the $\Theta^+$ width is made small, we have determined it in three independent ways: (i) from the pole position in the complex energy plane, (ii) from the axial constant and by using the Goldberger–Treiman relation, (iii) from the semiclassical radiation theory. All three calculations lead to the same expression for the width $\Gamma$.

### 3.7.2. *Finite $m_\Theta - m_N$, vanishing $\Gamma_\Theta$*

The analytical equations of the previous subsection remain accurate as long as the imaginary part of the pole position is much less than the real part, that is insofar as $\Gamma \ll \Delta$. From Eq. (3.7.59) one infers that actually this condition is $\gamma e^2 N_c \ll 1$ where $e^2$ is the inverse coefficient in front of the Skyrme term in the action, and $\gamma$ is the coefficient of the Wess–Zumino term. If one likes to fix once and forever $\gamma = 1$ (say, from topology arguments), one is still able to support the regime $\Gamma \to 0$ but $\Delta = {\rm const}$ by rescaling the other two constants of the Skyrme model. Namely, we consider the following regime:

$$F_\pi = F_0 \beta^{-3}, \quad e = e_0 \beta, \qquad \beta \to 0, \qquad \gamma = \gamma_0 = 1. \qquad (3.7.63)$$

The Skyrmion mass, its size and moments of inertia scale then as

$$m_N \sim \frac{F_\pi}{e} \propto \frac{1}{\beta^4} \to \infty, \qquad r_0 \sim \frac{1}{F_\pi e} \propto \beta^2 \to 0, \qquad I_{1,2} \sim \frac{1}{F_\pi e^3} \propto {\rm const.},$$

such that the splittings between rotational $SU(3)$ multiplets remain fixed, $m_\Delta - m_N \sim {\rm const.}$, and

$$\Delta = m_\Theta - m_N = \frac{N_c}{4I_2} \sim N_c F_\pi e^3 \propto {\rm const.}, \qquad {\rm but} \ \ \Gamma_\Theta \sim N_c^2 F_\pi e^5 \propto \beta^2 \to 0.$$

$$(3.7.64)$$

We stress that $\Theta^+$ becomes stable in this regime not because the decay phase volume tends to zero (which would have been trivial but it is not the case here) but because the Skyrmion size $r_0$ is small. Taking $r_0$ to zero we mimic to some extent the limit of nonrelativistic quarks in the Skyrme model, where we expect a narrow width. Since the Skyrme model is opposite in spirit to the nonrelativistic quarks (see Sections 1 and 6) it is difficult to achieve this limit. Indeed, the regime (3.7.63) is not too realistic. However it serves well to illustrate the point: when $\Gamma$ is small, the real part of the pole position coincides with the rotational frequency

$$m_\Theta - m_N = \omega_{\rm rot} = \frac{N_c}{4I_2} \qquad (3.7.65)$$

as it follows from the quantization of the $SU(3)$ rotations, Eq. (3.3.24).

Is it a coincidence? Probably, not: $\Theta^+$ *is* an $SU(3)$ rotational excitation of the nucleon. (At large $N_c$ the rotation is more like a precession near the "North pole" but nevertheless ) It remains a (deformed) rotational state even in the worse case scenario provided by the Skyrme model where at "realistic" parameters it becomes a broad and hence strong resonance but then what should be called the resonance mass becomes ambiguous. Its precise determination is then from the pole position which is away from the real axis, such that the real part of the pole position does not need to coincide with the rotational splitting just for the trivial reason that the imaginary part is large.

Therefore, the key issue is the resonance width. On the one hand, a rotating body must radiate, in this case the kaon field. Since in the $\frac{1}{2}^+ \rightarrow \frac{1}{2}^+$ transition the kaon is in the $p$ wave, the $\Theta^+$ width is entirely due to dipole radiation. The dipole radiation intensity is proportional to $(\ddot{d})^2 = \omega^4 d^2$ at small frequencies where $d$ is the dipole moment. Generically, $d \sim r_0$ where $r_0$ is the characteristic size of the system. The Skyrme model illustrates the generic case, therefore the only way to suppress the dipole radiation at fixed $\omega = \Delta$ is to shrink the size $r_0$ as we have done above. On the other hand, in our case it is the *transition* dipole moment $d$ corresponding to strangeness emission, which can be, in principle, much less than $r_0$, even zero. We have argued in Section 6 that in the real world the transition dipole moment is small as the nucleon is essentially nonrelativistic and hence has a small $5Q$ component. If the width is small, we see no reasons why would not the real part of the $\Theta^+$ pole coincide with the rotational splitting of the $SU(3)$ multiplets.[a]

## 3.8. Goldberger–Treiman Relation and the $\Theta^+$ Width

In this section we show that the $\Theta^+$ width can be expressed through the transition axial constant *provided* the width is small. We reaffirm the validity of a modified Goldberger–Treiman relation between the axial and pseudoscalar $\Theta KN$ constants in the chiral limit. We derive these relations in the framework of the Skyrme model where all equations are explicit. However, these relations are, of course, of a general nature.

If $\Theta^+$ is a narrow and well-defined state one can define the transitional axial $g_A = g_A(\Theta \rightarrow KN)$ and pseudoscalar $G = G(\Theta \rightarrow KN)$ constants as

$$\frac{\varepsilon_{\alpha\beta}}{\sqrt{2}} \langle \Theta_k^+ | j_{\mu 5}^\alpha(x) | N^{\beta,i} \rangle = e^{iq \cdot x} \bar{u}(\Theta, k) \left( \frac{\gamma_\mu}{2} g_A - \frac{q_\mu}{q^2} F_\pi G \right) \gamma_5 \, u(N, i) \qquad (3.8.66)$$

where $i, k = 1, 2$ are the nucleon and $\Theta^+$ spin projections, $\alpha, \beta = 1, 2$ are the isospin projections of the nucleon and of the kaon current; we are interested in the isospin $T = 0$ channel. Finally, $u, \bar{u}$ are $N$ and $\Theta^+$ 4-spinors. We assume that they obey the non-relativistic normalization $\bar{u}(i)u(k) = \delta_{ik}$. In the non-relativistic limit appropriate at large $N_c$ one has:

$$\bar{u}(k)\gamma_5 u(i) = \psi^*(k) \frac{\mathbf{q} \cdot \sigma}{2m} \psi(i) = \frac{1}{2m} (\mathbf{q} \cdot \sigma)_k^i \qquad (3.8.67)$$

where $\psi(i)$ is a non-relativistic 2-spinor with polarization $i$.

---

[a] A quantum-mechanical counter-example by T. Cohen[42] does not seem to capture the necessary physics as the spectrum there is discrete and there is no radiation.

The modified Goldberger–Treiman relation follows immediately from the conservation of the axial current, $\partial_\mu j^\alpha_{\mu 5} = 0$:

$$g_A(\Theta \to KN)(m_N + m_\Theta) = 2G(\Theta \to KN)F_\pi.$$

It should be stressed that it holds true even if $m_\Theta$ differs significantly from $m_N$. In the large $N_c$ limit however one can put $m_\Theta \approx m_N$.

Let us consider now the nucleon matrix element of the product of two strangeness-changing axial currents $j^\alpha_{\mu 5}(x)$ and expand it in intermediate states $|n\rangle$:

$$\Pi^{T=0}(\omega, \mathbf{q})^i_k = \int d^4x\, e^{iq\cdot x}\, \frac{\varepsilon^{\alpha_1\beta_1}}{\sqrt{2}} \langle N_{\beta_1 k}|j^\dagger_{\nu 5\,\alpha_1}(x)j^{\alpha_2}_{\mu 5}(0)|N^{\beta_2 i}\rangle \frac{\varepsilon_{\alpha_2\beta_2}}{\sqrt{2}} \tag{3.8.68}$$

$$= \sum_n \frac{\varepsilon^{\alpha_1\beta_1}}{\sqrt{2}} \langle N_{\beta_1 k}|j^\dagger_{\nu 5\,\alpha_1}(0)|n(\mathbf{q})\rangle\, 2\pi\delta\left(\omega - E_n(\mathbf{q})\right)\, \langle n(\mathbf{q})|j^{\alpha_2}_{\mu 5}(0)|N^{\beta_2 i}\rangle \frac{\varepsilon_{\alpha_2\beta_2}}{\sqrt{2}}.$$

Here $E_n(\mathbf{q})$ is the kinetic energy of the intermediate state. Since the nucleon is infinitely heavy at large $N_c$ the energy $\omega$ and the 3-momentum $\mathbf{q}$ are conserved. We write relativistic equations for the kaon field, however.

The correlation function (3.8.68) can be calculated *e.g.* in the Skyrme model. We wish to isolate the kaon pole contribution to the strange axial current, that is we have to consider $\omega^2 \approx \mathbf{q}^2 + m_K^2$ where we temporarily take the chiral limit, $m_K = 0$, for the current to be conserved. In fact, this requirement can be relaxed. The singular contribution to the current arises from the asymptotics of the kaon scattering wave (3.5.50):

$$j^\alpha_{\mu 5}(\omega, \mathbf{q}) = iq_\mu\, F_\pi\, K^\alpha(\omega, \mathbf{q}) = -q_\mu\, F_\pi\, \frac{\sin\delta(\omega)}{\omega^2 - \mathbf{q}^2}\, (\mathbf{q}\cdot\boldsymbol{\tau})^\alpha_\beta\, \mathbf{b}^{\dagger\,\beta}(\omega)\, \sqrt{\frac{4\pi}{\omega^3}}. \tag{3.8.69}$$

The last factor arises here in accordance with the commutation relation for the creation-annihilation operators $[\mathbf{b}^{\dagger\,\beta}(\omega_1)\mathbf{b}_\alpha(\omega_2)] = 2\pi\,\delta^\beta_\alpha\,\delta(\omega_1 - \omega_2)$.[25] Substituting Eq. (3.8.69) into Eq. (3.8.68) we obtain

$$\Pi^{T=0}_{\mu\nu}(\omega, \mathbf{q})^i_k = q_\mu q_\nu\, \frac{4\pi F_\pi^2}{\omega^3}\, \mathbf{q}^2\, \delta^i_k\, \frac{\sin^2\delta(\omega)}{(\omega^2 - \mathbf{q}^2)^2}. \tag{3.8.70}$$

The correlation function is therefore expressed through the phase shift $\delta(\omega)$! The conservation of the axial current implies that there is also a contact term in the correlation function, proportional to $g_{\mu\nu}$; the coefficient in front of it must be exactly the coefficient in front of $q_\mu q_\nu/q^2$, with the minus sign.

Let us now assume that one of the intermediate states in Eq. (3.8.68) is a *narrow* $\Theta^+$ resonance. Then, on the one hand, at $\omega \approx \Delta$ the phase shift $\delta(\omega)$ must exhibit the Breit–Wigner behaviour as it follows from unitarity:

$$\sin^2\delta(\omega) = \frac{\Gamma^2/4}{(\omega - \Delta)^2 + \frac{\Gamma^2}{4}} \xrightarrow{\Gamma \to 0} \frac{\Gamma}{4} 2\pi\delta(\omega - \Delta). \tag{3.8.71}$$

On the other hand, one can extract the contribution of the $\Theta^+$ intermediate state using the definition of the matrix elements of the axial current (3.8.66). Taking there the contribution that is singular near the kaon pole and recalling Eq. (3.8.67) we get

$$\Pi_{\mu\nu}^{T=0}(\omega,\mathbf{q})_k^i = 2\pi\delta(\omega - \Delta)\frac{q_\mu q_\nu}{(q^2)^2}F_\pi^2 G^2 \frac{\mathbf{q}^2}{4m_N^2}\delta_k^i. \tag{3.8.72}$$

We now compare Eq. (3.8.72) and Eqs. (3.8.70,3.8.71) and immediately obtain the already familiar equation for the $\Theta^+$ width through the $\Theta KN$ pseudoscalar coupling $G$ (*cf.* Section 7):

$$\Gamma = \frac{G^2(\Theta \to KN)}{4\pi m_N^2}\mathbf{p}^3 \tag{3.8.73}$$

where $\mathbf{p}$ is the kaon momentum, equal to $\Delta$ in the chiral limit. We stress that the Born graph for the $KN$ scattering with pseudoscalar Yukawa coupling arises automatically — through unitarity — from the $KN$ scattering phase, provided it corresponds to a narrow resonance.

To conclude, if $\Theta^+$ happens to be a narrow resonance, one can find its width from the $\Theta KN$ transition axial coupling or, thanks to the Goldberger–Treiman relation, from the transition pseudoscalar coupling (it is contrary to the recent claim of Refs. 29 and 43). This is how the narrow $\Theta^+$ has been first predicted[5] and how a more stringent estimate of the width $\Gamma_\Theta \sim 1\,\mathrm{MeV}$ has been recently performed.[16,37,38]

## 3.9. Finite-$N_c$ Effects in the $\Theta^+$ Width

In any chiral soliton model of baryons, the baryon-baryon-meson coupling can be written in terms of the rotational coordinates given by the $SU(3)$ matrix R as[28]

$$L = -ip_i\frac{3G_0}{2m_N}\tfrac{1}{2}\mathrm{Tr}(R^\dagger\lambda^a R\tau^i) \tag{3.9.74}$$

where $\lambda^a$ is the Gell-Mann matrix for the pseudoscalar meson of flavour $a$, and $p_i$ is its 3-momentum. The pseudoscalar coupling $G_0$ is directly related to the asymptotics of the Skyrmion profile function $P(r) \approx r_0^2/r^2$ :[28]

$$G_0 = \frac{8\pi}{3}F_\pi m_N r_0^2. \tag{3.9.75}$$

Note that $G_0 = \mathcal{O}(N_c^{\frac{3}{2}})$. In a generic case there are baryon-baryon-meson couplings other than (3.9.74), labeled in Ref.[5] by $G_1$ and $G_2$. It is the interplay of these constants that leads to a small $\Theta^+$ width. In the nonrelativistic limit the combination of $G_{0,1,2}$ is such that the width goes to zero strictly, however each of the constants remain finite being then determined solely by valence quarks. Unfortunately, in the Skyrme model $G_{1,2}$ are altogether absent, related to the fact that there are no valence quarks in the Skyrme model. For example, $G_2$ is proportional to the fraction of nucleon spin carried by valence quarks which is known to be exactly zero in the Skyrme model.[40] Since only the coupling $G_0$ is present in the Skyrme model, we are forced to mimic the nonrelativistic limit there by taking the size $r_0$ to zero, which leads to unrealistic parameters. In any chiral model with explicit valence quarks there are less traumatic ways to obtain a very small $\Theta^+$ width.

In the chiral limit $SU(3)$ symmetry is exact, therefore Eqs. (3.9.74) and (3.9.75) determine also the leading term in the $\Theta \to KN$ decay width, provided $\Theta^+$ is understood as an excited rotational state of a nucleon.[5] For arbitrary $N_c$ the appropriate Clebsch–Gordan coefficient has been computed by Praszalowicz[44]:

$$\Gamma_\Theta(N_c) = \frac{3(N_c + 1)}{(N_c + 3)(N_c + 7)} \frac{3}{8\pi m_N^2} G_0^2 |\mathbf{p}|^3. \qquad (3.9.76)$$

To compare it with the width computed in Section 7 from the imaginary part of the pole in the kaon-Skyrmion scattering amplitude, one has to take the limit $N_c \to \infty$, as only in this limit the use of the Callan–Klebanov linearized scattering approach is legal. Using (3.9.75) we find

$$\Gamma_\Theta(N_c \to \infty) = \frac{8\pi F_\pi^2 r_0^4}{N_c} \Delta^3 = \mathcal{O}(N_c^0), \qquad (3.9.77)$$

which coincides exactly with the width obtained in Section 7 by other methods, in particular from the resonance pole position, where one has to put the coefficient $\gamma = 1$. To guarantee the validity of this result one has to make sure that the width is small, $\Gamma \ll \Delta$, for example, by taking the limit considered in subsection 7.2. In more realistic models the condition $\Gamma \ll \Delta$ can be achieved not by taking small $G_0$ but due to the cancelation of several pseudoscalar coupling $G_{0,1,2}$ as it in fact must happen in the nonrelativistic limit. Then, as shown from unitarity in Section 8, Eq. (3.9.76) modified to incorporate other couplings[5,44] remains valid.

Looking into Eq. (3.9.76) we can discuss what happens with the width as one goes from the idealized case of $N_c = \infty$ to the real world with $N_c = 3$. Unfortunately, at finite $N_c$ the whole Skyrmion approach becomes problematic since quantum corrections to the saddle point are then not small. Quantum corrections to a saddle point in general and here in particular are of two kinds: coming from zero and nonzero modes. Corrections from nonzero modes can be viewed as a meson loop in the Skyrmion background. As any other quantum loop in 4 dimensions, it has a typical additional suppression by $1/(2\pi)$ arising from the integral over loop momenta $\int d^4p/(2\pi)^4$. We remind the reader that in QED radiative corrections are not of the order of $\alpha$, the fine structure constant, but rather $\alpha/(2\pi) \approx 10^{-3}$. Therefore, quantum corrections from nonzero modes are expected to be of the order of $1/(2\pi N_c) \approx 1/20$ and look as if they can be neglected. As to zero modes, which are the translations and the rotations of the Skyrmion as a whole, they do not lead to the additional $1/(2\pi)$ suppression. On the contrary, they lead to "kinematical" factors like the Clebsch–Gordan coefficient in Eq. (3.9.76), which bear huge $1/N_c$ corrections. Hence it is desirable to take rotations into account exactly for any $N_c$.

We are therefore inclined to take Eq. (3.9.76) at face value for any $N_c$ and claim that it is the leading effect in accounting for finiteness of $N_c$. At $N_c = 3$ it leads to the relation

$$\Gamma_\Theta(N_c = 3) = \frac{1}{5}\Gamma_\Theta(N_c \to \infty). \tag{3.9.78}$$

The Clebsch–Gordan coefficient "1/5" was actually used in the original paper[5] predicting a narrow pentaquark. A large suppression of $\Gamma_\Theta$ as compared to its asymptotic value at $N_c = \infty$ has been also noticed in Ref. 29 in another estimate of the finite $N_c$ effects. Whatever is the width found from the pole position in the kaon-Skyrmion scattering amplitude, the real $\Theta^+$ width is expected to be at least 5 times less! Estimates for the real-world $N_c = 3$ in Refs. 16, 37 and 38 demonstrate that it can easily by obtained at the level of 1 MeV or even less, without any fitting parameters.

## 3.10. Conclusions

The remarkable idea of Skyrme that baryons can be viewed as nonlinear solitons of the pion field, finds a justification from the modern QCD point of view. However, the concrete realization of this idea — the use of the two- and four-derivative Skyrme Lagrangian supplemented by the four-derivative Wess–Zumino term — is an oversimplification of reality. Therefore, the Skyrme model as it is, may work reasonably well for certain baryon observables but may fail qualitatively for other.

To understand where the Skyrme model fails, one has to keep in mind that the model implies that the valence quarks are close to the negative-energy Dirac sea and cannot be separated from the sea that is strongly distorted. The number of $Q\bar{Q}$ pairs in a baryon, corresponding to a strongly polarized sea is $\mathcal{O}(1)$, times $N_c$. Exotic baryons are then not distinguishable from ordinary ones as they differ only by one additional $Q\bar{Q}$ pair as compared to the infinite $\mathcal{O}(N_c)$ number of pairs already present in the nucleon in that model, hence the exotic decays are not suppressed. In principle, it does not contradict QCD at strong coupling, however in reality we know that the octet and decuplet baryons are mainly 'made of' $N_c = 3$ constituent quarks with only a small (order of 30%) admixture of the $N_c + 2 = 5$ quark Fock component. In fact there is an implicit small parameter in baryon physics that may be called "relativism" $\epsilon \ll 1$ such that valence quark velocities are $v^2/c^2 \sim \epsilon$ and the number of $Q\bar{Q}$ pairs is $\epsilon N_c$.[16] For observables where the "nonrelativism" is essential one expects a qualitative disagreement with the Skyrme model predictions. For computing such observables it is better, while preserving the general and correct Skyrme's idea, to use a model that interpolates between the two extremes: the Skyrme model and the nonrelativistic quark model where there are no antiquarks at all.

Quantization of the $SU(3)$ zero rotational modes of the Skyrmion, whatever is its dynamical realization, leads to the spectrum of baryons forming a sequence of bands: each band is characterized by "exoticness", *i.e.* the number of additional $Q\bar{Q}$ pairs minimally needed to form a baryon multiplet. Inside the band, the splittings are $\mathcal{O}(1/N_c)$ whereas the splittings between bands with increasing exoticness is $\mathcal{O}(1)$, see Fig. 3.5. At large $N_c$ the lowest-mass baryons with nonzero exoticness (like the $\Theta^+$ baryon) have rotational wave functions corresponding to a small-angle precession. Therefore, the $\Theta^+$ and other exotic baryons can be, at asymptotically large $N_c$, studied *à la* Callan–Klebanov by considering the small oscillations of the kaon field about a Skyrmion, or the kaon-Skyrmion scattering in the linear order.

This problem has been solved numerically by Klebanov *et al.* [27] who have found that there is no resonance or bound state with the $\Theta^+$ exotic quantum numbers at least in the large $N_c$ limit, and suggested that it therefore could be an artifact of the rigid rotator approximation. In this paper, we study this scattering in more detail and come to the opposite conclusion. While reproducing numerically the phase shifts found in[27] we find, both analytically and numerically, that there is a pole in the complex energy plane, corresponding to a strong $\Theta^+$ resonance which would have definitely revealed itself in $KN$ scattering. Moreover, its origin is precisely the $SU(3)$ rotational mode. By varying the Skyrme model parameters, we are able to make $\Theta^+$ as narrow as one likes, as compared to the resonance excitation energy which can be held arbitrary. Being arbitrary it nevertheless coincides with the rotational excitation energy. To understand better the origin of the $\Theta^+$ width, we have computed it in five different ways yielding the same result. The problem is not the existence of $\Theta^+$ which *is* predicted by the Skyrme model and is a rotational excitation there, but what dynamics makes it narrow.

Although we can deform the parameters of the Skyrme model to make a finite-energy $\Theta^+$ narrow, they are not natural. It is precisely a problem where the deficiency of the Skyrme model mentioned above becomes, unfortunately, critical. To get a chance of explaining the narrow width, one needs a model that interpolates between the Skyrme model and the nonrelativistic quark models. The narrow $\Theta^+$ is near the nonrelativistic end of this interpolation. Fortunately, the Chiral Quark Soliton Model makes the job and indeed estimates of the $\Theta^+$ width there appear naturally with no parameter fitting at the 1 MeV level.

It is exciting and challenging to write this paper at the time when experimental evidence in favour of the exotic pentaquark $\Theta^+$ is still controversial. We hope that we have waived certain theoretical prejudice against $\Theta$'s existence and its small width, so it must be there.


## Acknowledgements

We have benefited from discussions with many people but most importantly from conversations and correspondence with Tom Cohen and Igor Klebanov. We are grateful to Klaus Goeke and Maxim Polyakov for hospitality at Bochum University where this work has been finalized. D.D. gratefully acknowledges Mercator Fellowship by the Deutsche Forschungsgemeinschaft. This work has been supported in part by Russian Government grants RFBR-06-02-16786 and RSGSS-3628.2008.2.

# References

1. T.H.R. Skyrme, *Proc. Roy. Soc. Lond. A* **260** (1961) 127; *Nucl. Phys.* **31** (1962) 556.
2. E. Witten, *Nucl. Phys. B* **223** (1983) 422, 433.
3. J. Wess and B. Zumino, *Phys. Lett. B* **37** (1971) 95.
4. E. Guadagnini, *Nucl. Phys. B* **236** (1984) 35;
   L.C. Biedenharn, Y. Dothan and A. Stern, *Phys. Lett. B* **146** (1984) 289;
   P.O. Mazur, M.A. Nowak and M. Praszalowicz, *Phys. Lett. B* **147** (1984) 137;
   A.V. Manohar, *Nucl. Phys. B* **248** (1984) 19;
   M. Chemtob, *Nucl. Phys. B* **256** (1985) 600;
   S. Jain and S.R. Wadia, *Nucl. Phys. B* **258** (1985) 713;
   D. Diakonov and V. Petrov, *Baryons as solitons*, preprint LNPI-967 (1984), a write-up of the lectures presented by D. Diakonov at the 12th ITEP Winter School (Feb. 1984), published in: Elementary Particles, Energoatomizdat, Moscow (1985) pp. 50–93.
5. D. Diakonov, V. Petrov and M. Polyakov, *Z. Phys. A* **359** (1997) 305, arXiv:hep-ph/9703373.
6. D. Diakonov and M. Eides, *Sov. Phys. JETP Lett.* **38** (1983) 433.
7. A. Dhar and S. Wadia, *Phys. Rev. Lett.* **52** (1984) 959;
   A. Dhar, R. Shankar and S. Wadia, *Phys. Rev. D* **31** (1985) 3256.
8. D. Diakonov and V. Petrov, *Nucl. Phys. B* **272** (1986) 457.
9. D. Diakonov, V. Petrov and P. Pobylitsa, *Nucl. Phys. B* **306** (1988) 809.
10. M. Rho, A.S. Goldhaber and G.E. Brown, *Phy. Rev. Lett.* **51** (1983) 747;
    G.E. Brown, A.D. Jackson, M. Rho and V. Vento, *Phys. Lett. B* **140** (1984) 285.
11. A. Hosaka and H. Toki, *Phys. Rep.* **277** (1996) 65.
12. A. Manohar and H. Georgi, *Nucl. Phys. B* **234** (1984) 189.
13. S. Kahana, G. Ripka and V. Soni, *Nucl. Phys. A* **415** (1984) 351;
    S. Kahana and G. Ripka, *Nucl. Phys. A* **429** (1984) 462.
14. M.S. Birse and M.K. Banerjee, *Phys. Lett. B* **136** (1984) 284.
15. D. Diakonov and V. Petrov, *Sov. Phys. JETP Lett.* **43** (1986) 57;
    D. Diakonov, in *Skyrmions and Anomalies*, eds. M. Jeżabek and M. Praszalowicz, (World Scientific, Singapore, 1987) p. 27.
16. D. Diakonov and V. Petrov, *Phys. Rev. D* **72** (2005) 074009, arXiv:hep-ph/0505201.
17. D. Diakonov, V. Petrov, P. Pobylitsa, M. Polyakov and C. Weiss, *Nucl. Phys. B* **480** (1996) 341, arXiv:hep-ph/9606314; *Phys. Rev. D* **56** (1997) 4069.
18. V. Petrov and M. Polyakov, arXiv:hep-ph/0307077.
19. G. Ripka *Quarks bound by chiral fields*, Clarendon Press, Oxford (1997).
20. D. Diakonov and V. Petrov, in *At the Frontiers of Particle Physics (Handbook of QCD)*, ed. M. Shifman, World Scientific (2001) Vol. 1, p. 359, arXiv:hep-ph/0009006.
21. A. Blotz *et al.*, *Nucl. Phys. A* **355** (1993) 765.
22. D. Diakonov and V. Petrov, *Phys. Rev. D* **69** (2004) 056002, arXiv:hep-ph/0309203.
23. Z. Dulinski and M. Praszalowicz, *Acta Phys. Polon. B* **18** (1987) 1157.
24. T. Cohen, *Phys. Lett. B* **581** (2004) 175, arXiv:hep-ph/0309111; *Phys. Rev. D* **70** (2004) 014011, arXiv:hep-ph/0312191.
25. C. Callan and I. Klebanov, *Nucl. Phys. B* **262** (1985) 365; C. Callan, K. Hornbostel and I. Klebanov, *Phys. Lett. B* **202** (1988) 269.

26. B.-Y. Park, M. Rho and D.-P. Min, *Phys. Rev. D* **70** (2004) 114026, arXiv:hep-ph/0405246.
27. N. Itzhaki, I.R. Klebanov, P. Ouyang and L. Rastelli, *Nucl. Phys. B* **684** (2004) 264, arXiv:hep-ph/0309305v5.
28. G. Adkins, C. Nappi and E. Witten, *Nucl. Phys. B* **228** (1983) 552.
29. H. Walliser and H. Weigel, *Eur. Phys. J. A* **26** (2005) 361, arXiv:hep-ph/0510055.
30. L.D. Landau and E.M. Lifshitz, Quantum Mechanics (Non-Relativistic Theory), 3rd edition, Butterworth-Heinemann, Oxford (1999).
31. J.B. Hyslop *et al.*, *Phys. Rev. D* **46** (1992) 961.
32. A.S. Caroll *et al.*, *Phys. Lett. B* **45** (1973) 531;
C.B. Dover and G.E. Walker, *Phys. Rep.* **89** (1982) 1.
33. R. A. Arndt, I. I. Strakovsky and R. L. Workman, *Phys. Rev. C* **68** (2003) 042201 [Erratum-ibid. **69** (2004) 019901], arXiv:nucl-th/0308012;
A. Sibirtsev, J. Haidenbauer, S. Krewald and U. G. Meissner, *Phys. Lett. B* **599** (2004) 230, arXiv:hep-ph/0405099.
34. D. Diakonov, V. Petrov and M. Praszalovicz, *Nucl. Phys. B* **323** (1989) 53.
35. M. Wakamatsu and H. Yoshiki, *Nucl. Phys. A* **524** (1991) 561.
36. D. Diakonov and V. Petrov, *Phys. Rev. D* **69** (2004) 094011, arXiv:hep-ph/0310212.
37. C. Lorcé, *Phys. Rev. D* **74** (2006) 054019, arXiv:hep-ph/0603231.
38. T. Ledwig, H.-C. Kim and K. Goeke, *Phys. Rev. D* **78** (2008) 054005, arXiv:0805.4063; arXiv:0803.2276.
39. M. Polyakov and A. Rathke, *Eur. Phys. J. A* **18** (2003) 691, arXiv:hep-ph/0303138.
40. S.J. Brodsky, J.R. Ellis and M. Karliner, *Phys. Lett. B* **206** (1988) 309.
41. D. Diakonov and V. Petrov, arXiv:hep-ph/0312144;
D. Diakonov, *Acta Phys. Pol. B* **25** (1994) 17.
42. A. Cherman, T. Cohen, A. Nellore, *Phys. Rev. D* **70** (2004) 096003.
43. H. Weigel, *Phys. Rev. D* **75** (2007) 114018, arXiv:hep-ph/0703072.
44. M. Praszalowicz, *Phys. Lett. B* **583** (2004) 96, arXiv:hep-ph/0311230.

# Chapter 4

# Heavy-Quark Skyrmions

N.N. Scoccola

*Departmento de Física, Comisión Nacional de Energía Atómica,*
*(1429) Buenos Aires, Argentina*
*CONICET, Rivadavia 1917, (1033) Buenos Aires, Argentina*
*Universidad Favaloro, Solís 453, (1078) Buenos Aires, Argentina*

The description of the heavy baryons as heavy-meson–soliton bound systems is reviewed. We outline how such bound systems arise from effective lagrangians that respect both chiral symmetry and heavy quark symmetry. Effects due to finite heavy quark masses are also discussed, and the resulting heavy baryon spectra are compared with existing quark model and empirical results. Finally, we address some issues related to a possible connection between the usual bound state approach to strange hyperons and that for heavier baryons.

## Contents

4.1  Introduction . . . . . . . . . . . . . . . . . . . . . . . . . . . . . . . . . . . .  91
4.2  Heavy Baryons as Skyrmions in the Heavy Quark Limit . . . . . . . . . . . . . . .  92
    4.2.1  Effective chiral lagrangians and heavy quark symmetry . . . . . . . . . . . . .  93
    4.2.2  Heavy-meson–soliton bound states in the heavy quark limit and their collective
           quantization . . . . . . . . . . . . . . . . . . . . . . . . . . . . . . . . .  94
4.3  Beyond the Heavy Quark Limit . . . . . . . . . . . . . . . . . . . . . . . . . . .  101
4.4  Relation with the Bound State Approach to Strangeness . . . . . . . . . . . . . . .  107
4.5  Summary and Conclusions . . . . . . . . . . . . . . . . . . . . . . . . . . . . .  111
References . . . . . . . . . . . . . . . . . . . . . . . . . . . . . . . . . . . . . . .  112

## 4.1.  Introduction

During the last quarter of a century it has become clear that the applicability of the Skyrme's topological soliton model for light baryon structure[1,2] goes far beyond all the original expectations. In fact, as described in other chapters of this book the underlying ideas have found applications in other areas of physics, notably in the physics of complex nuclei and dense matter, condensed matter physics and gauge/string duality. The purpose of the present contribution is to summarize the work done on the extension of the skyrmion picture to the study of the

"

structure of baryons containing heavy quarks. In this scheme, such baryons are described as bound systems of heavy mesons and a soliton. This so-called "bound state approach" was first developed to describe strange hyperons[3,4] and was later shown[5] to be applicable to baryons containing one or more charm (c) and bottom (b) quarks. In these early works only pseudoscalar fields were taken as explicit degrees of freedom with their interactions given by a flavor symmetric Skyrme lagrangian supplemented by explicit flavor symmetric terms to account for the effect of the heavy quark mass. The results on the mass spectra[6] and magnetic moments[7] for charm baryons were found to be strikingly close to the quark model description which is expected to work better as the heavy quark involved becomes heavier. However, it was then realized that this description in terms of only pseudoscalar fields was at odds with the heavy quark symmetry[8] which states that in the heavy quark limit the heavy pseudoscalar and vector fields become degenerate and, thus, should be treated on an equal footing. This difficulty was resolved in Ref. 9 where it was proposed to apply the bound state approach to the heavy meson effective lagrangian[10-13] which simultaneously incorporates chiral symmetry and heavy quark symmetry. Such observation led to a quite important number of works where various properties of heavy baryons have been studied within this framework. Here, we present a short review of those studies pointing out their main results as well as the relationship between some different approaches used in the literature. Some still remaining open questions are also mentioned.

This contribution is organized as follows. In Sec. 4.2 we outline how heavy baryons can be described within soliton models in the heavy quark limit. In particular, in Sec. 4.2.1 we introduce the type of lagrangian that describes the interactions between light and heavy mesons, and which simultaneously respect chiral and heavy quark symmetries, while in Sec. 4.2.2 we show how bound states of a soliton and heavy mesons are obtained and the system quantized. In Sec. 4.3 we show how departures from the heavy quark limit can be taken into account. In Sec. 4.4 we discuss some issues related to the connection between the usual bound state approach to strange hyperons with that for heavier baryons given in the previous section. Finally, in Sec. 4.5 a summary with some conclusions is given.

## 4.2. Heavy Baryons as Skyrmions in the Heavy Quark Limit

In this section we outline how a heavy baryon can be described within topological soliton models in the limit in which the heavy quarks are assumed to be infinitely heavy. Corrections due to finite heavy quark masses will be discussed in the following section. In Sec. 4.2.1 we introduce a type of lagrangian for a system of Goldstone bosons and the heavy mesons, which possesses both chiral symmetry and heavy quark symmetry. Next, in Sec. 4.2.2 we show how a heavy-meson–soliton bound state can arise at the classical level, and the way in which such bound system can be quantized.

### 4.2.1. *Effective chiral lagrangians and heavy quark symmetry*

For the light sector, the simplest lagrangian that supports stable soliton configuration is the Skyrme model lagrangian[1]

$$\mathcal{L}_l^{Sk} = \frac{f_\pi^2}{4} \operatorname{Tr} \left[ \partial_\mu U^\dagger \partial^\mu U \right] + \frac{1}{32e^2} \operatorname{Tr} \left[ [U^\dagger \partial_\mu U, U^\dagger \partial_\nu U]^2 \right], \tag{4.2.1}$$

where $f_\pi$ is the pion decay constant ($\approx 93$ MeV empirically) and $U$ is an $SU(2)$ matrix of the chiral field, i.e.

$$U = \exp\left[ iM/f_\pi \right], \tag{4.2.2}$$

with $M$ being a $2 \times 2$ matrix of the pion triplet

$$M = \vec{\tau} \cdot \vec{\pi} = \begin{pmatrix} \pi^0 & \sqrt{2}\pi^+ \\ \sqrt{2}\pi^- & -\pi^0 \end{pmatrix}. \tag{4.2.3}$$

Here, the chiral $SU(2)_L \times SU(2)_R$ symmetry is realized nonlinearly under the transformation of $U$

$$U \longrightarrow L\, U\, R^\dagger, \tag{4.2.4}$$

with $L \in SU(2)_L$ and $R \in SU(2)_R$. Due to the presence of the Skyrme term with the Skyrme parameter $e$, the lagrangian $\mathcal{L}_l^{Sk}$ supports stable soliton solutions.

When discussing the interaction of the Goldstone fields $M(x)$ with other fields it is convenient to introduce $\xi(x)$ such that

$$U = \xi^2, \tag{4.2.5}$$

and which transforms under the $SU(2)_L \times SU(2)_R$ as

$$\xi \to \xi' = L\, \xi\, \vartheta^\dagger = \vartheta\, \xi\, R^\dagger, \tag{4.2.6}$$

where $\vartheta$ is a local unitary matrix depending on $L$, $R$, and $M(x)$.

Consider now heavy mesons containing a heavy quark $Q$ and a light antiquark $\bar{q}$. Here, the light antiquark in a heavy meson is assumed to form a point-like object with the heavy quark, endowing it with appropriate color, flavor, angular momentum and parity. Let $\Phi$ and $\Phi_\mu^*$ be the field operators that annihilate $j^\pi=0^-$ and $1^-$ heavy mesons with $C = +1$ or $B = -1$. They form $SU(2)$ antidoublets: for example, when the heavy quark constituent is the $c$-quark,

$$\Phi = \left( D^0, D^+ \right), \qquad \Phi^* = \left( D^{*0}, D^{*+} \right). \tag{4.2.7}$$

In the limit of infinite heavy quark mass, the heavy quark symmetry implies that the dynamics of the heavy mesons depends trivially on their spin and mass. Such a trivial dependence can be eliminated by introducing a redefined $4 \times 4$ matrix field $H(x)$ as

$$H = \frac{1 + \slashed{v}}{2} \left( \Phi_v \gamma_5 - \Phi_{v\mu}^* \gamma^\mu \right). \tag{4.2.8}$$

94                                    *N.N. Scoccola*

Here, we use the conventional Dirac $\gamma$-matrices and $\not v$ denotes $v_\mu\gamma^\mu$. The fields $\Phi_v$ and $\Phi^*_{v\mu}$, respectively, represent the heavy pseudoscalar field and heavy vector fields in the moving frame with a four velocity $v_\mu$. They are related to the $\Phi$ and $\Phi^*_\mu$ as[14]

$$\Phi = e^{-iv\cdot x m_\Phi}\frac{\Phi_v}{\sqrt{2m_\Phi}} \; , \qquad \Phi^*_\mu = e^{-iv\cdot x m_{\Phi^*}}\frac{\Phi^*_{v\mu}}{\sqrt{2m_{\Phi^*}}} \; . \tag{4.2.9}$$

Under $SU(2)_L \times SU(2)_R$ chiral symmetry operations $H$ transforms as

$$H \to H\,\vartheta \; , \tag{4.2.10}$$

while under the heavy quark spin rotation,

$$H \to S\,H \; , \tag{4.2.11}$$

with $S \in SU(2)_v$, i.e. the heavy quark spin symmetry group boosted by the velocity $v$. Taking this into account it is possible to write down a lagrangian that describes the interactions of heavy mesons and Goldstone bosons, and which possesses both chiral symmetry and heavy quark symmetry. To leading order in derivatives acting on the Goldstone fields, the most general form of such lagrangian is given by[10–13]

$$\mathcal{L}_{lh} = -iv_\mu \,\mathrm{Tr}\left[D^\mu H\bar H\right] - g\,\mathrm{Tr}(\left[H\gamma_5 A_\mu \gamma^\mu \bar H\right] \; , \tag{4.2.12}$$

where $\bar H = \gamma_0 H^\dagger \gamma_0$, and

$$V_\mu = \frac{1}{2}(\xi^\dagger \partial_\mu \xi + \xi\partial_\mu \xi^\dagger) \; , \qquad A_\mu = \frac{i}{2}(\xi^\dagger \partial_\mu \xi - \xi\partial_\mu \xi^\dagger) \; . \tag{4.2.13}$$

Here, $g$ is a universal coupling constant for the $\Phi\Phi^*\pi$ and $\Phi^*\Phi^*\pi$ interactions. The nonrelativistic quark model provides the naive estimation[12] $g = -3/4$. On the other hand, for the case of the $D^* \to \pi D$ decay this lagrangian leads to a width given by

$$\Gamma(D^{*+} \to D^0\pi^+) = \frac{1}{6\pi}\frac{g^2}{f_\pi^2}|\vec p_\pi|^3 \; . \tag{4.2.14}$$

Recent experimental results for this width imply $|g|^2 \approx 0.36$.[15]

### 4.2.2. *Heavy-meson–soliton bound states in the heavy quark limit and their collective quantization*

Following the discussions in the previous subsection we consider here the chiral and heavy quark symmetric effective lagrangian given by

$$\mathcal{L} = \mathcal{L}_l^{Sk} + \mathcal{L}_{lh} \; , \tag{4.2.15}$$

where $\mathcal{L}_l^{Sk}$ and $\mathcal{L}_{lh}$ are given by Eqs. (4.2.1) and (4.2.12), respectively.

In what follows we will discuss how to obtain heavy baryons following a procedure in which a heavy-meson–soliton bound state is first found and then quantized by rotating the whole system in the collective coordinate quantization scheme.[16,17] An alternative method[9] will be briefly discussed at the end of this subsection.

The non-linear lagrangian $\mathcal{L}_l^{Sk}$ supports a classical soliton solution

$$U_0(\vec{r}) = \exp[i\vec{\tau} \cdot \hat{r} F(r)] ,$$ (4.2.16)

with the boundary conditions

$$F(0) = \pi \quad \text{and} \quad F(\infty) = 0 ,$$ (4.2.17)

which, due to its nontrivial topological structure, carries a winding number identified as the baryon number $B = 1$. It also has a finite mass $M_{sol}$ whose explicit expression in terms of the soliton profile function $F(r)$ can be found in e.g. Refs. 1 and 2.

In order to look for possible heavy-meson–soliton bound states we have to find the eigenstates of the heavy meson fields interacting with the static potentials

$$V^\mu = \left(0, \vec{V}\right) = \left(0, i \, v(r) \, \hat{r} \times \vec{\tau}\right) ,$$

$$A^\mu = \left(0, \vec{A}\right) = \left(0, \frac{1}{2} \, a_1(r) \, \vec{\tau} + \frac{1}{2} \, a_2(r) \, \hat{r} \, \vec{\tau} \cdot \hat{r}\right),$$ (4.2.18)

where

$$v(r) = \frac{\sin^2(F/2)}{r} , \qquad a_1(r) = \frac{\sin F}{r} , \qquad a_2(r) = F' - \frac{\sin F}{r} .$$ (4.2.19)

These expressions result from the soliton configuration (4.2.16) sitting at the origin. In the rest frame $v_\mu = (1,0,0,0)$, it follows from Eq. (4.2.8) that $H(x)$ can be expressed in terms of 2 x 2 blocks as

$$H(x) = \begin{pmatrix} 0 & h(x) \\ 0 & 0 \end{pmatrix} .$$ (4.2.20)

Here we have used that, in that frame, $\Phi^*_{v,0}$ is identically zero due to the condition $v \cdot \Phi^*_v = 0$. Thus, the lagrangian Eq. (4.2.12) takes the form

$$L_0 = -M_{sol} + \int d^3r \left(-i \operatorname{Tr}\left[\dot{h} \, \bar{h}\right] + g \operatorname{Tr}\left[h \, \vec{A} \cdot \vec{\sigma} \, \bar{h}\right]\right) ,$$ (4.2.21)

where $\bar{h} = -h^\dagger$. The corresponding equation of motion for the $h$-field is[17,18]

$$i \, \dot{h} = g \, h \, \vec{A} \cdot \vec{\sigma} .$$ (4.2.22)

In the "hedgehog" configuration (4.2.16), and consequently in the static potentials (4.2.18), the isospin and the angular momentum are correlated in such a way that neither of them is separately a good quantum number, but their sum (the so-called "grand spin") $\vec{K}$ is. Here

$$\vec{K} = \vec{J} + \vec{I} \equiv (\vec{L} + \vec{S}) + \vec{I} .$$ (4.2.23)

Thus, the equation of motion Eq. (4.2.22) is invariant under rotations in $K$-space, and the wavefunctions of the heavy meson eigenmodes can be written as the product of a radial function and the eigenfunction of the grand spin $\mathcal{K}^{(a)}_{kk_3}(\hat{r})$. Namely,

$$h(\vec{r}, t) = \sum_a \alpha_a \, h_k^{(a)}(r) \, \mathcal{K}^{(a)}_{kk_3}(\hat{r}) \, e^{-i\varepsilon t} ,$$ (4.2.24)

where the sum over $a$ accounts for the possible ways of constructing the eigenstates of the same grand spin and parity by combining the eigenstates of the spin, isospin and orbital angular momentum, and the expansion coefficients $\alpha_a$ are normalized by $\sum_a |\alpha_a|^2 = 1$. Since we are assuming here that both the soliton and the heavy mesons are infinitely heavy in the lowest energy state they should be sitting one on top of the other at the same spatial point, just propagating in time. That is, the radial functions $h_k^{(a)}(r)$ of the lowest energy eigenstate can be approximated by a delta-function-like one, say $f(r)$, which is strongly peaked at the origin and normalized as $\int dr\, r^2 |f(r)|^2 = 1$. Thus, using orthonormalized eigenfunctions $\mathcal{K}_{kk_3}^{(a)}(\hat{r})$ of the grand spin which satisfy

$$\int d\Omega \, \mathrm{Tr}\left[ \mathcal{K}_{kk_3}^{(a)}(\hat{r}) \bar{\mathcal{K}}_{k'k_3'}^{(a')}(\hat{r}) \right] = -\delta_{aa'}\delta_{kk'}\delta_{k_3 k_3'} \,, \tag{4.2.25}$$

the field $h$ is normalized as

$$-\int d^3 r \, \mathrm{Tr}[h\bar{h}] = 1 \,. \tag{4.2.26}$$

Replacing Eq. (4.2.24) in Eq. (4.2.22) and integrating out the radial part, we obtain

$$\varepsilon \, \mathcal{K}_{kk_3}(\hat{r}) = \frac{gF'(0)}{2} \, \mathcal{K}_{kk_3}(\hat{r}) \left( 2\vec{\sigma} \cdot \hat{r} \, \vec{\tau} \cdot \hat{r} - \vec{\sigma} \cdot \hat{r} \right) \,, \tag{4.2.27}$$

with $\mathcal{K}_{kk_3} \equiv \sum_a \alpha_a \, \mathcal{K}_{kk_3}^{(a)}$. Here, we have used that, near the origin, $F(r) \sim \pi + F'(0)\, r$ and consequently $\vec{A} \cdot \vec{\sigma} \sim \frac{1}{2}F'(0)(2\vec{\sigma} \cdot \hat{r}\vec{\tau} \cdot \hat{r} - \vec{\sigma} \cdot \hat{r})$.

Thus, our problem is reduced to finding $\mathcal{K}_{kk_3}$. For this purpose it is convenient to construct the grand spin eigenstates $\mathcal{K}_{kk_3}^{(a)}(\hat{r})$ by combining the eigenstates of the spin, isospin and orbital angular momentum. Here, we construct first the eigenfunctions of $\vec{\Lambda} = \vec{L} + \vec{I}$ by combining orbital angular momentum and isospin eigenstates, and then couple the resulting states to the spin eigenstates. Since we are interested here in the lowest energy eigenmode of positive parity, we can restrict the angular momentum $\ell$ to be 1. This statement requires some explanation. In general, when departures from a delta-like behavior are considered the differential equations for the heavy meson radial functions have a centrifugal term with a singularity $\ell_{eff}(\ell_{eff}+1)/r^2$ near the origin. Here, $\ell_{eff}$ is the "effective" angular momentum[3] given by $\ell_{eff} = \ell \pm 1$ if $\lambda = \ell \pm 1/2$. That behavior is due to the presence of a vector potential from the soliton configuration $\vec{V}(\sim i(\hat{r} \times \vec{\tau})/r$, near the origin), which alters the singular structure of $\vec{D}^2 = (\vec{\nabla} - \vec{V})^2$ from $\ell(\ell+1)/r^2$ of the usual $\vec{\nabla}^2$ to $\ell_{eff}(\ell_{eff}+1)/r^2$. Thus, the state with $\ell_{eff} = 0$ can have most strongly peaked radial function and become the lowest eigenstate. Note that $\ell_{eff} = 0$ can be achieved only when $\ell = 1$. It is important to notice that combining the negative parity resulting from this orbital wavefunction with the heavy meson intrinsic negative parity we obtain that ground state heavy baryons have positive parity, as expected. For $\ell = 1$ two values of $\lambda$, $\frac{1}{2}$ and $\frac{3}{2}$, are possible. Moreover, from the experience of the bound-state approach to strange hyperons, where a similar situation arises,[3] the lowest energy state is expected to correspond to the lowest possible value of $k$, i.e. $k = \frac{1}{2}$. Since we have $s = 0, 1$ and $\lambda = \frac{1}{2}, \frac{3}{2}$, we can construct three different grand spin states of $k = \frac{1}{2}$. Explicitly,[17]

$$\mathcal{K}^{(1)}_{\frac{1}{2}, \pm\frac{1}{2}}(\hat{r}) = \frac{1}{\sqrt{8\pi}} \, \chi_\pm \, \vec{\tau} \cdot \hat{r} \, ,$$

$$\mathcal{K}^{(2)}_{\frac{1}{2}, \pm\frac{1}{2}}(\hat{r}) = \frac{1}{\sqrt{24\pi}} \, \chi_\pm \, \vec{\sigma} \cdot \vec{\tau} \, \vec{\tau} \cdot \hat{r} \, , \qquad (4.2.28)$$

$$\mathcal{K}^{(3)}_{\frac{1}{2}, \pm\frac{1}{2}}(\hat{r}) = \frac{1}{\sqrt{48\pi}} \, \chi_\pm \, (\vec{\sigma} \cdot \vec{\tau} \, \vec{\tau} \cdot \hat{r} - 3 \, \vec{\sigma} \cdot \hat{r}) \, .$$

Here, $\chi_+ = (0, -1)$ and $\chi_- = (+1, 0)$ are the isospin states corresponding to $\bar{u}$ and $\bar{d}$, respectively. The eigenstates $\mathcal{K}_{\frac{1}{2}, \pm\frac{1}{2}}(\hat{r})$ of Eq. (4.2.27) can be expanded in terms of these states

$$\mathcal{K}_{\frac{1}{2}, \pm\frac{1}{2}}(\hat{r}) = \sum_{a=1}^{3} \alpha_a \, \mathcal{K}^{(a)}_{\frac{1}{2}, \pm\frac{1}{2}}(\hat{r}) \, , \qquad (4.2.29)$$

with the expansion coefficients given by the solution of the secular equation

$$\sum_{b=1}^{3} \mathcal{M}_{ab} \, \alpha_b = -\, \varepsilon \, \alpha_a \, , \qquad (4.2.30)$$

where the matrix elements $\mathcal{M}_{ab}$ are defined by

$$\mathcal{M}_{ab} = \frac{gF'(0)}{2} \int d\Omega \, \mathrm{Tr} \left[ \mathcal{K}^{(a)}(\hat{r}) \, (2 \, \vec{\sigma} \cdot \hat{r} \, \vec{\tau} \cdot \hat{r} - \vec{\sigma} \cdot \hat{r}) \, \bar{\mathcal{K}}^{(b)}(\hat{r}) \right] \, . \qquad (4.2.31)$$

Note that the minus sign in Eq. (4.2.30) is due to the fact that the basis states $\mathcal{K}^{(a)}_{\frac{1}{2}, \pm\frac{1}{2}}(\hat{r})$ are normalized as indicated in Eq. (4.2.25). With the explicit form of $\mathcal{K}^{(a)}_{\frac{1}{2}, \pm\frac{1}{2}}(\hat{r})$ given by Eq. (4.2.29), these matrix elements can be easily calculated. The secular equation (4.2.29) yields three eigenstates. Since $g < 0$ and $F'(0) < 0$ (in the case of the baryon-number-1 soliton solution), there is a heavy-meson–soliton bound state of binding energy $-\frac{3}{2}gF'(0)$. The two *unbound* eigenstates with positive eigenenergy $+\frac{1}{2}gF'(0)$ are not consistent with the strongly peaked radial functions. They are improper solutions of Eq. (4.2.27).

In terms of the eigenmodes, the hamiltonian of the system in the body fixed (i.e. soliton) frame has the diagonal form

$$H_{\mathrm{bf}} = M_{sol} + \sum_{nkk_3} \varepsilon_{nk} \, a_{nkk_3} \, a^\dagger_{nkk_3} =$$

$$= M_{sol} + \varepsilon_{bs} \left( a^\dagger_{+1/2} \, a_{+1/2} + a^\dagger_{-1/2} \, a_{-1/2} \right) + \dots \, , \qquad (4.2.32)$$

where $n$ represents the extra quantum numbers needed to completely specify a given eigenstate. Moreover, $a_{nkk_3}$ ($a^\dagger_{nkk_3}$) are the usual meson annihilation (creation) operators. In the second line of Eq. (4.2.32) we have explicitly written the contribution of the bound state with $\varepsilon_{gs} = -\frac{3}{2}gF'(0)$ found above, using the subscript $\pm 1/2$ to indicate the grand spin projection $k_3$.

What we have obtained so far is the heavy-meson–soliton bound state which carries a baryon number and a heavy flavor. Therefore, up to order $O(m_Q^0 N_c^0)$ baryons containing a heavy quark such as $\Lambda_Q$, $\Sigma_Q$ and $\Sigma_Q^*$ are degenerate in mass. However, to extract physical heavy baryons of correct spin and isospin, we have to go to the next order in $1/N_c$, while remaining in the same order in $m_Q$, i.e. $O(m_Q^0 N_c^{-1})$. This can be done by introducing time dependent $SU(2)$ collective variables $C(t)$ associated with the degeneracy under simultaneous $SU(2)$ rotation of the soliton configuration and the heavy meson fields

$$\xi(\vec{r}, t) = C(t)\, \xi_{\mathrm{bf}}(\vec{r})\, C^\dagger(t) \qquad \text{and} \qquad h(\vec{r}, t) = h_{\mathrm{bf}}(\vec{r}, t)\, C^\dagger(t) \ , \qquad (4.2.33)$$

where $\xi_{\mathrm{bf}}^2 \equiv U_0$, and then performing the quantization by elevating them to the corresponding quantum mechanical operators. In Eq. (4.2.33) and in what follows, $h_{\mathrm{bf}}$ refers to the heavy meson field in the (isospin) soliton frame, while $h$ refers to that in the laboratory frame, *i.e.*, the heavy quark rest frame. Inserting Eq. (4.2.33) in Eq. (4.2.15) we obtain an extra collective contribution of $O(m_Q^0 N_c^{-1})$ to the lagrangian

$$L_{coll} = \frac{1}{2}\, \mathcal{I}\, \omega^2 + \vec{Q} \cdot \vec{\omega}\ , \qquad (4.2.34)$$

where the "angular velocity" $\vec{\omega}$ of the collective rotation is defined by

$$C^\dagger \dot{C} \equiv \frac{i}{2}\vec{\tau} \cdot \vec{\omega}\ , \qquad (4.2.35)$$

$\mathcal{I}$ is the moment of inertia of the rotating soliton, whose explicit expression in terms of the soliton profile function $F(r)$ can be found in e.g. Refs. 1 and 2, and

$$\vec{Q} = -\frac{1}{4}\int d^3r\, \mathrm{Tr}\left[ h_{\mathrm{bf}} \left( \xi_{\mathrm{bf}}^\dagger\, \vec{\tau}\, \xi_{\mathrm{bf}} + \xi_{\mathrm{bf}}\, \vec{\tau}\, \xi_{\mathrm{bf}}^\dagger \right) \bar{h}_{\mathrm{bf}} \right]\ . \qquad (4.2.36)$$

Taking the Legendre transform of the lagrangian we obtain the collective hamiltonian as

$$H_{coll} = \frac{1}{2\mathcal{I}} \left( \vec{R} - \vec{Q} \right)^2\ , \qquad (4.2.37)$$

where $\vec{R}$ is the spin of the rotor given by $\vec{R} = \mathcal{I}\,\vec{\omega} + \vec{Q}$.

With the collective variable introduced as in Eq. (4.2.33), the isospin of the fields $U(x)$ and $h(x)$ is entirely shifted to $C(t)$. To see this, consider the isospin rotation

$$U \to \mathcal{A}\, U\, \mathcal{A}^\dagger\ , \qquad h \to h\, \mathcal{A}^\dagger\ , \qquad (4.2.38)$$

with $\mathcal{A} \in SU(2)_V$, under which the collective variables and fields in the soliton frame transform as

$$C(t)\ \to \mathcal{A}\, C(t), \qquad h_{\mathrm{bf}}(x)\ \to\ h_{\mathrm{bf}}(x)\ . \qquad (4.2.39)$$

Thus, the $h$-field is isospin-blind in the (isospin) soliton frame. The conventional Noether construction gives the isospin of the system,

$$I^a = \frac{1}{2} \operatorname{Tr}\left[\tau^a C \tau^b C^\dagger\right]\left(\mathcal{I}\,\omega^b + Q^b\right) = D^{ab}(C)R^b \;, \tag{4.2.40}$$

where $D^{ab}(C)$ is the adjoint representation of the $SU(2)$ transformation associated with the collective variables $C(t)$.

The eigenfunctions of the rotor-spin operator are the usual Wigner $\mathcal{D}$-functions. In terms of these eigenfunctions and the heavy meson bound states $|\pm 1/2\rangle_{bs}$, the heavy baryon state of isospin $i_3$ and spin $s_3$ containing a heavy quark can be constructed as

$$i; i_3, s_3\rangle = \sqrt{2i+1} \sum_{m=\pm 1/2} (i, s_3 - m, 1/2, m | 1/2, s_3)\, \mathcal{D}^{(i)}_{i_3,-s_3+m}(C)\, |m\rangle_{bs}\;, \tag{4.2.41}$$

where $i = 0$ for $\Lambda_Q$ and $i = 1$ for $\Sigma_Q$ and $\Sigma_Q^*$.

Treating the collective Hamiltonian (4.2.37) to first order in perturbation theory we obtain

$$m_i = M_{sol} + \varepsilon_{bs} + \frac{1}{2\mathcal{I}}\left(i(i+1) + 3/4\right)\;. \tag{4.2.42}$$

Here, we have used that explicit evaluation shows[18]

$$_{bs}\langle m|\vec{Q}|m\rangle_{bs} = 0\;, \tag{4.2.43}$$

$$_{bs}\langle m|\vec{Q}^2|m\rangle_{bs} = 3/4\;. \tag{4.2.44}$$

These two results deserve some comments. First we note that general use of the Wigner-Eckart theorem implies

$$< n, k, k_3|\vec{Q}|n, k, k_3' > = c_{nk} < n, k, k_3|\vec{K}|n, k, k_3' > \;. \tag{4.2.45}$$

The constants $c_{nk}$ are usually called "hyperfine splitting" constants. Eq. (4.2.43) implies that for the ground state $c_{gs} = 0$ in the heavy quark limit. As a consequence of this, the Hamiltonian depends only on the rotor-spin so that $\Sigma_Q$ and $\Sigma_Q^*$ become degenerate as expected from the heavy quark symmetry. It is clear that corrections that imply departures from heavy quark limit could lead to non-vanishing values of $c_{gs}$. It is also important to notice that to obtain the result Eq. (4.2.44) one should take into account all possible intermediate states.

In order to compare the results with experimental heavy baryon masses, we have to add the heavy meson masses subtracted so far from the eigenenergies. The mass formulas to be compared with data are

$$m_{\Lambda_Q} = M_{sol} + \overline{m}_\Phi - \frac{3}{2}gF'(0) + \frac{3}{8\mathcal{I}}\;,$$

$$m_{\Sigma_Q} = m_{\Sigma_Q^*} = M_{sol} + \overline{m}_\Phi - \frac{3}{2}gF'(0) + \frac{11}{8\mathcal{I}}\;, \tag{4.2.46}$$

where $\overline{m}_\Phi$ is the weighted average mass of the heavy meson multiplets, $\overline{m}_\Phi = (3m_{\Phi^*} + m_\Phi)/4$. In the case of $Q = c$, we have $\overline{m}_\Phi = 1973$ MeV while for

$Q = b$, $\overline{m}_\Phi = 5314$ MeV. The $SU(2)$ quantities $M_{sol}$ and $\mathcal{I}$ are obtained from the nucleon and $\Delta$ masses

$$M_{sol} = 866 \text{ MeV}, \qquad \text{and} \qquad 1/\mathcal{I} = 195 \text{ MeV} . \tag{4.2.47}$$

Finally, the unknown value of $gF'(0)$ can be adjusted to fit the observed value of the $\Lambda_c$ mass,

$$m_{\Lambda_c} = 2286 \text{ MeV} = M_{sol} + \overline{m}_\Phi - \frac{3}{2}gF'(0) + \frac{3}{8\mathcal{I}} , \tag{4.2.48}$$

which implies that

$$gF'(0) = 417 \text{ MeV} . \tag{4.2.49}$$

This leads to a prediction on the $\Lambda_b$ mass and the average masses of the $\Sigma_Q$-$\Sigma_Q^*$ multiplets, $\overline{m}_{\Sigma_Q}[\equiv \frac{1}{3}(2m_{\Sigma_Q^*} + m_{\Sigma_Q})]$,

$$m_{\Lambda_b} = M_{sol} + \overline{m}_B - \frac{3}{2}gF'(0) + 3/8\mathcal{I} = 5627 \text{ MeV} , \tag{4.2.50}$$

$$\overline{m}_{\Sigma_c} = M_{sol} + \overline{m}_D - \frac{3}{2}gF'(0) + 11/8\mathcal{I} = 2481 \text{ MeV} , \tag{4.2.51}$$

$$\overline{m}_{\Sigma_b} = M_{sol} + \overline{m}_B - \frac{3}{2}gF'(0) + 11/8\mathcal{I} = 5822 \text{ MeV} . \tag{4.2.52}$$

These are comparable with the experimental masses[19] of $\Lambda_b$ (5620 MeV), $\Sigma_c$ (2454 MeV), $\Sigma_c^*$ (2518 MeV), $\Sigma_b$ (5811 MeV) and $\Sigma_b^*$ (5833 MeV). Furthermore, with the Skyrme lagrangian (with the quartic term for stabilization), the wavefunction has a slope $F'(0) \sim -2ef_\pi \approx -700$ MeV near the origin, which implies $g \sim -0.6$. This is also consistent with the values given at the end of the previous subsection.

The role of light vector mesons in the description of the heavy-meson–soliton system was analyzed in Ref. 16. In fact, using effective heavy quark symmetric lagrangians that incorporate light vector mesons,[21,22] it was shown that the effect of these light degrees of freedom could be relevant. Within this scheme the extension of the light flavor group to SU(3) was also considered.[23]

Up to now, we have discussed how one can obtain the heavy baryon states containing a heavy quark, $\Sigma_Q$, $\Sigma_Q^*$ and $\Lambda_Q$, as heavy-meson–soliton bound states treated in the standard way: a heavy-meson–soliton bound state is first found and then quantized by rotating the whole system in the collective coordinate quantization scheme. This amounts to proceeding systematically in a decreasing order in $N_c$; i.e. in the first step only terms up to $N_c^0$ order are considered, in the next step terms of order $1/N_c$ are also taken into account, etc. In this way of proceeding, the heavy mesons first lose their quantum numbers (such as the spin and isospin), with only the grand spin preserved. The good quantum numbers are recovered when the whole system is quantized properly. An alternative approach was adopted in Ref. 9. In this approach, the soliton is first quantized to produce the light baryon states such as nucleons and $\Delta$'s with correct quantum numbers. Then, the heavy mesons with explicit spin and isospin are coupled to the light baryons to form heavy baryons as a bound state. Compared with the traditional one which is a "soliton body-fixed" approach, this approach may be interpreted as a "laboratory-frame" approach. It has been shown,[17] however, that both approaches lead to the same results in the heavy quark limit.

It should be stressed that in the heavy quark limit discussed so far one cannot account for the experimentally observed hyperfine splittings, like e.g. the $\Sigma_c^*$-$\Sigma_c$ mass difference. Another consequence of taking such limit is the existence of parity doublets in the spectrum of the low-lying excited states.[18,20] This follows from the fact that in the heavy quark limit the centrifugal barrier that would affect states with $\ell_{eff} > 0$ plays no role. It is clear that finite heavy quark mass corrections have to be taken into account in order to have a more realistic description of the heavy baryon properties in the present topological soliton framework. How to account for such corrections will be discussed in the following section.

## 4.3. Beyond the Heavy Quark Limit

In the previous section, we have limited ourselves to the heavy quark limit. Thus, heavy baryon masses have been computed to leading order in $1/m_Q$, that is to $O(m_Q^0)$. Here, we will consider the corrections implied by the use of finite heavy quark masses.

The $\Sigma_Q^*$-$\Sigma_Q$ mass difference due to the leading heavy quark symmetry breaking was first computed in Ref. 24 using the alternative method mentioned at the end of Sec. 4.2.2. As an illustration of the equivalence of the two approaches, we briefly discuss how the corresponding results can be obtained using the soliton body fixed approach described at length in that subsection. The leading order lagrangian in the derivative expansion that breaks the heavy quark symmetry is[10]

$$\mathcal{L}_1 = \frac{\lambda_2}{m_Q} \operatorname{Tr} \left[ \sigma^{\mu\nu} H \sigma_{\mu\nu} \bar{H} \right] , \qquad (4.3.53)$$

which leads to a $\Phi^*$-$\Phi$ mass difference

$$m_{\Phi^*} - m_\Phi = -\frac{8\lambda_2}{m_Q} . \qquad (4.3.54)$$

Assuming as in Sec. 4.2.2 that the radial functions are peaked strongly at the origin, the inclusion of this heavy quark symmetry breaking lagrangian implies that the equation of motion Eq. (4.2.22) gets an additional term. Namely, one obtains

$$i\,\dot{h} = g\,h\,\vec{A}\cdot\vec{\sigma} + \frac{2\lambda_2}{m_Q}\,\vec{\sigma}\cdot(h\,\vec{\sigma}) . \qquad (4.3.55)$$

One can now consider the last term as a perturbation and compute its effect on the $k = 1/2$ bound state. Since $\mathcal{L}_1$ breaks only the heavy quark spin symmetry the grand spin is still a good symmetry of the equation of motion. Thus, the eigenstates can be classified by the corresponding quantum numbers. Expanding in terms of the three possible basis states $\mathcal{K}^{(a)}_{\frac{1}{2}k_3}$ given in Eq. (4.2.29) the problem reduces to finding the solution of the secular equation

$$\sum_{b=1}^{3} \left(\mathcal{M}_{ab} + \delta\mathcal{M}_{ab}\right) \alpha_b = -\varepsilon\, \alpha_a \,,\qquad (4.3.56)$$

with $\mathcal{M}_{ab}$ given by Eq. (4.2.31) and

$$\delta\mathcal{M}_{ab} = \frac{2\lambda_2}{m_Q} \int d\Omega\, \mathrm{Tr}\left[\vec{\sigma}\cdot\left(\mathcal{K}^{(a)}_{\frac{1}{2}k_3}\,\vec{\sigma}\right)\bar{\mathcal{K}}^{(b)}_{\frac{1}{2}k_3}\right]\,.\qquad (4.3.57)$$

It turns out that up to first order in perturbation, the bound state energy remains unchanged while the corresponding eigenstate $\mathcal{K}^{bs}_{\frac{1}{2}k_3}$ is perturbed to

$$\mathcal{K}^{bs}_{\frac{1}{2}k_3} = \frac{1}{2}\left(1+3\,\kappa\right)\mathcal{K}^{(1)}_{\frac{1}{2}k_3} - \frac{\sqrt{3}}{2}\left(1-\kappa\right)\mathcal{K}^{(2)}_{\frac{1}{2}k_3}\,,\qquad (4.3.58)$$

with

$$\kappa = -\frac{\lambda_2}{m_Q}\frac{1}{gF'(0)}\,.\qquad (4.3.59)$$

The heavy baryons can be obtained by quantizing the heavy-meson–soliton bound state in the same way as explained in Sec. 4.2.2. It leads to the heavy baryon states of Eq. (4.2.41) with $|m\rangle_{bs}$ replaced by the perturbed state of Eq. (4.3.58). Due to the perturbation, the expectation value of $\vec{Q}$ defined by Eq. (4.2.36) with respect to the bound states does not vanish. In fact, one gets that the hyperfine constant is given by

$$c = 2\epsilon = -\frac{2\lambda_2}{m_Q}\frac{1}{gF'(0)}\,.\qquad (4.3.60)$$

With the help of Eq. (4.2.45), one can compute the expectation value of the collective hamiltonian (4.2.37)

$$m_{i,j} = M_{sol} + \varepsilon_{bs} + \frac{1}{2\mathcal{I}}\left((1-c)i(i+1) + cj(j+1) - ck(k+1) + \frac{3}{4}\right)\,.\qquad (4.3.61)$$

Thus, the $\Sigma_Q^*$-$\Sigma_Q$ mass difference is obtained as

$$m_{\Sigma_Q^*} - m_{\Sigma_Q} = \frac{3c}{2\mathcal{I}} = \frac{(m_\Delta - m_N)(m_{\Phi^*} - m_\Phi)}{4gF'(0)}\,,\qquad (4.3.62)$$

where Eqs. (4.3.54) and (4.3.60) together with the resulting expression for the $\Delta$-$N$ mass splitting in terms of the moment of inertia $\mathcal{I}$ have been used. Note that the mass splittings have the dependence on $m_Q$ and $N_c$ that agrees with the constituent quark model. The $\Phi^*$-$\Phi$ mass difference is of order $1/m_Q$ and the $\Delta$-$N$ mass difference is of order $1/N_c$. This implies that the $\Sigma_Q^*$-$\Sigma_Q$ mass difference is of order $1/(m_Q N_c)$. Substituting $gF'(0) = 417$ MeV, we obtain

$$m_{\Sigma_c^*} - m_{\Sigma_c} = 25\text{ MeV}\qquad\text{and}\qquad m_{\Sigma_b^*} - m_{\Sigma_b} = 8\text{ MeV}\,.\qquad (4.3.63)$$

The experimentally measured $\Sigma_c^*$-$\Sigma_c$ mass difference $\sim 64$ MeV is about three times larger than this Skyrme model prediction. Something similar happens in the case of the $\Sigma_b^*$-$\Sigma_b$ mass difference, the empirical value of which is $\sim 21$ MeV.

This failure to reproduce the observed hyperfine splittings naturally suggests the need for including additional heavy-spin violating terms, of higher order in derivatives. However, since there are many possible terms with unknown coefficients such a systematic perturbative approach turns out not to be very predictive. To overcome this problem some relativistic lagrangian models written in terms of the ordinary pseudoscalar and vector fields (rather than the heavy fluctuation field multiplet Eq. (4.2.8)) have been used. A typical model of this type which only includes pseudoscalar fields in the light sector is given by

$$\mathcal{L} = \mathcal{L}_l^{Sk} + D_\mu \Phi (D^\mu \Phi)^\dagger - m_\Phi^2 \Phi \Phi^\dagger - \frac{1}{2} \Phi^{*\mu\nu} \Phi_{\mu\nu}^{*\dagger} + m_{\Phi^*}^2 \Phi^{*\mu} \Phi_\mu^{*\dagger}$$

$$+ f_Q (\Phi A^\mu \Phi_\mu^{*\dagger} + \Phi_\mu^* A^\mu \Phi^\dagger) + \frac{g_Q}{2} \varepsilon^{\mu\nu\lambda\rho}(\Phi_{\mu\nu}^* A_\lambda \Phi_\rho^{*\dagger} + \Phi_\rho^* A_\lambda \Phi_{\mu\nu}^{*\dagger}) \, , \quad (4.3.64)$$

where $D_\mu \Phi = \partial_\mu \Phi - \Phi V_\mu^\dagger$, $\varepsilon_{0123} = +1$, and $f_Q$ and $g_Q$ are the $\Phi^* \Phi M$ and $\Phi^* \Phi^* M$ coupling constants, respectively. The field strength tensor is defined in terms of the covariant derivative $D_\mu \Phi_\nu^* = \partial_\mu \Phi_\nu^* - \Phi_\nu^* V_\mu^\dagger$ as

$$\Phi_{\mu\nu}^* = D_\mu \Phi_\nu^* - D_\nu \Phi_\mu^* \, , \quad (4.3.65)$$

and the vector $V_\mu$ and axial vector $A_\mu$ have been defined in Eq. (4.2.13). In principle, Eq. (4.3.64) has two independent coupling constants $f_Q$ and $g_Q$. However, in order to respect heavy quark symmetry they should be related to each other as[12]

$$\lim_{m_Q \to \infty} f_Q / 2m_{\Phi^*} = \lim_{m_Q \to \infty} g_Q = g \, , \quad (4.3.66)$$

where $g$ is the universal coupling constant appearing in Eq. (4.2.12). It should be noted that even to order $1/m_Q$, Eq. (4.3.64) leads to extra contributions to the hyperfine splittings.[25]

The interacting heavy-meson–soliton system described by the lagrangian Eq. (4.3.64) can be treated following a procedure similar to the one described at length in Sec. 4.2.2. It should be noted, however, that the need to treat the finite mass corrections non-perturbatively implies that departures from a $\delta$-like behaviour of the heavy meson radial wavefunctions should be taken into account. Thus, the equations of motions which describe the dynamics of the heavy mesons moving in the static soliton background field should be solved numerically. It turns out that, for a given value of $g$, the binding energies are somewhat smaller than the ones obtained in the heavy quark limit.[26] Concerning the hyperfine splittings, although the use of the effective lagrangian Eq. (4.3.64) leads to some improvement, it is not still sufficient to bring the predicted $\Sigma_Q^*$-$\Sigma_Q$ mass splitting into agreement with experiment. The prediction for such a splitting is actually correlated to those for the $\Sigma_Q - \Lambda_Q$ and $\Delta$-$N$ splittings according to

$$m_{\Sigma_Q^*} - m_{\Sigma_Q} = m_\Delta - m_N - \frac{3}{2}(m_{\Sigma_Q} - m_{\Lambda_Q}) \, . \quad (4.3.67)$$

This formula follows from Eq. (4.3.61), and depends only on the collective quantization procedure being used rather than on the detailed structure of the model. If $m_\Delta - m_N$ and $m_{\Sigma_c} - m_{\Lambda_c}$ are taken to agree with their empirical value, Eq. (4.3.67) predicts 42 MeV rather than the empirical value 64 MeV. In the case of the bottom baryons one gets 6 MeV to be compared to the empirical value 21 MeV. This means that, within the present quantization framework, it is not possible to exactly predict all the three mass differences appearing in Eq. (4.3.67). Thus, the goodness of the approach must be judged by looking at the overall predictions for the heavy baryon masses.

In this context, the study of possible excited states turns out to be of great interest. As already mentioned, in the heavy quark limit degenerate doublets of excited states are obtained. However, such limit implies that both the soliton and the heavy mesons are infinitely heavy sitting one on top of the other. It is evident that, due to the ignorance of any kinetic effects, this approximation is not expected to work well for the orbitally and/or radially excited states. In Ref. 27 the kinetic effects due to the finite heavy meson masses were estimated by approximating their static potentials by a quadratic form with the curvature determined at the origin. Such a harmonic oscillator approximation is valid only when the heavy mesons are sufficiently massive so that their motions are restricted to a very small range. The situation was somewhat improved in Ref. 20 by solving an approximate Schrödinger-like equation and incorporating the light vector mesons. In the context of the model defined by Eq. (4.3.64), in which only pseudoscalar degrees of freedom are present in the light sector, the exact solution of the equations of motion of the heavy meson bound states were first obtained in Ref. 28 and their collective coordinate quantization performed in Ref. 29. The typical resulting excitation spectra for the low-lying charm and bottom baryons obtained from these calculations (SM) are shown in Figs. 4.1 and 4.2, respectively.

For comparison, we also include in these figures the results of the quark model (QM) calculation of Ref. 30 (more recent quark model calculations[31] lead to qualitatively similar results), those resulting from naive extension[6] of the bound state approach to the strangeness (NSM) and the empirically known values[19] (EXP). Note that the excitation energies are taken with respect to the mass (also indicated in the figures) of the lowest $\Lambda_c$ and $\Lambda_b$, respectively. Finally, in order to see the impact of including the light vector mesons in the effective lagrangian, the excitation spectra resulting from the calculations of Ref. 32 (VMM) are also displayed.

In the case of the charm sector, we observe that the predictions for the absolute values of the ground state $\Lambda_c$ mass are similar in all soliton models calculations, and are in reasonable agreement with its empirical value and the QM prediction. As for the low lying spectra, we see that they are all qualitatively similar. From a more quantitative point of view, the SM version of the skyrmion models seems to provide a more accurate description of the splitting between the two lowest lying negative parity excited $\Lambda_c$ baryons, although the corresponding centroid is somewhat underestimated as compared with present experimental results. In any case, for these particular states the soliton models based on heavy quark symmetry certainly do better than the QM of Ref. 30 and the soliton calculation NSM. For the $\Sigma_c$ baryons, the predictions of the SM and VMM results are very similar with the main difference, with respect to the QM and NSM predictions, being the position of the second $1/2^-$ state. Concerning the bottom sector, looking at the absolute value of the ground state $\Lambda_b$, we clearly see that the NSM tends to grossly overestimate the bottom meson binding energy. In this sense, although as discussed below the inclusion of other effects might still be required, the soliton models based in heavy quark symmetry (SM and VMM) lead to predictions which are in much better agreement with the empirical values. As for the excitation spectra, we see that all the models predict a similar ordering of low-lying states. However, the only two excitation energies that can be compared with existing empirical data, i.e. those

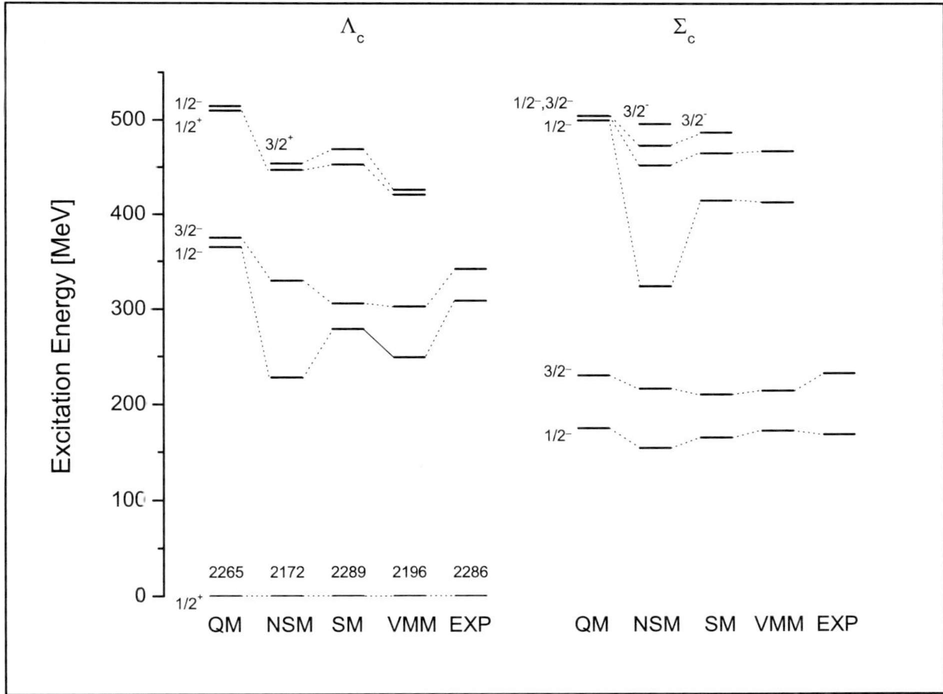


Fig. 4.1. Excitation spectra of charm baryons in soliton models as compared to the results of the quark model (QM) of Ref. 30 and the present empirical data[19] (EXP). NSM corresponds to the soliton model calculation of Ref. 6 where heavy quark symmetry has not been explicitly implemented. SM and VMM refer to soliton models which incorporate heavy quark symmetry. SM corresponds to a calculation[29] where only pseudoscalars have taken into account in the light sector, while VMM to the calculation of Ref. 32 where light vector mesons have been also explicitly included. The numbers above the lowest $\Lambda_c$ state correspond to the absolute masses (in MeV) of this state.

corresponding to the $\Sigma_b$ and $\Sigma_b^*$, are also much better reproduced by the SM and VMM results. It should be noticed that those models also predict rather small excitation energies ($\approx 200$ MeV) for the lowest lying negative $1/2^-$ and $3/2^-$ states as compared with the QM prediction (above 300 MeV).

Another kinetic correction that has to be taken into account is related to the recoil effects due to the finite soliton mass. This type of effect has been considered in several works.[20,28,32–34] As expected, they tend to decrease the heavy-meson–soliton binding energies leading to predictions which, particularly in the case of bottom baryons, are in better agreement with empirical data.

It should be mentioned that in the combined heavy quark and large $N_c$ limit a dynamical symmetry connecting excited heavy baryon states with the corresponding ground states exists.[43] Assuming that such symmetry holds as an approximate symmetry at finite values of $m_Q$ and $N_c$ one can develop an effective theory formulated in terms of the expansion parameter $\lambda \sim 1/m_Q, 1/N_c$. Within such scheme, up to next-to-leading order an average excitation energy of $\sim 300$ MeV is obtained for the first negative parity $\Lambda_b$ excited states. Such value is somewhat larger than the one obtained within heavy-meson–soliton bound state models, as it can be seen from Fig. 4.2.

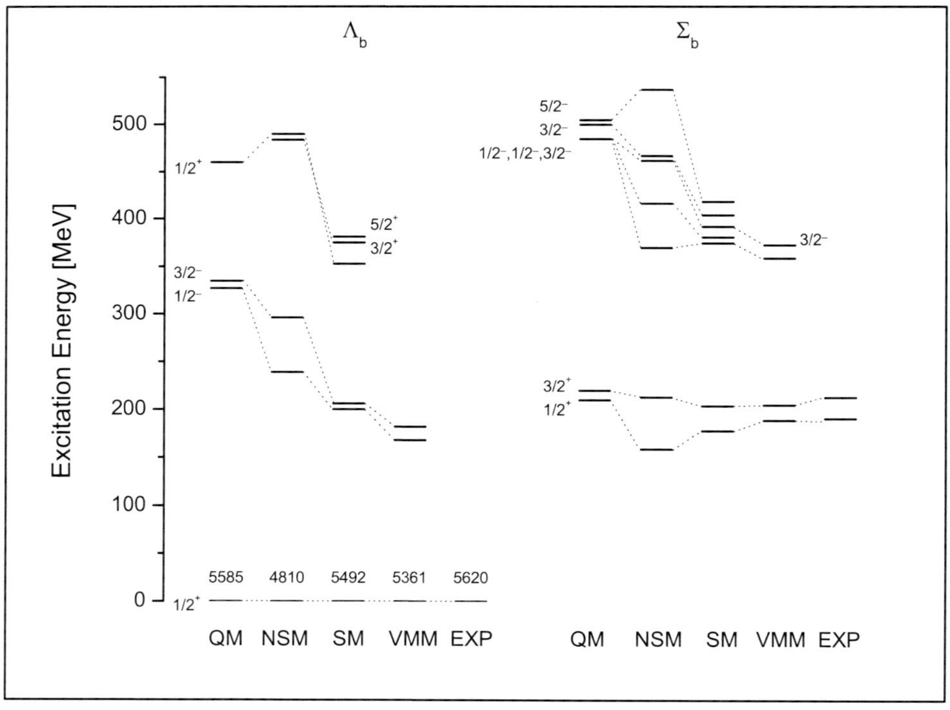

Fig. 4.2.   Excitation spectra of bottom baryons. Notation as in Fig. 4.1.

We conclude this section by mentioning that, in addition to the masses, other heavy baryon properties have been studied using the heavy-meson–soliton bound state picture. For example, magnetic moments have been analyzed in the heavy quark limit[35] and beyond it.[36] The radiative decays of excited $\Lambda_Q$ have been also considered.[37] Finally, the possible existence of multibaryons with heavy flavors[38,39] and other exotic states[40–42] have also been investigated.

## 4.4. Relation with the Bound State Approach to Strangeness

Thus far, we have discussed in detail a description of heavy baryons in which one begins from the heavy quark symmetry limit and then consider deviations from such a limit which start with order $1/m_Q$ corrections. However, as mentioned in the introduction, the picture proposed in Ref. 6 in which the heavy quark regime is approached from below, i.e. starting form a chiral invariant lagrangian and accounting for the heavy meson mass effects by the inclusion of suitable symmetry breaking terms, also turns out to be, at least qualitatively, successful. Therefore, it is interesting to see whether it is possible to find a dynamical scheme which allows to go continuously from the chiral regime to the heavy quark regime.

Suppose that one starts with three massless quarks, assuming the spontaneous breaking of chiral $SU(3)_L \times SU(3)_R$ down to the $SU(3)_V$ vector symmetry. The chiral field can be written as

$$U = \exp\left[\frac{i}{f_\pi}\begin{pmatrix} \pi^0 + \frac{1}{\sqrt{3}}\Psi & \sqrt{2}\pi^+ & \Phi^+ \\ \sqrt{2}\pi^- & -\pi^0 + \frac{1}{\sqrt{3}}\Psi & \Phi^0 \\ \Phi^- & \bar{\Phi}^0 & -\frac{2}{\sqrt{3}}\Psi \end{pmatrix}\right]. \tag{4.4.68}$$

Here, $\Phi^+$, $\Phi^0$, $\Phi^-$, $\bar{\Phi}^0$ and $\Psi$ denote the mesons with the quantum numbers of $\bar{h}\gamma_5 u$, $\bar{h}\gamma_5 d$, $\bar{u}\gamma_5 h$ and $\bar{d}\gamma_5 h$ and $\bar{u}\gamma_5 u + \bar{d}\gamma_5 d - 2\bar{h}\gamma_5 h$, respectively. For example, if $h=s$, they correspond to $K^+$, $K^0$, $K^-$, $\bar{K}^0$ and $\eta_8$. The effective action can be obtained by adding the Wess-Zumino term[44] $\Gamma_{WZ}$ to the lagrangian for interactions among the Goldstone bosons given by generalizing Eq. (4.2.1) to three flavors. Namely,

$$\Gamma = \int d^4x \, \mathcal{L}_l^{Sk} + \Gamma_{WZ}. \tag{4.4.69}$$

The Wess-Zumino term cannot be written as a local lagrangian density in $(3+1)$ dimensions. However, it can be expressed as a local action in five-dimensions,[45]

$$\Gamma_{WZ} = -\frac{iN_c}{240\pi^2}\int_{M_5} d^5x \, \varepsilon^{\mu\nu\rho\sigma\lambda} \, \mathrm{Tr}\left[U^\dagger\partial_\mu U U^\dagger\partial_\nu U U^\dagger\partial_\rho U U^\dagger\partial_\sigma U U^\dagger\partial_\lambda U\right], \tag{4.4.70}$$

where the integration is over a five-dimensional disk whose boundary is the ordinary space-time $M_4$ and $U$ is extended so that $U(\vec{r}, t, s = 0) = 1$ and $U(\vec{r}, t, s = 1) = U(\vec{r}, t)$. This term is non-vanishing for $N_f \geq 3$. When the soliton is built in $SU(2)$ space, this term does not contribute. However, we shall be considering $(2+1)$ flavors where one flavor can be heavy, in which case the dynamics can be influenced by the

Wess-Zumino term as in the Callan-Klebanov (CK) model.[3] What we are interested in is the situation where the symmetry $SU(3)_L \times SU(3)_R$ is *explicitly* broken to $SU(2)_L \times SU(2)_R \times U(1)$ by an $h$-quark mass, thereby making the $\Phi$-meson massive and its decay constant $f_\Phi$ different from that of the pion. These two symmetry breaking effects can be effectively incorporated into the lagrangian by a term of the form[6]

$$\mathcal{L}_{sb} = \frac{1}{6} f_\Phi^2 m_\Phi^2 \, \mathrm{Tr}[(1 - \sqrt{3}\lambda_8)(U + U^\dagger - 2)]$$

$$+ \frac{1}{12}(f_\Phi^2 - f_\pi^2) \, \mathrm{Tr}[(1 - \sqrt{3}\lambda_8)(U\partial_\mu U^\dagger \partial^\mu U + U^\dagger \partial_\mu U \partial^\mu U^\dagger)] \,, \qquad (4.4.71)$$

where, for simplicity, we turn off the light quark masses. The appropriate ansatz for the chiral field is the CK-type which we shall take in the form

$$U = N_\pi \, N_\Phi \, N_\pi \,, \qquad (4.4.72)$$

where $N_\pi = \mathrm{diag}\,(\xi, 1)$, with the $SU(2)$ matrix $\xi$ defined by Eq. (4.2.5), and

$$N_\Phi = \exp\left[ \frac{i\sqrt{2}}{f_\pi} \begin{pmatrix} \mathbf{0} & \Phi^\dagger \\ \Phi & 0 \end{pmatrix} \right] \,, \qquad (4.4.73)$$

with the $\Phi$-meson anti-doublets $\Phi = (\Phi^-, \bar{\Phi}^0)$ and doublets $\Phi^\dagger = (\Phi^+, \Phi^0)^T$.

Substituting the CK ansatz (4.4.72) into the action (4.4.69) with the symmetry breaking term (4.4.71) and expanding up to second order in the $\Phi$-meson field, we obtain

$$\mathcal{L} = \mathcal{L}_l^{Sk} + D_\mu \Phi (D_\mu \Phi)^\dagger - M_\Phi^2 \Phi \Phi^\dagger - \Phi A_\mu^\dagger A^\mu \Phi^\dagger - \frac{iN_c}{4f_P^2} B_\mu \left( D^\mu \Phi \Phi^\dagger - \Phi (D^\mu \Phi)^\dagger \right),$$

$$(4.4.74)$$

where we have rescaled the $\Phi$-meson fields as $\Phi/\kappa$ with $\kappa = f_\Phi/f_\pi$. The covariant derivative $(D_\mu \Phi)^\dagger$ is $(\partial_\mu + V_\mu)\Phi^\dagger$, the vector field $V_\mu$ and the axial-vector field $A_\mu$ are the same as in the lagrangian (4.3.64), and $B_\mu$ is the topological current

$$B^\mu = \frac{1}{24\pi^2} \varepsilon^{\mu\nu\lambda\rho} \, \mathrm{Tr} \left[ U^\dagger \partial_\nu U U^\dagger \partial_\lambda U U^\dagger \partial_\rho U \right] \,, \qquad (4.4.75)$$

which is the baryon number current in the Skyrme model.

With the identification $\Phi = K$, the lagrangian Eq. (4.4.74) has been successfully used in the strange sector. In fact, using the empirical values for $m_K$ and the $f_K/f_\pi$ ratio this lagrangian leads to a kaon-soliton bound state which allows for a very good description of the strange hyperon spectrum,[6] once an $SU(2)$ collective quantization similar to the one described in Sec. 4.2.2 is performed. Moreover, the existence of an excited $\ell = 0$ state provides a natural explanation for the negative parity $\Lambda(1405)$ hyperon.[3,46] The results displayed at the end of Sec. 4.3 (those labelled NSM in Figs. 1 and 2) show that the straightforward extension of this approach[5,6] leads to reasonable results in the charm sector, while it certainly fails to provide a quantitative good description of the bottom baryons. This clearly indicates that new explicit degrees of freedom have to be included in the effective lagrangian in order to have the correct heavy quark limit.

To proceed it is important to observe that, to the lowest order in derivatives on the Goldstone boson fields, Eq. (4.4.74) is the same as the lagrangian Eq. (4.3.64) when only the heavy pseudoscalars are considered. Furthermore, as argued in Refs. 47–49, as the $h$ quark mass increases above the chiral scale $\Lambda_\chi$, the Wess-Zumino term is expected to vanish, thereby turning off the last term of Eq. (4.4.74). Thus, the two lagrangians are indeed equivalent as far as the pseudoscalars are concerned. However, as discussed in the previous sections, in order to have the correct heavy quark limit one should explicitly take into account the heavy vector degrees of freedom, which become degenerate with the pseudoscalars as one approaches that limit. From an effective lagrangian point of view, the vector mesons can be viewed as "matter fields". There are several ways of introducing vector matter fields. Here we follow the hidden gauge symmetry (HGS) approach[50] in which case the non-anomalous effective lagrangian is

$$\mathcal{L}_0 = -\frac{f_\pi^2}{4}\,\mathrm{Tr}[\mathcal{D}_\mu\xi_L\xi_L^\dagger - \mathcal{D}_\mu\xi_R\xi_R^\dagger]^2 - a\frac{f_\pi^2}{4}\,\mathrm{Tr}[\mathcal{D}_\mu\xi_L\xi_L^\dagger + \mathcal{D}_\mu\xi_R\xi_R^\dagger]^2 - \frac{1}{2}\,\mathrm{Tr}(F_{\mu\nu}F^{\mu\nu}).$$

$$(4.4.76)$$

Here, $\mathcal{D}_\mu = \partial_\mu + ig_* U_\mu$ with

$$U_\mu = \frac{1}{2}\begin{pmatrix} \omega_\mu + \rho_\mu & \sqrt{2}\Phi_\mu^{*\dagger} \\ \sqrt{2}\Phi_\mu^* & \Psi_\mu^* \end{pmatrix}, \qquad (4.4.77)$$

and $g_*$ is a gauge coupling constant to be specified later. The field strength tensor of the vector mesons is $F_{\mu\nu} = \mathcal{D}_\mu U_\nu - \mathcal{D}_\nu U_\mu$, and the fields $\xi_L$ and $\xi_R$ are related to the chiral field by $U(x) = \xi_L^\dagger \xi_R$. The vector meson mass $M_{\rho,\omega}$ and the $\rho\pi\pi$ coupling constant can be read off from the lagrangian,

$$M_{\rho,\omega}^2 = ag_*^2 f_\pi^2; \qquad g_{\rho\pi\pi} = \frac{a}{2}g_*. \qquad (4.4.78)$$

The usual KSRF relation $m_\rho^2 = 2g_*^2 f_\pi^2$, and the universality of the vector-meson coupling $g_{\rho\pi\pi} = g_*$, can be used[50] to fix the arbitrary parameter $a$ to 2.

The effective action should satisfy the same anomalous Ward identities as does the underlying fundamental theory, QCD[44]. In the presence of vector mesons $A_{L,R}^\mu$ associated with the external (e.g. electroweak) gauge transformations, the general form of the anomalous lagrangian is given by a special solution of the anomaly equation plus general solutions of the homogeneous equation.[51] The former is the so-called gauged Wess-Zumino action $\Gamma_{WZ}^g$ (see e.g. Ref. 52 for details) and the latter, the anomaly free terms, can be made of four independent blocks $\mathcal{L}_i$ whose explicit forms can be found in Ref. 50. Thus, for the anomalous processes we have

$$\Gamma_{an} = \Gamma_{WZ}^g[\xi_L^\dagger\xi_R, A_L, A_R] + \sum_{i=1}^4 \gamma_i \int_{M_4} d^4x\,\mathcal{L}_i, \qquad (4.4.79)$$

with four arbitrary constants $\gamma_i$, which are determined by experimental data. Vector meson dominance (VMD) in processes like $\pi^0 \to 2\gamma$ and $\gamma \to 3\pi$ is very useful in determining these constants.

As for the symmetry breaking one can take the form[53]

$$\mathcal{L}_{sb} = -\frac{f_\pi^2}{4} \operatorname{Tr} \left\{ (\mathcal{D}_\mu \xi_L \xi_L^\dagger - \mathcal{D}_\mu \xi_R \xi_R^\dagger)^2 (\xi_R \varepsilon_A \xi_L^\dagger + \xi_L \varepsilon_A \xi_R^\dagger) \right\}$$
$$- \frac{a f_\pi^2}{4} \operatorname{Tr} \left\{ (\mathcal{D}_\mu \xi_L \xi_L^\dagger + \mathcal{D}_\mu \xi_R \xi_R^\dagger)^2 (\xi_R \varepsilon_V \xi_L^\dagger + \xi_L \varepsilon_V \xi_R^\dagger) \right\} \ . \quad (4.4.80)$$

The matrix $\varepsilon_{A(V)}$ is taken to be $\varepsilon_{A(V)} = \operatorname{diag}(0, 0, c_{A(V)})$, where $c_{A(V)}$ are the SU(3)-breaking real parameters to be determined. In terms of them one obtains

$$m_{\Phi^*}^2 = (1 + c_V)\, m_{\rho,\omega}^2, \qquad f_\Phi^2 = (1 + c_A)\, f_\pi^2 \ . \quad (4.4.81)$$

Finally, we substitute the CK ansatz Eq. (4.4.72), (that is, $\xi_L^\dagger = N_\pi \sqrt{U_\Phi}$ and $\xi_R = \sqrt{U_\Phi} N_\pi$) into the total effective action

$$\Gamma = \Gamma_0 + \Gamma_{an} + \Gamma_{sb} \ , \quad (4.4.82)$$

where $\Gamma_0$ and $\Gamma_{sb}$ are obtained from the lagrangians Eq. (4.4.76) and Eq. (4.4.80), respectively, and the action $\Gamma_{an}$ is given in Eq. (4.4.79). One may check that the resulting lagrangian contains all the terms of Eq. (4.3.64). Explicitly, one gets[17]

$$\mathcal{L} = \mathcal{L}_l^{Sk} + D_\mu \Phi D_\mu \Phi^\dagger - m_\Phi^2 \Phi \Phi^\dagger - \frac{1}{2} \Phi^{*\mu\nu} \Phi_{\mu\nu}^{*\dagger} + m_{\Phi^*}^2 \Phi^{*\mu} \Phi_\mu^{*\dagger}$$
$$- \sqrt{2} m_{\Phi^*} (\Phi A^\mu \Phi_\mu^{*\dagger} + \Phi_\mu^* A^\mu \Phi^\dagger) + \frac{i}{2} c_4 g_*^2 \varepsilon^{\mu\nu\lambda\rho} (\Phi_{\mu\nu}^* A_\lambda \Phi_\rho^{*\dagger} + \Phi_\lambda^* A_\rho \Phi_{\mu\nu}^{*\dagger}) + \cdots \ ,$$
$$(4.4.83)$$

where the light vector meson fields $\rho_\mu$ and $\omega_\mu$ have been replaced by $2i\, V_\mu/g_*$ and $(c_1 - c_2) i 6\pi^2 B_\mu / g_* f_\pi^2$, respectively, and terms with higher derivatives acting on the pion fields have not been explicitly written. Comparing Eq. (4.4.83) with Eq. (4.3.64), we obtain two relations

$$f_Q = -\sqrt{2} m_{\Phi^*}, \qquad \text{and} \qquad g_Q = i\gamma_4 g_*^2 \ . \quad (4.4.84)$$

The first relation implies that

$$\frac{f_Q}{2 m_{\Phi^*}} = -\frac{1}{\sqrt{2}} \ , \quad (4.4.85)$$

which is quite close to the expected heavy quark limit result Eq. (4.3.66) with $g = -0.75$ evaluated with the NRQM in Sec. 2. Using this relation and assuming that the VMD works in the heavy meson sector, in which case $\gamma_4 = i N_c / 16\pi^2$, one obtains $g_*$ in the heavy quark limit, i.e.

$$g_* = \sqrt{\frac{16\pi^2}{\sqrt{2} N_c}} \simeq 6 \qquad \text{(with } N_c = 3 \text{)} \ . \quad (4.4.86)$$

which is close to $g_* = g_{\rho\pi\pi}$ found in the light sector. These results seem to indicate that, in principle, it might be possible to construct an effective soliton model which could be used to describe both the strange sector and the heavier sectors. Of course, further work is definitely required in order to test in detail the feasibility of this ambitious program.

To conclude this section, we note that there is an alternative method[54] to describe strange hyperons within topological soliton models (for reviews see e.g. Ref. 55). That method is based on treating strange degrees of freedom as light and, thus, on the introduction of rotational $SU(3)$ collective quantization. It is clear that this treatment becomes better the closer one is to the limit $m_K \rightarrow 0$. It has been suggested,[56] however, that even in such a limit the bound state picture is applicable. In the present context this brings in the very interesting question concerning the possibility of having a unified framework that may allow to smoothly interpolate between the chiral symmetry limit and the heavy quark limit.

## 4.5. Summary and Conclusions

Heavy baryons represent an extremely interesting problem since they combine the dynamics of the heavy and light sectors of the strong interactions. In this contribution we have reviewed the work done on the description of heavy baryons as heavy-meson–soliton bound systems. We have first discussed how these bound systems can be obtained in the infinite heavy quark limit using effective lagrangians that respect both chiral symmetry and heavy quark symmetry. Next, we have shown how the effects due to finite heavy quark masses can be accounted for, and compared the resulting heavy baryon spectra with existing quark model and empirical results. This comparison indicates that, even though room for improvement is certainly left, the bound heavy-meson–soliton models are reasonably successful in reproducing those results. Finally, we have addressed some issues related to a possible connection between the usual bound state approach to strange hyperons and that for heavier baryons. We have shown that there are some indications that it might be possible to construct an effective soliton model which could be used to describe baryons formed by quarks of any flavor. Of course, further work is definitely required in order to test in detail the feasibility of this ambitious program. We finish by recalling that, although in recent years there has been an enormous progress in both the theoretical and experimental aspects of the heavy baryon physics, many problems still remain to be resolved. For example, most of the $J^P$ quantum numbers of the heavy baryons have not been yet determined experimentally, but are assigned on the basis of quark model predictions. In this sense, the insight obtained from alternative models such as the bound state soliton model discussed in the present contribution might be particularly useful.

## Acknowledgements

I would like to thank J.L. Goity, B-Y. Park, M. Rho and D.O. Riska for useful comments. This work was supported in part by CONICET (Argentina) grant # PIP 6084 and by ANPCyT (Argentina) grants # PICT04 03-25374 and # PICT07 03-00818.

## References


1. T.H.R. Skyrme, *Proc. R. Soc. London A* **260** (1961) 127; *Proc. R. Soc. London A* **262** (1961) 237; *Nucl. Phys.* **31** (1962) 556.
2. I. Zahed and G.E. Brown, *Phys. Rept.* **142** (1986) 1; G. Holzwarth and B. Schwesinger, *Rept. Prog. Phys.* **49** (1986) 825.
3. C.G. Callan and I. Klebanov, *Nucl. Phys. B* **262** (1985) 365.
4. N.N. Scoccola, H. Nadeau, M. Nowak and M. Rho, *Phys. Lett. B* **201** (1988) 425; C.G. Callan, K. Hornbostel and I. Klebanov, *Phys. Lett. B* **202** (1988) 269; U. Blom, K. Dannbom and D.O. Riska, *Nucl. Phys. A* **493** (1989) 384; N.N. Scoccola, D.P. Min, H. Nadeau and M. Rho, *Nucl. Phys. A* **505** (1989) 497.
5. M. Rho, D.O. Riska and N.N. Scoccola, *Phys. Lett. B* **251** (1990) 597.
6. D.O. Riska and N.N. Scoccola, *Phys. Lett. B* **265** (1991) 188; M. Rho, D.O. Riska and N.N. Scoccola, *Z. Phys. A* **341** (1992) 343.
7. Y. Oh, D.P. Min, M. Rho and N.N. Scoccola, *Nucl. Phys. A* **534** (1991) 493.
8. See e.g. the following monographs where references to the original work can be found: A.V. Manohar and M.B. Wise, *Camb. Monogr. Part. Phys. Nucl. Phys. Cosmol.* **10** (2000) 1; A.G. Grozin, *Springer Tracts Mod. Phys.* **201** (2004) 1.
9. E.E. Jenkins, A.V. Manohar and M.B. Wise, *Nucl. Phys. B* **396** (1993) 27; Z. Guralnik, M.E. Luke and A.V. Manohar, *Nucl. Phys. B* **390** (1993) 474.
10. M.B. Wise, *Phys. Rev.* **45** (1992) R2188.
11. G. Burdman and J.F. Donoghue, *Phys. Lett. B* **280** (1992) 287.
12. T.-M. Yan, H.-Y. Cheng, C.-Y.Cheung, G.-L. Lin, Y.C. Lin and H.-L. Yu, *Phys. Rev. D* **46** (1992) 1148.
13. J.L. Goity, *Phys. Rev. D* **46** (1992) 3929.
14. H. Georgi, *Phys. Lett. B* **240** (1990) 447; H. Georgi, in *Proc. of the Theoretical Advanced Study Institute*, eds. R.K. Ellis *et al.* (World Scientific, Singapore, 1992) and references therein.
15. A. Anastassov *et al.* [CLEO Collaboration], *Phys. Rev. D* **65** (2002) 032003.
16. K.S. Gupta, M. Arshad Momen, J. Schechter and A. Subbaraman, *Phys. Rev. D* **47** (1993) 4835.
17. D.P. Min, Y. Oh, B.Y. Park and M. Rho, *Int. J. Mod. Phys. E* **4** (1995) 47.
18. Y. Oh, B.Y. Park and D.P. Min, *Phys. Rev. D* **50** (1994) 3350.
19. C. Amsler *et al.* [Particle Data Group], *Phys. Lett. B* **667** (2008) 1.
20. J. Schechter and A. Subbaraman, *Phys. Rev. D* **51** (1995) 2311.
21. R. Casalbuoni, A. Deandrea, N. Di Bartolomeo, R. Gatto, F. Feruglio and G. Nardulli, *Phys. Lett. B* **292** (1992) 371; R. Casalbuoni, A. Deandrea, N. Di Bartolomeo, R. Gatto, F. Feruglio and G. Nardulli, *Phys. Rept.* **281** (1997) 145.
22. J. Schechter and A. Subbaraman, *Phys. Rev. D* **48** (1993) 332.
23. A. Momen, J. Schechter and A. Subbaraman, *Phys. Rev. D* **49** (1994) 5970.
24. E.E. Jenkins and A.V. Manohar, *Phys. Lett. B* **294** (1992) 273.

25. M. Harada, A. Qamar, F. Sannino, J. Schechter and H. Weigel, *Phys. Lett. B* **390** (1997) 329.
26. Y. Oh, B.Y. Park and D. P. Min, *Phys. Rev. D* **49** (1994) 4649.
27. C.K. Chow and M.B. Wise, *Phys. Rev. D* **50** (1994) 2135.
28. Y. Oh and B.Y. Park, *Phys. Rev. D* **51** (1995) 5016.
29. Y. Oh and B.Y. Park, *Phys. Rev. D* **53** (1996) 1605.
30. S. Capstick and N. Isgur, *Phys. Rev. D* **34** (1986) 2809.
31. S. Migura, D. Merten, B. Metsch and H.R. Petry, *Eur. Phys. J. A* **28** (2006) 41; H. Garcilazo, J. Vijande and A. Valcarce, *J. Phys. G* **34** (2007) 961; D. Ebert, R.N. Faustov and V.O. Galkin, *Phys. Lett. B* **659** (2008) 612; W. Roberts and M. Pervin, *Int. J. Mod. Phys. A* **23** (2008) 2817.
32. J. Schechter, A. Subbaraman, S. Vaidya and H. Weigel, *Nucl. Phys. A* **590** (1995) 655 [Erratum-ibid. **598** (1996) 583]; M. Harada, A. Qamar, F. Sannino, J. Schechter and H. Weigel, *Nucl. Phys. A* **625** (1997) 789.
33. Y. Oh and B.Y. Park, *Z. Phys. A* **359** (1997) 83.
34. T.D. Cohen and P.M. Hohler, *Phys. Rev. D* **75** (2007) 094007.
35. Y. Oh and B.Y. Park, *Mod. Phys. Lett. A* **11** (1996) 653.
36. S. Scholl and H. Weigel, *Nucl. Phys. A* **735** (2004) 163.
37. C.K. Chow, *Phys. Rev. D* **54** (1996) 3374.
38. C.L. Schat and N.N. Scoccola, *Phys. Rev. D* **61** (2000) 034008.
39. V.B. Kopeliovich and W.J. Zakrzewski, *Eur. Phys. J. C* **18** (2000) 369.
40. D.O. Riska and N.N. Scoccola, *Phys. Lett. B* **299** (1993) 338.
41. Y. Oh, B.Y. Park and D.P. Min, *Phys. Lett. B* **331** (1994) 362.
42. M. Bander and A. Subbaraman, *Phys. Rev. D* **50** (1994) 5478.
43. C.K. Chow and T.D. Cohen, *Phys. Rev. Lett.* **84** (2000) 5474; Z. Aziza Baccouche, C.K. Chow, T.D. Cohen and B.A. Gelman, *Phys. Lett. B* **514** (2001) 346; C.K. Chow and T.D. Cohen, *Nucl. Phys. A* **688** (2001) 842; C.K. Chow, T.D. Cohen and B. Gelman, *Nucl. Phys. A* **692** (2001) 521; Z. Aziza Baccouche, C.K. Chow, T.D. Cohen and B.A. Gelman, *Nucl. Phys. A* **696** (2001) 638.
44. J. Wess and B. Zumino, *Phys. Lett. B* **37** (1971) 95.
45. E. Witten, *Nucl. Phys. B* **223** (1983) 422; *Nucl. Phys. B* **223** (1983) 433.
46. C.L. Schat, N.N. Scoccola and C. Gobbi, *Nucl. Phys. A* **585** (1995) 627.
47. M.A. Nowak, M. Rho and I. Zahed, *Phys. Rev. D* **48** (1993) 4370.
48. H.K. Lee and M. Rho, *Phys. Rev. D* **48** (1993) 2329.
49. H.K. Lee, M.A. Nowak, M. Rho and I. Zahed, *Annals Phys.* **227** (1993) 175.
50. M. Bando, T. Kugo and K. Yamawaki, *Phys. Rept.* **164** (1988) 217.
51. T. Fujiwara, T. Kugo, H. Terao, S. Uehara and K. Yamawaki, *Prog. Theor. Phys.* **73** (1985) 926.
52. U.G. Meissner, *Phys. Rept.* **161** (1988) 213.
53. A. Bramon, A. Grau and G. Pancheri, *Phys. Lett. B* **345** (1995) 263.
54. H. Yabu and K. Ando, *Nucl. Phys. B* **301** (1988) 601.
55. H. Weigel, *Int. J. Mod. Phys. A* **11** (1996) 2419; *Lect. Notes Phys.* **743** (2008) 1.
56. D.B. Kaplan and I.R. Klebanov, *Nucl. Phys. B* **335** (1990) 45.

# Chapter 5

# Skyrmion Approach to Finite Density and Temperature

Byung-Yoon Park[*] and Vicente Vento[*,†]

*Department of Physics, Chungnam National University
Daejon 305-764, Korea
bypark@cnu.ac.kr
†Departament de Física Teòrica and Institut de Física Corpuscular
Universitat de València and Consejo Superior de Investigaciones Científicas
E-46100 Burjassot (València), Spain
Vicente.Vento@uv.es

We review an approach, developed over the past few years, to describe hadronic matter at finite density and temperature, whose underlying theoretical framework is the Skyrme model, an effective low energy theory rooted in large $N_c$ QCD. In this approach matter is described by various crystal structures of skyrmions, classical topological solitons carrying baryon number, from which conventional baryons appear by quantization. Chiral and scale symmetries play a crucial role in the dynamics as described by pion, dilaton and vector meson degrees of freedom. When compressed or heated skyrmion matter describes a rich phase diagram which has strong connections with the confinement/deconfinement phase transition.

## Contents

5.1   Introduction . . . . . . . . . . . . . . . . . . . . . . . . . . . . . . . . . .   116
5.2   Matter at Finite Density . . . . . . . . . . . . . . . . . . . . . . . . . . . .   119
       5.2.1   Skyrmion matter . . . . . . . . . . . . . . . . . . . . . . . . . . .   119
       5.2.2   Pions in skyrmion matter . . . . . . . . . . . . . . . . . . . . . .   123
5.3   Implementing Scale Invariance . . . . . . . . . . . . . . . . . . . . . . . .   125
       5.3.1   Dilaton dynamics . . . . . . . . . . . . . . . . . . . . . . . . . .   125
       5.3.2   Dynamics of the single skyrmion . . . . . . . . . . . . . . . . . .   127
       5.3.3   Dense skyrmion matter and chiral symmetry restoration . . . . . . . .   128
       5.3.4   Pions in a dense medium with dilaton dynamics . . . . . . . . . . . .   130
5.4   Skyrmion Matter at Finite Temperature . . . . . . . . . . . . . . . . . . . .   133
5.5   Vector Mesons and Dense Matter . . . . . . . . . . . . . . . . . . . . . . .   136
       5.5.1   Dynamics of the single skyrmion . . . . . . . . . . . . . . . . . . .   137
       5.5.2   Skyrmion matter: an FCC skyrmion crystal . . . . . . . . . . . . . .   138
       5.5.3   A resolution of the $\omega$ problem . . . . . . . . . . . . . . . . .   140
5.6   Conclusions . . . . . . . . . . . . . . . . . . . . . . . . . . . . . . . . .   142
References . . . . . . . . . . . . . . . . . . . . . . . . . . . . . . . . . . . . .   144

## 5.1. Introduction

An important issue at present is to understand the properties of hadronic matter under extreme conditions, e.g., at high temperature as in relativistic heavy-ion physics and/or at high density as in compact stars. The phase diagram of hadronic matter turns out richer than what has been predicted by perturbative Quantum Chromodynamics (QCD).[1] Two approaches have been developed thus far to discuss this issue: on the one hand, lattice QCD which deals directly with quark and gluon degrees of freedom, and on the other, effective field theories which are described in terms of hadronic fields. We shall describe in here a formalism for the second approach based on the topological soliton description of hadronic matter firstly introduced by Skyrme.[2,3]

Lattice QCD, the main computational tool accessible to highly nonperturbative QCD, has provided much information on the the finite temperature transition, such as the value of the critical temperature, the type of equation of state, etc.[4] However, due to a notorious 'sign problem', lattice QCD could not be applied to study dense matter. Only in the last few years, it has become possible to simulate QCD with small baryon density.[5] Chiral symmetry is a flavor symmetry of QCD which plays an essential role in hadronic physics. At low temperatures and densities it is spontaneously broken leading to the existence of the pion. Lattice studies seem to imply that chiral symmetry is restored in the high temperature and/or high baryon density phases and that it may go hand-in-hand with the confinement/deconfinement transition. The quark condensate $\langle \bar{q}q \rangle$ of QCD is an order parameter of this symmetry and decreases to zero when the symmetry is restored.

The Skyrme model is an effective low energy theory rooted in large $N_c$ QCD,[6,7] which we have applied to dense and hot matter studies.[8–15] The model does not have explicit quark and gluon degrees of freedom, and therefore one can not investigate the confinement/deconfinement transition directly, but we may study the chiral symmetry restoration transition which occurs close by. The schemes which aim at approaching the phase transition from the hadronic side are labelled 'bottom up' schemes. The main ingredient associated with chiral symmetry is the pion, the Goldstone boson associated with the spontaneously broken phase. The various patterns in which the symmetry is realized in QCD will be directly reflected in the in-medium properties of the pion and consequently in the properties of the skyrmions made of it.

The most essential ingredients of the Skyrme model are the pions, Goldstone bosons associated with the spontaneous breakdown of chiral symmetry. Baryons arise as topological solitons of the meson Lagrangian. The pion Lagrangian can be realized non-linearly as $U = \exp(i\vec{\tau} \cdot \vec{\pi}/f_\pi)$, which transforms as $U \to g_L U g_R^\dagger$ under the global chiral transformations $SU_L(N_f) \times SU_L(N_f)$; $g_L \in SU_L(N_f)$ and $g_R \in SU_R(N_f)$. Hereafter, we will restrict our consideration to $N_f = 2$. In the case of $N_f = 2$, the meson field $\pi$ represents three pions as

$$\vec{\tau} \cdot \vec{\pi} = \begin{pmatrix} \pi^0 & \sqrt{2}\pi^+ \\ \sqrt{2}\pi^- & -\pi^0 \end{pmatrix}. \tag{5.1.1}$$

The Lagrangian for their dynamics can be expanded in powers of the right and left invariant currents $R_\mu = U\partial_\mu U^\dagger$ and $L_\mu = U^\dagger \partial_\mu U$, which transforms as $R_\mu \to g_L R_\mu g_L^\dagger$ and $L_\mu \to g_R L_\mu g_R^\dagger$. The lowest order term is

$$\mathcal{L}_\sigma = \frac{f_\pi^2}{4} \mathrm{tr}(\partial_\mu U^\dagger \partial^\mu U). \tag{5.1.2}$$

Here, $f_\pi = 93$ MeV is the pion decay constant.

Throughout this paper, we take the following convention for the indices: (i) $a, b, \cdots = 1, 2, 3$ (Euclidean metric) for the isovector fields; (ii) $i, j, \cdots = 1, 2, 3$ (Euclidean metric) for the spatial components of normal vectors; (iii) $\mu, \nu, \cdots = 0, 1, 2, 3$ (Minkowskian metric) for the space-time 4-vectors; (iv) $\alpha, \beta, \cdots = 0, 1, 2, 3$ (Euclidean metric) for isoscalar(0)+ isovectors(1,2,3).

In the next order, one may find three independent terms consistent with Lorentz invariance, parity and $G$-parity as

$$\mathcal{L}_4 = \alpha \mathrm{tr}[L_\mu, L_\nu]^2 + \beta \mathrm{tr}\{L_\mu, L_\nu\}_+ + \gamma \mathrm{tr}(\partial_\mu L_\nu)^2. \tag{5.1.3}$$

In his original work,[2,3] Skyrme introduced only the first term to be denoted as

$$\mathcal{L}_{\mathrm{sk}} = \frac{1}{32e^2} \mathrm{tr}[L_\mu, L_\nu]^2, \tag{5.1.4}$$

which it is still second order in the time derivatives. The value of the "Skyrme parameter" may be evaluated by using $\pi\pi$ data. In the Skyrme model, it is also determined, for example, as $e = 5.45$ [16] to fit the nucleon-Delta masses, or as $e = 4.75$ [17] to fit the axial coupling constant of nucleon.

One may build up higher order terms with more and more phenomenological parameters. However, this naive derivative expansion leads to a Lagrangian which has an excessive symmetry; that is, it is invariant under $U \leftrightarrow U^\dagger$, which is not a genuine symmetry of QCD. To break it, we need the Wess-Zumino-Witten term.[18] The corresponding action can be written locally as

$$S_{WZW} = -\frac{iN_c}{240\pi^2} \int d^5 x \varepsilon^{\mu\nu\lambda\rho\sigma} \mathrm{tr}(L_\mu L_\nu \cdots L_\sigma), \tag{5.1.5}$$

i.e. in a five dimensional space whose boundary is the ordinary space and time. For $N_f = 2$ this action vanishes trivially, but for $N_f = 3$ it provides a hypothesized process $KK \to \pi^+\pi^0\pi^-$. When the action is $U(1)$ gauged for the pions to interact with photons, this term plays a nontrivial role even with two flavors.

Chiral symmetry is explicitly broken by the quark masses, which provides the masses to the Goldstone bosons. The mass term can be incorporated in the same way as chiral symmetry is broken in QCD; that is,

$$\mathcal{L}_{\mathrm{m}} = \frac{f_\pi^2 m_\pi^2}{4} \mathrm{tr}((U + U^\dagger - 2)) \sim -\frac{\langle \bar{q}q \rangle}{4} \mathrm{tr}(\mathcal{M}(U + U^\dagger - 2)), \tag{5.1.6}$$

where

$$\mathcal{M} = \begin{pmatrix} m_u & 0 \\ 0 & m_d \end{pmatrix}. \tag{5.1.7}$$

We neglect the u- and d-quark mass difference.

The approach has been generalized to more sophisticated meson Lagrangians which are constructed by implementing the symmetries of QCD.[19] The scale dilaton has been incorporated into the effective scheme to describe in hadronic language the scale anomaly.[20,21] The vector mesons $\rho$ and $\omega$ with masses $m_{\rho,\omega} \sim 780$ MeV can be incorporated into the Lagrangian by using the hidden local symmetry (HLS)[22] and guided by the matching of this framework to QCD in what is called 'vector manifestation' (VM).[23] We shall discuss these generalizations, when required in the discussion of skyrmion matter, later on.

The classical nature of skyrmions enables us to construct a dense system quite conveniently by putting more and more skyrmions into a given volume. Then, skyrmions shape and arrange themselves to minimize the energy of the system. The ground state configuration of skyrmion matter are crystals. At low density it is made of well-localized single skyrmions.[24] At a critical density, the system undergoes a structural phase transition to a new kind of crystal. It is made of 'half-skyrmions' which are still well-localized but carry only half winding number. In the half-skyrmion phase, the system develops an additional symmetry which leads to a vanishing average value of $\sigma = \frac{1}{2} Tr(U)$, the normalized trace of the $U$ field.[25] In the studies of the late 80's,[26] the vanishing of this average value $\langle \sigma \rangle$ was interpreted as chiral symmetry restoration by assuming that $\langle \sigma \rangle$ is related to the QCD order parameter $\langle \bar{q}q \rangle$. However, in Ref. 8, it was shown that the vanishing of $\langle \sigma \rangle$ cannot be an indication of a genuine chiral symmetry restoration, because the decay constant of the pion fluctuating in such a half-skyrmion matter does not vanish. This was interpreted as a signal of the appearance of a pseudogap phase similar to what happens in high $T_c$ superconductors.[27]

The puzzle was solved in Ref. 11 by incorporating a suitable degree of freedom, the dilaton field $\chi$, associated to the scale anomaly of QCD. The dilaton field takes over the role of the order parameter for chiral symmetry restoration. As the density of skyrmion matter increases, both $\langle \sigma \rangle$ and $\langle \chi \rangle$ vanish (not necessarily at the same critical density). The effective decay constant of the pion fluctuation vanishes only when $\langle \chi \rangle$ becomes zero. It is thus the dilaton field which provides the mechanism for chiral symmetry restoration.

Contrary to lattice QCD, there are few studies on the temperature dependence of skyrmion matter. Skyrmion matter has been heated up to melt the crystal into a liquid to investigate the crystal-liquid phase transition,[28,29] a phenomenon which is irrelevant to the restoration of chiral symmetry. We have studied skyrmion matter at *finite density and temperature* and have obtained the phase diagram describing the realization of the chiral symmetry.[15]

The contents of this review are as follows. Section 5.2 deals with the history of skyrmion matter and how our work follows from previous investigations. We also study the pion properties inside skyrmion matter at finite density. To confront the results with reality, in Sec. 5.3 we show that the scale dilaton has to be incorporated and we discuss how the properties of the pion change thereafter. Section 5.4 is devoted to the study of the temperature dependence and the description of the phase diagram. In Sec. 5.5 we incorporate vector mesons to the scheme and discuss the problem that arises due to the coupling of the $\omega$ meson and our solution to it. Finally the last section is devoted to a summary of our main results and to some conclusions we can draw from our study.

## 5.2. Matter at Finite Density

### 5.2.1. *Skyrmion matter*

The Skyrme model describes baryons, with arbitrary baryon number, as static soliton solutions of an effective Lagrangian for pions.[2,3] The model has been used to describe not only single baryon properties,[16,30] but also has served to derive the nucleon-nucleon interaction,[3,31] the pion-nucleon interaction,[32] properties of light nuclei and of nuclear matter. In the case of nuclear matter, most of the developments[24,25,33–35] done in late 80's involve a crystal of skyrmions.

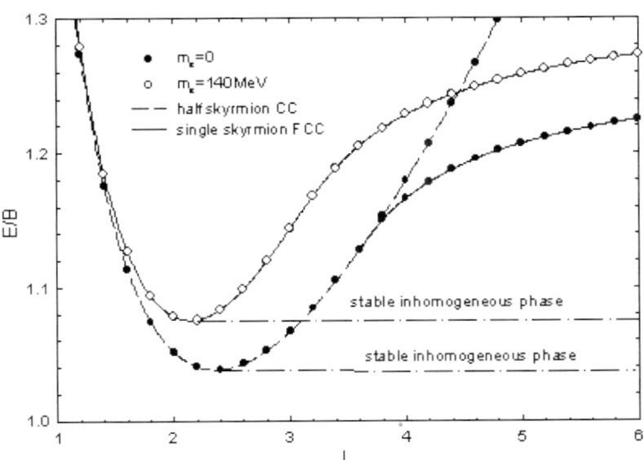


Fig. 5.1. Energy per single skyrmion as a function of the size parameter $L$. The solid circles show the results for massless pions and the open circles are those for massive pions. Note the rapid phase transition around $L \sim 3.8$ for massless pions.

The first attempt to understand the dense skyrmion matter was made by Kutchera *et al.*[36] These authors proceeded by introducing a single skyrmion into a spherical Wigner-Seitz cell without incorporating explicit information on the interaction. The presently considered conventional approaches were developed later. In them one assumes that the skyrmions form a crystal with a specific symmetry and then performs numerical simulations using this symmetry as a constraint. The first guess at this symmetry was made by Klebanov.[24] He considered a system where the skyrmions are located in the lattice site of a cubic crystal (CC) and have relative orientations in such a way that the pair of nearest neighbors attract maximally. Goldhaber and Manton [25] suggested that contrary to Klebanov's findings, the high density phase of skyrmion matter is to be described by a body-centered crystal (BCC) of half skyrmions. This suggestion was confirmed by numerical calculations.[26] Kugler and Shtrikman,[37] using a variational method, investigated the ground state of the skyrmion crystal including not only the single skyrmion CC and half-skyrmion BCC but also the single skyrmion face-centered-cubic crystal (FCC) and half-skyrmion CC. In their calculation a phase transition from the single FCC to half-skyrmion CC takes place and the ground state is found in the half-skyrmion CC configuration. Castillejo *et al.*[33] obtained similar conclusions.

In Fig. 5.1 we show the energy per baryon $E/B$ as a function of the FCC box size parameter $L$.[a] Each point in the figure denotes a minimum of the energy for the classical field configuration associated with the Lagrangians (5.1.2), (5.1.4) and (5.1.6) for a given value of $L$. The solid circles correspond to the zero pion mass calculation and reproduce the results of Kugler and Shtrikman.[34] The quantities $L$ and $E/B$, appearing in the figure, are given in units of $(ef_\pi)^{-1}$ ($\sim 0.45$fm with $f_\pi = 93$ MeV and $e = 4.75$) and $E/B$ in units of $(6\pi^2 f_\pi)/e$ ($\sim 1160$ MeV), respectively. The latter enable us to compare the numerical results on $E/B$ easily with its Bogolmoln'y bound for the skyrmion in the chiral limit, which can be expressed as $E/B = 1$ in this convention.

In the chiral limit, as we squeeze the system from $L = 6$ to around $L = 3.8$, one sees that the skyrmion system undergoes a phase transition from the FCC single skyrmion configuration to the CC half-skyrmion configuration. The system reaches a minimum energy configuration at $L = L_{\min} \sim 2.4$ with the energy per baryon $E/B \sim 1.038$. This minimum value is close to the Bogolmoln'y bound for the energy associated to Eqs. (5.1.2) and (5.1.4).

On the other hand, the configuration found at $L > L_{\min}$ with the constrained symmetry may not be the genuine low energy configuration of the system for that given $L$. Note that the pressure $P \equiv \partial E/\partial V$ is negative, which implies that the system in that configuration is unstable. Some of the skyrmions may condense to form dense lumps in the phase leaving large empty spaces forming a stable inhomogeneous as seen in Fig. 5.1 for $L = L_{\min}$. Only the phase to the left of the minimum, $L < L_{\min}$, may be referred to as "*homogeneous*" and there the background field is described by a crystal configuration.

The open circles are the solutions found with a nonvanishing pion mass, $m_\pi = 140$ MeV.[b] Comparing to the skyrmion system for massless pions, the energy per

---

[a] A single FCC is a cube with a side length $2L$, so that there are 4 single skyrmions in a volume of $8L^3$, that is, the baryon number density is related to $L$ as $\rho_B = 1/2L^3$.

[b] Incorporating the pion mass into the problem introduces a new scale in the analysis and therefore we are forced to give specific values to the parameters of the chiral effective Lagrangian, the pion decay constant and the Skyrme parameter, a feature which we have avoided in the chiral limit.

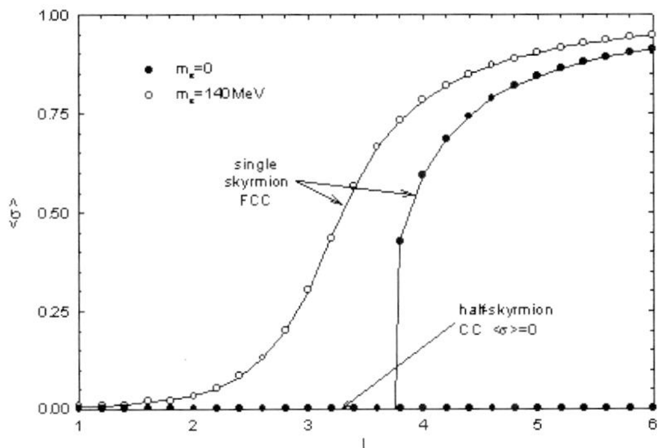

Fig. 5.2. $\langle\sigma\rangle$ as a function of the size parameter $L$. The notation is the same as in Fig. 5.1.

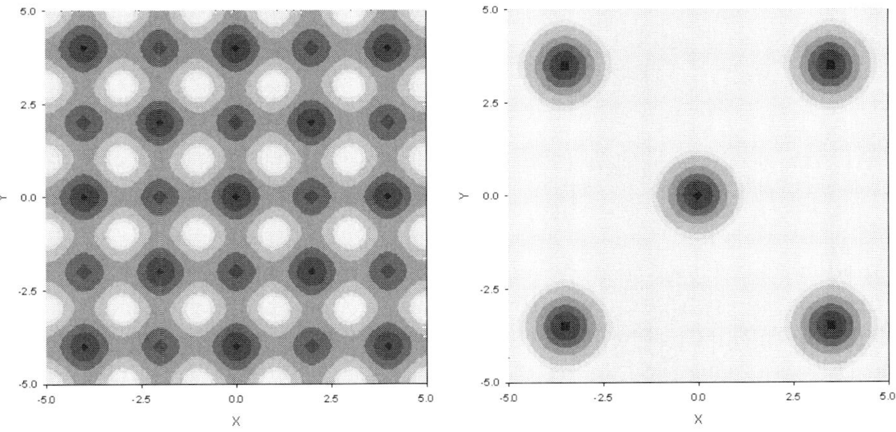

Fig. 5.3. Local baryon number densities at $L = 3.5$ and $L = 2.0$ with massive pions. For $L = 2.0$ the system is (almost) a half-skyrmion in a CC crystal configuration.

baryon is somewhat higher. Furthermore, there is no first order phase transition at low densities.

In Fig. 5.2, we show $\langle\sigma\rangle$, i.e. the space average value of $\sigma$ as a function of $L$. In the chiral limit, $\langle\sigma\rangle$ rapidly drops as the system shrinks and reaches zero at $L \sim 3.8$, where the system goes to the half-skyrmion phase. This phase transition was interpreted[38] as a signal for chiral symmetry restoration. However, as we sall see in the next section, this is not the expected transition. In the case of massive pions, the transition in $\langle\sigma\rangle$ is soft. Its value decreases monotonically and reaches

---

In order to proceed, we simply take their empirical values, that is, $f_\pi = 93$ MeV and $e = 4.75$. Although the numerical results depend on these values, their qualitative behavior will not.

zero asymptotically, as the density increases. Furthermore, as we can see in Fig. 5.3, where the local baryon number density is shown, for $L = 2$ (left) and $L = 3.5$ (right) in the $z = 0$ plane, the system becomes a half-skyrmion crystal at high density.

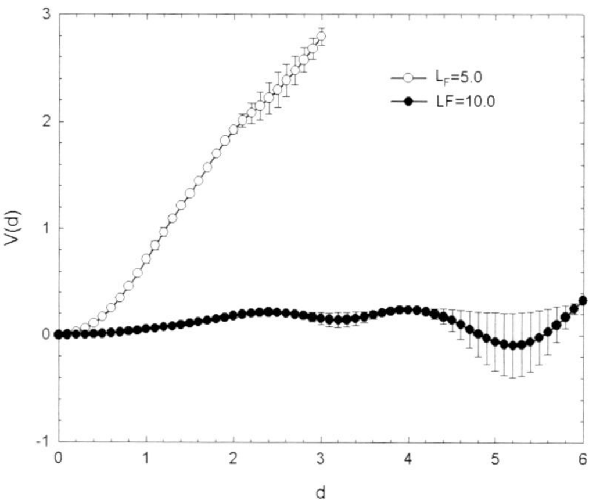


Fig. 5.4.   The energy cost to shift a single skyrmion from its stable position by an amount $d$ in the direction of the $z$-axis.

Another scheme used to study multi-skyrmion systems is the procedure based on the Atiyah-Manton Ansatz.[39] In this scheme, skyrmions of baryon number $N$ are obtained by calculating the holonomy of Yang-Mills instantons of charge $N$. This Ansatz has been successful in describing few-nucleon systems.[40–42] This procedure has been also applied to nuclear matter with the instanton solution on a four torus.[35] The energy per baryon was found to be $(E/B)_{min} = 1.058$ at $L_{\min} = 2.47$, which is comparable to the variational result of Kugler and Shtrikman.[37]

In Ref. 43, the Atiyah-Maton Ansatz is employed to get skyrmion matter from the 't Hooft's multi-instanton solution, which is modified to incorporate dynamical variables such as the positions and relative orientations of the single skyrmions. This description provides information on the dynamics of a single skyrmion in skyrmion matter. Shown in Fig. 5.4 is the energy change of the system when a single skyrmion is shifted from its FCC lattice site by an amount $d$ in the direction of the z-axis. Two extreme cases are shown. In the case of a dense system ($L_F \equiv 2L = 5.0$), the energy changes abruptly. For small $d$, it is almost quadratic in $d$. It implies that the dense system is in a crystal phase. On the other hand, in the case of a dilute system ($L_F = 10.0$), the system energy remains almost constant up to some large $d$, which implies that the system is in a gas (or liquid) phase. If we let all the variables vary freely, the system will prefer to change to a disordered or inhomogeneous phase in which some skyrmions will form clusters, as we have discussed before.

### 5.2.2. *Pions in skyrmion matter*

The Skyrme model also provides the most convenient framework to study the pion properties in dense matter. The basic strategy is to take the static configuration $U_0(\vec{x})$ discussed in Sec. 5.2.1 as the background fields and to look into the properties of the pion fluctuating on top of it. This is the conventional procedure used to find single particle excitations when one has solitons in a field theory.[44]

The fluctuating *time-dependent* pion fields can be incorporated on top of the static fields through the Ansatz[45]

$$U(\vec{x}, t) = \sqrt{U_\pi} U_0(\vec{x}) \sqrt{U_\pi},$$
(5.2.8)

where

$$U_\pi = \exp\left( i\vec{\tau} \cdot \vec{\phi}(x)/f_\pi \right),$$
(5.2.9)

with $\vec{\phi}$ describing the fluctuating pions.

When $U_0(\vec{r}) = 1(\rho_B = 0)$, the expansion in power of $\phi$'s leads us to

$$\mathcal{L}(\phi) = \frac{1}{2}\partial_\mu \phi^a \partial^\mu \phi^a + \frac{1}{2}m_\pi^2 \sigma(\vec{x})\phi^a \phi^a + \cdots,$$
(5.2.10)

which is just a Lagrangian for the self-interacting pion fields without any interactions with baryons. Here, we have written only the kinetic and mass terms relevant for further discussions. With a non trivial $U_0(\vec{r})$ describing dense skyrmion matter, the Lagrangian becomes,

$$\mathcal{L} = \frac{1}{2}G^{ab}(\vec{x})\partial_\mu \phi_a \partial^\mu \phi_b + \frac{1}{2}m_\pi^2 \sigma(\vec{x})\phi^a \phi^a + \cdots,$$
(5.2.11)

with

$$G^{ab}(\vec{x}) = \sigma^2 \delta_{ab} + \pi_a \pi_b.$$
(5.2.12)

The structure of our Lagrangian is similar to that of chiral perturbation theory Eq. (5.2.13) of Refs. 46 and 47. These authors start with a Lagrangian containing all the degrees of freedom, including nucleon fields, and free parameters. They integrate out the nucleons in and out of an *à priori* assumed Fermi sea and in the process they get a Lagrangian density describing the pion in the medium. Their result corresponds to the above Skyrme Lagrangian except that the quadratic (current algebra) and the mass terms pick up a density dependence of the form

$$-\frac{f_\pi^2}{4}\left( g^{\mu\nu} + \frac{D^{\mu\nu}\rho}{f_\pi^2} \right) \mathrm{Tr}(U^\dagger \partial_\mu U U^\dagger \partial^\nu U) + \frac{f_\pi^2 m_\pi^2}{4}\left( 1 - \frac{\Sigma_{\pi N}}{f_\pi^2 m_\pi^2}\rho \right) \mathrm{Tr}(U + U^\dagger - 2),$$
(5.2.13)

where $\rho$ is the density of the nuclear matter and $D^{\mu\nu}$ and $\sigma$ are physical quantities obtained from the pion-nucleon interactions. Note that in this scheme, nuclear matter is assumed *ab initio* to be a Fermi sea devoid of the intrinsic dependence mentioned above.

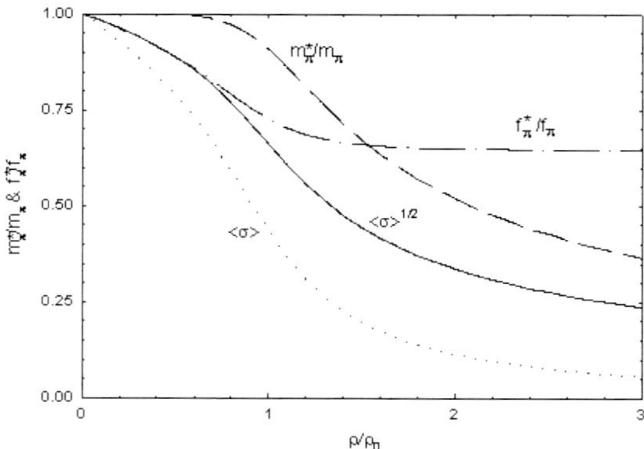

Fig. 5.5.   Estimates of $f_\pi^*/f_\pi$ and $m_\pi^*/m_\pi$ as functions of the baryon number density of skyrmion matter.

We proceed via a mean field approximation consisting in averaging the background modifications $G^{ab}(\vec{x})$ and $\sigma(\vec{x})$ appearing in the Lagrangian which are reduced to constants, $\langle G^{ab}\rangle = G\delta_{ab}$ and $\langle\sigma\rangle$. Then, the Lagrangian can be rewritten as

$$\mathcal{L}(\phi^*) = \frac{1}{2}\partial_\mu\phi_a^*\partial^\mu\phi_a^* + \frac{1}{2}m_\pi^*\phi_a^*\phi_a^* + \cdots, \qquad (5.2.14)$$

where we have carried out a wavefunction renormalization, $\phi_a^* = \sqrt{G}\phi_a$, which leads to a medium modified pion decay constant and mass as

$$\frac{f_\pi^*}{f_\pi} = \sqrt{G}, \qquad (5.2.15)$$

$$\frac{m_\pi^*}{m_\pi} = \frac{\langle\sigma\rangle}{\sqrt{G}}. \qquad (5.2.16)$$

In Fig. 5.5 we show the estimates of $f_\pi^*/f_\pi$ and $m_\pi^*/m_\pi$ as a function of the density. As the density increases, $f_\pi^*$ decreases only to $\sim 0.65 f_\pi$ and then it remains constant at that value. Our result is different from what was the general believe[38]: *the vanishing of $\langle\sigma\rangle$ is not an indication of chiral symmetry restoration since the pion decay constant does not vanish.*

Note that $\langle\sigma\rangle^{\frac{1}{2}}$ has the same slope at low densities, which leads to $m_\pi^*/m_\pi \sim 1$ at low densities. Since at higher densities $G$ becomes a constant, $m_\pi^*/m_\pi$ decreases like $\langle\sigma\rangle^{1/2}$ with a factor which is greater than 1. As the density increases, higher order terms in $\rho$ come to play important roles and $m_\pi^*/m_\pi$ decreases. A more rigorous derivation of these quantities can be obtained using perturbation theory.[8]

The slope of $\langle\sigma\rangle$ at low density is approximately $1/3$. If we expand $\langle\sigma\rangle$ about $\rho = 0$ and compare it with Eq. (5.2.13), we obtain

$$\langle\sigma\rangle \sim 1 - \frac{1}{3}\frac{\rho}{\rho_0} + \cdots \sim 1 - \frac{\Sigma_{\pi N}}{f_\pi^2 m_\pi^2}\rho + \cdots, \qquad (5.2.17)$$

which yields $\Sigma_{\pi N} \sim m_\pi^2 f_\pi^2/(3\rho_0) \sim 42$ MeV, which is comparable with the experimental value 45 MeV.[c] This comparison is fully justified from the point of view of the $\frac{1}{N}$ expansion since both approaches should produce the same result to leading order in this expansion. The liner term is $\mathcal{O}(1)$.

The length scale is strongly dependent on our choice of the parameters $f_\pi$ and $e$. Thus one should be aware that the $\rho$ scale in Fig. 5.5 could change quantitatively considerably if one chooses another parameter set, however the qualitative behavior will remain unchanged.

Note that the density dependence of the background is taken into account to all orders. No low-density approximation, whose validity is in doubt except at very low density, is ever made in the calculation. The power of our approach is that the dynamics of the background and its excitations can be treated in a unified way on the same footing with a single Lagrangian.

## 5.3. Implementing Scale Invariance

### 5.3.1. *Dilaton dynamics*

The dynamics introduced in Sec. 5.1 as an effective theory for the hadronic interactions is probably incomplete. In fact, it is not clear that the intrinsic density dependence required by the matching to QCD is fully implemented in the model. One puzzling feature is that the Wigner phase represented by the half-skyrmion matter with $\langle\sigma\rangle = 0$ supports a non-vanishing pion decay constant. This may be interpreted as a possible signal for a pseudogap phase. However, at some point, the chiral symmetry should be restored and there the pion decay constant should vanish.

This difficulty can be circumvented in our framework by incorporating in the standard skyrmion dynamics the trace anomaly of QCD in an effective manner.[20] The end result is the skyrmion Lagrangian introduced by Ellis and Lanik[21] and employed by Brown and Rho[49] for nuclear physics which contains an additional scalar field, the so called scale dilaton.

The classical QCD action of scale dimension 4 in the chiral limit is invariant under the scale transformation

$$x \to {}^\lambda x = \lambda^{-1}x, \quad \lambda \geq 0, \tag{5.3.18}$$

under which the quark field and the gluon fields transform with the scale dimension 3/2 and 1, respectively. The quark mass term of scale dimension 3 breaks scale invariance. At the quantum level, scale invariance is also broken by dimensional transmutation even for massless quarks, as signaled by the non-vanishing of the trace of the energy-momentum tensor. Equivalently, this phenomenon can be formulated

---

by the non-vanishing divergence of the dilatation current $D_\mu$, the so called trace anomaly,

$$\partial^\mu D_\mu = \theta^\mu_\mu = \sum_q m_q \bar{q}q + \frac{\beta(g)}{g} \mathrm{Tr} G_{\mu\nu} G^{\mu\nu}, \tag{5.3.19}$$

where $\beta(g)$ is the beta function of QCD.

Broken scale invariance can be implemented into large $N_c$ physics by modifying the standard skyrmion Lagrangian, introduced in Sec. 5.1, to

$$\mathcal{L} = \frac{f_\pi^2}{4} \left(\frac{\chi}{f_\chi}\right)^2 \mathrm{Tr}(\partial_\mu U^\dagger \partial^\mu U) + \frac{1}{32e^2} \mathrm{Tr}([U^\dagger \partial_\mu U, U^\dagger \partial_\nu U])^2$$

$$+ \frac{f_\pi^2 m_\pi^2}{4} \left(\frac{\chi}{f_\chi}\right)^3 \mathrm{Tr}(U + U^\dagger - 2)$$

$$+ \frac{1}{2}\partial_\mu \chi \partial^\mu \chi - \frac{1}{4}\frac{m_\chi^2}{f_\chi^2}\left[\chi^4\left(\ln(\chi/f_\chi) - \frac{1}{4}\right) + \frac{1}{4}\right]. \tag{5.3.20}$$

We have denoted the non vanishing vacuum expectation value of $\chi$ as $f_\chi$, a constant which describes the decay of the scalar into pions. The second term of the trace anomaly (5.3.19) can be reproduced by the potential energy $V(\chi)$, which is adjusted in the Lagrangian (5.3.20) so that $V = dV/d\chi = 0$ and $d^2V/d\chi^2 = m_\chi^2$ at $\chi = f_\chi$.[20]

The vacuum state of the Lagrangian at zero baryon number density is defined by $U = 1$ and $\chi = f_\chi$. The fluctuations of the pion and the scalar fields about this vacuum, defined through

$$U = \exp(i\vec{\tau} \cdot \vec{\phi}/f_\pi), \quad \text{and} \quad \chi = f_\chi + \tilde{\chi} \tag{5.3.21}$$

give physical meaning to the model parameters: $f_\pi$ as the pion decay constant, $m_\pi$ as the pion mass, $f_\chi$ as the scalar decay constant, and $m_\chi$ as the scalar mass. For the pions, we use their empirical values as $f_\pi = 93\,\mathrm{MeV}$ and $m_\pi = 140\,\mathrm{MeV}$. We fix the Skyrme parameter $e$ to 4.75 from the axial-vector coupling constant $g_A$ as in Ref. 50. However, for the scalar field $\chi$, no experimental values for the corresponding parameters are available.

In Ref. 51, the scalar field is incorporated into a relativistic hadronic model for nuclear matter not only to account for the anomalous scaling behavior but also to provide the mid-range nucleon-nucleon attraction. Then, the parameters $f_\chi$ and $m_\chi$ are adjusted so that the model fits finite nuclei. One of the parameter sets is $m_\chi = 550$ MeV and $f_\chi = 240$ MeV (Set A). On the other hand, Song *et al.*[52] obtain the "best" values for the parameters of the effective chiral Lagrangian with the "soft" scalar fields so that the results are consistent with "Brown-Rho" scaling,[49] explicitly, $m_\chi = 720$ MeV and $f_\chi = 240\,\mathrm{MeV}$ (Set B). For completeness, we consider also a parameter set of $m_\chi = 1$ GeV and $f_\chi = 240$ MeV (Set C) corresponding to a mass scale comparable to that of chiral symmetry $\Lambda_\chi \sim 4\pi f_\pi$.

### 5.3.2. *Dynamics of the single skyrmion*

The procedure one has to follow can be found in Ref. 11 and is similar to the one discussed in Sec. 5.2.1. The first step is to find the solution for the single skyrmion which includes the dilaton dynamics. The skyrmion with the baryon number $B = 1$ can be found by generalizing the spherical hedgehog Ansatz of the original Skyrme model as

$$U_0(\vec{r}) = \exp(i\vec{\tau} \cdot \hat{r} F(r)), \text{ and } \chi_0(\vec{r}) = f_\chi C(r), \tag{5.3.22}$$

with two radial functions $F(r)$ and $C(r)$. Minimization of the mass equation leads to a coupled set of equations of motion for these functions. In order for the solution to carry a baryon number, $U_0$ has the value $-1$ at the origin, that is, $F(x = 0) = \pi$, while there is no such topological constraint for $C(x = 0)$. All that is required is that it be a positive number below 1. At infinity, the fields $U_0(\vec{r})$ and $\chi_0(\vec{r})$ should reach their vacuum values.

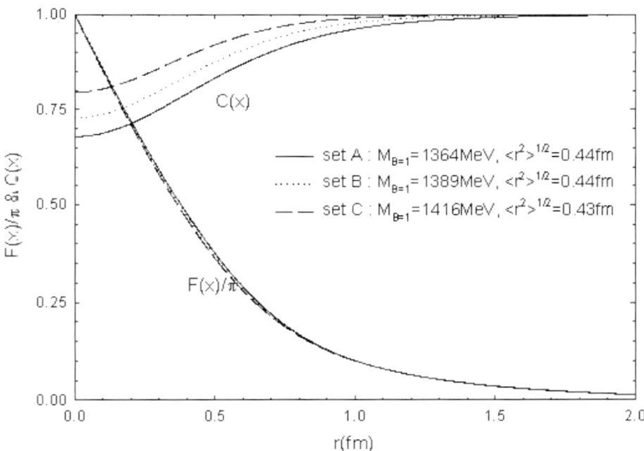


Fig. 5.6. Profile functions $F(x)$ and $C(x)$ as a function of $x$.

Shown in Fig. 5.6 are profile functions as a function of $x(= ef_\pi r)$. $F(r)$ and consequently the root mean square radius of the baryon charge show little dependence on $m_\chi$. On the other hand, the changes in $C(r)$ and the skyrmion mass are recognizable. Inside the skyrmion, especially at the center, $C(r)$ deviates from its vaccum value 1. Note that this change in $C(r)$ is multiplied by $f_\pi^2$ in the current algebra term of the Lagrangian. Thus, $C(r) \leq 1$ reduces the *effective* $f_\pi$ inside the single skyrmion, which implies a partial restoration of the chiral symmetry there. The reduction in the effective pion decay constant is reflected in the single skyrmion mass.

The larger the scalar mass is, the smaller its coupling to the pionic field and the less its effect on the single skyrmion. In the limit of $m_\chi \rightarrow \infty$, the scalar field is completely decoupled from the pions and the model returns back to the original one, where $C(r) = 1$, $M_{sk} = 1479$ MeV and $\langle r^2 \rangle^{1/2} = 0.43$ fm.

### 5.3.3. *Dense skyrmion matter and chiral symmetry restoration*

The second step is to construct a crystal configuration made up of skyrmions with a minimal energy for a given density.

Referring to Refs. 11 and 12 for the full details, we emphasize here the role the dilaton field in the phase transition scenario for skyrmion matter. Let the dilaton field $\chi(\vec{r})$ be a constant throughout the whole space as

$$\chi/f_\chi = X. \tag{5.3.23}$$

Then the energy per baryon number of the system for a given density can be calculated and conveniently expressed as [11]

$$E/B(X, L) = X^2(E_2/B) + (E_4/B) + X^3(E_m/B) + (2L^3)\left(X^4(\ln X - \tfrac{1}{4}) + \tfrac{1}{4}\right), \tag{5.3.24}$$

where $E_2$, $E_4$ and $E_m$ are, respectively, the contributions from the current algebra term, the Skyrme term and the pion mass term of the Lagrangian to the energy of the skyrmion system, described in Sec. 5.1, and $(2L^3)$ is the volume occupied by a single skyrmion

The quantity $E/B(X, L)$ can be taken as an *in medium* effective potential for $X$, modified by the coupling of the scalar to the background matter. Using the parameter values of Ref. 12 for the Skyrme model without the scalar field, the effective potential $E/B(X)$ for a few values of $L$ behaves as shown in Fig. 5.7(a). At low density (large $L$), the minimum of the effective potential is located close to $X = 1$. As the density increases, the quadratic term in the effective potential $E/B(X)$ develops another minimum at $X = 0$ which is an unstable extremum of the potential $V(X)$ in free space. At $L \sim 1$ fm, the newly developed minimum competes with the one near $X \sim 1$. At higher density, the minimum shifts to $X = 0$ where the system stabilizes.

In Fig. 5.7(b), we plot $E/B(X_{min}, L)$ as a function of $L$, which is obtained by minimizing $E/B(X, L)$ with respect to $X$ for each $L$. The figure in the small box is the corresponding value of $X_{min}$ as function of $L$. There we see the explicit manifestation of a first-order phase transition. Although the present discussion is based on a simplified analysis, it essentially encodes the same physics as in the more rigorous treatment of $\chi$ given in Ref. 11.

We show in Fig. 5.8 the average values $\langle \sigma \rangle$ and $\langle \chi/f_\chi \rangle$ over space for the minimum energy crystal configurations obtained by the complete numerical calculation without any approximation for $\chi$. These data show that a 'structural' phase transition takes place, characterized by $\langle \sigma \rangle = 0$, at lower density then the *genuine* chiral phase transition which occurs when $\langle \chi \rangle = 0$. The value of $\langle \sigma \rangle$ becomes 0 when the structure of the skyrmion crystal undergoes a change from the single skyrmion

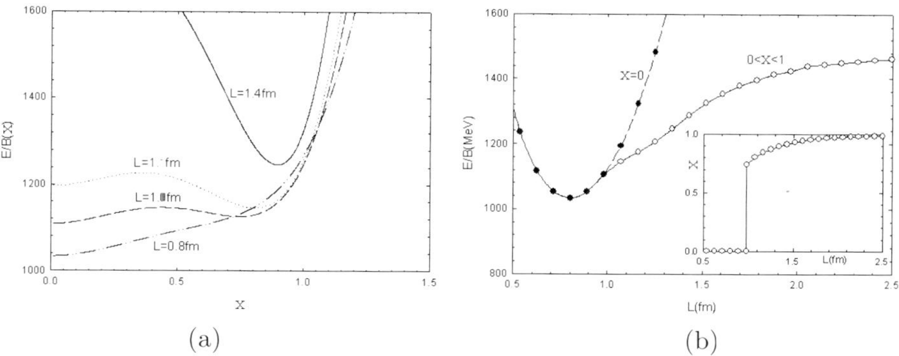

Fig. 5.7. (a) Energy per single skyrmion as a function of the scalar field $X$ for a given $L$. The results are obtained with the $(E_2/B)$, $(E_4/B)$, and $(E_m/B)$ of Ref. 12 and with the parameter sets B, (b) Energy per single skyrmion as a function of $L$.

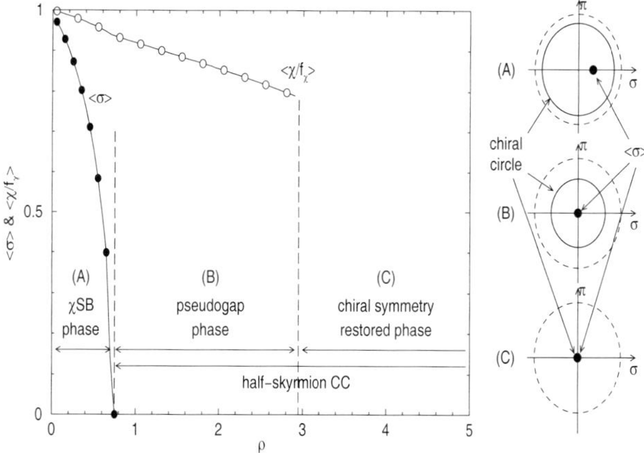

Fig. 5.8. Average values of $\sigma = \frac{1}{2}\mathrm{Tr}(U)$ and $\chi/f_\chi$ of the lowest energy crystal configuration at a given baryon number density.

FCC to the half-skyrmion CC. Thus, the pseudogap phase persists in an intermediate density region, where the $\langle \chi/f_\chi \rangle$ does not vanish while $\langle \sigma \rangle$ does.[53] A similar pseudogap structure has been also proposed in hot QCD.[54]

The two step phase transition is schematically illustrated in Fig. 5.8. Let $\rho_p$ and $\rho_c$ be the density at which $\langle \sigma \rangle$ and $\langle \chi \rangle$ vanish, repectively.

(A) At low density ($\rho < \rho_p$), matter slightly reduces the vacuum value of the dilaton field from that of the baryon free vacuum. This implies a shrinking of the radius of the chiral circle by the same ratio. Since the skyrmion takes all the values on the chiral circle, the expectation value of $\sigma$ is not located on the circle but inside the circle. Skyrmion matter at this density is in the chiral symmetry broken phase.

(B) At some intermediate densities ($\rho_p < \rho < \rho_c$), the expectation value of $\sigma$ vanishes while that of the dilaton field is still nonzero. The skyrmion crystal is in a CC configuration made of half skyrmions localized at the points where $\sigma = \pm 1$. Since the average value of the dilaton field does not vanish, the radius of the chiral circle is still finite. Here, $\langle \sigma \rangle = 0$ does not mean that chiral symmetry is completely restored. We interpret this as a pseudogap phase.

(C) At higher density ($\rho > \rho_c$), the phase characterized by $\langle \chi/f_\chi \rangle = 0$ becomes energetically favorable. Then, the chiral circle, describing the fluctuating pion dynamics, shrinks to a point.

The density range for the occurrence of a pseudogap phase strongly depends on the parameter choice of $m_\chi$. For small $m_\chi$ below 700 MeV, the pseudogap has almost zero size.

In the case of massive pions, the chiral circle is tilted by the explicit (mass) symmetry breaking term. Thus, the exact half-skyrmion CC, which requires a symmetric solution for points with value $\sigma = +1$ and those with $\sigma = -1$ cannot be constructed and consequently the phase characterized by $\langle \sigma \rangle = 0$ does not exist for any density. Thus no pseudogap phase arises. However, $\langle \sigma \rangle$ is always inside the chiral circle and its value drops much faster than that of $\langle \chi/f_\chi \rangle$. Therefore, only if the pion mass is small a pseudogap phase can appear in the model.

### 5.3.4. *Pions in a dense medium with dilaton dynamics*

Since we have achieved, via dilaton dynamics, a reasonable scenario for chiral symmetry restoration, it is time to revisit the properties of pions in a dense medium. As was explained in Sec. 5.2.2 and in Ref. 8, we proceed to incorporate the fluctuations on top of the static skyrmion crystal. (We refer to Refs. 11 and 12 for details.)

Using a mean field approximation we calculate the in-medium pion mass $m_\pi^*$ and decay constant $f_\pi^*$ obtaining,

$$Z_\pi^2 = \left\langle \left( \frac{\chi_0(\vec{x})}{f_\chi} \right)^2 \left( 1 - \tfrac{2}{3}\pi^2(\vec{x}) \right) \right\rangle \equiv \left( \frac{f_\pi^*}{f_\pi} \right)^2, \tag{5.3.25}$$

$$m_\pi^{*2} Z_\pi^2 = \left\langle \left( \frac{\chi_0(\vec{x})}{f_\chi} \right)^3 \sigma(\vec{x}) \, m_\pi^2 \right\rangle. \tag{5.3.26}$$

The wave function renormalization constant $Z_\pi$ gives the ratio of the in-medium pion decay constant $f_\pi^*$ to the free one, and the above expression arises from the current algebra term in the Lagrangian. The explicit calculation of $m_\chi^*$ is given in Ref. 12.

In Fig. 5.9 we show the (exact) ratios of the in-medium parameters relative to their free-space values. Only the results obtained with the parameter set B are shown. The parameter set A yields similar results while set C shows a two step structure with an intermediate pseudogap phase. Not only the average value of $\chi_0$

over the space but also $\chi_0(\vec{r})$ itself vanishes at any point in space. This is the reason for the vanishing of $m_\pi^*$ and $f_\pi^*$. That is, $f_\pi^*$ really vanishes when $\rho < \rho_c$ in the Skyrme model with dilaton dynamics.

At low matter density, the ratio $f_\pi^*/f_\pi$ can be fitted to a linear function

$$\frac{f_\pi^*}{f_\pi} \sim 1 - 0.24(\rho/\rho_0) + \cdots \tag{5.3.27}$$

At $\rho = \rho_0$, this yields $f_\pi^*/f_\pi = 0.76$, which is comparable to the other predictions.

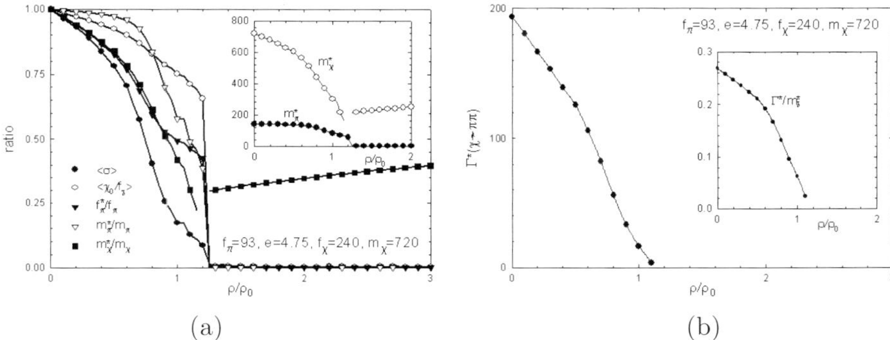


(a)        (b)

Fig. 5.9.   (a) The ratios of the in-medium parameters to the free space parameters. The graph in a small box shows the masses of the pion and the scalar, (b) the in-medium decay width $\Gamma^*(\chi \to \pi\pi)$ as a function of $\rho$.

In Ref. 12, the in-medium modification of the $\chi$ decay into two pions is also studied using the mean field approximation. Gathering the terms with a fluctuating scalar field and two fluctuating pion fields, we get the Lagrangian density for the process $\chi \to \pi\pi$

$$\mathcal{L}_{M,\chi\pi^2} = \frac{\chi_0}{f_\chi^2}(\delta_{ab} + g_{ab})\chi\partial_\mu\phi_a\partial^\mu\phi_b, \tag{5.3.28}$$

where only the term from $\mathcal{L}_\sigma$ is used.   Averaging the space dependence of the background field configuration modifies the coupling constant by a factor $\langle(\chi_0/f_\chi)(1 + g_{11})\rangle = \langle(\chi_0/f_\chi)(1 - \frac{2}{3}\pi^2)\rangle$.   Taking into account the appropriate wave function renormalization factors, $Z_\pi$, and the change in the scalar mass, we obtain the in-medium decay width as

$$\Gamma^*(\chi \to \pi\pi) = \frac{3m_\chi^{*3}}{32\pi f_\chi^2}\left|\frac{\langle(\chi_0/f_\chi)(1 - \frac{2}{3}\pi^2)\rangle}{\langle(\chi_0/f_\chi)^2(1 - \frac{2}{3}\pi^2)\rangle}\right|^2 \approx \frac{3m_\chi^{*3}}{32\pi f_\chi^{*2}}. \tag{5.3.29}$$

We show in Fig. 5.9 the in-medium decay width predicted with the parameter set B. In the region $\rho \geq \rho_{pt}$ where $\chi_0 = 0$, $\Gamma^*$ cannot be defined to this order. Near the critical point, the scalar becomes an extremely narrow-width excitation, a feature which has been discussed in the literature as a signal for chiral restoration.[55,56]

Another interesting change in the properties of the pion in the medium is associated with the in-medium pion dispersion relation. This relation requires, besides the mass, the so-called in medium pion velocity, $v_\pi$. This property allows us to gain more insight into the real time properties of the system under extreme conditions and enables us to analyze how the phase transition from normal matter to deconfined QCD takes place from the hadronic side, the so called 'bottom up' approach.

At nonzero temperature and/or density, the Lorentz symmetry is broken by the medium. In the dispersion relation for the pion modes (in the chiral limit)

$$p_0^2 = v_\pi^2 |\vec{p}|^2, \tag{5.3.30}$$

the velocity $v_\pi$ which is 1 in free-space must depart from 1. This may be studied reliably, at least at low temperatures and at low densities, via chiral perturbation theory.[57] The in-medium pion velocity can be expressed in terms of the time component of the pion decay constant, $f_\pi^t$ and the space component, $f_\pi^s$,[58,59]

$$\begin{aligned}
\langle 0|A_a^0|\pi^b(p)\rangle_{\text{in-medium}} &= i f_\pi^t \delta^{ab} p^0, \\
\langle 0|A_a^i|\pi^b(p)\rangle_{\text{in-medium}} &= i f_\pi^s \delta^{ab} p^i.
\end{aligned} \tag{5.3.31}$$

The conservation of the axial vector current leads to the dispersion relation (5.3.30) with the pion velocity given by

$$v_\pi^2 = f_\pi^s / f_\pi^t. \tag{5.3.32}$$

In Ref. 60 two decay constants, $f_t$ and $f_s$, are defined differently from those of Eq. (5.3.31)), through the effective Lagrangian,

$$\mathcal{L}_{\text{eff}} = \frac{f_t^2}{4}\text{Tr}(\partial_0 U^\dagger \partial_0 U) - \frac{f_s^2}{4}\text{Tr}(\partial_i U^\dagger \partial_i U) + \cdots, \tag{5.3.33}$$

where $U$ is an SU(2)-valued chiral field whose phase describes the in-medium pion. In terms of these constants, the pion velocity is defined by

$$v_\pi = f_s / f_t. \tag{5.3.34}$$

In Ref. 12, it is shown that local interactions with background skyrmion matter lead to a breakdown of Lorentz symmetry in the dense medium and to an effective Lagrangian for pion dynamics in the form of Eq. (5.3.33). The results are shown in Figs. 5.10. Both of the pion decay constants change significantly as a function of density and vanish — in the chiral limit — when chiral symmetry is restored. However, the second-order contributions to the $f_s$ and $f_\pi$, which break Lorentz symmetry, turn out to be rather small, and thus their ratio, the pion velocity, stays $v_\pi \sim 1$. The lowest value found is $\sim 0.9$. Note, however, the drastic change in its behavior at two different densities. At the lower density, where skyrmion matter is in the chiral symmetry broken phase, the pion velocity decreases and has the minimum at $\rho = \rho_p$. If one worked only at low density in a perturbative scheme, one would conclude that the pion velocity decreases all the way to zero. However, the presence of the pseudogap phase transition changes this behavior. In the pseudogap phase, the pion velocity not only stops decreasing but starts increasing with increasing density. In the chiral symmetry restored phase both $f_t$ and $f_s$ vanish. Thus their ratio makes no sense.

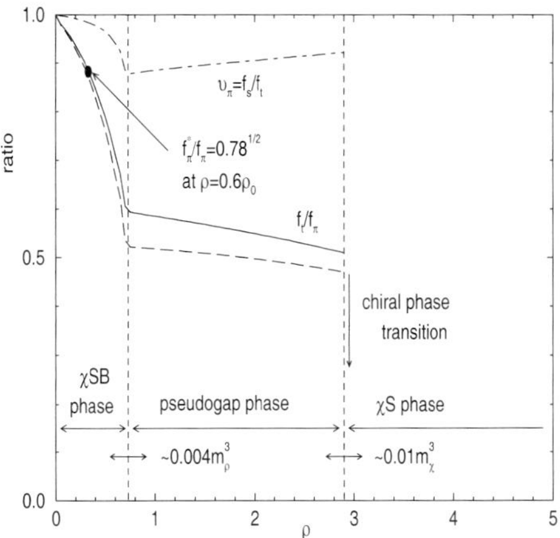


Fig. 5.10.   In-medium pion decay constants and their ratio, the pion velocity.

In Refs. 13 and 14, the in-medium modification of the neutral pion decay process into two $\gamma$s and neutrino-anti-neutrino pair are studied in the same manner. The $\pi^0 \to \gamma\gamma$ process is shown to be strongly suppressed in dense medium, while the process $\pi^0 \to \nu\bar{\nu}$ forbidden in free space becomes possible by the Lorentz symmetry breaking effect of the medium.

## 5.4. Skyrmion Matter at Finite Temperature

There are many studies of lattice QCD at finite temperature. The situation is completely different for skyrmion matter where the number of studies is limited. For example, skyrmion matter has been heated up to melt the crystal into a liquid to study the crystal-liquid phase transition.[28,29] However this phenomenon is irrelevant for the restoration of chiral symmetry, which interests us here for the reasons discussed in previous sections.

What happens if we heat up the system? Naively, as the temperature increases, the kinetic energy of the skyrmions increases and the skyrmion crystal begins to melt. The kinetic energy associated with the translations, vibrations and rotations of the skyrmions is proportional to $T$. This mechanism leads to a solid-liquid-gas phase transition of the skyrmion system. However, we are interested in the chiral symmetry restoration transition, which is not related to the melting. Therefore, a new mechanism must be incorporated to describe chiral symmetry restoration. We show in what follows that the thermal excitation of the pions in the medium is the appropriate mechanism, since this phenomenon is proportional to $T^4$ and therefore dominates the absorption of heat.

The pressure of non-interacting pions is given by[61]

$$P = \frac{\pi^2}{30} T^4,$$ (5.4.35)

where we have taken into account the contributions from three species of pion, $\pi^+, \pi^0, \pi^-$. This term contributes to the energy per single skyrmion volume as $3PV(\chi/f_\chi)^2$. The kinetic energy of the pions arises from $\mathcal{L}_\sigma$ (5.1.2), and therefore scale symmetry implies that it should carry a factor $\chi^2$. The factor 3 comes from the fact that our pions are massless.

To estimate the properties of skyrmion matter at finite temperature let us take $\chi$ as a constant field as we did in Sec. 5.3.3. After including thermal pions, Eq. (5.3.24) can be rewritten as

$$E/B(\rho, T, X) = \left( E_2/B)(\rho) + \frac{\pi^2}{10} T^4 V \right) X^2 + (E_4/B)(\rho) + X^4 (\ln X - \tfrac{1}{4}) + \tfrac{1}{4}),$$ (5.4.36)

where we have dropped the pion mass term.

As in Sec. 5.3.3, chiral restoration will occur when the value of $X_{\min}$ that minimizes $E/B$ vanishes. By minimizing $E/B$ with respect to $X$, we observe that the phase transition takes place from a non-vanishing $X = e^{-1/4}$ to $X = 0$. Thus, the nature of the phase transition is of the first order.

After a straightforward calculation we obtain,

$$\rho^c (E_2/B) + \frac{\pi^2}{10} T_c^4 = \frac{f_\chi^2 m_\chi^2}{8 e^{1/2}}.$$ (5.4.37)

which leads to

$$T_c = \left( \frac{10}{\pi^2} \left( \frac{f_\chi^2 m_\chi^2}{8 e^{1/2}} - \rho^c (E_2/B)(\rho^c) \right) \right)^{1/4}$$ (5.4.38)

For $\rho = 0$ (zero density), our estimate for the critical temperature is

$$T_c = \left( \frac{10}{\pi^2} \frac{f_\chi^2 m_\chi^2}{8 e^{1/2}} \right)^{1/4} \sim 205 \text{ MeV},$$ (5.4.39)

where we have used the following values for the parameters. $f_\chi = 210$ MeV and $m_\chi = 720$ MeV. It is remarkable that our model leads to $T_c \sim 200$ MeV, which is close to that obtained by lattice QCD[4] and in agreement with the data.[62] To us this is a confirmation that the mechanism chosen for the absorption of heat plays a fundamental role in the hadronic phase.

The numerical results on $E_2/B$ that minimize the energy of the system for a given $\rho_B$ can be approximated by

$$E_2/B = \begin{cases} 10 f_\pi^2 / \rho^{1/3}, & \rho > \rho_0 \\ 36 f_\pi / e_{\text{sk}}, & \rho < \rho_0, \end{cases}$$ (5.4.40)

where $\rho_0 = (e_{\text{sk}} f_\pi / 3.6)^3$.

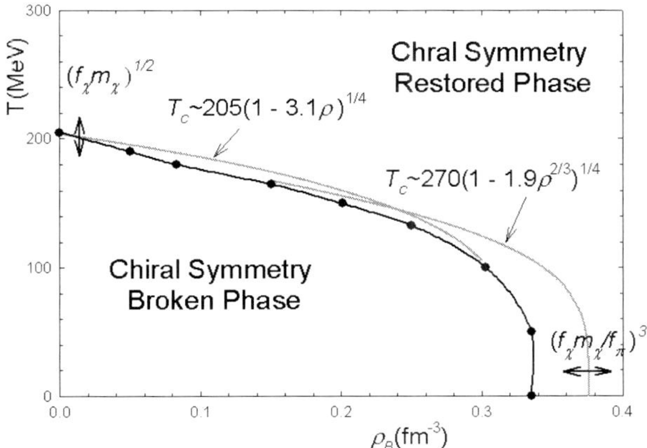



Fig. 5.11. The chiral phase transition. The solid line shows the exact calculation, while the gray lines two approximate estimates.

Using Eq. (5.4.40) for $E_2/B$, we obtain the critical density for chiral symmetry restoration at zero temperature as

$$\rho^c(T=0) = \left( \frac{f_\chi^2 m_\chi^2}{8e^{1/2}} \frac{1}{10 f_\pi^2} \right)^{3/2} \sim 0.37 \text{ fm}^{-3}. \tag{5.4.41}$$

Since $\rho_0 = 0.24$ fm$^{-3}$ $< \rho^c(T=0)$ our result is consistent with the high density formula for $E_2/E$ used.

The resulting critical density $\rho^c(T=0) \sim 0.37$ fm$^{-3}$ is only twice normal nuclear matter density and it is low with respect to the expected values. This result does not represent a problem since $\rho^c(T=0)$ scales with $(f_\chi m_\chi/f_\pi)^3$ and $T_c^{\rho=0}$ with $(f_\chi m_\chi)^{1/2}$ and small changes in the parameters lead to larger values for the critical density without changing the critical temperature too much.

For a finite density smaller than $\rho^c(T=0)$, we obtain the corresponding critical temperature by substituting the asymptotic formulas (5.4.40) for $E_2/B$,

$$T_c = \begin{cases} T_c(\rho=0)\,(1 - 3.09\,\rho_c)^{1/4} & \text{for } \rho < \rho_0, \\ T_c(\rho=0)\,(1 - 1.92\,\rho_c^{2/3})^{1/4} & \text{for } \rho > \rho_0, \end{cases} \tag{5.4.42}$$

where the density is measured in fm$^{-3}$. The gray lines in Fig. 5.11 show these two curves. The results from the *exact* calculations obtained by minimization of the energy (5.4.36) are shown by black dots connected by black line in Fig. 5.11. The resulting phase diagram has the same shape but the values of the temperatures and densities are generally smaller than in the approximate estimates.

## 5.5. Vector Mesons and Dense Matter

In our effort to approach the theory of the hadronic interactions and inspired by Weinberg's theorem[19] we proceed to incorporate to the model the lowest-lying vector mesons, namely the $\rho$ and the $\omega$. In this way we also do away with the ad hoc Skyrme quartic term. It is known that these vector mesons play a crucial role in stabilizing the single nucleon system[30,63] as well as in the saturation of normal nuclear matter.[64]

We consider a skyrmion-type Lagrangian with vector mesons possessing hidden local gauge symmetry,[22] spontaneously broken chiral symmetry and scale symmetry.[11,21] Such a theory might be considered as a better approximation to reality than the extreme large $N_c$ approximation to QCD represented by the Skyrme model. Specifically, the model Lagrangian, which we investigate, is given by[65]

$$
\mathcal{L} = \frac{f_\pi^2}{4}\left(\frac{\chi}{f_\chi}\right)^2 \text{Tr}(\partial_\mu U^\dagger \partial^\mu U) + \frac{f_\pi^2 m_\pi^2}{4}\left(\frac{\chi}{f_\chi}\right)^3 \text{Tr}(U + U^\dagger - 2)
$$

$$
-\frac{f_\pi^2}{4}a\left(\frac{\chi}{f_\chi}\right)^2 \text{Tr}[\ell_\mu + r_\mu + i(g/2)(\vec{\tau}\cdot\vec{\rho}_\mu + \omega_\mu)]^2 - \tfrac{1}{4}\vec{\rho}_{\mu\nu}\cdot\vec{\rho}^{\mu\nu} - \tfrac{1}{4}\omega_{\mu\nu}\omega^{\mu\nu}
$$

$$
+\tfrac{3}{2}g\omega_\mu B^\mu + \tfrac{1}{2}\partial_\mu\chi\partial^\mu\chi - \frac{m_\chi^2 f_\chi^2}{4}\left[(\chi/f_\chi)^4(\ln(\chi/f_\chi) - \tfrac{1}{4}) + \tfrac{1}{4}\right], \tag{5.5.43}
$$

where, $U = \exp(i\vec{\tau}\cdot\vec{\pi}/f_\pi) \equiv \xi^2$, $\ell_\mu = \xi^\dagger\partial_\mu\xi$, $r_\mu = \xi\partial_\mu\xi^\dagger$, $\vec{\rho}_{\mu\nu} = \partial_\mu\vec{\rho}_\nu - \partial_\nu\vec{\rho}_\mu + g\vec{\rho}_\mu\times\vec{\rho}_\nu$, $\omega_{\mu\nu} = \partial_\mu\omega_\nu - \partial_\nu\omega_\mu$, and $B^\mu = \frac{1}{24\pi^2}\varepsilon^{\mu\nu\alpha\beta}\text{Tr}(U^\dagger\partial_\nu U U^\dagger\partial_\alpha U U^\dagger\partial_\beta U)$. Note that the Skyrme quartic term is not present. The vector mesons, $\rho$ and $\omega$, are incorporated as dynamical gauge bosons for the local hidden gauge symmetry of the non-linear sigma model Lagrangian and the dilaton field $\chi$ is introduced so that the Lagrangian has the same scaling behavior as QCD. The physical parameters appearing in the Lagrangian are summarized in Table 5.1.

Table 5.1.  Parameters of the model Lagrangian

| notation | physical meaning | value |
|---|---|---|
| $f_\pi$ | pion decay constant | 93 MeV |
| $f_\chi$ | $\chi$ decay constant | 210 MeV |
| $g$ | $\rho\pi\pi$ coupling constant | 5.85* |
| $m_\pi$ | pion mass | 140 MeV |
| $m_\chi$ | $\chi$ meass | 720 MeV |
| $m_V$ | vector meson masses | 770 MeV† |
| $a$ | vector meson dominance | 2 |

* obtained by using the KSFR relation $m_V^2 = m_\rho^2 = m_\omega^2 = af_\pi^2 g^2$ with $a = 2$. cf. $g_{\rho\pi\pi} = 6.11$ from the decay width of $\rho \to \pi\pi$.
† experimentally measured values are $m_\rho$=768 MeV and $m_\omega$=782 MeV.

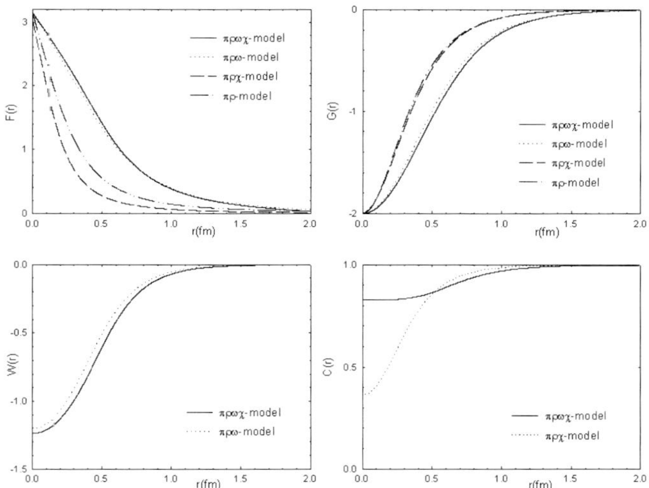

Fig. 5.12.   Profile functions — $F(r)$, $G(r)$, $W(r)$ and $C(r)$.

### 5.5.1. *Dynamics of the single skyrmion*

The spherically symmetric hedgehog Ansatz for the $B = 1$ soliton solution of the standard Skyrme model can be generalized to

$$U^{B=1} = \exp(i\vec{\tau} \cdot \hat{r} F(r)), \qquad (5.5.44)$$

$$\rho^{a,B=1}_{\mu=i} = \varepsilon^{ika}\hat{r}^k \frac{G(r)}{gr}, \qquad \rho^{a,B=1}_{\mu=0} = 0, \qquad (5.5.45)$$

$$\omega^{B=1}_{\mu=i} = 0, \qquad \omega^{B=1}_{\mu=0} = f_\pi W(r), \qquad (5.5.46)$$

$$\chi^{B=1} = f_\chi C(r). \qquad (5.5.47)$$

The boundary conditions that the profile functions satisfy at infinity are

$$F(\infty) = G(\infty) = W(\infty) = 0, \qquad C(\infty) = 1, \qquad (5.5.48)$$

and at the center $(r = 0)$ are

$$F(0) = \pi, \qquad G(0) = -2, \qquad W'(0) = C'(0) = 0. \qquad (5.5.49)$$

The profile functions are obtained numerically by minimizing the soliton mass with the boundary conditions (see Ref. 9 for the technical details). The results are summarized in Table 5.2 and the corresponding profile functions are given in Fig. 5.12. The role of the $\omega$ meson that provides a strong repulsion is prominent. Comparing the $\pi\rho$ model with the $\pi\rho\omega$ model, the presence of the $\omega$ increases the mass by more than 415 MeV and the size, i.e. $\langle r^2 \rangle$, by more than 0.28 $\text{fm}^2$.

Table 5.2. Single skyrmion mass and various contributions to it.

| Model | $\langle r^2 \rangle$ | $E^{B=1}$ | $E_\pi^{B=1}$ | $E_{\pi\rho}^{B=1}$ | $E_\rho^{B=1}$ | $E_\omega^{B=1}$ | $E_{WZ}^{B=1}$ | $E_\chi^{B=1}$ |
|---|---|---|---|---|---|---|---|---|
| $\pi\rho$-model | 0.27 | 1054.6 | $400.2 + 9.2$ | 110.4 | 534.9 | 0.0 | 0.0 | 0.0 |
| $\pi\rho\chi$-model | 0.19 | 906.5 | $103.1 + 1.4$ | 155.1 | 504.1 | 0.0 | 0.0 | 142.8 |
| $\pi\rho\omega$-model | 0.49 | 1469.0 | $767.6 + 39.9$ | 33.2 | 370.7 | $-257.6$ | 515.1 | 0.0 |
| $\pi\rho\omega\chi$-model | 0.51 | 1408.3 | $646.0 + 29.2$ | 34.9 | 355.7 | $-278.3$ | 556.7 | 64.2 |

How does the dilaton affect this calculation? The $\pi\rho$ model with much smaller skyrmion has a larger baryon density near the origin and this affects the dilaton, significantly changing its mean-field value from its vacuum one. The net effect of the dilaton mean field on the mass is a reduction of $\sim 150$ MeV, whereas for the $\pi\rho\omega$ model it is only of 50 MeV. The details can be seen in Table 5.2. The effect on the soliton size is, however, different: while the dilaton in the $\pi\rho$ model produces an additional localization of the baryon charge and hence reduces $\langle r^2 \rangle$ from 0.21 fm$^2$ to 0.19 fm$^2$, in the $\pi\rho\omega$ model, on the contrary, the dilaton produces a *delocalization* and increases $\langle r^2 \rangle$ from 0.49 fm$^2$ to 0.51 fm$^2$. We will see, however, that this strong repulsion provided by $\omega$ causes a somewhat serious problem in the chiral restoration of the skyrmion matter at higher density.

### 5.5.2. *Skyrmion matter: an FCC skyrmion crystal*

Again, the lowest-energy configuration is obtained when one of the skyrmions is rotated in isospin space with respect to the other by an angle $\pi$ about an axis perpendicular to the line joining the two.[9] If we generalize this Ansatz to many-skyrmion matter, we obtain that the configuration at the classical level for a given baryon number density is an FCC crystal where the nearest neighbor skyrmions are arranged to have the attractive relative orientations.[8] Kugler's Fourier series expansion method[37] can be generalized to incorporate the vector mesons, although some subtleties associated with the vector fields have to be implemented. The details can be found in Ref. 9.

The figures in Fig. 5.13 are the the numerical results of the energy per baryon $E/B$, $\langle \chi \rangle$ and $\langle \sigma \rangle$ in various models as a function of the FCC lattice parameter $L$. In the $\pi\rho\chi$ model, as the density of the system increases ($L$ decreases), $E/B$ changes little. It is close to the energy of a $B = 1$ skyrmion up to a density greater than $\rho_0$ ($L \sim 1.43$). This result is easy to interpret. As we discussed before the size of the skyrmion in this model is very small and therefore the skyrmions in the lattice will interact only at very high densities, high enough for their tails to overlap.

In the absence of the $\omega$, the dilaton field plays a dramatic role. A skyrmion matter undergoes an abrupt phase transition at high density at which the expectation value of the dilaton field vanishes $\langle \chi \rangle = 0$. (In general, $\langle \chi \rangle = 0$ does not necessarily require $\langle \chi^2 \rangle = 0$. However, since $\chi \geq 0$, $\langle \chi \rangle = 0$ always accompanies $\chi = 0$ in the whole space.) The $\rho$ meson on the other hand is basically a spectator

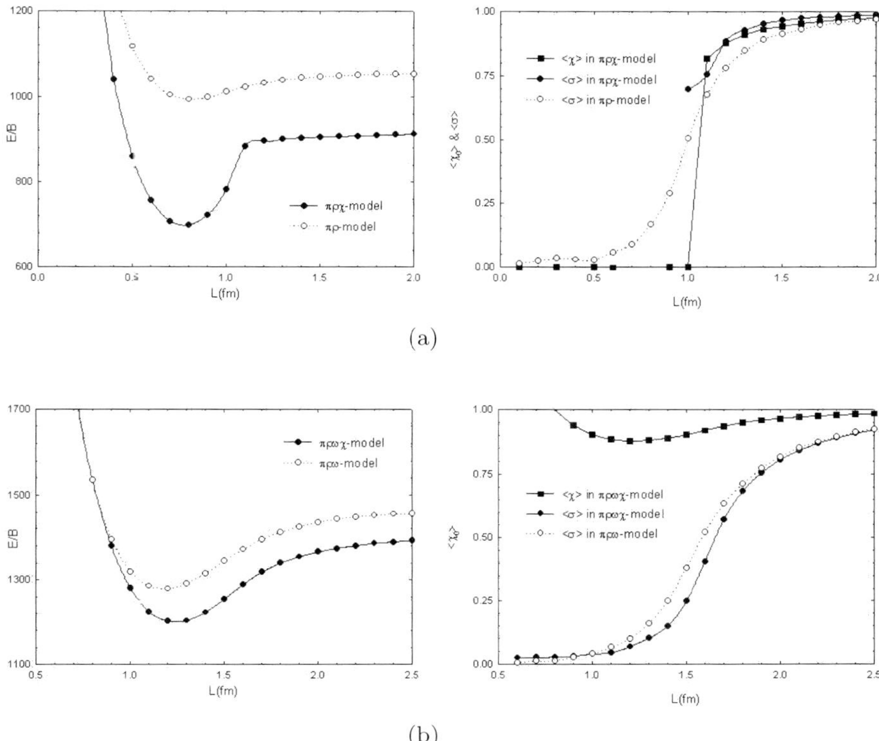

Fig. 5.13. $E/B$, $\langle\chi\rangle$ and $\langle\sigma\rangle$ as a function of $L$ in the models (a) without the $\omega$ and (b) with omega.

at the classical level, producing little change with respect to our previously studied $\pi\chi$ model except that at high densities, once the $\rho$ starts to overlap, the energy of nuclear matter increases due to the repulsive effect at short distances. The densities have to be quite high since these skyrmions are very small. Since $\chi$ vanishes at the phase transition, we recover the standard behavior, namely, $f_\pi^* = 0$ and $m_\rho^* = 0$.

In the $\pi\rho\omega\chi$ model, the situation changes dramatically. The reason is that the $\omega$ provides not only a strong repulsion among the skyrmions, but somewhat surprisingly, also an intermediate range attraction. Note the different mass scales between Figs. 5.13(a) and 5.13(b). In both the $\pi\rho\omega$ and the $\pi\rho\omega\chi$ models, at high density, the interaction reduces $E/B$ to 85% of the $B = 1$ skyrmion mass. This value should be compared with 94% in the $\pi\rho$ model. In the $\pi\rho\chi$-model, $E/B$ goes down to 74% of the $B = 1$ skyrmion mass, but in this case it is due to the dramatic behavior of the dilaton field.

In the $\pi\rho\omega\chi$ model the role of the dilaton field is suppressed. It provides a only a small attraction at intermediate densities. Moreover, the phase transition towards its vanishing expectation value, $\langle\chi\rangle = 0$, does not take place. Instead, its value grows at high density!

The problem involved is associated with the Lagrangian (5.5.43) which includes an anomalous part known as Wess-Zumino term, namely the coupling of the $\omega$ to the baryon current, $B_\mu$. To see that this term is the one causing the problem, consider the energy per baryon contributed by this term.[9]

$$\left(\frac{E}{B}\right)_{WZ} = \frac{1}{4}\left(\frac{3g}{2}\right)^2 \int_{Box} d^3x \int d^3x' B_0(\vec{x})\frac{\exp(-m_\omega^*|\vec{x}-\vec{x}'|)}{4\pi|\vec{x}-\vec{x}'|}B_0(\vec{x}') \qquad (5.5.50)$$

where "Box" corresponds to a single FCC cell. Note that while the integral over $\vec{x}$ is defined in a single FCC cell, that over $\vec{x}'$ is not. Thus, unless it is screened, the periodic source $B_0$ filling infinite space will produce an infinite potential $w$ which leads to an infinite $(E/B)_{WZ}$. The screening is done by the omega mass, $m_\omega^*$. Thus the effective $\omega$ mass cannot vanish. Our numerical results reflect this fact: at high density the $B_0$-$B_0$ interaction becomes large compared to any other contribution. In order to reduce it, $\chi$ has to increase, and thereby the effective screening mass $m_\omega^* \sim m_\omega\langle\chi\rangle$ becomes larger. In this way we run into a phase transition where the expectation value of $\chi$ does not vanish and therefore $f_\pi$ does not vanish but instead increases.

### 5.5.3. *A resolution of the $\omega$ problem*

Assuming that there is nothing wrong with (5.5.43), we focus on the Wess-Zumino term in the Lagrangian. Our objective is to find an alternative to (5.5.43) that leads to a behavior consistent with the expected behavior. In the absence of any reliable clue, we try the simplest, admittedly *ad hoc*, modification of the Lagrangian (5.5.43) that allows a reasonable and appealing way-out.[10] Given our ignorance as to how spontaneously broken scale invariance manifests in matter, we shall simply forego the requirement that the anomalous term be scale invariant and multiply the anomalous $\omega \cdot B$ term by $(\chi/f_\chi)^n$ for $n \geq 2$. We have verified that it matters little whether we pick $n = 2$ or $n = 3$.[10] We therefore take $n = 3$:

$$L'_{an} = \tfrac{3}{2}g(\chi/f_\chi)^3\omega_\mu B^\mu \qquad (5.5.51)$$

This additional factor has two virtues:

i) It leaves meson dynamics in free space (i.e. $\chi/f_\chi = 1$) unaffected, since chiral symmetry is realized à la sigma model as required by QCD.

ii) It plays the role of an effective density-dependent coupling constant so that at high density, when scale symmetry is restored and $\chi/f_\chi \to 0$, there will be no coupling between the $\omega$ and the baryon density as required by hidden local symmetry with the vector manifestation.

The properties of this Lagrangian for the meson ($B = 0$) sector are the same as in our old description. The parameters of the Lagrangian are determined by meson physics as given in Table 5.1.

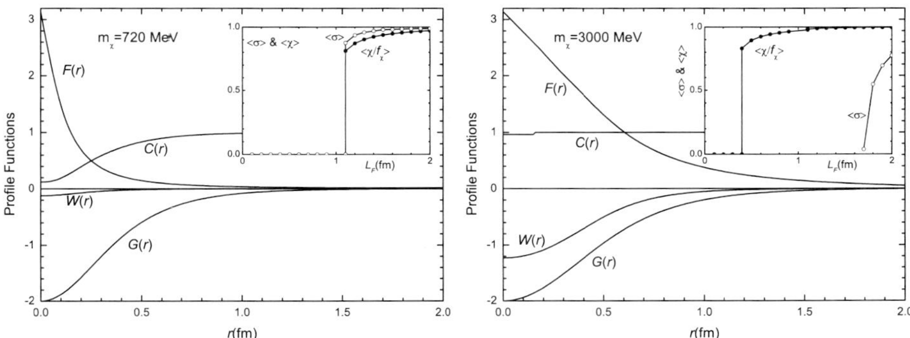


Fig. 5.14.   A small and large skyrmion obtained with $m_\chi = 720$MeV (left) and $m_\chi = 3000$ MeV (right). Shown in small boxes are $\langle \chi \rangle$ and $\langle \sigma \rangle$ as a function of the FCC lattice size $L_F$.

Figure 5.14 summarizes the consequences of the modification. Depending on the dilaton mass, the properties of a single skyrmion show distinguished characters and consequently undergoes different phase transition. A small dilaton mass, say $m_\chi < 1$ GeV, leads to a very small skyrmion with an rms radius about 0.1 fm. The light dilaton seems to react quite sensitively to the presence of the matter. One can see that at the center of the single skyrmion the chiral symmetry is almost restored. It weakens most of the repulsion from $\omega - B$ coupling, which leads us to such a small sized skyrmion. Since these small skyrmions are already chiral-symmetry-restored objects, simply filling the space with them restores the symmetry. As shown in the small box, chiral symmetry is restored simultaneously when $\langle \sigma \rangle$ vanishes. In case of having a large mass, the dilaton does not play any significant role in the structure of a single skyrmion. This scenario leads, as the density of skyrmion matter increases, first to a pseudogap phase transition where $\langle \sigma \rangle = 0$ and thereafter, at higher density, to a genuine chiral symmetry restoration phase transition where $\langle \chi/f_\chi \rangle = 0$. Anyway, whether the dilaton is light or heavy, we finally have a reasonable phase transition scenario that at some critical density chiral symmetry restoration occurs where $\langle \chi/f_\chi \rangle$ vanishes.

Under the same mean field approximation, this skyrmion approach to the dense matter leads us to the scaling behaviors of the vector mesons

$$\frac{m_\rho^*}{m_\rho} = \frac{m_\omega^*}{m_\omega} = \sqrt{\langle \left(\frac{\chi}{f_\chi}\right)^2 \rangle}, \tag{5.5.52}$$

while that of the pion decay constant is

$$\frac{f_\pi^*}{f_\pi} = \sqrt{\langle \left(\frac{\chi}{f_\chi}\right)^2 (1 + (a-1)\tfrac{2}{3}\pi^2) \rangle}. \tag{5.5.53}$$

With $a = 1$, a remarkably simple BR scaling law is obtained. These scaling laws imply that as the density of the matter increases the effecive quantities in medium scale down.

We have shown how a slight modification of the Lagrangian resolves the $\omega$ problem. However, in modifying the Lagrangian we have taken into account only the phenomenological side of the problem. Multiplying the Wess-Zumino term by the factor $(\chi/f_\chi)^n$ has no sound theoretical support. It breaks explicitly the scale invariance of the Lagrangian. Recall that the dilaton field was introduced into the model to respect scale symmetry. Furthermore, we don't have any special reason for choosing $n = 3$, except that it works well. Recently, a more fundamental explanation for the behavior in Eq. (5.5.51) has been found.[d]

## 5.6. Conclusions

In trying to understand what happens to hadrons under extreme conditions, it is necessary that the theory adopted for the description be consistent with QCD. In terms of effective theories this means that they should match to QCD at a scale close to the chiral scale $\Lambda_\chi \sim 4\pi f_\pi \sim 1$ GeV. It has been shown that this matching can be effectuated in the framework of hidden local symmetry (HLS) and leads to what is called 'vector manifestation' (VM)[23] which provides a theoretical support for a low-energy effective field theory for hadrons and which gives, in the chiral limit, an elegant and unambiguous prediction of the behavior of light-quark hadrons at high temperature and/or at high density. Following the indications of the HLS theory, we have described a Skyrme model in which the dilaton field $\chi$, whose role in dense matter was first pointed out by Brown and Rho,[49] and the vector meson fields $\rho$ and $\omega$ were incorporated into the Skyrme Lagrangian to construct dense skyrmion matter.

We have presented an approach to hadronic physics based on Skyrme's philosophy, namely that baryons are solitons of a theory described in terms of meson fields, which can be justified from QCD in the large $N_c$ expansion. We have adopted the basic principles of effective field theory. Given a certain energy domain we describe the dynamics by a Lagrangian defined in terms of the mesonic degrees of freedom active in that domain, we thereafter implement the symmetries of QCD and VM, and describe the baryonic sectors as topological winding number sectors and solve in these sectors the equations derived from the Lagrangian with the appropriate boundary conditions for the sector. In this way one can get all of Nuclear Physics out of a single Lagrangian. We have studied the B=1 sector to obtain the properties of the single skyrmion, the B=2 sector to understand the interaction between skyrmions, and our main effort has been to study skyrmion matter, as a model for hadronic matter, investigating its behavior at finite density and temperature and the description of meson properties in that dense medium.

Skyrme models have been proven successful in describing nuclei, the nucleon-nucleon interaction and pion-nucleon interactions. It turns out that Skyrme models also represent a nice tool for understanding low density cold hadronic matter and

---

[d]Private communication by M. Rho on work in progress by H.K. Lee and M. Rho.

the behavior of the mesons, in particular, the pion inside matter. We have shown in here that when hadronic matter is compressed and/or heated Skyrme models provide useful information on the chiral phase transitions. Skyrmion matter is realized as a crystal and we have seen that at low densities it is an FCC crystal made of skyrmions. The phase transition occurs when the FCC crystal transforms into a half skyrmion CC one. In our study we have discovered the crucial role of the scale dilaton in describing the expected phase transition towards a chiral symmetry restored phase. We have also noticed the peculiar behavior of the $\omega$ associated to its direct coupling to the baryon number current and we have resolved the problem by naturally scaling the coupling constant using the scale dilaton.

Another aspect of our review has been the study of the properties of elementary mesons in the medium, in particular those involved in the model, the pion and the dilaton. Moreover we have described how their properties change when we move from one phase to another.

A description of the chiral restoration phase transition in the temperature-density plane has been presented, whose main ingredient is that the dominant scenario is the absorption of heat by the fluctuating pions in the background of crystal skyrmion matter. This description leads to a phase transition whose dynamical structure is parameter independent and whose shape resembles much the conventional confinement/deconfiment phase transition. We obtain, for parameter values close to the conventional ones, the expected critical temperatures and densities.

For clarity, the presentation has been linear, in the sense, that given the Lagrangian we have described its phenomenology, and have made no effort to interpret the mechanisms involved and the results obtained from QCD. In this way we have taken a 'bottom up' approach: the effective theory represents confined QCD and it should explain the hadronic phenomenology in its domain of validity.

The main result of our calculation is the realization that the phase transition scenario is not as simple as initially thought but contains many features which make it highly interesting and phenomenologically appealing. It is now time to try to collect ideas based on fundamental developments and see how our effective theory and the principles that guide it realize these ideas. In this line of thought, it is exciting to have unveiled scenarios near the phase transition of unexpected interesting phenomenology in line with recent proposals.[66,67]

## Acknowledgements

We would like to thank our long time collaborators Dong-Pil Min and Hee-Jung Lee whose work is reflected in these pages and who have contributed greatly to the effort. We owe inspiration and gratitude to Mannque Rho, who during many years has been a motivating force behind our research. Skyrmion physics had a boom in the late 80's and thereafter only a few groups have maintained this activity obtaining very beautiful results, which however, have hardly influenced the community. We hope that this book contributes to make skyrmion physics more widely appreciated. Byung-Yoon Park thanks the members of Departamento de Física Teórica of the University of Valencia for their hospitality. Byung-Yoon Park and Vicente Vento were supported by grant FPA2007-65748-C02-01 from Ministerio de Ciencia e Innovación.

## References


1. J.-E. Alam, S. Chattopadhyay, T. Nayak, B. Sinha and Y.P. Viyogi, Quark Matter 2008, *J. Phys. G* **35** (2008) 100301.
2. T.H.R. Skyrme, A Nonlinear field theory, *Proc. Roy. Soc. Lond. A* **260** (1961) 127.
3. T.H.R. Skyrme, A Unified Field Theory of Mesons and Baryons, *Nucl. Phys.* **31** (1962) 556.
4. F. Karsch, Recent lattice results on finite temerature and density QCD, part II, *PoS.* **LAT2007** (2007) 015.
5. M. Fromm and P. de Forcrand, Revisiting strong coupling QCD at finite temperature and baryon density (2008).
6. G. 't Hooft, A Planar Diagram Theory For Strong Interactions, *Nucl. Phys. B* **72** (1974) 461.
7. E. Witten, Current Algebra, Baryons, and Quark Confinement, *Nucl. Phys. B* **223** (1983) 433.
8. H.-J. Lee, B.-Y. Park, D.-P. Min, M. Rho and V. Vento, A unified approach to high density: Pion fluctuations in skyrmion matter, *Nucl. Phys. A* **723** (2003) 427.
9. B.-Y. Park, M. Rho and V. Vento, Vector mesons and dense skyrmion matter, *Nucl. Phys. A* **736** (2004) 129.
10. B.-Y. Park, M. Rho and V. Vento, The Role of the Dilaton in Dense Skyrmion Matter, *Nucl. Phys. A* **807** (2008) 28.
11. H.-J. Lee, B.-Y. Park, M. Rho and V. Vento, Sliding vacua in dense skyrmion matter, *Nucl. Phys. A* **726** (2003) 69.
12. H.-J. Lee, B.-Y. Park, M. Rho and V. Vento, The Pion Velocity in Dense Skyrmion Matter, *Nucl. Phys. A* **741** (2004) 161.
13. A.C. Kalloniatis, J.D. Carroll and B.-Y. Park, Neutral pion decay into nu anti-nu in dense skyrmion matter, *Phys. Rev. D* **71** (2005) 114001.
14. A.C. Kalloniatis and B.-Y. Park, Neutral pion decay in dense skyrmion matter, *Phys. Rev. D* **71** (2005) 034010.
15. B.-Y. Park, H.-J. Lee, and V. Vento, Skyrmions at finite density and temperature: the chiral phase transition (2008).
16. G.S. Adkins, C.R. Nappi, and E. Witten, Static Properties of Nucleons in the Skyrme Model, *Nucl. Phys. B* **228** (1983) 552.
17. A.D. Jackson and M. Rho, Baryons as Chiral Solitons, *Phys. Rev. Lett.* **51** (1983) 751.
18. E. Witten, Global Aspects of Current Algebra, *Nucl. Phys. B* **223** (1983) 422.
19. S. Weinberg, Phenomenological Lagrangians, *Physica A* **96** (1979) 327.
20. A.A. Migdal and M.A. Shifman, Dilaton Effective Lagrangian in Gluodynamics, *Phys. Lett. B* **114** (1982) 445.
21. J. R. Ellis and J. Lanik, Is scalar gluonium observable?, *Phys. Lett. B* **150** (1985) 289.
22. M. Bando, T. Kugo and K. Yamawaki, Nonlinear Realization and Hidden Local Symmetries, *Phys. Rept.* **164** (1988) 217.

23. M. Harada and K. Yamawaki, Hidden local symmetry at loop: A new perspective of composite gauge boson and chiral phase transition, *Phys. Rept.* **381** (2003) 1.

24. I.R. Klebanov, Nuclear matter in the skyrme model, *Nucl. Phys. B* **262** (1985) 133.

25. A.S. Goldhaber and N.S. Manton, Maximal symmetry of the skyrme crystal, *Phys. Lett. B* **198** (1987) 231.

26. A.D. Jackson and J.J.M. Verbaarschot, Phase structure of the skyrme model, *Nucl. Phys. A* **484** (1988) 419.

27. Z. Tesanovic, O. Vafek and M. Franz, Chiral symmetry breaking and phase fluctuations: A QED-3 theory of the pseudogap state in cuprate superconductors, *Phys. Rev. B* **65** (2002) 180511.

28. G. Kaelbermann, Nuclei as skyrmion fluids, *Nucl. Phys. A* **633** (1998) 331.

29. O. Schwindt and N.R. Walet, Soliton systems at finite temperatures and finite densities (2002).

30. I. Zahed and G.E. Brown, The Skyrme Model, *Phys. Rept.* **142** (1986) 1.

31. A. Jackson, A.D. Jackson and V. Pasquier, The Skyrmion-Skyrmion Interaction, *Nucl. Phys. A* **432** (1985) 567.

32. B. Schwesinger, H. Weigel, G. Holzwarth and A. Hayashi, The skyrme soliton in pion, vector and scalar meson fields: pi n scattering and photoproduction, *Phys. Rept.* **173** (1989) 173.

33. L. Castillejo, P.S.J. Jones, A.D. Jackson, J.J.M. Verbaarschot and A. Jackson, Dense Skyrmion Systems, *Nucl. Phys. A* **501** (1989) 801.

34. M. Kugler and S. Shtrikman, A new skyrmion crystal, *Phys. Lett. B* **208** (1988) 491.

35. N.S. Manton and P.M. Sutcliffe, Skyrme crystal from a twisted instanton on a four torus, *Phys. Lett. B* **342** (1995) 196.

36. M. Kutschera, C.J. Pethick and D.G. Ravenhall, Dense matter in the chiral soliton model, *Phys. Rev. Lett.* **53** (1984) 1041.

37. M. Kugler and S. Shtrikman, Skyrmion crystals and their symmetries, *Phys. Rev. D* **40** (1989) 3421.

38. H. Forkel et al., Chiral symmetry restoration and the skyrme model, *Nucl. Phys. A* **504** (1989) 818.

39. M.F. Atiyah and N.S. Manton, Skyrmions from Instantons, *Phys. Lett. B* **222** (1989) 438.

40. M.F. Atiyah and N.S. Manton, Geometry and kinematics of two skyrmions, *Commun. Math. Phys.* **153** (1993) 391.

41. R.A. Leese and N.S. Manton, Stable instanton generated Skyrme fields with baryon numbers three and four, *Nucl. Phys. A* **572** (1994) 575.

42. N.R. Walet, Quantising the B=2 and B=3 Skyrmion systems, *Nucl. Phys. A* **606** (1996) 429.

43. B.-Y. Park, D.-P. Min, M. Rho and V. Vento, Atiyah-Manton approach to Skyrmion matter, *Nucl. Phys. A* **707** (2002) 381.

44. R. Jackiw, Quantum meaning of classical field theory, *Rev. Mod. Phys.* **49** (0000) 681.

45. S. Saito, T. Otofuji and M. Yasino, Pion Fluctuations about the Skyrmion, *Prog. Theor. Phys.* **75** (1986) 68.

46. H. Yabu, F. Myhrer and K. Kubodera, Meson condensation in dense matter revisited, *Phys. Rev. D* **50** (1994) 3549.

47. V. Thorsson and A. Wirzba, S-wave Meson-Nucleon Interactions and the Meson Mass in Nuclear Matter from Chiral Effective Lagrangians, *Nucl. Phys. A* **589** (1995) 633.

48. W.R. Gibbs and W.B. Kaufmann, The Contribution of the Quark Condensate to the pi N Sigma Term, *nucl-th/0301095* (2003).

49. G.E. Brown and M. Rho, Scaling effective Lagrangians in a dense medium, *Phys. Rev. Lett.* **66** (1991) 2720.

50. G.E. Brown, A.D. Jackson, M. Rho and V. Vento, The nucleon as a topological chiral soliton, *Phys. Lett. B* **140** (1984) 285.

51. R.J. Furnstahl, H.-B. Tang and B.D. Serot, Vacuum contributions in a chiral effective Lagrangian for nuclei, *Phys. Rev. C* **52** (1995) 1368.

52. C. Song, G.E. Brown, D.-P. Min and M. Rho, Fluctuations in 'Brown-Rho scaled' chiral Lagrangians, *Phys. Rev. C* **56** (1997) 2244.

53. H. Reinhardt and B. V. Dang, Modified Skyrme Model with correct QCD scaling behavior on S3, *Phys. Rev. D* **38** (1988) 2881.

54. K. Zarembo, Possible pseudogap phase in qcd, *JETP Lett.* **75** (2002) 59.

55. T. Hatsuda and T. Kunihiro, The sigma-meson and pi pi correlation in hot/dense medium: Soft modes for chiral transition in QCD (2001).

56. H. Fujii, Scalar density fluctuation at critical end point in NJL model, *Phys. Rev. D* **67** (2003) 094018.

57. R.D. Pisarski and M. Tytgat, Propagation of Cool Pions, *Phys. Rev. D* **54** (1996) 2989.

58. H. Leutwyler, Nonrelativistic effective Lagrangians, *Phys. Rev. D* **49** (1994) 3033.

59. M. Kirchbach and A. Wirzba, In-medium chiral perturbation theory and pion weak decay in the presence of background matter, *Nucl. Phys. A* **616** (1997) 648.

60. D.T. Son and M. A. Stephanov, Real-time pion propagation in finite-temperature QCD, *Phys. Rev. D* **66** (2002) 076011.

61. A. Bochkarev and J.I. Kapusta, Chiral symmetry at finite temperature: linear vs nonlinear $\sigma$-models, *Phys. Rev. D* **54** (1996) 4066.

62. I. Arsene et al., Quark-gluon plasma and color glass condensate at RHIC? The perspective from the BRAHMS experiment, *Nuclear Physics A* **757**(2005) 1.

63. U. G. Meissner, Low-energy hadron physics from effective chiral lagrangians with vector mesons, *Phys. Rept.* **161** (1988) 213.

64. B.D. Serot and J.D. Walecka, The relativistic nuclear many body problem, *Adv. Nucl. Phys.* **16** (1986) 1.

65. U.-G. Meissner, A. Rakhimov and U.T. Yakhshiev, The nucleon nucleon interaction and properties of the nucleon in a pi rho omega soliton model including a dilaton field with anomalous dimension, *Phys. Lett. B* **473** (2000) 200.

66. L. McLerran, Quarkyonic Matter and the Phase Diagram of QCD (2008).

67. L. McLerran and R.D. Pisarski, Phases of Cold, Dense Quarks at Large $N_c$, *Nucl. Phys. A* **796** (2007) 83.

Chapter 6

# Half-Skyrmion Hadronic Matter at High Density


Hyun Kyu Lee* and Mannque Rho*,†

*Department of Physics, Hanyang University, Seoul 133-791, Korea
†Institut de Physique Théorique, CEA Saclay, 91191 Gif-sur-Yvette, France



The hadronic matter described as a skyrmion matter embedded in an FCC crystal is found to turn into a half-skyrmion matter with vanishing (in the chiral limit) quark condensate and *non-vanishing* pion decay constant at a density $n_{1/2S}$ lower than or at the critical density $n_c^{\chi SR}$ at which hadronic matter changes over to a chiral symmetry restored phase with deconfined quarks. When hidden local gauge fields and dilaton scalars — one "soft" and one "hard" — are incorporated, this phase is characterized by $a = 1$, $f_\pi \neq 0$ with the hidden gauge coupling $g \neq 0$ but $\ll 1$. While chiral symmetry is restored in this region in the sense that $\langle \bar{q}q \rangle = 0$, quarks are still confined in massive hadrons and massless pions. This phase seems to correspond to the "quarkyonic phase" predicted in large $N_c$ QCD. It also represents the "hadronic freedom" regime relevant to kaon condensation at compact-star density. As $g \to 0$ (in the chiral limit), the symmetry "swells" — as an emergent symmetry due to medium — to $SU(N_f)^4$ as proposed by Georgi for the "vector limit." The fractionization of skyrmion matter into half-skyrmion matter is analogous to what appears to happen in condensed matter in (2+1) dimensions where half-skyrmions or "merons" enter as relevant degrees of freedom at the interface. Finally the transition from baryonic matter to color-flavor-locked quark matter can be bridged by a half-skyrmion matter.


## Contents


6.1 Introduction . . . . . . . . . . . . . . . . . . . . . . . . . . . . . . . . . . . . . . 148
6.2 Vector Mesons and Dilatons in Skyrmion Matter . . . . . . . . . . . . . . . . 150
    6.2.1 Dilatons in hidden local symmetry . . . . . . . . . . . . . . . . . . . . . 150
    6.2.2 The 1/2-skyrmion matter . . . . . . . . . . . . . . . . . . . . . . . . . . 152
    6.2.3 The effect of the $\omega$ meson . . . . . . . . . . . . . . . . . . . . . . . . . 154
6.3 Vector Symmetry at High Density . . . . . . . . . . . . . . . . . . . . . . . . . 155
    6.3.1 Hadronic freedom . . . . . . . . . . . . . . . . . . . . . . . . . . . . . . . 156
    6.3.2 The fate of neutron stars . . . . . . . . . . . . . . . . . . . . . . . . . . 158
6.4 Transition from Nuclear Matter to CFL Phase . . . . . . . . . . . . . . . . . . 159
6.5 Further Remarks . . . . . . . . . . . . . . . . . . . . . . . . . . . . . . . . . . . 160
References . . . . . . . . . . . . . . . . . . . . . . . . . . . . . . . . . . . . . . . . . . 161

## 6.1. Introduction

Hadronic matter at high density is presently poorly understood and the issue of the equation of state (EOS) in the density regime appropriate for the interior of compact stars remains a wide open problem. Unlike at high temperature where lattice QCD backed by relativistic heavy ion experiments is providing valuable insight into hot medium, the situation is drastically different for cold hadronic matter at a density a few times that of the ordinary nuclear matter relevant for compact stars. While asymptotic freedom should allow perturbative QCD to make well-controlled predictions at superhigh densities, at the density regime relevant for compact stars, there are presently neither reliable theoretical tools nor experimental guides available to make clear-cut statements. The lattice method, so helpful in high-T matter, is hampered by the sign problem and cannot as yet handle the relevant density regime.

What is generally accepted at the moment is that effective field theories formulated in terms of hadronic variables, guided by a wealth of experimental data, can accurately describe baryonic matter up to nuclear matter density $n_0 \approx 0.16$ fm$^{-3}$ and perturbative QCD unambiguously predicts that color superconductivity should take place in the form of color flavor locking (CFL) at some asymptotically high density $n_{CFL}$.[1] In between, say, $n_0 \leq n \leq n_{CFL}$, presently available in the literature are a large variety of model calculations which however have not been checked by first-principle theories or by experiments. The model calculations so far performed paint a complex landscape of phases from $n_0$ to $n_{CFL}$, starting with kaon condensation at $n_c^K \sim 3n_0$,[2] followed by a plethora of color superconducting quark matter with or without color flavor locking near and above the chiral restoration $n_c^{\chi SR}$ and ultimately CFL with or without kaon condensation. It is unclear which of the multitude of the phases could be realized and how they would manifest themselves in nature.

In this note, we would like to zero in on the vicinity of the chiral restoration density denoted $n_c^{\chi SR}$ at which both the quark condensate $\langle \bar{q}q \rangle$ and the physical pion decay constant $f_\pi$ go to zero in the chiral limit and explore a hitherto unsuspected novel phenomenon that could take place very near $n_c^{\chi SR}$. This can be efficiently done by putting skyrmions in a crystal lattice.[a] While the skyrmion structure has been extensively studied in hadronic physics as a description of a baryon in QCD at large $N_c$, one expects it to equally provide a powerful approach to many-body systems: A skyrmion with winding number $B$ is to encode entire strong interactions of QCD at large $N_c$ for systems with $B$ baryons. Thus the skyrmion description has the potential to provide a *unified approach* to baryonic dynamics, not only that of elementary baryons but also the structure of complex nuclei as well as infinite matter at any density, both below and above the deconfinement point. Perhaps

---

[a]No formulation for the phenomenon described in this note is available in continuum but we expect the topological structure will remain intact in the continuum limit.

academic but theoretically fascinating is the possibility that the CFL phase can also be described as a skyrmion matter of different form — to be referred to as "superqualiton" matter. Thus the transition from normal matter to CFL matter can be considered as a skyrmion-superqualiton transition, with a half-skyrmion phase as the border between the two.[3]

Treating dense nuclear matter in terms of skyrmion matter, we will argue that at a density denoted $n_{1/2S}$ lying at $\lesssim n_c^{\chi SR}$, a skyrmion in dense matter fractionizes into two half skyrmions with chiral $SU(N_f) \times SU(N_f)$ symmetry restored but with a *non-vanishing* pion decay constant $f_\pi \neq 0$. Phrased in terms of hidden local symmetry Lagrangian where the lowest-lying vector mesons $\rho$ and $\omega$ are introduced in addition to the Nambu-Goldstone pions, the chiral symmetry is restored in the sense that $f_\pi = f_\sigma \neq 0$ where

$$\langle \pi^b(q)|A_\mu^a|0\rangle = iq_\mu \delta^{ab} f_\pi, \quad \langle \sigma^b(q)|V_\mu^a|0\rangle = iq_\mu \delta^{ab} f_\sigma \tag{6.1.1}$$

but with $\langle \bar{q}q \rangle = 0$, where $\sigma^b$ is the Nambu-Goldstone boson to be higgsed to become the longitudinal component of the $\rho$ meson. Here chiral symmetry is restored but the phase transition scenario differs from the standard Nambu-Goldstone-to-Wigner Weyl transition in that $f_\pi \neq 0$ in this phase. To distinguish this phase from the usual chiral symmetry restored phase with $\langle \bar{q}q \rangle \sim f_\pi = 0$, we shall refer to it as "1/2-skyrmion phase." This phase could be identified with the "vector symmetry" of Georgi.[4,b]

It should be noted that in this half-skyrmion phase, quarks are still confined although chiral symmetry is restored. Thus it resembles the "quarkyonic phase" predicted[6] in the large $N_c$ limit of QCD characterized by the order parameter $B_0$ in which chiral symmetry is restored but the quarks are confined. In this phase baryons in which the quarks are confined are massive, so cannot enter in the 't Hooft anomaly condition.[c] The 't Hooft anomaly matching could be assured in the 1/2-skyrmion phase by the massless pion which is present.

We suggest that the phase $n_{1/2S} \lesssim n \lesssim n_c^{\chi SR}$ can also be identified with the "hadronic freedom" regime and $n_{1/2S}$ as the "flash density" $n_{flash}$,[2] both of which play an important role in describing dense matter near and just below the chiral transition point.[d] There is also a tantalizing analogy between the half-skyrmion phase present in dense matter and the meron phases in (2+1) dimensions encountered in condensed matter.

---

[b]It was argued by Harada and Yamawaki[5] that the vector limit with $g = 0$ and $f_\pi = f_\sigma \neq 0$ does not satisfy the axial Ward identity and that it is the limit $g = f_\pi = 0$, called "vector manifestation," that does. A comment will be made on this point later.

[c]The anomaly matching condition states that a composite particle has to reproduce exactly the anomaly present in the fundamental theory, that is to say that the fundamental anomaly and the anomalies in the composite theory must match. For this matching to be satisfied by the composite system, there must exist massless excitations. It has been shown that this matching condition holds in the presence of chemical potential.[7]

[d]The corresponding temperature in hot medium is called "flash temperature." More on this below in connection with heavy ion collisions.[8]

## 6.2. Vector Mesons and Dilatons in Skyrmion Matter

Up to date, most of the works done on skyrmions relied on the Skyrme Lagrangian that contains the current algebra term and the Skyrme term,[e] viz.

$$\mathcal{L} = \frac{F_\pi^2}{4}\mathrm{Tr}(\partial_\mu U \partial^\mu U^\dagger) + \frac{1}{32e^2}\mathrm{Tr}[U^\dagger \partial_\mu U, U^\dagger \partial^\nu U]^2 \qquad (6.2.2)$$

implemented with mass terms. But there are compelling reasons to believe that other degrees of freedom than the pions are essential for reliably describing systems with $B > 1$. It seems certain that both vector and scalar excitations are essential. It has in fact been argued since sometime that vector mesons must figure in the topological structure of elementary baryon as well as baryonic matter.[10,f] Indeed the recent development in holographic dual QCD (hQCD)[14] indicates that not just the lowest vector mesons but the infinite tower of vector mesons encapsulated in five-dimensional (5D) Yang-Mills Lagrangian can drastically modify the structure of baryons arising as instantons.[15–17] This suggests that dense matter described with a hidden local symmetric Lagrangian with the infinite tower would be drastically different from the picture given by the pion-only skyrmion description. This point will be addressed below. Furthermore it has become evident that certain scalar degrees of freedom associated with the trace anomaly of QCD could also figure crucially.[12,13] This development came about in implementing broken scale invariance in the skyrmion structure of dense matter built in the presence of vector mesons. A remarkable structure arises in the presence of the $\omega$ meson and two scalar mesons corresponding to the dilatons of spontaneously broken scale invariance as we will describe.

### 6.2.1. *Dilatons in hidden local symmetry*

To start with, let us describe the Lagrangian with which we will develop our arguments. To bring out the notion that hidden local symmetry in low-energy dynamics is quite generic, it is instructive to see how hidden local fields "emerge" naturally from a low-energy theory.[18] As will be noted, the same structure can be obtained top-down from string theory.

The idea is simply that the chiral field $U = e^{2i\pi/F_\pi}$ — which represent the coordinates for the symmetry $SU(N_f)_L \times SU(N_f)_R / SU(N_f)_{L+R}$ — can be written

----

[e]We reserve $f_\pi$ for the physical pion decay constant while $F_\pi$ stands for a parameter in the Lagrangian. In the mean field approximation used below, they are equivalent.

[f]A glaring defect of the skyrmion with pion fields but with no other fields (such as the vector mesons $\rho$, $\omega$ etc) is that when applied to nuclei, the parameters needed to even approximately fit nature are totally unnatural. For instance, the parameter $f_\pi$ is much too small compared with the physical value $f_\pi \approx 93$ MeV — this is so even for a single nucleon — and the pion mass parameter $m_\pi$ is much too large compared with its free-space value $m_\pi \approx 140$ MeV. See e.g.[11] When the parameters are taken to be close to their physical values, the resulting structure at the mean field level of complex nuclei, e.g., shape, comes out to be completely different from what is known in nature.

in terms of the left and right coset-space coordinates as

$$U = \xi_L^\dagger \xi_R \qquad (6.2.3)$$

with transformation under $SU(N_f)_L \times SU(N_f)_R$ as $\xi_L \to \xi_L L^\dagger$ and $\xi_R \to \xi_R R^\dagger$ with $L(R) \in SU(N_f)_{L(R)}$. Now the redundancy that is hidden, namely, the invariance under the *local* transformation

$$\xi_{L,R} \to h(x)\xi_{L,R} \qquad (6.2.4)$$

where $h(x) \in SU(N_f)_{V=L+R}$ can be elevated to a local gauge invariance[5] with the corresponding gauge field $V_\mu \in SU(N_f)_V$ that transforms

$$V_\mu \to h(x)(V_\mu + i\partial_\mu)h^\dagger(x). \qquad (6.2.5)$$

The resulting hidden local symmetry (HLS) Lagrangian given in terms of the covariant derivative $D_\mu$ takes the form[4] (with $V_\mu = g\rho_\mu$):

$$\mathcal{L} = \frac{F_\pi^2}{4}\mathrm{Tr}\{|D_\mu\xi_L|^2 + |D_\mu\xi_R|^2 + \gamma|D_\mu U|^2\} - \frac{1}{2}\mathrm{Tr}\left[\rho_{\mu\nu}\rho^{\mu\nu}\right] + \cdots \qquad (6.2.6)$$

where the ellipsis stands for higher derivative and other higher dimension terms including the gauged Skyrme term. If one parameterizes $\xi_{L,R} = e^{i\sigma/F_\sigma}e^{\mp i\pi/F_\pi}$, gauge-fixing with $\sigma = 0$ corresponds to unitary gauge, giving the usual gauged non-linear sigma model with a mass term for the gauge field. Clearly one can extend such a construction to an infinite tower of vector mesons spread in energy in the fifth dimension  Such a construction has been made and led to the so-called "dimensionally deconstructed QCD" encapsulated in a 5D Yang-Mills theory.[19] The latter is essentially equivalent in form to the 5D Yang-Mills theory of holographic dual QCD that comes from string theory.[14] This infinite-tower HLS theory will be denoted as $\mathrm{HLS}_\infty$. As noted by Harada, Matsuzaki and Yamawaki,[20] the Lagrangian (6.2.6) — denoted in an obvious notation as $\mathrm{HLS}_1$ — can be thought of as a truncated version of $\mathrm{HLS}_\infty$ where *all* other than the lowest vector mesons $\rho$ and $\omega$ are integrated out.

For studying the properties of dense hadronic matter, the scaling behavior of the effective Lagrangian is crucial. In fact the early description of how hadron properties change in hot/dense medium was anchored on the role played by the scalar dilaton associated with the trace anomaly of QCD.[21] It was clear then that the spontaneous breaking of chiral symmetry which leads to the generation of hadron masses and the explicit breaking of scale invariance by the quantum anomaly in QCD, which brings a length scale, must be connected. How to introduce scalar degrees of freedom to the HLS Lagrangian (6.2.6) is, however, not so straightforward since both chiral symmetry breaking ($\chi$SB) and confinement are intricately involved. In Ref. 13, this problem was solved by introducing two dilatons, one "soft" and the other "hard," with the soft dilaton $\chi_s$ intervening in $\chi$SB and the hard dilaton $\chi_h$ intervening in confinement-deconfinement. By integrating out the latter to focus on the chiral symmetry properties of hadrons, a suitable $\mathrm{HLS}_1$ Lagrangian was obtained in Ref. 13. Written in unitary gauge and with some harmless simplifications, it takes the form (including the pion mass term) for two light flavors (up and down)[g]:

$$\mathcal{L} = \mathcal{L}_{\chi_s} + \mathcal{L}_{hWZ} \tag{6.2.7}$$

where

$$\mathcal{L}_{\chi_s} = \frac{F_\pi^2}{4}\kappa^2\mathrm{Tr}(\partial_\mu U^\dagger \partial^\mu U) + \kappa^3 v^3 \mathrm{Tr}M(U + U^\dagger)$$
$$- \frac{F_\pi^2}{4}a\kappa^2\mathrm{Tr}[\ell_\mu + r_\mu + i(g/2)(\vec{\tau}\cdot\vec{\rho}_\mu + \omega_\mu)]^2$$
$$- \tfrac{1}{4}\vec{\rho}_{\mu\nu}\cdot\vec{\rho}^{\mu\nu} - \tfrac{1}{4}\omega_{\mu\nu}\omega^{\mu\nu} + \tfrac{1}{2}\partial_\mu\chi_s\partial^\mu\chi_s + V(\chi_s) \tag{6.2.8}$$
$$\mathcal{L}_{hWZ} = \tfrac{3}{2}g\kappa^3\omega_\mu B^\mu \tag{6.2.9}$$

where $\kappa = \chi_s/f_{\chi_s}$ with $f_{\chi_s} = \langle 0|\chi_s|0\rangle$ and

$$B^\mu = \frac{1}{24\pi^2}\varepsilon^{\mu\nu\alpha\beta}\mathrm{Tr}(U^\dagger\partial_\nu U U^\dagger\partial_\alpha U U^\dagger\partial_\beta U) \tag{6.2.10}$$

is the baryon current and

$$V(\chi_s) = B\chi_s^4\ln\frac{\chi_s}{f_{\chi_s}e^{1/4}} \tag{6.2.11}$$

is the dilaton potential.

### 6.2.2. *The 1/2-skyrmion matter*

For understanding a generic feature of dense skyrmion matter, it is illuminating to first consider the Skyrme Lagrangian coupled to the dilaton $\chi_s$ which is gotten from (6.2.7) by setting $\rho_\mu = \omega_\mu = 0$ and putting a quartic Skyrme term to assure the topological stability. There have been a series of works on dense matter treated with this Skyrme-dilaton Lagrangian[22] on which we will base our beginning arguments.

In Ref. 22, following the seminal work of Klebanov,[23] density effect is simulated by putting skyrmions in a crystal and squeezing the crystal. In (3+1) dimensions, it is found to be energetically favorable to arrange the skyrmions as a face-centered cubic crystal (FCC) lattice.[24] One should however recognize that there is no proof that this is indeed the absolute minimal configuration. There may be other configurations that are more favorable. Indeed, it has been recently shown that in baby-skyrmion systems,[25] of all possible crystalline structures, it is the hexagonal, not the cubic, that gives the minimal energy. This caveat notwithstanding, we will base our discussions on the FCC crystalline structure. We will say more on this below, in particular concerning certain qualitative features that could be different for different crystalline structures.

---

[g]For flavor number $N_f < 3$, the well-known 5D topological Wess-Zumino term is absent. However in the presence of vector mesons as in hidden local symmetry formulation, there are in general four terms — that we shall call "hWZ terms" — in the anomalous parity sector that satisfy homogenous anomaly equation. It turns out that if one requires vector dominance in photon-induced processes involving the hWZ terms and use the equation of motion for a heavy $\rho$ field, then the hWZ terms can be reduced to one term in the regime we are concerned with as given in this formula.

Briefly, what is done in Ref. 22 is as follows. The crystal configuration made up of skyrmions has each FCC lattice site occupied by a single skyrmion centered with $U_0 = -1$ with each nearest neighbor pair relatively rotated in isospin space by $\pi$ with respect to the line joining the pair. In order to have the Skyrme Lagrangian possess the correct scaling under scale change of the crystalline, the dilaton scalar $\chi$ associated with the trace anomaly of QCD has to be implemented as suggested in.[21] The energy density of the lattice skyrmions is then given by[22]

$$\epsilon = \frac{1}{4} \int_{Box} d^3x \left\{ \frac{f_\pi^2}{4} \left( \frac{\chi^*}{f_\chi} \right)^2 \mathrm{Tr}(\partial_i U_0^\dagger \partial_i U_0) + \cdots \right.$$
$$\left. + \frac{1}{2} \partial_i \chi^* \partial_i \chi^* + V(\chi) \right\} \tag{6.2.12}$$

where $f_\pi$ is the physical pion decay constant (which is equal to the parametric constant $F_\pi$ at the tree order). We have dropped the subscript "s" since we are dealing only with $\chi_s$. Here, the ellipsis stands for the familiar Skyrme quartic term and quark mass terms which need not be explicited, the subscript 'box' denotes that the integration is over a single FCC box and the factor $1/4$ in front appears because the box contains baryon number four. The asterisk "*" denotes the mean field (a density-dependent object for $n \neq 0$), $f_\chi$ is the $\chi$ decay constant and $V(\chi)$ is the dilaton potential for $\chi_s$.[13,26] The field $\chi$ is coupled to the chiral field $U$, so the mean field $\chi^* = \langle \chi \rangle_n$ (for a given density $n$) scales with the background provided by the crystal configuration. The minimization of this energy density with respect to the coefficients of the Fourier expansion of the (mean) fields taken as variational parameters reveals that at some minimum size of the box corresponding to a density, say, $n_{1/2S}$ of the matter, there is a phase transition from the FCC crystal configuration of skyrmions into a body-centered cubic crystal (BCC) configuration of half skyrmions as predicted on symmetry grounds.[24,27] We should point out two aspects here that characterize the transition. One is that what is involved here is a topology change, also observed in (2+1) dimensions. Therefore it has the possibility of being stable against quantum fluctuations. The other is that in terms of the mean chiral field $U_0(x) = \sigma(x) + i\tau \cdot \pi$, the expectation value $\langle \sigma \rangle \propto \langle \bar{q}q \rangle$ is zero at $n_{1/2S}$, so the transition is indeed a chiral restoration phase transition.

The result of the calculation in Ref. 22 uncovers several striking features in skyrmions at dense matter. The most prominent among them is that while chiral symmetry is restored at $n_{1/2S}$, the pion decay constant given by $f_\pi^*/f_\pi \propto \langle \chi \rangle_{n \geq n_p}/f_\chi \neq 0$. In terms of the chiral order parameter $\langle \bar{q}q \rangle \sim \langle c\mathrm{Tr}(U + U^\dagger) \rangle$, the 1/2-skyrmion phase has $\langle \mathrm{Tr}(U + U^\dagger) \rangle = 0$ but $c \sim f_\pi^* \neq 0$. A similar property has been proposed for high temperature and identified with a "pseudogap phase" in analogy to high T superconductivity.[28] Note that this phase is distinct from the standard $\chi$SB phase where the pion decay constant is directly proportional to the quark condensate. Although the connection is not clear, the phase between $n_p$ and the density denoted $n_c^{\chi SR}$ at which $f_\pi^* = 0$ is called "pseudogap phase." The density range of the pseudogap phase depends on the mass of the scalar $\chi$. As will be seen, the range can be shrunk to a point for certain value of the mass for the dilaton but there is always some region in which the 1/2-skyrmion phase is the lowest energy state.

### 6.2.3. *The effect of the $\omega$ meson*

In the presence of vector-meson fields, particularly the $\omega$ meson field, the phase structure is dramatically different from the one without vector mesons. With vector mesons, in particular, with the $\omega$, the dilaton $\chi$ plays a crucial role. There is a close interplay between the $\omega$ which supplies repulsion between skyrmions (nucleons) and the dilaton which provides attraction that leads to the binding in nuclei.[h]

How the dilaton influences dense matter depends on the mass of the dilaton that enters in the dynamics. At present the structure of the dilaton — in fact the structure of low-energy scalars in general — is not well understood.[29] In Ref. 13, two extreme cases were taken, a low mass object at $\sim 700$ MeV and a high mass at $\sim 3$ GeV, the two giving drastically different scenarios. Given that the "hard" dilaton whose excitation could be of the order of the high mass object taken, the latter may not be relevant to the phenomenon concerned whereas the low mass object $f_0(600)$ is most likely to be relevant.

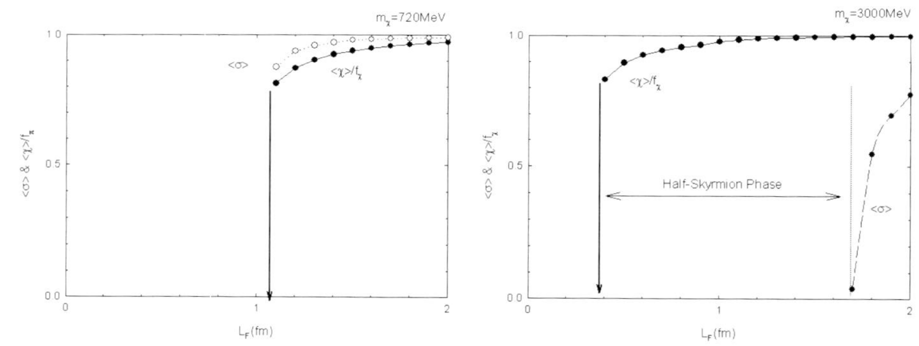

Fig. 6.1.  Behavior of $\langle \chi \rangle$ and $\langle \sigma \rangle \propto \langle \bar{q}q \rangle$ where $\sigma = \frac{1}{2}\mathrm{Tr}U$ as a function of lattice size for "light" dilaton mass $m_\chi = 720$ MeV (left figure) and for "heavy" dilaton mass $m_\chi = 3000$ MeV (right figure).

The result of the Lagrangian (6.2.7) put on an FCC crystal[12] is shown in Fig. 6.1. For a heavy dilaton with mass $m_\chi \gg 1$ GeV, there is a distinctive phase in which $\langle \bar{q}q \rangle^* \propto \langle \mathrm{Tr}U \rangle^* = 0$ but $f_\pi^* \sim \langle \chi \rangle^* \neq 0$. This phase has the skyrmions fractionized into half-skyrmions. However if the dilaton is light, say, $m_\chi \sim 700$ MeV, the 1/2-skyrmion phase shrinks to a point. The model cannot describe the confinement-

---

[h]It is worth mentioning here that the scalar "$\sigma$" in Walecka mean-field theory corresponds to this dilaton, not to the fourth component of the chiral four-vector in linear sigma model.

deconfinement transition but one expects that above the transition point, there could be a deconfined quark phase or a color superconducting phase. In either case, the pion decay constant $f_\pi^*$ and the $\omega$ mass $m_\omega^*$ are predicted to decrease as density increases, both going to zero at the critical point (in the chiral limit).

The important role of the dilaton in the presence of the $\omega$ meson can be seen by modifying the homogeneous Wess-Zumino term (6.2.9). As mentioned, this is a special form gotten in certain approximations but its property is expected to be generic. Suppose that one sets $\kappa = 1$ in (6.2.9) and take

$$\mathcal{L}'_{hWZ} = \tfrac{3}{2}g\omega_\mu B^\mu. \tag{6.2.13}$$

In fact, one naively expects this to be of the correct form from the point of view of the scaling dimension of the hWZ term which is 4 if $\kappa = 1$. However as argued in Ref. 13, with the two dilatons $\chi_{s,h}$, one can construct (6.2.9) — with the exponent 3 on $\kappa$ — valid for the soft-dilaton sector without violating the scale invariance. Now what happens with the skyrmion matter simulated on FCC crystal using the hWZ Lagrangian (6.2.13) is a disastrous result totally at odds with nature[30]: Both $f_\pi^*$ and $m_\omega^*$ are found to *increase* with increasing density rather than decrease as desired.

This feature is not difficult to understand. The key point is that the $\omega$ meson gives rise to a Coulomb potential. The hWZ term then leads to the repulsive interaction, contributing to the energy per baryon, $E/B$, of the form

$$(E/B)_{hWZ} = \frac{9g^2}{16} \int_{\text{Box}} d^3x \int d^3x' B_0(\vec{x}) \frac{e^{-m_\omega^*|\vec{x} - \vec{x}'|}}{4\pi|\vec{x} - \vec{x}'|} B_0(\vec{x}'). \tag{6.2.14}$$

What is important is that this repulsive interaction turns out to dominate over other terms as density increases. Now while the integral over $\vec{x}$ is defined in a single lattice (FCC) cell, that over $\vec{x}'$ is not, so will lead to a divergence unless tamed. In order to prevent the $(E/B)_{hWZ}$ from diverging, $m_\omega^*$ has to *increase* sufficiently fast. And since $m_\omega^* \sim f_\pi^* g$ in this HLS model, for a fixed $g$, $f_\pi^*$ must therefore *increase*. In fact this phenomenon is a generic feature associated with the role that the vector mesons in the $\omega$ channel play in dense medium. This feature, however, is at variance with nature: QCD predicts that the pion decay constant tied to the chiral condensate should *decrease* and go to zero (in the chiral limit) at the chiral transition.

The suppression by the soft dilaton of the repulsion at high density has an important consequence on the maximum stable mass of neutron stars as described below.


## 6.3. Vector Symmetry at High Density

What is perhaps the most significant for dense matter near chiral restoration is that the 1/2-skyrmion (or pseudogap) state exhibits an emerging or "enhanced" symmetry. In $\mathrm{HLS}_1$ theory, the 1/2-skyrmion state has the chiral $SU(N_f)_L \times SU(N_f)_R$ symmetry restored. Thus for $n_{1/2S} \leq n < n_c^{\chi SR}$,

$$(F_\sigma/F_\pi)^2 \equiv a = 1, \quad F_\pi \neq 0, \tag{6.3.15}$$

which corresponds to $\gamma = 0$ in Eq. (6.2.6). Note however that the gauge coupling $g \neq 0$, so the vector meson remains massive. Since the vector meson is massive, $F_\sigma$ is the decay constant for the longitudinal component of the vector meson, not of a free scalar. The gauge coupling $g$ goes to zero, however, at chiral restoration, $n = n_c^{\chi SR}$. This corresponds to Georgi's "vector limit." As noted by Georgi,[4] at this point the symmetry "swells" to $SU(N_f)^4$, with $\xi_L$ and $\xi_R$ transforming under independent $SU(N_f) \times SU(N_f)$ symmetries,

$$\xi_L \rightarrow h_L(x)\xi_L L^\dagger, \quad \xi_R \rightarrow h_R(x)\xi_R R^\dagger, \tag{6.3.16}$$

where $L, R$ and $h_{L,R}$ are the unitary matrices generating the corresponding global and local $SU(N_f)$ groups. The hidden local symmetry is the diagonal sum of $SU(N_f)_{h_L}$ and $SU(N_f)_{h_R}$. Away from the vector limit, the non-zero gauge couplings break the vector symmetry explicitly producing the nonzero vector meson mass and couplings for the transverse components of the vector mesons. In terms of this symmetry pattern, we see that the pseudogap phase is the regime where one has $a = 1$ ($\gamma = 0$) and the gauge coupling $g$ is weak but non-zero.

It is noteworthy that while chiral symmetry is restored, the quarks are confined in hadrons. This suggests to identify the hadronic freedom (or pseudogap) regime to be "quarkyonic" as predicted in large $N_c$ QCD.

### 6.3.1. *Hadronic freedom*

The matter between the 1/2-skyrmion threshold density $n_{1/2S}$ and $n_{\chi SR}$ with the gauge coupling $g \rightarrow 0$ has been referred to as "hadronic freedom" region with $n_{1/2S}$ identified as a "flash density" $n_{flash}$ in analogy to the "flash temperature" in hot medium as defined below. This pseudogap region has an important astrophysical implication. With $\gamma \rightarrow 0$ ($a \rightarrow 1$), the gauge coupling $g$ goes to zero as density approaches $n_c^{\chi SR}$, so hadrons interact weakly in that regime. In Ref. 39, this reasoning was used to predict kaon condensation at a density $\sim 3n_0$. There the assumption was that kaons must condense somewhere between the flash density $n_{flash}$ and $n_c^{\chi SR}$. Therefore one can start from the vector manifestation fixed point of HLS theory with $a = 1$ and $g = 0$ but $F_\pi \neq 0$. This calculation reinforced the previous conclusion that kaons must condense *before* any other phase changes can take place and hence determine the fate of compact stars. This is reviewed in Ref. 2.

In relativistic heavy-ion collisions, one is dealing with high temperature and relatively low density. There is no equivalent hadronic description comparable to dense skyrmions for what happens between the chiral transition temperature $T_{\chi SR}$

and the "flash temperature" $T_{flash} \sim 120$ MeV at which hadrons, in particular, the $\rho$ meson, go nearly on-shell. There are however lattice calculations with dynamical quarks[31] on thermodynamic properties of hot matter which indicate that at a temperature corresponding to $T_{flash}$ at which the condensate of the soft dilaton $\langle \chi_s \rangle^*$ (the asterisk here denotes temperature dependence) starts "melting," vanishing at the chiral transition temperature $T_{\chi SR} \sim 200$ MeV. Between $T_{flash}$ and $T_{\chi SR}$, the gauge coupling must be small according to the hidden local symmetry in the vector manifestation.[5] Since temperature induces violation of the vector dominance in the photon-pion coupling, $a$ must be approaching 1. Thus with $g \approx 0$ and $a \approx 1$, this region can be considered as the temperature counter part of the "hadronic freedom" established above in density. In Ref. 8, it has been proposed that dileptons decouple from the $\rho$ meson in this hadronic freedom region, which could explain the recent dilepton measurements at CERN and RHIC where no evidences for precursors to chiral symmetry restoration are seen.

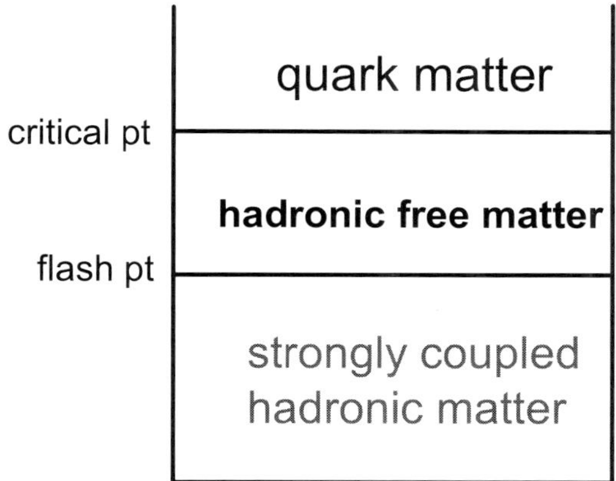



Fig. 6.2.   A schematic picture of the proposed phase diagram modified by the hidden local symmetry in the vector manifestation that is conjectured to lead to "hadronic freedom." The "critical (flash) pt" corresponds to $n_{\chi SR}$ ($n_{flash}$) in density and $T_{\chi SR}$ ($T_{flash}$) in temperature.

In sum, a new hitherto unsuspected phase — that is drastically different from the currently accepted one — emerges from the above observations, i.e., the "hadronic freedom phase," that connects a possible quark gluon phase to the normal hadronic phase, that ranges in the $(T, n)$ plane from $T_{flash} \lesssim T \lesssim T_{\chi SR}$ and $n = 0$ to $n_{flash} \lesssim n \lesssim n_{\chi SR}$ and $T = 0$ — which may be identified with the quarkyonic matter at large $N_c$. A schematic form of the new phase structure is given in Fig. 6.2. Here the "quark matter" stands for what might be identified with "sQGP"[i] at

---

[i]The state just above $T_{\chi SR}$ (and at low density) is not understood. In fact, it may have nothing to do with sQGP. All one can say at present is that it is most likely in the Wigner phase.

$T \gtrsim T_{\chi SR}$ and $n \approx 0$ and mnemonically for the variety of color superconducting states with chiral symmetry either restored or broken or true quark matter with chiral symmetry restored. The case where the CFL phase figures there presents an interesting, though academic, case as discussed below.

### 6.3.2. *The fate of neutron stars*

The suppression by the dilaton $\chi_s$ of the repulsion brought about by the $\omega$ exchange between baryons has an interesting consequence on the fate of compact stars that are more massive than some critical mass $M_{crit}$ of the star. Phrased in terms of an effective field theory (EFT) for nuclear matter, the taming of the $\omega$ repulsion in (6.2.14) can be understood as follows. In nuclear EFT, (6.2.14) represents the contribution to $(E/B)$ from the mean field of a four-Fermi interaction in the effective Lagrangian of the form

$$\frac{1}{2}C_\omega^{*\,2}(\overline{N}\gamma_\mu N)^2 \qquad\qquad (6.3.17)$$

where $N$ is the nucleon field and $C_\omega^{*\,2}$ is a constant proportional to $g^2/m_\omega^2$ coming from the $\omega$ exchange between two nucleons. It was observed[30] that if in (6.2.14), the coupling constant $g$ is taken to be a constant, the repulsion grows in the skyrmion matter as density increases. This is translated in (6.3.17) as the coefficient $C_\omega^{*\,2}$ growing with density. This means that the vector meson mass $m_\omega$ is decreasing at increasing density with the vector coupling held constant. The remedy to this disease discussed above and given in the references[12,13] correspond to the vector coupling decreasing in some proportion to the mass $m_\omega$. In fact, in $\mathrm{HLS}_1$ theory, the coupling constant $g$ is to drop proportionally to the quark condensate, and this circumvents the necessity for the $\omega$ mass to increase to counter the increasing repulsion. The intrinsic density dependence of the gauge coupling constant in $\mathrm{HLS}_1$, when truncated at the four-Fermi interaction level, subsumes, among others, three- and more-body forces and hence the repulsion that is generated when $g$ is held constant can be considered as an effect of repulsive many-body forces. This is indeed what is found in specific model studies in many-body nuclear physics approaches[32] where the many-body repulsion would lead to the maximum neutron star mass $\gtrsim 2M_\odot$ while it leaves unaffected the equation of state at the equilibrium density of normal nuclear matter. Such a repulsion sourced by many-body forces, if unsuppressed, would push kaon condensation to a density $n \gtrsim 7n_0$,[33] so that the maximum stable neutron star mass $M_{max} \simeq 1.56 M_\odot$ conjectured by Brown and Bethe[2,34] would be ruled out. Thus the role of the light dilaton which renders the skyrmion matter consistent with $\mathrm{HLS}_1$ theory is found to be crucial for the physics of compact stars. This issue will be addressed elsewhere.

## 6.4. Transition from Nuclear Matter to CFL Phase

At asymptotically high density, diquarks will condense to a form where color and flavor get locked. Here chiral symmetry is again spontaneously broken by the diquark condensate. Although it is not clear whether the color-flavor locked state is relevant for compact stars, so this phenomenon could be purely academic, it is nonetheless a theoretically interesting object. Now it is tempting to think of the phase transition from normal baryonic matter to quark matter going via the 1/2-skyrmion phase at some high density in analogy to the condensed matter case. To see whether this analogy can be made closer, let us consider the CFL phase of quark matter. In the real world of two (u and d) light flavors and one heavy (s) flavor, a variety of model calculations predict a multitude of superconducting states, some unstable and some others (such as LOFF crystalline) presumably stable, but we are going to consider, for simplicity, the CFL configuration which is favored for degenerate quark masses. Furthermore, there is also a possibility that the CFL phase can come down in density all the way to the nuclear matter density for a light-enough s-quark mass, say, in the presence of strong $U(1)_A$ anomaly.[36]

Since in the CFL phase, the global color symmetry $SU(3)$ is completely broken and chiral $SU(3)_L \times SU(3)_R$ (for $N_f = 3$) is broken down to the diagonal subgroup $SU(3)_V$,[1] low-energy excitations can be described by the coordinates $\xi_{C+L} \in SU(3)_{C+L}$ and $\xi_{C+R} \in SU(3)_{C+R}$ given in terms of the octet pseudoscalar $\pi$ and the octet scalar $s$. The scalars are eaten up by the gluons which become massive and are mapped one-to-one to the vector mesons present in the hadronic sector. The Lagrangian that describes low-energy excitations is of the same local gauge invariant form as (6.2.6). The gauge symmetry here is explicit, not hidden as in the hadronic sector but we will nonetheless call it HLS'. Now as in the hadronic sector, the HLS' Lagrangian supports solitons that carry fermion number $B$, which are nothing but skyrmions.[37] The CFL soliton is called "superqualiton" to be distinguished from the soliton in the hadronic phase. It is actually a quark excitation on top of the vacuum with condensed Cooper pairs, effectively color singlet with spin 1/2. But in this formulation, it is a topological object.

Given the skyrmion matter for $n \lesssim n_{1/2S}$ and the superqualiton matter for $n \gtrsim n_c^{\chi SR}$, the transition from nuclear matter to the CFL matter can be considered as a skyrmion-superqualiton transition with half skyrmions figuring in between. In both the skyrmion phase and the superqualiton phase, chiral symmetry is spontaneously broken and quarks are confined. The order parameters are however different, the former with $\langle \bar{q}q \rangle$ and the latter with $\langle qq \rangle$. The 1/2-skyrmion phase sandwiched by the two Nambu-Goldstone phases has chiral symmetry restored but quarks are still confined inside hadrons. Therefore the conjectured phase change takes place in a confined phase. This is the analogy to the Néel-VBS transition with half-skyrmions (spinons) at the boundary.[9] Independently of whether this analogy is just a coincidence or has a non-trivial meaning, what is significant is that the pseudogap region can deviate strongly from the Fermi-liquid state that is usually assumed in studying color superconductivity.

## 6.5. Further Remarks

The main assumption made in this note is that the dense skyrmion matter simulated in a crystal using $HLS_1$ Lagrangian represents dense baryonic matter. There are several questions one can raise here.

The first is whether there are no other crystal configurations that could (1) give a lower ground state and (2) induce different skyrmion fractionization. The answer to this is not known. It is unquestionably an important question to address. For instance, in (2+1) dimensions, while a baby-skyrmion fractionizes into two half-skyrmions for the known square-cell configuration, it is the hexagonal configuration that has the minimal energy and induces the fractionization of a baby-skyrmion into four *quarter-skyrmions*.[25]

Given that the skyrmion-1/2skyrmion transition scenario is anchored on the crystalline structure at the mean field level, one wonders whether quantum fluctuations would not wash out the soliton structure of the 1/2-skyrmion matter. As mentioned in,[22] since nuclear matter is known to be a liquid, not a crystal, it might be that quantum fluctuations would "melt" the crystal. The phase change could then be merely a lattice artifact although at high density baryonic matter is favored to be in the form of a crystal. In addition, the spin and statistics of the 1/2-skyrmion would require quantization. It seems highly plausible however that given that the transition involved here is a topology change, the phase change be robust against quantum fluctuations. Similar issues are raised in condensed matter physics where the concept of "topological order" is invoked for robustness of topology-changing phase transitions.

It should be stressed that the half-skyrmions "live" in the confined phase, i.e., hadrons, so need not have to be identified with the QCD degrees of freedom, i.e., the quarks with color and fractional electric charges.

The next unanswered question is the mechanism for the fractionization of a skyrmion to two half-skyrmions at $n_{1/2S}$. The fractionization under certain external conditions seems generic, taking place both in (2+1) and (3+1) dimensions. The treatment made in this note was based on energetics considerations but the mechanism was left unclarified. In the condensed matter case discussed in Ref. 9, the key role for the fractionization is played by the emergent $U(1)$ gauge field and its monopole structure. The pair of half-skyrmions (referred to as "up-meron" and "down-anti-meron" in Ref. 9) are confined — or bound — to a single skyrmion in both the initial Néel state and the final VBS state but the skyrmion fractionizes into half skyrmions at the boundary due to the "irrelevance" of the monopole tunneling, with an emergent global symmetry not present in the many-body Hamiltonian. It would be exciting to see a similar mechanism at work in the present case. It could elucidate what the hadronic phase could be at the doorway to color superconductivity, should the latter survive the black-hole collapse following kaon condensation.[2] In this regard, it would be interesting to investigate the skyrmion-1/2skyrmion transition in terms of the instantons and merons of 5D Yang-Mills Lagrangian of hQCD which would reveal the role of the infinite tower.

If the pseudogap phase is indeed the hadronic freedom region, how can one exploit the background provided by the half-skyrmion matter for describing kaon condensation? One may embed and bind $K^-$'s in dense half-skyrmion matter where $a \approx 1$ and $g \sim 0$ and exploit that in compact stars, electrons with high chemical potential decay into $K^-$'s once the kaon mass falls sufficiently low and the kaons Bose-condense. To do this calculation, it may be necessary to know what the quantum structure of the half-skyrmion phase is.

## Acknowledgments

This work was supported in part by the WCU project of Korean Ministry of Education, Science and Technology (R33-2008-000-10087-0)

## References

1. For review, see K. Rajagopal and F. Wilczek, *At the frontier of particle physics: Handbook of QCD* ed by M. Shifman (World Scientific, Singapore, 2001) Vol. 3, p.2061.
2. For review, see G.E. Brown, C.-H. Lee and M. Rho, "Recent developments on kaon condensation and its astrophysical implications," *Phys. Rept.* **462** (2008) 1 [arXiv:0708.3137 [hep-ph]].
3. M. Rho, "Hidden local symmetry and dense half-skyrmion matter," arXiv:0711.3895 [nucl-th].
4. H. Georgi, "New realization of chiral symmetry," *Phys. Rev. Lett.* **63** (1989) 1917; "Vector realization of chiral symmetry," *Nucl. Phys. B* **331** (1990) 311.
5. M. Harada and K. Yamawaki, "Hidden local symmetry at loop: A new perspective of composite gauge bosons and chiral phase transition," *Phys. Rept.* **381** (2003) 1.
6. L. McLerran and R.D. Pisarski, "Phases of cold, dense quarks at large $N_c$," *Nucl. Phys. A* **796** (2007) 83.
7. S.D.H. Hsu, F. Sannino and M. Schwetz, "Anomaly matching in gauge theories at finite matter density," *Mod. Phys. Lett. A* **16** (2001) 1871 [arXiv:hep-ph/0006059].
8. G.E. Brown, M. Harada, J.W. Holt, M. Rho and C. Sasaki, "Hidden local field theory and dileptons in relativistic heavy ion collisions," *Prog. Theor. Phys.*, in press, arXiv:0901.1513 [hep-ph].
9. T. Senthil et al., "Deconfined quantum critical points," *Nature* **303** (2004) 1490.
10. M. Rho, "Hidden local symmetry and the vector manifestation of chiral symmetry in hot and/or dense matter," *Prog. Theor. Phys. Suppl.* **168** (2007) 519.
11. R. Battye and P. Sutcliffe, *Phys. Rev. C* **73** (2006) 055205; R. Battye et al, hep-th/0605284.
12. B.-Y. Park, M. Rho and V. Vento, "The role of the dilaton in dense skyrmion matter", *Nucl. Phys. A* **807** (2008) 28.
13. H.K. Lee and M. Rho, "Dilatons in hidden local symmetry for hadrons in dense matter," arXiv:0902.3361 [hep-ph].

14. T. Sakai and S. Sugimoto, "Low energy hadron physics in holographic QCD," *Prog. Theor. Phys.* **113** (2005) 843; "More on a holographic dual of QCD," *Prog. Theor. Phys.* **114** (2005) 1083.

15. D.-K. Hong, M. Rho, H.-U. Yee and P. Yi, "Chiral dynamics of baryons from string theory," *Phys. Rev. D* **76** (2007) 061901; "Dynamics of baryons from string theory and vector dominance," *JHEP* **09**, 063 (2007); "Nucleon form factors and hidden symmetry in holographic QCD," arXiv:0710.4615 [hep-ph].

16. K. Hashimoto, T. Sakai and S. Sugimoto, "Holographic baryons : Static properties and form factors from gauge/string duality," arXiv:0806.3122 [hep-th]; H. Hata, T. Sakai, S. Sugimoto and S. Yamato, "Baryons from instantons in holographic QCD," arXiv:hep-th/0701280.

17. K.Y. Kim and I. Zahed, "Electromagnetic baryon form factors from holographic QCD," *JHEP* **0809** (2008) 007 [arXiv:0807.0033 [hep-th]].

18. For a lucid exposition, see N. Arkani-Hamed, H. Georgi and M.D. Schwartz, "Effective field theory for massive gravitons and gravity in theory space," *Ann. Phys.* **305** (2003) 96.

19. D.T. Son and M.A. Stephanov, "QCD and dimensional deconstruction," *Phys. Rev. D* **69** (2004) 065020.

20. M. Harada, S. Matsuzaki and K. Yamawaki, "Implications of holographic QCD in ChPT with hidden local symmetry," *Phys. Rev. D* **74** (2006) 076004.

21. G.E. Brown and M. Rho, "Scaling effective Lagrangians in a dense medium," *Phys. Rev. Lett.* **66** (1991) 2720.

22. B.-Y. Park, D.-P. Min, M. Rho and V. Vento, *Nucl. Phys. A* **707** (2002) 381; H.-J. Lee, B.-Y. Park, D.-P. Min, M. Rho and V. Vento, *Nucl. Phys. A* **723** (2003) 427.

23. I.R. Klebanov, "Nuclear matter in the Skyrme model," *Nucl. Phys. B* **262** (1985) 133.

24. M. Kugler and S. Shtrikman, "Skyrmion crystals and their symmetries," *Phys. Rev. D* **40** (1989) 3421; L. Castillejo, P.S.J. Jones, A.D. Jackson and J.J.M Verbaarschot, "Dense skyrmion systems," *Nucl. Phys. A* **501** (1989) 801.

25. I. Hen and M. Karliner, "Hexagonal structure of baby skyrmion lattice," arXiv:0711.2387 [hep-th].

26. G.E. Brown, J.W. Holt, C-H. Lee and M. Rho, "Late hadronization and matter formed at RHIC: Vector manifestation, BR scaling and hadronic freedom," *Phys. Rept.* **439** (2006) 161.

27. A.S. Goldhaber and N.S. Manton, " Maximal symmetry of the Skyrme crystal," *Phys. Lett.B* **198** (1987) 231.

28. K. Zarembo, "Possible pseudogap phase in QCD," *JETP Lett.* **75** (2002) 59.

29. For a recent discussion, see A. H. Fariborz, R. Jora and J. Schechter, "Global aspects of the scalar meson puzzle," arXiv:0902.2825 [hep-ph].

30. B.-Y. Park, M. Rho and V. Vento, "Vector mesons and dense skyrmion matter," *Nucl. Phys. A* **736** (2004)129.

31. D.E. Miller, "Lattice QCD calculations for the physical equation of state," *Phys. Rept.* **443** (2007) 55

32. A. Akmal, V.R. Pandharipande and D.G. Ravenhall, "The equation of state for nucleon matter and neutron star structure," *Phys. Rev. C* **58** (1998) 1804 [arXiv:nucl-th/9804027].

33. V.R. Pandharipande, C.J. Pethick and V. Thorsson, "Kaon energies in dense matter," *Phys. Rev. Lett.* **75** (1995) 4567 [arXiv:nucl-th/9507023].

34. G.E. Brown and H.A. Bethe, "A scenario for a large number of low-mass black holes in the galaxy," *Astrophys. J.* **423** (1994) 659.

35. Y. Oh, B.-Y. Park and M. Rho, work in progress.

36. T. Hatsuda. M. Tachibana, N. Yamamoto and G. Baym, "New critical point induced by the axial anomaly in dense QCD," *Phys. Rev. Lett.* **97** (2006) 122001.
37. D.K. Hong, M. Rho and I. Zahed, "Qualitons at high density," *Phys. Lett. B* **468** (1999) 261.
38. G.E. Brown, C.-H. Lee and M. Rho, "Vector manifestation of hidden local symmetry, hadronic freedom, and the STAR $\rho^0/\pi^-$ ratio," *Phys. Rev. C* **74** (2006) 024906.
39. G.E. Brown, C.-H. Lee, H.-J. Park and M. Rho, "Study of strangeness condensation by expanding about the fixed point of the Harada-Yamawaki vector manifestation," *Phys. Rev. Lett.* **96** (2006) 062303.

## Chapter 7

# Superqualitons: Baryons in Dense QCD

Deog Ki Hong

*Department of Physics, Pusan National University, Busan 609-735, Korea*

QCD predicts matter at high density should exhibit color superconductivity. We review briefly several pertinent properties of color superconductivity and then discuss how baryons are realized in color superconductors. Especially, we explain an attempt to describe the color-flavor locked quark matter in terms of bosonic degrees of freedom, where the gapped quarks and Fermi sea are realized as Skyrmions, called superqualitons, and $Q$-matter, respectively.



## Contents



7.1   Introduction . . . . . . . . . . . . . . . . . . . . . . . . . . . . . . . . . . . . . . . . . 165
7.2   Color Superconducting Quark Matter . . . . . . . . . . . . . . . . . . . . . . 166
7.3   Color-Flavor Locked Quark Matter . . . . . . . . . . . . . . . . . . . . . . . . 167
7.4   Superqualitons and Gapped Quarks . . . . . . . . . . . . . . . . . . . . . . . 170
7.5   Bosonization of QCD at High Density . . . . . . . . . . . . . . . . . . . . . 174
7.6   Conclusion . . . . . . . . . . . . . . . . . . . . . . . . . . . . . . . . . . . . . . . . . . . 176
References . . . . . . . . . . . . . . . . . . . . . . . . . . . . . . . . . . . . . . . . . . . . . . . 176



## 7.1.  Introduction

Quantum chromodynamics (QCD) is now widely accepted as an undisputed theory of strong interactions. The QCD prediction on how its coupling changes at different energies is thoroughly tested, and well confirmed, for the wide range of energy from order of 1 GeV to a few hundred GeV by numerous and independent experiments. QCD is however extremely difficult to solve, since it is highly non-linear and strongly coupled at the same time, offering no apparant expansion parameters. So far it has precluded any analytic solutions.

One of the reasons why QCD is hard to solve is that quarks and gluons, the basic degrees of freedom of QCD, become less relevant at low energy, where they are strongly coupled. Since the right degrees of freedom of strong interactions at low energy are hadrons rather than quarks and gluons, one may try to solve QCD in terms of hadrons. The Skyrme picture based on chiral Lagrangian is such an attempt.[1,a] One writes down the effective Lagrangian for the pions in pow-

---

[a]Holographic QCD[2] is one of the recent attempts to solve QCD directly with hadrons, especially with pions and vector mesons.

ers of momentum in accord with the QCD realization of (chiral) symmetry and then determines the couplings in the effective Lagrangian by experimental data. One interesting aspect of chiral Lagrangian is that it admits a topological solition, which is stable if one allows the so-called Skyrme term only for quartic couplings. One can also show that the topological current associated with the soliton is nothing but the baryon number current which arises from the Wess-Zumino-Witten term.[3] The baryons are therefore realized as topological solitions, known as Skyrmions in the chiral Lagrangian. The phenomenology of Skyrme Lagrangian was quite successful.[4]

Recently QCD at high density[5,6] has been studied intensively not only because it is relevant to dense matter, found in compact stars like neutron stars or in heavy ion collisions, but it may shed light on the nonperturbative behavior of QCD like chiral symmetry breaking and color confinement. Furthermore, it is an interesting question to ask how the Skyrme picture changes as one increases baryon density, which will be addressed in this article.

The study of dense matter is ultimately related to the properties of quarks, the basic building blocks of atomic nuclei; how QCD and electroweak interaction of quarks behave at high baryon density. QCD predicts because of the asymptotic freedom a phase transition at baryon density around the QCD scale, $1/\Lambda^3_{\mathrm{QCD}}$, that dense hadronic matter become quark matter.[7] The wave function of quarks inside a nucleon overlaps with those of quarks in other nucleons as nucleons pack closely at high density, liberating quarks from nucleons.

QCD also predicts that quark matter should be color superconducting at high baryon density[8] since it is energetically preferred for quarks to form Cooper-pairs rather than to form quark-anti-quark condensates. Though color superconductivity has not been observed yet, one expects however to find it in the core of compact stars like neutron stars or quark stars. Finding color superconductivity will be a great challenge for QCD.

## 7.2. Color Superconducting Quark Matter

Unlike ordinary electron superconductors, color-superconducting quark matter has a rather rich phase structure because quarks have not only three different color charges but also come in several flavors, which makes it extremely interesting to find color superconductors. The number of quark flavors in quark matter depends on its density because the mass gap is flavor-dependent.

At intermediate density where the strange quarks are too heavy to populate, only up and down quarks participate in Cooper-pairing. Since the color anti-triplet channel provides attraction among quarks, the quark Cooper pairs are flavor singlet but transform as anti-fundamental under $SU(3)_c$ color:

$$\langle q^q_{Li}(\vec{p})q^b_{Lj}(-\vec{p})\rangle = -\langle q^q_{Ri}(\vec{p})q^b_{Rj}(-\vec{p})\rangle = \epsilon_{ij}\epsilon^{ab3}\Delta_2, \qquad (7.2.1)$$

where $i, j = 1, 2$ and $a, b = 1, 2, 3$ are flavor and color indices, respectively, and $\Delta_2$ is the gap opened at the Fermi surface of two-flavor quark matter. (We will call the 3 direction in color space as blue.) In the ground state of two flavor quark matter the Cooper pairs form condensates, breaking $SU(3)_c$ down to its subgroup $SU(2)_c$. Since the Cooper pairs are flavor singlet, the ground state preserves all the global symmetries of QCD except the U(1) baryon number, which is broken down to $Z_2$ by the condensation of the Cooper pairs. In two-flavor color superconductors five among eight gluons are coupled to the Cooper-pairs, becoming massive due to Higgs mechanism, and the Cooper-pair gap opens at the Fermi surface of green and red quarks.

The confinement scale, $\Lambda_C$, of unbroken $SU(2)_c$ is much smaller than the QCD scale[9] and also parametrically much smaller than the gap, $\Delta_2$. At energy lower than $\Lambda_C$ (and also lower than the Cooper-pair gap, $\Delta_2$), the particle spectrum of two-flavor quark matter consists of a massless Nambu-Goldstone boson associated with broken $U(1)_B$ and gapless up and down blue quarks, which should remain gapless as the chiral symmetry is unbroken, while four massive gluons and gapped quarks (red and green quarks), which are fundamental under $SU(2)_c$, are confined in bound states like baryons or glueballs.[10,11] (The massive 8th gluon is neutral under $SU(2)_c$ and thus decoupled from the rest of particles.) Baryons in two-flavor color superconductors are like a heavy (blue) quarkonium made of red and green quarks.

## 7.3. Color-Flavor Locked Quark Matter

As nucleons pack closely together, they will eventually form quark matter. The critical density or critical chemical potential for the phase transition to quark matter is rather difficult to estimate due to the nature of strong interactions. While the lattice calculation for the critical temperature to form quark matter has been quite successful,[12] lattice is of not much help at finite density due to the notorious sign problem associated with the complex measure, barring the Monte Carlo method.[13,14] One expects, however, the phase transition at finite density presumably occurs around at the quark chemical potential, $\mu \sim \Lambda_{QCD}$, solely on dimensional grounds, which corresponds to about 5 to 10 times the nuclear density, $n_0 \simeq 0.17 \ \mathrm{fm}^{-3}$.

In the previous section we assumed strange quarks are decoupled at intermediate density. However, matter at density close to the critical density strange quarks are not completely decoupled as the quark chemical potential is comparable to the strange quark mass, $m_s \simeq 100 \ \mathrm{MeV}$. Significant fraction of quark matter is therefore composed of strange quarks. Whether strange quarks participate in Cooper-paring with up and down quarks near the critical density is still an open issue, because we do not know yet whether the pairing gap is bigger than the stress to break pairing, $m_s^2/(2\mu)$, due to the mismatch of the Fermi surfaces of pairing quarks. On the other hand at density much higher than the critical density one surely expects that all of three light flavors do participate in Cooper-pairing. In fact one can show rigorously that Cooper-pairs take a so-called color-flavor locked (CFL) form[15] at asymptotic density,[16]

$$\langle q_{Li}^a(\vec{p}) q_{Lj}^b(-\vec{p}) \rangle = - \langle q_{Ri}^a(\vec{p}) q_{Rj}^b(-\vec{p}) \rangle = \Delta\, \epsilon^{ab\alpha} \epsilon_{ij\alpha} \,, \qquad (7.3.2)$$

where the flavor indices $i, j$ now run from 1 to 3 and we neglected the color-sextet components, since the instanton effect is negligible.

The color-flavor locked phase of quark matter turns out to be quite stable against various stress[17] and also theoretically very interesting. The particle spectrum of CFL phase maps one-to-one onto that of low density (hypernuclear) hadron matter, as if there is no phase boundary between them.[18] The chiral symmetry is spontaneously broken because the rotations of both left and right-handed quarks are locked to the same color-rotations. If one rotates both color and flavor simultaneously, the Cooper pairs remain invariant. The condensate of CFL Cooper-pairs also breaks the U(1)$_{\rm em}$ electromagnetism. Since the quarks transform under U(1)$_{\rm em}$ as

$$q \mapsto e^{i\varphi\, Q_{\rm em}}\, q \qquad (7.3.3)$$

where $Q_{\rm em} = {\rm diag}\,(2/3, -1/3, -1/3)$, the U(1)$_{\rm em}$ transformation on quarks can be undone by U(1)$_Y$ color hyper-charge transformation. A linear combination of photon and hyper-charge component of gluons, $A_\mu^Y$, remain un-Higgsed. The modified photon of unbroken U(1)$_{\tilde Q}$ is given as

$$\tilde{A}_\mu = A_\mu \cos\theta + A_\mu^Y \sin\theta, \quad \cos\theta = \frac{g_s}{\sqrt{e^2 + g_s^2}} \,, \qquad (7.3.4)$$

where $e$ is the electromagnetic coupling and $g_s$ is the QCD coupling. The symmetry breaking pattern in CFL phase is therefore given as

$$SU(3)_c \times U(1)_{\rm em} \times SU(3)_L \times SU(3)_R \times U(1)_B \to SU(3)_{c+L+R} \times Z_2 \times U(1)_{\tilde Q} \,. \quad (7.3.5)$$

At high baryon density antiquarks are highly suppressed, since it takes energy bigger than the chemical potential to excite them. An effective theory of modes near the Fermi surface, called High Density Effective Theory, has been derived by integrating out the modes far away from the Fermi surface.[19,20] The Dirac mass term, which breaks the chiral symmetry explicitly, is suppressed in dense medium and gives mass operators as, once antiquarks are integrated out,

$$m\bar{q}q = \frac{mm^\dagger}{2\mu} q_+^\dagger q_+ + \frac{mm^T}{4\mu^2}\left( q_+^\dagger \Delta\, q_{c+} + {\rm h.c.} \right) + \cdots \,, \qquad (7.3.6)$$

where $q_+$ denotes quarks near the Fermi surface and $q_{c+}$ their charge-conjugate fields.[21] Therefore, mass of pseudo NG bosons becomes suppressed in dense medium as $m^2/(2\mu)$. The particle spectrum of CFL phase consists of 8 pseudo Nambu-Goldstone (NG) bossons of mass $m^2/(2\mu)$ and one massless Nambu-Goldstone boson, corresponding to the baryon superfluid, and 8 massive gluons, and 9 massive quarks. Under the unbroken global symmetry, $SU(3)_{c+L+R} \times Z_2$, the particles transform as in Table 7.1.[b]

Table 7.1.  Particle spectrum of CFL phase.

| Particles | Spin | Mass | $SU(3)_{c+L+R}$ | $U(1)_B$ | $Z_2$ |
|---|---|---|---|---|---|
| NG bosons | 0 | $O(\frac{m^2}{2\mu})$ | $8 \oplus 1$ | 0 | $+1$ |
| Gapped quarks | 1/2 | $\Delta$ | 8 | $\frac{1}{3}$ | $-1$ |
| Gapped quarks | 1/2 | $2\Delta$ | 1 | $\frac{1}{3}$ | $-1$ |
| Gluons | 1 | $O(g_s\Delta)$ | 8 | 0 | $+1$ |

The ground state of the CFL phase is nothing but the Fermi sea with gap opened at Fermi surfaces of all nine quarks by Cooper-pairing; the octet under $SU(3)_{c+L+R}$ has a gap $\Delta$ while the singlet has $2\Delta$. The collective excitations of Cooper-pairs are possible without any energy gap at arbitrarily low energy, exhibiting superflows of mass and color charges.[c] To describe the low-energy excitations of the CFL quark matter, we introduce composite (diquark) fields $\phi_L$ and $\phi_R$ as

$$\phi_{L(R)\,ai}(x) \equiv \lim_{y \to x} \frac{|x - y|^{\gamma_m}}{\kappa} \, \epsilon_{abc}\epsilon_{ijk} q_{L(R)}^{bj}(-\vec{v}_F, x) q_{L(R)}^{ck}(\vec{v}_F, y), \qquad (7.3.7)$$

where $\gamma_m$ is the anomalous dimension of the diquark fields and $q(\vec{v}_F, x)$ denotes a quark field with momentum close to a Fermi momentum $p_F$.[19] A dimensional quantity $\kappa$ is introduced in (7.3.7) so that the expectation value of diquark fields in the ground state becomes identity,

$$\langle \phi_L \rangle = - \langle \phi_R \rangle = I \,. \qquad (7.3.8)$$

Under the color and flavor symmetry the diquark fields transform as

$$\phi_L \mapsto g_c^T \, \phi_L \, g_L \,, \quad \phi_R \mapsto g_c^T \, \phi_R \, g_R \,, \quad g \in SU(3) \,. \qquad (7.3.9)$$

The low-lying excitations of condensates are then described by the following unitary matrices,

$$U_L(x) = g_c^T(x) g_L(x), \quad U_R(x) = g_c^T(x) g_R(x) \,, \qquad (7.3.10)$$

which may be parameterized by Nambu-Goldstone fields as

$$U_{L(R)}(x) = \exp\left(2i\Pi_{L(R)}^A(x) T^A / F_\pi\right) \,, \qquad (7.3.11)$$

where $T^A$ are the SU(3) generators, normalized as $\mathrm{Tr}\, T^A T^B = \frac{1}{2}\delta^{AB}$. The parity-even combination of the Nambu-Goldstone bosons, constituting the longitudinal

[b]The baryon number is spontaneously broken by a condensate of Cooper pairs, which carry $B = 2/3$. So, it is defined modulo 2/3.

[c]The Cooper-pair breaks $U(1)$ electromagnetism but leaves $U(1)_{\tilde{Q}}$ unbroken, a linear combination of $U(1)$ electromagnetism and $U(1)_Y$ color-hypercharge. Therefore CFL quark matter will have a finite resistance for $\tilde{Q}$ charges.

components of gluons, generates the color supercurrents in quark matter, while the parity-odd combination becomes pseudo Nambu-Goldstone bosons, neutral under color, which can be written as

$$\Sigma_i^j(x) \equiv U_{Lai}(x) U_R^{*aj}(x) = \exp\left(2i\,\Pi^A(x) T^A / F\right) , \qquad (7.3.12)$$

where $\Pi^A(x) = \Pi_L^A(x) - \Pi_R^A(x)$ are correlated excitations of $\phi_L$ and $\phi_R$, having same quantum numbers as pions, kaons, and eta in hadronic phase.

Expanding in powers of derivatives, the low-energy effective Lagrangian density for the (colored) Nambu-Goldstone bosons is given as

$$\mathcal{L}_0 = \frac{F^2}{4}\,\mathrm{Tr}\left(\partial_0 U_L \partial_0 U_L^\dagger - v^2\,\vec{\nabla} U_L \cdot \vec{\nabla} U_L^\dagger\right) + n_L \mathcal{L}_{\mathrm{WZW}} + (L \leftrightarrow R) + \cdots (7.3.13)$$

where $v = 1/\sqrt{3}$ is the speed of Nambu-Goldstone bosons in medium and the Wess-Zumino-Witten (WZW) term $\mathcal{L}_{\mathrm{WZW}}$ is added to correctly reproduce the symmetries of dense QCD. The colored NG bosons will couple to (massive) gluons through minimal coupling, replacing the ordinary derivatives with covariant derivatives, $D = \partial + i g_s A$, which amounts to adding to the effective Lagrangian, (7.3.13), the minimal gauge coupling and the gluon mass

$$\mathcal{L}_1 = -g_s A_\mu J^\mu - m_g^2\,\mathrm{Tr}\,A_\mu^2 \qquad (7.3.14)$$

and also replacing with covariant ones the plain derivatives in Wess-Zumino-Witten term to reproduce the anomalous coupling of NG bosons with vector mesons.

The coefficient of the WZW term in the effective Lagrangian should be chosen to match the global anomalies of microscopic theory. For instance the $\mathrm{SU}(3)_L$ anomaly is given at the quark level as

$$\partial J_{L\mu}^A = \frac{\tilde{e}^2}{32\pi^2}\,\mathrm{Tr}\left(T^A \tilde{Q}^2\right)\,\epsilon_{\mu\nu\rho\sigma}\tilde{F}^{\mu\nu}\tilde{F}^{\rho\sigma} , \qquad (7.3.15)$$

where $\tilde{F}$ is the field strength tensor of the modified photon. On the other hand the WZW term contains a term, if one gauges $\mathrm{U}(1)_{\tilde{Q}}$,

$$\mathcal{L}_{WZW} \ni \frac{n_L \tilde{e}^2}{64\pi^2 F}\Pi^0 \epsilon_{\mu\nu\rho\sigma}\tilde{F}^{\mu\nu}\tilde{F}^{\rho\sigma} , \qquad (7.3.16)$$

which agrees with (7.3.15) if $n_L = 1$ since $\mathrm{Tr}\left(T^3 \tilde{Q}^2\right) = 1/2$.[23] Similarly, one can show that $n_R = 1$.


## 7.4. Superqualitons and Gapped Quarks

In vacuum chiral symmetry breaking occurs due to the condensation of quark-antiquark bilinear at strong coupling. The coefficients of operators in the chiral effective Lagrangian therefore contain the physics of strong dynamics and are hence very difficult to calculate directly from QCD. However, the chiral symmetry breaking in the color-flavor locked phase of quark matter occurs even at asymptotic density where the QCD coupling is extremely small, because it is due to the

Cooper-paring of quarks near the Fermi surface which can occur at arbitrarily weak attraction due to Cooper theorem.

The coefficients of operators in the low-energy effective Lagrangian of CFL matter are calculable at asymptotic density, using perturbation, called hard dense-loop approximation, which appropriately incorporates the medium effects. Similarly, the CFL gap, which characterizes the properties of CFL matter, can be also calculated precisely, using perturbation. However at not-so-high density where CFL matter is strongly coupled we do not have well-developed tools to study either the gap or the low-energy constants of the effective theory, as in the vacuum QCD. To study the properties of CFL matter at intermediate density the Skyrme's idea may be useful, which correctly captures the large $N_c$ behavior of baryons as topological solitons made of pions. In the case of CFL matter where quarks are deconfined, though gapped the topological solitons made out of (colored) NG bosons, called superqualitons, should be identified as gapped quarks, similar to Kaplan's qualiton[22] which realizes the constituent quarks inside nucleons. In this section we study the CFL gap of strongly interacting quark matter à la Skyrme.[23]

We first note that gapped quarks of each chirality should be treated independently, since Cooper-pairing occurs between quarks with same chirality. (We will concentrate on the left-handed gapped quark. But, the argument below applies equally to the right-handed gapped quark.) The manifold of NG bosons, $\Pi_L$, associated with the condensation of Cooper pairs of left-handed quarks has a nontrivial third homotopy,

$$\Pi_3 \left( \mathrm{SU}(3)_c \times \mathrm{SU}(3)_L / \mathrm{SU}(3)_{c+L} \right) = Z_n \,, \tag{7.4.17}$$

and thus the low-energy effective Lagrangian of $\Pi_L$ admits a topological soliton associated with a topological current,

$$J_L^\mu = \frac{1}{24\pi^2} \epsilon^{\mu\nu\rho\sigma} \, \mathrm{Tr} \, U_L^{-1} \partial_\nu U_L U_L^{-1} \partial_\rho U_L U_L^{-1} \partial_\sigma U_L \,, \tag{7.4.18}$$

whose charge counts the number of winding of the map $U_L$ from $S^3$, the boundary of space at infinity to the manifold of NG bosons. Since the sigma model description of the $\mathrm{SU}(3)_L$ quark current $J_{L\mu}^A = \bar{q}_{L+} T^A \gamma_\mu q_{L+}$ contains an anomalous piece from the WZW term,

$$J_{L\mu}^A \ni \frac{1}{24\pi^2} \epsilon^{\mu\nu\rho\sigma} \, \mathrm{Tr} \, T^A U_L^{-1} \partial_\nu U_L U_L^{-1} \partial_\rho U_L U_L^{-1} \partial_\sigma U_L \tag{7.4.19}$$

the topological current should be interpreted as the (left-handed) quark number current, $J_{L\mu} = \bar{q}_{L+} \gamma_\mu q_{L+}$ and the soliton of unit winding number should be identified as (left-handed) gapped quark, carrying a baryon number $1/3$ as $n_L = 1$ rather 3 in the case of vacuum QCD.

Once we identify the soliton as a gapped quark, it is straightforward to estimate the magnitude of the gap as a function of the low energy constants like the NG boson decay constant, $F$, or the QCD coupling, as the soliton is stabilized by the Coulomb repulsion due to color charges at the core. Following Skyrme[1] we seek a static configuration for the field $U_L$ in $SU(3)$ by embedding an $SU(2)$ hedgehog in color-flavor in $SU(3)$, with

$$U_{Lc}(x) = e^{i\vec{\tau}\cdot\hat{r}\theta(r)}, \quad U_R = 0, \quad A_0^Y = \omega(r), \quad \text{all other } A_\mu^A = 0, \qquad (7.4.20)$$

where $\tau$'s are Pauli matrices. The radial function $\theta(r)$ is monotonous and satisfies

$$\theta(0) = \pi, \quad \theta(\infty) = 0 \qquad (7.4.21)$$

for a soliton of winding number one. (Note that we can also look for a right-handed soliton by switching off the $U_L$ field. The solution should be identical because QCD is invariant under parity.) This configuration has only non-vanishing color charge in the color-hypercharge $Y$ direction

$$J_0^Y = \frac{\sin^2\theta\,\theta'}{2\pi r^2} \qquad (7.4.22)$$

and all others are zero. As shown in,[24] the energy of the configuration (7.4.20) is given as

$$E[\omega,\theta] = \int 4\pi r^2 dr \left[ -\frac{1}{2}\omega'^2 + F^2\left(\theta'^2 + 2\frac{\sin\theta^2}{r^2}\right) + \frac{g_s}{2\pi^2}\frac{\omega}{r^2}\sin^2\theta\,\theta' \right] \qquad (7.4.23)$$

and the total charge within a radius $r$ is

$$Q^Y(r) = g_s \int_0^r \text{Tr } Y J_0^Y(r')4\pi r'^2 dr' = -g_s\left(\frac{\theta(r) - \sin\theta(r)\cos\theta(r) - \pi}{\pi}\right). \qquad (7.4.24)$$

Using the Gauss law with screened charge density, we can trade $\omega$ in terms of $\theta(r)$,

$$\omega' = \frac{Q^Y(r)}{4\pi r^2}e^{-m_E r}, \qquad (7.4.25)$$

where $m_E = \sqrt{6\alpha_s/\pi}\,\mu$ is the electric screening mass for the gluons. Hence, the energy functional simplifies to

$$E[\theta] = \int_0^\infty \mathcal{E}(r)\,dr$$
$$= \int_0^\infty 4\pi dr \left[ F^2\left(r^2\theta'^2 + 2\sin^2\theta\right) + \frac{\alpha_s}{2\pi}\left(\frac{\theta - \sin\theta\cos\theta - \pi}{2r}\right)^2 e^{-2m_E r} \right], \qquad (7.4.26)$$

where $\alpha_s = g_s^2/(4\pi)$. The squared size of the superqualiton is $R_S^2 = \langle r^2 \rangle$ where the averaging is made using the (weight) density $\mathcal{E}(r)$.

The equation of motion for the superqualiton profile $\theta(r)$ is

$$\left(r^2\theta'\right)' = \sin 2\theta + \frac{\alpha_s}{4\pi}\frac{e^{-2m_E r}}{(F r)^2}\sin^2\theta\left(\pi - \theta + \frac{1}{2}\sin 2\theta\right) \qquad (7.4.27)$$

subject to the boundary conditions (7.4.21). We solve the equation Eq. (7.4.27) numerically for several values of $m_E$ and $\alpha_s$. The profile function of the superqualiton

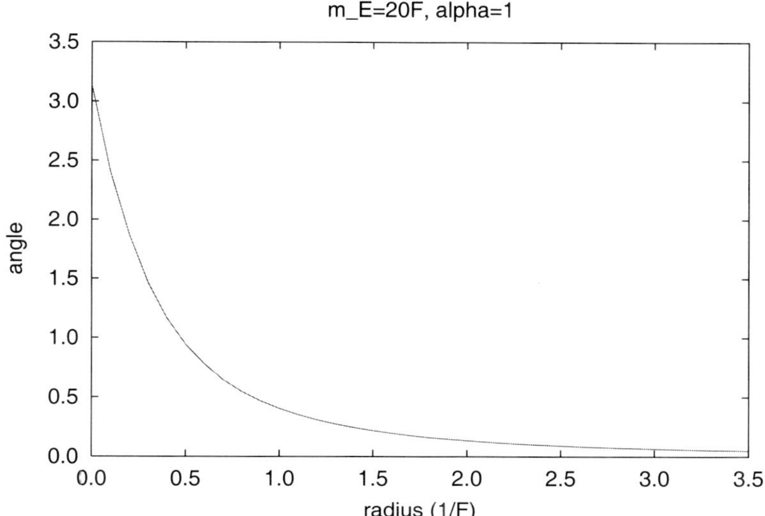


Fig. 7.1.  The qualiton profile for $m_E = 20F$ and $\alpha_s = 1$.

for $m_E = 20F$ and $\alpha_s = 1$ is shown in Fig. 7.1. For $m_E/F = 1, 10, 100$ and $\alpha_s = 1$ we find the soliton mass $M_S = 2.41$, $2.08$, $2.07 \times 4\pi F$ and $R_S = 1.36$, $1.35$, $1.347\, F^{-1}$, respectively. By varying the coupling for a fixed screening mass, $m_E = F$, we find $R_s = 1.25$, $1.30$, $1.58/F$ for $\alpha_s = 0.1$, $1$, $10$, respectively, showing that the soliton gets bigger for the stronger coupling, since the color-electric force, which balances the kinetic energy of the soliton, is more repulsive.[24]

To access the quantum numbers and the spectrum of the superqualiton, we note as usual that for any static solution to the equations of motion, one can generate another solution by a rigid $SU(3)$ rotation,

$$U(x) \mapsto AU(x)A^{-1}, \quad A \in SU(3). \tag{7.4.28}$$

The matrix $A$ corresponds to the zero modes of the superqualiton. Note that two $SU(3)$ matrices are equivalent if they differ by a matrix $h \in U(1) \subset SU(3)$ that commutes with $SU(2)$ generated by $\vec{\tau} \otimes I$. The Lagrangian for the zero modes is given by substituting $U(\vec{x}, t) = A(t)U_c(\vec{x})A(t)^{-1}$.[25] Hence,

$$L[A] = -M_s + \frac{1}{2}I_{\alpha\beta}\mathrm{Tr}\; T^\alpha A^{-1}\dot{A}\mathrm{Tr}\; T^\beta A^{-1}\dot{A} - i\frac{1}{2}\mathrm{Tr}\; Y A^{-1}\dot{A}, \tag{7.4.29}$$

where $I_{\alpha\beta}$ is an invariant tensor on $\mathcal{M} = SU(3)/U(1)$ and the hypercharge $Y$ is

$$Y = \frac{1}{3}\begin{pmatrix} 1 & 0 & 0 \\ 0 & 1 & 0 \\ 0 & 0 & -2 \end{pmatrix}. \tag{7.4.30}$$

Under the transformation $A(t) \mapsto A(t)h(t)$ with $h \in U(1)_Y$

$$L \mapsto L - \frac{i}{2} \mathrm{Tr}\, Y h^{-1}\dot{h}. \tag{7.4.31}$$

Therefore, if we rotate adiabatically the soliton by $\theta$ in the hypercharge space in $SU(3)$, $h = \exp(iY\theta)$, for time $T \to \infty$, then the wave function of the soliton changes by a phase in the semiclassical limit;

$$\psi(T) \sim e^{i\int dt L}\psi(0) = e^{i\theta/3}\psi(0), \tag{7.4.32}$$

where we neglected the irrelevant phase $-M_S T$ due to the rest mass energy. The simplest and lowest energy configuration that satisfies Eq. (7.4.32) is the fundamental representation of $SU(3)$. In a similar way, under a spatial (adiabatic) rotation by $\theta$ around the $z$ axis, $h = \exp(i\tau^3\theta)$, the phase of the wave function changes by $\theta/2$. Therefore, the ground state of the soliton is a spin-half particle transforming under the fundamental representation of both the flavor and the color group, which leads us to conclude that the soliton is a gapped left-handed (or right-handed) quark in the CFL phase.

## 7.5. Bosonization of QCD at High Density

The gap in superconductors can be estimated by measuring the energy needed to excite a pair of particle and hole, breaking a Cooper-pair. If one decreases the total number of particles by $\delta N$, creating holes in Fermi Sea, the thermodynamic potential (or total energy at zero temperature) of ground state is reduced by $\delta E = \mu \delta N$. Therefore the gap in the superqualiton should be defined as[d]

$$\Delta = \frac{1}{2}\left(M_S - \delta E\right), \tag{7.5.33}$$

where $M_S$ is mass of soliton, calculated from (7.4.26).

Quark matter with finite baryon number is described by QCD with a chemical potential, which restricts the system to have a fixed baryon number on average;

$$\mathcal{L} = \mathcal{L}_{\mathrm{QCD}} - \mu \sum_{i=u,d,s} \bar{q}_i\gamma^0 q_i, \tag{7.5.34}$$

where $\bar{q}_i\gamma^0 q_i$ is the quark number density of the $i$-th flavor. The ground state in the CFL phase is nothing but the Fermi sea where all quarks are Cooper-paired. Equivalently, this system can be described in term of bosonic degrees of freedom, namely pions and kaons, which are small fluctuations of Cooper pairs.[26]

As the baryon number (or the quark number) is conserved, though spontaneously broken, the ground state in the bosonic description should have the same baryon (or quark) number as the ground state in the fermionic description. Under the $U(1)_Q$ quark number symmetry, the bosonic fields transform as

$$U_{L,R} \mapsto e^{i\theta Q}U_{L,R}e^{-i\theta Q} = e^{2i\theta}U_{L,R}, \tag{7.5.35}$$

---

[d]One needs energy, twice of the gap, to break a Cooper pair.

where $Q$ is the quark number operator, given in the bosonic description as

$$Q = i \int \mathrm{d}^3 x \, \frac{F^2}{4} \mathrm{Tr} \left[ U_L^\dagger \partial_t U_L - \partial_t U_L^\dagger U_L + (L \leftrightarrow R) \right]. \tag{7.5.36}$$

The energy in the bosonic description is

$$E = \int \mathrm{d}^3 x \frac{F^2}{4} \mathrm{Tr} \left[ |\partial_t U_L|^2 + \left| \vec{\nabla} U_L \right|^2 + (L \leftrightarrow R) \right] + E_m + \delta E, \tag{7.5.37}$$

where $E_m$ is the energy due to meson mass and $\delta E$ is the energy coming from the higher derivative terms. Assuming the meson mass energy is positive and $E_m + \delta E \geq 0$, which is reasonable because $\Delta/F \ll 1$, we can take, dropping the positive terms due to the spatial derivative,

$$E \geq \int \mathrm{d}^3 x \frac{F^2}{4} \mathrm{Tr} \left[ |\partial_t U_L|^2 + (L \leftrightarrow R) \right] (\equiv E_Q). \tag{7.5.38}$$

Since for any number $\alpha$

$$\int \mathrm{d}^3 x \, \mathrm{Tr} \left[ |U_L + \alpha i \partial_t U_L|^2 + (L \leftrightarrow R) \right] \geq 0, \tag{7.5.39}$$

we get a following Schwartz inequality,

$$Q^2 \leq I \, E_Q, \tag{7.5.40}$$

where we defined

$$I = \frac{F^2}{4} \int \mathrm{d}^3 x \, \mathrm{Tr} \left[ U_L U_L^\dagger + (L \leftrightarrow R) \right]. \tag{7.5.41}$$

Note that the lower bound in Eq. (7.5.40) is saturated for $E_Q = \omega Q$ or

$$U_{L,R} = e^{i\omega t} \quad \text{with} \quad \omega = \frac{Q}{I}. \tag{7.5.42}$$

The ground state of the color superconductor, which has the lowest energy for a given quark number $Q$, is nothing but a so-called $Q$-matter, or the interior of a very large $Q$-ball.[27,28] Since in the fermionic description the system has the quark number $Q = \mu^3/\pi^2 \int \mathrm{d}^3 x = \mu^3/\pi^2 \cdot I/F^2$, we find, using $F \simeq 0.209\mu$,[29]

$$\omega = \frac{1}{\pi^2} \left( \frac{\mu}{F} \right)^3 F \simeq 2.32\mu, \tag{7.5.43}$$

which is numerically very close to $4\pi F$. The ground state of the system in the bosonic description is a $Q$-matter whose energy per unit quark number is $\omega$.

Since, reducing the quark number of the $Q$-matter by one, the minimum energy we gain by creating a hole in Fermi sea is $\delta E = \omega$ and therefore the energy cost to create a gapped quark from the ground state in the bosonic description is

$$2\Delta = M_S - \omega, \tag{7.5.44}$$

where $M_S$ is the energy of the superqualiton configuration in (7.4.26). From (7.5.44) we can estimate the CFL gap of strongly interacting quark matter.[26]

## 7.6. Conclusion

Solving QCD is an outstanding problem in physics. We review an attempt to solve three-flavor QCD at finite density in terms of pions, kaons, and eta that occur as collective excitations of condensed Cooper-pairs, following Skyrme's idea that was applied to strong interactions. This attempt is promising and conceptually beautiful, since it deals with the correct degrees of freedom at low energy. The ground state of color-flavor locked phase is realized as a $Q$-matter, a collective excitation of Nambu-Golstone bosons, carrying a fixed baryon number. The gapped quarks are realized as topological solitions, made of NG bosons, similar to Kaplan's qualiton picture of constituent quarks.

The Skyrme's picture on baryons is used to estimate the color-flavor locked gap in strongly interacting quark matter, where perturbation fails, after correctly identifying the ground state of color-superconducting quark matter.

## Acknowledgments

The author thanks M. Rho for the invitation to contribute to this volume. He is also grateful to S. T. Hong, Y. J. Park, M. Rho, and I. Zahed for the collaborations upon which this review is based. This work was supported in part by KOSEF Basic Research Program with the grant No. R01-2006-000-10912-0 and also by the Korea Research Foundation Grant funded by the Korean Government (MOEHRD, Basic Research Promotion Fund) (KRF-2007-314- C00052)

## References

1. T.H.R. Skyrme, *Proc. R. Soc. A* **260** (1961) 127.
2. See, for instances, T. Sakai and S. Sugimoto, "Low energy hadron physics in holographic QCD," *Prog. Theor. Phys.* **113** (2005) 843; D.K. Hong, M. Rho, H.U. Yee and P. Yi, "Chiral dynamics of baryons from string theory," *Phys. Rev. D* **76** (2007) 061901; "Dynamics of baryons from string theory and vector dominance," *JHEP* **0709** (2007) 063.
3. E. Witten, "Global aspects Of current algebra," *Nucl. Phys. B* **223** (1983) 422.
4. G.S. Adkins, C.R. Nappi and E. Witten, "Static Properties of nucleons in The Skyrme Model," *Nucl. Phys. B* **228** (1983) 552.
5. For a recent review see M.G. Alford, A. Schmitt, K. Rajagopal and T. Schafer, "Color superconductivity in dense quark matter," *Rev. Mod. Phys.* **80** (2008) 1455 [arXiv:0709.4635 [hep-ph]].
6. For standard reviews, see D.K. Hong, "Effective theory of color superconductivity," *Prog. Theor. Phys. Suppl.* **168** (2007) 397; "Aspects of color superconductivity," *Acta Phys. Polon. B* **32** (2001) 1253; M.G. Alford, "Color superconducting quark matter," *Ann. Rev. Nucl. Part. Sci.* **51** (2001) 131; G. Nardulli, "Effective description of QCD at very high densities," *Riv. Nuovo Cim.* **25N3** (2002) 1; I.A. Shovkovy, "Two lectures on color superconductivity," *Found. Phys.* **35** (2005) 1309.
7. J.C. Collins and M.J. Perry, "Superdense matter: Neutrons or asymptotically free quarks?," *Phys. Rev. Lett.* **34** (1975) 1353.

8. B.C. Barrois, "Superconducting quark matter," *Nucl. Phys. B* **129** (1977) 390.

9. D.H. Rischke, D.T. Son and M.A. Stephanov, "Asymptotic deconfinement in high-density QCD," *Phys. Rev. Lett.* **87** (2001) 062001 [arXiv:hep-ph/0011379].

10. R. Casalbuoni, Z.Y. Duan and F. Sannino, "Low energy theory for 2 flavors at high density QCD," *Phys. Rev. D* **62** (2000) 094004 [arXiv:hep-ph/0004207].

11. R. Ouyed and F. Sannino, "The glueball sector of two-flavor color superconductivity," *Phys. Lett. B* **511** (2001) 66 [arXiv:hep-ph/0103168].

12. M. Cheng *et al.*, "The QCD Equation of State with almost Physical Quark Masses," *Phys. Rev. D* **77** (2008) 014511 [arXiv:0710.0354 [hep-lat]].

13. S. Hands, "Simulating dense matter," *Prog. Theor. Phys. Suppl.* **168** (2007) 253 [arXiv:hep-lat/0703017].

14. D.K. Hong and S.D.H. Hsu, "Positivity of high density effective theory," *Phys. Rev. D* **66** (2002) 071501 [arXiv:hep-ph/0202236].

15. M.G. Alford, K. Rajagopal and F. Wilczek, "Color-flavor locking and chiral symmetry breaking in high density QCD," *Nucl. Phys. B* **537** (1999) 443 [arXiv:hep-ph/9804403].

16. D.K. Hong and S.D.H. Hsu, "Positivity and dense matter," *Phys. Rev. D* **68** (2003) 034011 [arXiv:hep-ph/0304156].

17. M. Alford, C. Kouvaris and K. Rajagopal, "Gapless color-flavor-locked quark matter," *Phys. Rev. Lett.* **92** (2004) 222001 [arXiv:hep-ph/0311286].

18. T. Schafer and F. Wilczek, "Continuity of quark and hadron matter," *Phys. Rev. Lett.* **82** (1999) 3956 [arXiv:hep-ph/9811473].

19. D.K. Hong, "An effective field theory of QCD at high density," *Phys. Lett. B* **473** (2000) 118: "Aspects of high density effective theory in QCD," *Nucl. Phys. B* **582** (2000) 451.

20. T. Schafer, "Hard loops, soft loops, and high density effective field theory," *Nucl. Phys. A* **728** (2003) 251 [arXiv:hep-ph/0307074].

21. D.K. Hong, "Radiative mass in QCD at high density," *Phys. Rev. D* **62** (2000) 091501 [arXiv:hep-ph/0006105].

22. D.B. Kaplan, "Constituent quarks as collective excitations of QCD," *Phys. Lett. B* **235** (1990) 163; "Qualitons," *Nucl. Phys. B* **351** (1991) 137.

23. D.K. Hong, M. Rho and I. Zahed, "Qualitons at high density," *Phys. Lett. B* **468** (1999) 261 [arXiv:hep-ph/9906551].

24. D.K. Hong and S.G. Rajeev, "Towards a bosonization Of quantum electrodynamics," *Phys. Rev. Lett.* **64** (1990) 2475.

25. A.P. Balachandran, in *High Energy Physics 1985*, edited by M.J. Bowick and F. Gürsey (World Scientific, Singapore, 1986).

26. D.K. Hong, S.T. Hong and Y.J. Park, "Bosonization of QCD at high density," *Phys. Lett. B* **499** (2001) 125 [arXiv:hep-ph/0011027].

27. S. Coleman, "Q balls," *Nucl. Phys. B* **262** (1985) 263; G. Baym, "Pion condensation at finite temperature. Mean field theory," *Nucl. Phys. A* **352** (1981) 355.

28. D.K. Hong, "Q balls in superfluid He-3," *J. Low. Temp. Phys.* **71** (1988) 483.

29. D.T. Son and M.A. Stephanov, "Inverse meson mass ordering in color-flavor-locking phase of high density QCD," *Phys. Rev. D* **61** (2000) 074012 [hep-ph/9910491].

# Chapter 8

# Rotational Symmetry Breaking in Baby Skyrme Models

Marek Karliner and Itay Hen

*Raymond and Beverly Sackler School of Physics and Astronomy
Tel-Aviv University, Tel-Aviv 69978, Israel.
marek@proton.tau.ac.il*

We discuss one of the most interesting phenomena exhibited by baby skyrmions — breaking of rotational symmetry. The topics we will deal with here include the appearance of rotational symmetry breaking in the static solutions of baby Skyrme models, both in flat as well as in curved spaces, the zero-temperature crystalline structure of baby skyrmions, and finally, the appearance of spontaneous breaking of rotational symmetry in rotating baby skyrmions.

## Contents

8.1  Breaking of Rotational Symmetry in Baby Skyrme Models . . . . . . . . . . . . . . . .  180
    8.1.1  Results . . . . . . . . . . . . . . . . . . . . . . . . . . . . . . . . . . . . . . . .  181
8.2  The Lattice Structure of Baby Skyrmions . . . . . . . . . . . . . . . . . . . . . . . . .  185
    8.2.1  Baby skyrmions inside a parallelogram . . . . . . . . . . . . . . . . . . . . . .  188
    8.2.2  Results . . . . . . . . . . . . . . . . . . . . . . . . . . . . . . . . . . . . . . . .  189
    8.2.3  Semi-analytical approach . . . . . . . . . . . . . . . . . . . . . . . . . . . . . .  193
    8.2.4  Further remarks . . . . . . . . . . . . . . . . . . . . . . . . . . . . . . . . . . .  195
8.3  Baby Skyrmions on the Two-Sphere . . . . . . . . . . . . . . . . . . . . . . . . . . . .  196
    8.3.1  The baby Skyrme model on the two-sphere  . . . . . . . . . . . . . . . . . . . .  196
    8.3.2  Static solutions . . . . . . . . . . . . . . . . . . . . . . . . . . . . . . . . . . .  197
    8.3.3  Relation to the 3D Skyrme model . . . . . . . . . . . . . . . . . . . . . . . . .  199
    8.3.4  Results . . . . . . . . . . . . . . . . . . . . . . . . . . . . . . . . . . . . . . . .  200
    8.3.5  Further remarks . . . . . . . . . . . . . . . . . . . . . . . . . . . . . . . . . . .  202
8.4  Rotating Baby Skyrmions . . . . . . . . . . . . . . . . . . . . . . . . . . . . . . . . . .  202
    8.4.1  SBRS from a dynamical point of view  . . . . . . . . . . . . . . . . . . . . . .  203
    8.4.2  SBRS in baby Skyrme models . . . . . . . . . . . . . . . . . . . . . . . . . . .  204
    8.4.3  The baby Skyrme model on the two-sphere  . . . . . . . . . . . . . . . . . . . .  205
    8.4.4  Results . . . . . . . . . . . . . . . . . . . . . . . . . . . . . . . . . . . . . . . .  205
    8.4.5  The rational map ansatz . . . . . . . . . . . . . . . . . . . . . . . . . . . . . .  207
    8.4.6  Further remarks . . . . . . . . . . . . . . . . . . . . . . . . . . . . . . . . . . .  208
A.1  Obtaining Baby Skyrmion Solutions — The Relaxation Method . . . . . . . . . . . .  209
References  . . . . . . . . . . . . . . . . . . . . . . . . . . . . . . . . . . . . . . . . . . . . .  210

## 8.1. Breaking of Rotational Symmetry in Baby Skyrme Models

The Skyrme model[1,2] is an SU(2)-valued nonlinear theory for pions in (3+1) dimensions with topological soliton solutions called skyrmions. Apart from a kinetic term, the Lagrangian of the model contains a 'Skyrme' term which is of the fourth order in derivatives, and is used to introduce scale to the model.[3] The existence of stable solutions in the Skyrme model is a consequence of the nontrivial topology of the mapping $\mathcal{M}$ of the physical space into the field space at a given time, $\mathcal{M} : S^3 \rightarrow SU(2) \cong S^3$, where the physical space $\mathbb{R}^3$ is compactified to $S^3$ by requiring the spatial infinity to be equivalent in each direction. The topology which stems from this one-point compactification allows the classification of maps into equivalence classes, each of which has a unique conserved quantity called the topological charge.

The Skyrme model has an analogue in (2+1) dimensions known as the baby Skyrme model, also admitting stable field configurations of a solitonic nature.[4] Due to its lower dimension, the baby Skyrme model serves as a simplification of the original model, but nonetheless it has a physical significance in its own right, having several applications in condensed-matter physics,[5] specifically in ferromagnetic quantum Hall systems.[6-9] There, baby skyrmions describe the excitations relative to ferromagnetic quantum Hall states, in terms of a gradient expansion in the spin density, a field with properties analogous to the pion field in the 3D case.[10]

The target manifold in the baby model is described by a three-dimensional vector $\phi = (\phi_1, \phi_2, \phi_3)$ with the constraint $\phi \cdot \phi = 1$. In analogy with the (3+1)D case, the domain of this model $\mathbb{R}^2$ is compactified to $S^2$, yielding the topology required for the classification of its field configurations into classes with conserved topological charges. The Lagrangian density of the baby Skyrme model is given by:

$$\mathcal{L} = \frac{1}{2}\partial_\mu \phi \cdot \partial^\mu \phi - \frac{\kappa^2}{2}\left[(\partial_\mu \phi \cdot \partial^\mu \phi)^2 - (\partial_\mu \phi \cdot \partial_\nu \phi) \cdot (\partial^\mu \phi \cdot \partial^\nu \phi)\right] - U(\phi_3),$$

$$(8.1.1)$$

and consists of a kinetic term, a Skyrme term and a potential term.

While in (3+1) dimensions the latter term is optional,[11] its presence in the (2+1)D model is necessary for the stability of the solutions. However, aside from the requirement that the potential vanishes at infinity for a given vacuum field value (normally taken to be $\phi^{(0)} = (0, 0, 1)$), its exact form is arbitrary and gives rise to a rich family of possible baby-Skyrme models, several of which have been studied in detail in the literature. The simplest potential is the 'holomorphic' model with $U(\phi_3) = \mu^2(1 - \phi_3)^4$.[12-14] It is known to have a stable solution only in the charge-one sector (the name refers to the fact that the stable solution has an analytic form in terms of holomorphic functions). The model with the potential $U(\phi_3) = \mu^2(1 - \phi_3)$ (commonly referred to as the 'old' model) has also been extensively studied. This potential gives rise to very structured non-rotationally-symmetric multi-skyrmions.[4,15] Another model with $U(\phi_3) = \mu^2(1 - \phi_3^2)$ produces ring-like multi-skyrmions.[16] Other double-vacuum potentials which give rise to other types of solutions have also been studied.[17]

Clearly, the form of the potential term has a decisive effect on the properties of the minimal energy configurations of the model. It is then worthwhile to see how the multisolitons of the baby Skyrme model look like for the one-parametric family of potentials $U = \mu^2(1 - \phi_3)^s$ which generalizes the 'old' model ($s = 1$) and the holomorphic model ($s = 4$).[18] As it turns out, the value of the parameter $s$ has dramatic effects on the static solutions of the model, both quantitatively and qualitatively, in the sense that it can be viewed as a 'control' parameter responsible for the repulsion or attraction between skyrmions, which in turn determines whether or not the minimal-energy configuration breaks rotational symmetry.

The Lagrangian density is now:

$$\mathcal{L} = \frac{1}{2}\partial_\mu\phi \cdot \partial^\mu\phi - \frac{\kappa^2}{2}\left((\partial_\mu\phi \cdot \partial^\mu\phi)^2 - (\partial_\mu\phi \cdot \partial_\nu\phi) \cdot (\partial^\mu\phi \cdot \partial^\nu\phi)\right) - \mu^2(1 - \phi_3)^s,$$

$$(8.1.2)$$

and contains three free parameters, namely $\kappa, \mu$ and $s$. Since either $\kappa$ or $\mu$ may be scaled away, the parameter space of this model is in fact only two dimensional. Our main goal here is to study the effects of these parameters on the static solutions of the model within each topological sector.

The multi-skyrmions of our model are those field configurations which minimize the static energy functional within each topological sector. In polar coordinates the energy functional is given by

$$E = \int r\,dr\,d\theta \left(\frac{1}{2}(\partial_r\phi \cdot \partial_r\phi + \frac{1}{r^2}\partial_\theta\phi \cdot \partial_\theta\phi) + \frac{\kappa^2}{2}\frac{(\partial_r\phi \times \partial_\theta\phi)^2}{r^2} + \mu^2(1 - \phi_3)^s\right).$$

$$(8.1.3)$$

The Euler-Lagrange equations derived from the energy functional (8.1.3) are nonlinear *PDE*'s, so in most cases one must resort to numerical techniques in order to solve them. In our approach, the minimal energy configuration of a baby skyrmion of charge B and a given set of values $\mu, \kappa, s$ is found by a full-field relaxation method, which we describe in more detail in the Appendix.

### 8.1.1. *Results*

Applying the minimization procedure, we obtain the static solutions of the model for $1 \leq B \leq 5$. Since the parameter space of the model is effectively two dimensional (as discussed earlier), without loss of generality we fix the potential strength at $\mu^2 = 0.1$ throughout, and the $s$-$\kappa$ parameter space is scanned in the region $0 < s \leq 4$, $0.01 \leq \kappa^2 \leq 1$.

### 8.1.1.1. *Charge-one skyrmions*

In the charge-one sector, the solutions for every value of $s$ and $\kappa$ are stable rotationally-symmetric configurations. Figure 8.1(a) shows the obtained profile functions of the $B = 1$ solution for different values of $s$ with $\kappa$ fixed at $\kappa^2 = 0.25$. Interestingly, the skyrmion energy as a function of $s$ is not monotonic, but acquires a minimum at $s \approx 2.2$, as is shown in Fig. 8.2.

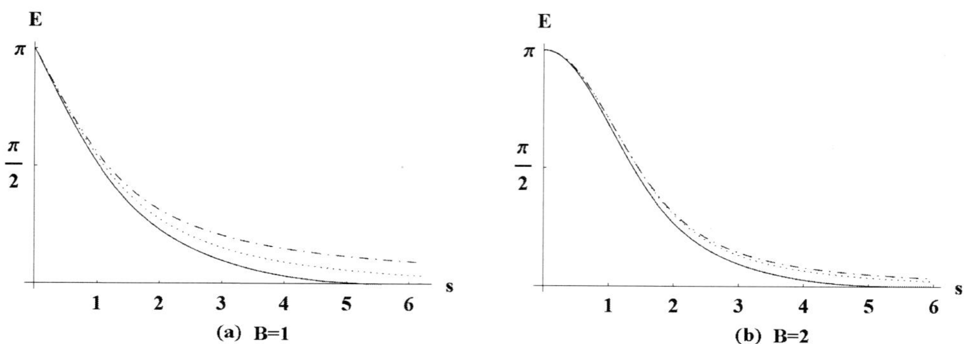

Fig. 8.1. Profile functions of the $B = 1$ (left) and $B = 2$ (right) skyrmions for $s = 0.5$ (solid), $s = 1$ (dotted) and $s = 2$ (dot-dashed). Here $\kappa$ is fixed at $\kappa^2 = 0.25$.

### 8.1.1.2. *Charge-two skyrmions*

Stable solutions also exist in the $B = 2$ sector, but only for $s < 2$. They are rotationally-symmetric and ring-like, corresponding to two charge-one skyrmions on top of each other. Figure 8.1(b) shows the profile functions of the stable solutions for different values of $s$ and $\kappa^2 = 0.25$.

As in the $B = 1$ case, the energy of the charge-two skyrmion as a function of $s$ is non-monotonic and has a minimum around $s = 1.3$. As shown in Fig. 8.2, at $s \approx 2$ the energy of the ring-like configuration reaches the value of twice the energy of the charge-one skyrmion and stable configurations cease to exist. At this point, the skyrmion breaks apart into its constituent charge-one skyrmions, which in turn start drifting away from each other, thus breaking the rotational symmetry of the solution. Contour plots of the energy distribution of the $B = 2$ skyrmion are shown in Fig. 8.3 for $\kappa^2 = 1$ and for two $s$ values. While for $s = 1.5$ a ring-like stable configuration exists (Fig. 8.3(a)), for $s = 2.6$ the skyrmion breaks apart. The latter case is shown in Fig. 8.3(b) which is a "snapshot" taken while the distance between the individual skyrmions kept growing.

These results are in accord with corresponding results from previously known studies of both the 'old' ($s = 1$) model in which ring-like configurations have been

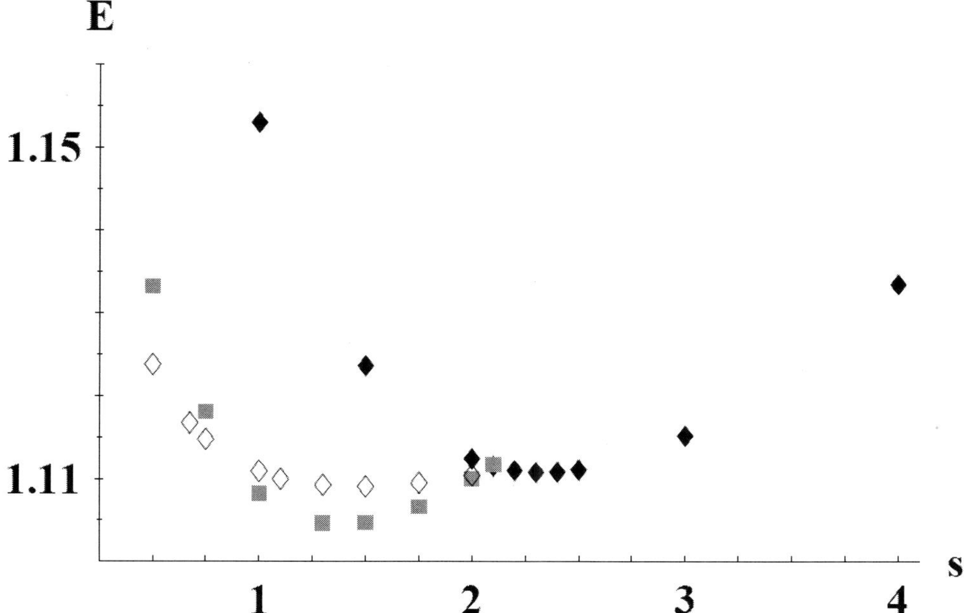

Fig. 8.2. Total energies (divided by $4\pi B$) of the charge-one ($\blacklozenge$) charge-two ($\blacksquare$) and charge-three ($\lozenge$) skyrmions as a function of the parameter $s$ for $\kappa^2 = 0.05$. Each of the energy graphs attains a minimal value at some $s$. At $s \approx 2$ the energy-per-topological-charge of the charge-two and charge-three solutions reaches the charge-one energy (from below), and stable solutions are no longer observed.

observed,[4,15] and the holomorphic model for which no stable solutions have been found.[12,13]

Rotationally-symmetric charge-two locally stable solutions may also be observed in the large $s$ regime, including the 'holomorphic' $s = 4$ case, in which case the global minimum in this regime corresponds to two infinitely separated charge-one skyrmions. The total energy of the rotationally symmetric solutions is larger than twice the energy of a charge-one skyrmion, indicating that the split skyrmion is an energetically more favorable configuration. We discuss this issue in more detail in the section 8.2 2.

### 8.1.1.3. *Charge-three and higher-charge skyrmions*

As with the $B = 2$ skyrmion, the existence of stable charge-three skyrmions was also found to be $s$ dependent. For any tested value of $\kappa$ in the range $0.01 \leq \kappa^2 \leq 1$, we have found that above $s \approx 2$, no stable charge-three solutions exist; in this region the skyrmion breaks apart into individual skyrmions drifting further and further away from each other.

In the $s < 2$ region, where stable solutions exist, the energy distribution of the

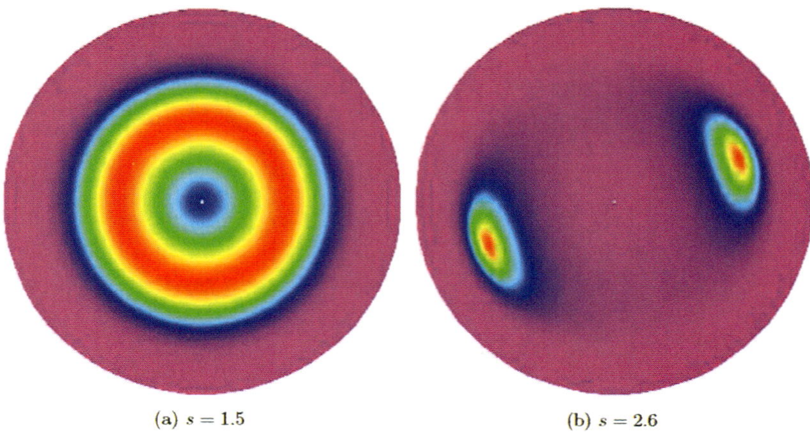

(a) $s = 1.5$        (b) $s = 2.6$

Fig. 8.3. Contour plots of the energy distributions of the $B = 2$ skyrmion for $\kappa^2 = 1$. In the $s < 2$ regime, ring-like rotationally-symmetric configurations exist, corresponding to two charge-one skyrmions on top of each other (left), whereas for $s > 2$, the charge-two skyrmion splits into two one-charge skyrmions drifting infinitely apart (right).

charge-three skyrmion turns out to be highly dependent on both $s$ and $\kappa$. Keeping $\kappa$ fixed and varying $s$, we find that in the small $s$ regime, ring-like rotationally-symmetric configurations exist. Increasing the value of $s$ results in stable minimal energy configurations with only $\mathbb{Z}(2)$ symmetry, corresponding to three partially-overlapping charge-one skyrmions in a row, as shown in Figs. 8.4(b) and 8.4(c). The energy of the charge-three skyrmion also has a minimum in $s$, at around $s = 1.5$ (as shown in Fig. 8.2). At $s \approx 2$ the energy of the skyrmion becomes larger than three times the energy of a charge-one skyrmion and stable configurations are no longer obtainable. This is illustrated in Fig. 8.4 which shows contour plots of the energy distribution of the $B = 3$ skyrmion for different values of $s$ and fixed $\kappa$. For $s = 0.5$ (Fig. 8.4(a)), the solution is rotationally symmetric and for $s = 0.75$ and $s = 1$ (Figs. 8.4(b) and 8.4(c) respectively) the rotational symmetry of the solution is broken and only $\mathbb{Z}(2)$ symmetry remains. At $s = 3$, no stable solution exists. The latter case is shown in Fig. 8.4(d) which is a "snapshot" taken while the distance between the individual skyrmions kept growing.

The dependence of the skyrmion solutions on the value of $\kappa$ with fixed $s$ show the following behavior: While for small $\kappa$ the minimal energy configurations are rotationally-symmetric, increasing its value results in an increasingly larger rotational symmetry breaking. This is illustrated in Fig. 8.5.

The $B = 4$ and $B = 5$ skyrmion solutions behave similarly to the $B = 3$ solutions. This is illustrated in Fig. 8.6, which shows the stable solutions that have been obtained in the $s = 0.9$ case and the splitting of these skyrmions into their constituents in the $s = 4$ case.

## 8.2. The Lattice Structure of Baby Skyrmions

The Skyrme model[1] may also be used to describe systems of a few nucleons, and has also been applied to nuclear and quark matter.[19–21] One of the most complicated aspects of the physics of hadrons is the behavior of the phase diagram of hadronic matter at finite density at low or even zero temperature. Particularly, the properties of zero-temperature skyrmions on a lattice are interesting, since the behavior of nuclear matter at high densities is now a focus of considerable interest. Within the standard zero-temperature Skyrme model description, a crystal of nucleons turns into a crystal of half nucleons at finite density.[22–26]

To study skyrmion crystals one imposes periodic boundary conditions on the Skyrme field and works within a unit cell.[11] The first attempted construction of a crystal was by Klebanov,[22] using a simple cubic lattice of skyrmions whose symmetries maximize the attraction between nearest neighbors. Other symmetries were proposed which lead to crystals with slightly lower, but not minimal energy.[23,24] It is now understood that it is best to arrange the skyrmions initially as a face-centered

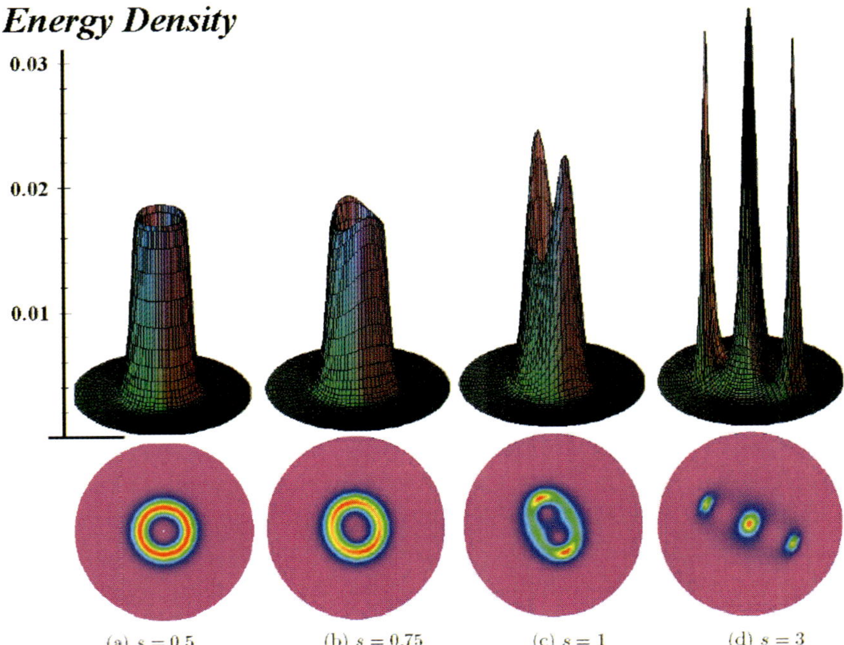


Fig. 8.4. Energy densities and corresponding contour plots of the $B = 3$ skyrmion for fixed $\kappa$ ($\kappa^2 = 0.01$) and varying $s$. In the $s = 0.5$ case, the minimal energy configuration is rotationally symmetric, corresponding the three one-skyrmions on top of each other. For $s = 0.75$ and $s = 1$ the solutions exhibit only $\mathbb{Z}(2)$ symmetry, corresponding to partially-overlapping one-skyrmions. For $s = 3$ no stable solution exists and the individual skyrmions are drifting apart.

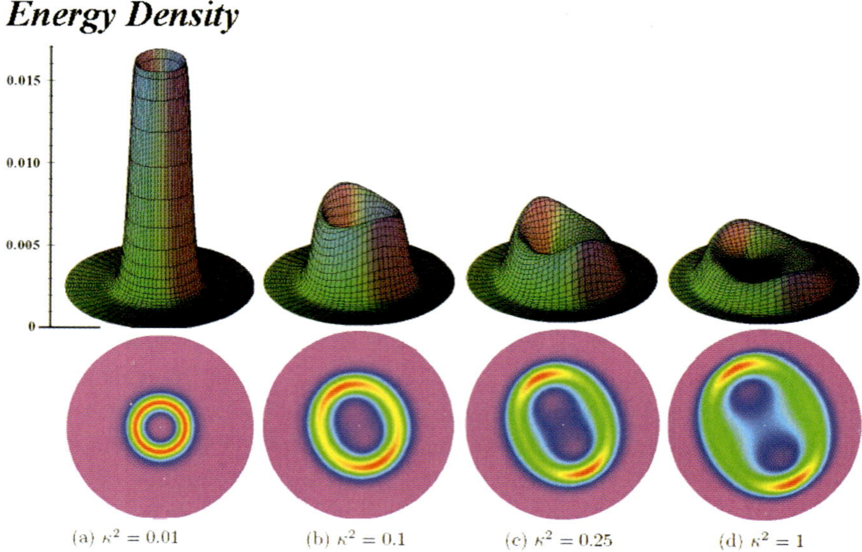



Fig. 8.5.  Energy densities and corresponding contour plots of the $B = 3$ skyrmion for fixed $s$ ($s = 0.5$) and varying $\kappa$. At low $\kappa$, the minimal energy configuration is rotationally symmetric. As $\kappa$ is increased, breaking of rotational symmetry appears, and only $\mathbb{Z}(2)$ symmetry remains.

cubic lattice, with their orientations chosen symmetrically to give maximal attraction between all nearest neighbors.[25,26]

The baby Skyrme model too has been studied under various lattice settings[27–31] and in fact, it is known that the baby skyrmions also split into half-skyrmions when placed inside a rectangular lattice.[29]  However, as we shall see, the rectangular periodic boundary conditions do not yield the true minimal energy configurations over all possible lattice types.[32] This fact is particularly interesting both because of its relevance to quantum Hall systems in two-dimensions, and also because it may be used to conjecture the crystalline structure of nucleons in three-dimensions.

In two dimensions there are five lattice types, as given by the crystallographic restriction theorem.[33] In in all of them the fundamental unit cell is a certain type of a parallelogram. To find the crystalline structure of the baby skyrmions, we place them inside different parallelograms with periodic boundary conditions and find the minimal energy configurations over all parallelograms of fixed area (thus keeping the skyrmion density fixed). As we show later, there is a certain type of parallelogram, namely the hexagonal, which yields the minimal energy configuration. In particular, its energy is lower than the known 'square-cell' configurations in which the skyrmion splits into half-skyrmions. As will be pointed out later, the hexagonal structure revealed here is not unique to the present model, but also arises in other solitonic models, such as Ginzburg-Landau vortices,[34] quantum Hall systems,[6,7] and even in the context of 3D skyrmions.[35] The reason for this will also be discussed later.

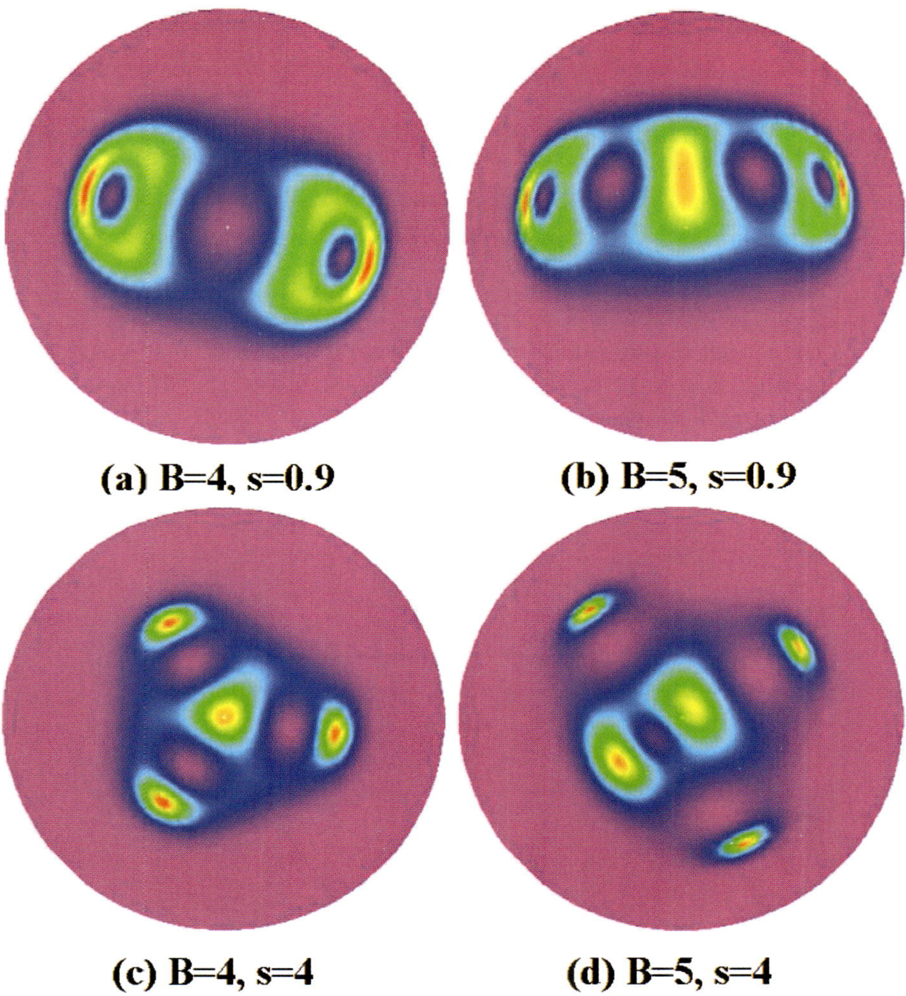

(a) B=4, s=0.9      (b) B=5, s=0.9

(c) B=4, s=4      (d) B=5, s=4

Fig. 8.6.   Contour plots of the energy distributions of the $B = 4$ and $B = 5$ skyrmions for $s = 0.9$ and $s = 4$ ($\kappa^2 = 0.1$). In the lower $s$ region stable solutions exist; the upper figures show a $B = 4$ skyrmion in a bound state of two charge-two skyrmions (left), and a $B = 5$ skyrmion in a two-one-two configuration. For values of $s$ higher than 2, the multi-skyrmions split into individual one-skyrmions constantly drifting apart (lower figures).

In what follows we review the setup of our numerical computations, introducing a systematic approach for the identification of the minimal energy crystalline structure of baby skyrmions. In section 8.2.2 we present the main results of our study and in section 8.2.3, a somewhat more analytical analysis of the problem is presented.

### 8.2.1. *Baby skyrmions inside a parallelogram*

We find the static solutions of the model by minimizing the static energy functional:

$$E = \frac{1}{2} \int_{\Lambda} \mathrm{d}x\mathrm{d}y \Big( (\partial_x \boldsymbol{\phi})^2 + (\partial_y \boldsymbol{\phi})^2 + \kappa^2 (\partial_x \boldsymbol{\phi} \times \partial_y \boldsymbol{\phi})^2 + 2\mu^2 (1 - \phi_3) \Big) , \quad (8.2.4)$$

within each topological sector. In this example, we use the 'old' model potential term. In our setup, the integration is over parallelograms, denoted here by $\Lambda$:

$$\Lambda = \{ \alpha_1(L, 0) + \alpha_2(sL \sin\gamma, sL \cos\gamma) : 0 \le \alpha_1, \alpha_2 < 1 \} . \quad (8.2.5)$$

Here $L$ is the length of one side of the parallelogram, $sL$ with $0 < s \le 1$ is the length of its other side and $0 \le \gamma < \pi/2$ is the angle between the '$sL$' side and the vertical to the '$L$' side (as sketched in Fig. 8.7).

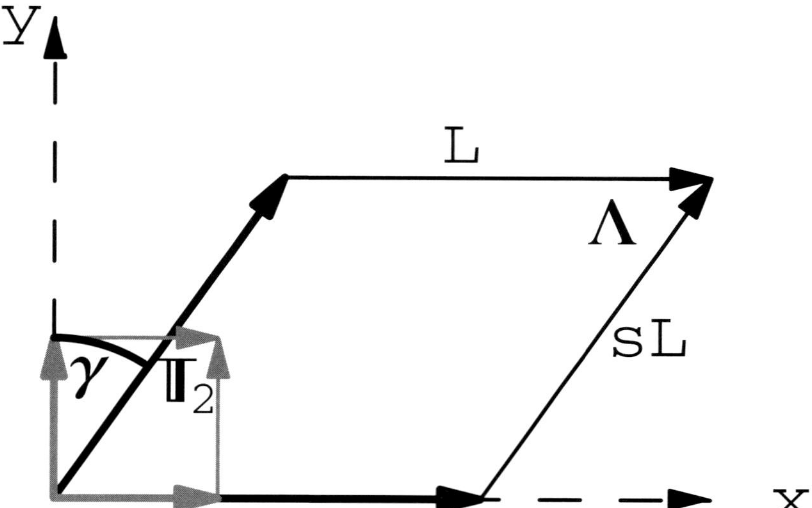


Fig. 8.7.   The parameterization of the fundamental unit-cell parallelogram $\Lambda$ (in black) and the two-torus $\mathbb{T}_2$ into which it is mapped (in gray).

Each parallelogram is thus specified by a set $\{L, s, \gamma\}$ and the skyrmion density inside a parallelogram is $\rho = B/(sL^2 \cos\gamma)$, where $B$ is the topological charge of the skyrmion. The periodic boundary conditions are taken into account by identifying each of the two opposite sides of a parallelogram as equivalent:

$$\boldsymbol{\phi}(\boldsymbol{x}) = \boldsymbol{\phi}(\boldsymbol{x} + n_1(L, 0) + n_2(sL \sin\gamma, sL \cos\gamma)) , \quad (8.2.6)$$

with $n_1, n_2 \in \mathbb{Z}$. We are interested in static finite-energy solutions, which in the language of differential geometry are $\mathbb{T}_2 \mapsto S_2$ maps. These are partitioned into homotopy sectors parameterized by an invariant integral topological charge $B$, the degree of the map, given by

$$B = \frac{1}{4\pi} \int_\Lambda \mathrm{d}x\mathrm{d}y \left( \boldsymbol{\phi} \cdot (\partial_x \boldsymbol{\phi} \times \partial_y \boldsymbol{\phi}) \right) . \tag{8.2.7}$$

The static energy $E$ can be shown to satisfy

$$E \geq 4\pi B , \tag{8.2.8}$$

with equality possible only in the 'pure' $O(3)$ case (i.e., when both the Skyrme and the potential terms are absent).[29] We note that while in the baby Skyrme model on $\mathbb{R}^2$ with fixed boundary conditions the potential term is necessary to prevent the solitons from expanding indefinitely, in our setup it is not required, due to the periodic boundary conditions.[29] We study the model both with and without the potential term.

The problem in question can be simplified by a linear mapping of the parallelograms $\Lambda$ into the unit-area two-torus $\mathbb{T}_2$. In the new coordinates, the energy functional becomes

$$E = \frac{1}{2s \cos \gamma} \int_{\mathbb{T}_2} \mathrm{d}x\mathrm{d}y \left( s^2 (\partial_x \boldsymbol{\phi})^2 - 2s \sin \gamma \partial_x \boldsymbol{\phi} \partial_y \boldsymbol{\phi} + (\partial_y \boldsymbol{\phi})^2 \right)$$

$$+ \frac{\kappa^2 \rho}{2B} \int_{\mathbb{T}_2} \mathrm{d}x\mathrm{d}y \left( \partial_x \boldsymbol{\phi} \times \partial_y \boldsymbol{\phi} \right)^2 + \frac{\mu^2 B}{\rho} \int_{\mathbb{T}_2} \mathrm{d}x\mathrm{d}y \left( 1 - \phi_3 \right) . \tag{8.2.9}$$

We note that the dependence of the energy on the Skyrme parameters $\kappa$ and $\mu$ and the skyrmion density $\rho$ is only through $\kappa^2 \rho$ and $\mu^2/\rho$.

In order to find the minimal energy configuration of skyrmions over all parallelograms with fixed area (equivalently, a specified $\rho$), we scan the parallelogram parameter space $\{s, \gamma\}$ and find the parallelogram for which the resultant energy is minimal over the parameter space. An alternative approach to this problem, which is of a more analytical nature, may also been implemented. We discuss it in detail in section 8.2.3.

### 8.2.2. *Results*

In what follows, we present the minimal energy static skyrmion configurations over all parallelograms, for various settings: The 'pure' $O(3)$ case, in which both $\kappa$, the Skyrme parameter, and $\mu$, the potential coupling, are set to zero, the Skyrme case for which only $\mu = 0$, and the general case for which neither the Skyrme term nor the potential term vanish.

In each of these settings, we scanned the parameter space of parallelograms, while the skyrmion density $\rho$ was held fixed, yielding for each set of $\{s, \gamma\}$ a minimal energy configuration. The choice as to how many skyrmions are to be placed inside the unit cells was made after some preliminary testing in which skyrmions of other charges (up to $B = 8$) were also examined. The odd-charge minimal-energy configurations turn out to have substantially higher energies than even-charge ones, where among the latter, the charge-two skyrmion is found to be the most fundamental, as it is observed that the charge-two configuration is a 'building-block' of the higher-charge configurations. This is illustrated in Fig. 8.8 in which the typical behavior of the multi-skyrmion energies as a function of topological charge is shown.

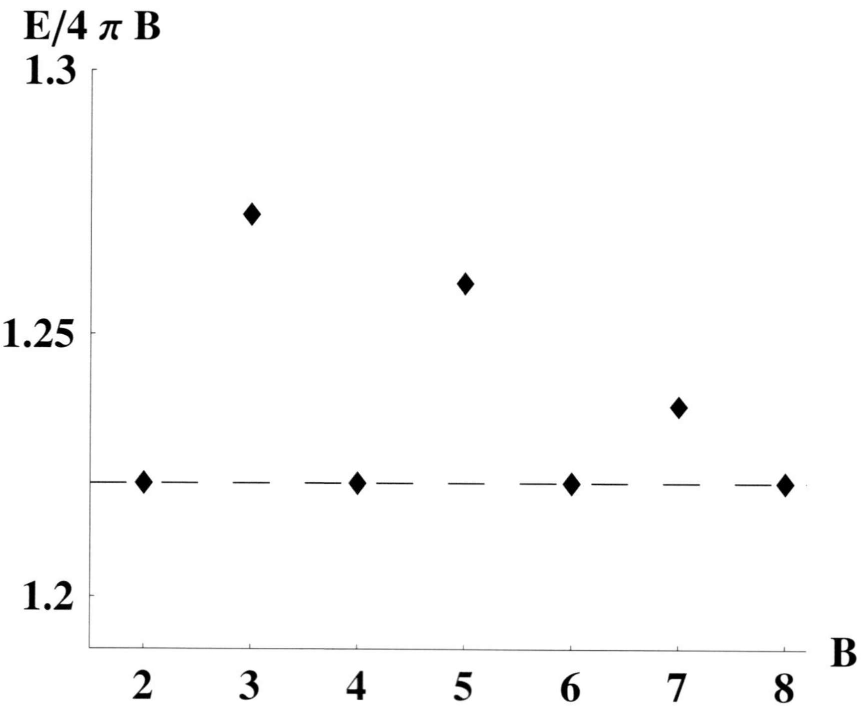

Fig. 8.8.  Energy per charge of the multi-skyrmion configurations as a function of topological charge. The horizontal dashed line was added to guide the eye. (Here, $\kappa^2 = 0.03$, $\mu = 0$, $\rho = 1$, $s = 1$ and $\gamma = \pi/6$.)

### 8.2.2.1. *The pure $O(3)$ case $(\kappa = \mu = 0)$*

The pure $O(3)$ case corresponds to setting both $\kappa$ and $\mu$ to zero. In this case, analytic solutions in terms of Weierstrass elliptic functions may be found[29-31] and the minimal energy configurations, in all parallelogram settings, saturate the energy bound in (8.2.8) giving $E = 4\pi B$. Thus, comparison of our numerical results with the analytic solutions serves as a useful check on the precision of our numerical procedure. The agreement is to 6 significant digits. Contour plots of the charge densities for different parallelogram settings for the charge-two skyrmions are shown in Fig. 8.9, all of them of equal energy $E/8\pi = 1$.

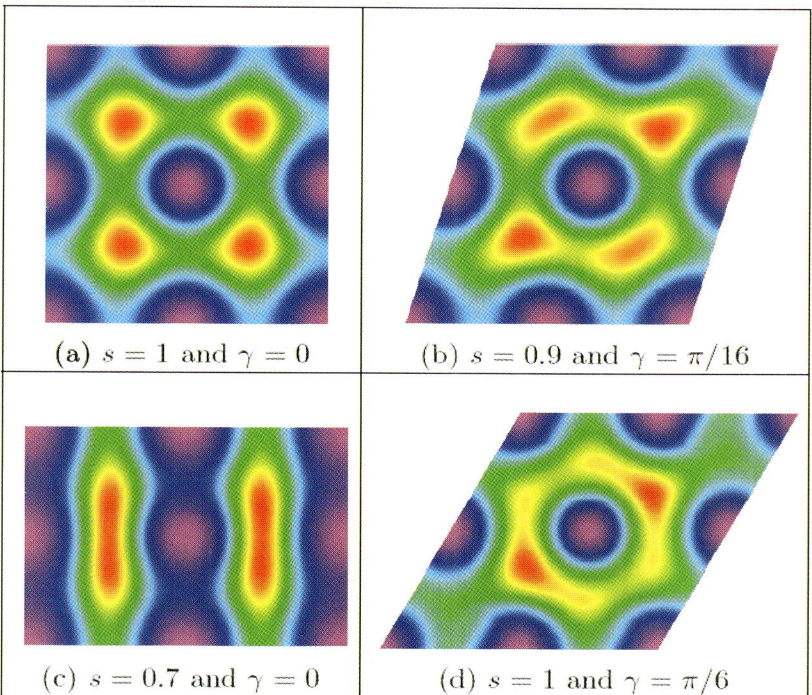

Fig. 8.9.  Charge-two skyrmions in the pure $O(3)$ case: Contour plots of the charge densities for various parallelogram settings, all of which saturate the energy bound $E = 4\pi B = 8\pi$.

### 8.2.2.2.  *The Skyrme case ($\kappa \neq 0, \mu = 0$)*

As pointed out earlier, for $\mu = 0$ the dependence of the energy functional on the Skyrme parameter $\kappa$ is only through $\kappa^2 \rho$, so without loss of generality we vary $\rho$ and fix $\kappa^2 = 0.03$ throughout (this particular choice for $\kappa$ was made for numerical convenience). Minimization of the energy functional (8.2.9) over all parallelograms yields the following. For any fixed density $\rho$, the minimal energy is obtained for $s = 1$ and $\gamma = \tau/6$. This set of values corresponds to the 'hexagonal' or 'equilateral triangular' lattice. In this configuration, any three adjacent zero-energy loci (or 'holes') are the vertices of equilateral triangles, and eight distinct high-density peaks are observed (as shown in Fig. 8.10b). This configuration can thus be interpreted as the splitting of the two-skyrmion into eight quarter-skyrmions. This result is independent of the skyrmion density $\rho$.

In particular, the well-studied square-cell minimal energy configuration (Fig. 8.10(a)), in which the two-skyrmion splits into four half-skyrmions, has higher energy than the hexagonal case. Figure 8.10 shows the total energies (divided by $8\pi$) and the corresponding contour plots of charge densities of the hexagonal, square and other configurations (for comparison), all of them with $\rho = 2$.

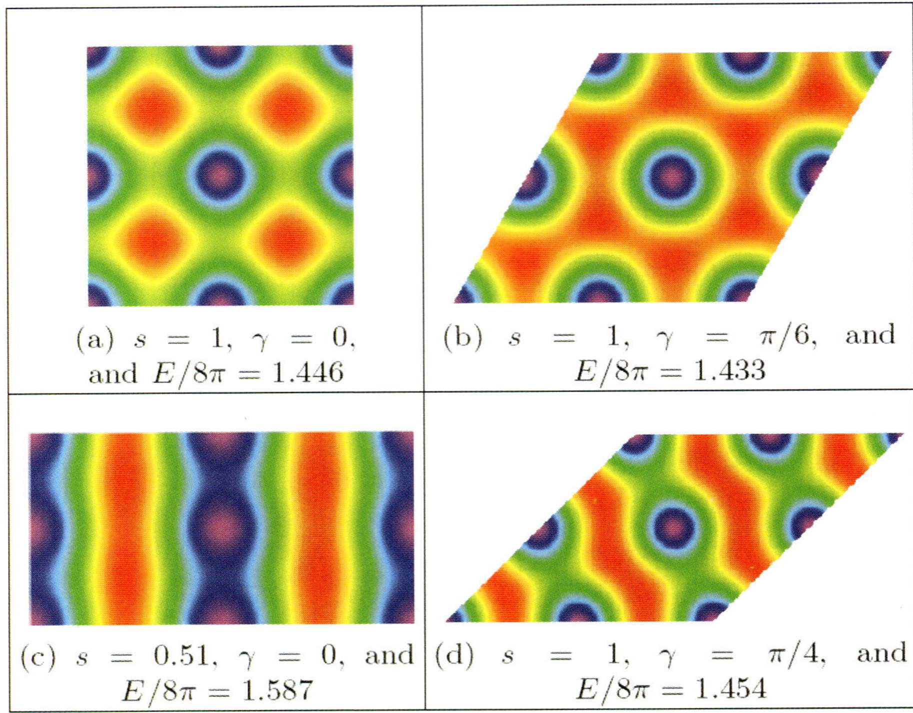

(a) $s = 1$, $\gamma = 0$, and $E/8\pi = 1.446$

(b) $s = 1$, $\gamma = \pi/6$, and $E/8\pi = 1.433$

(c) $s = 0.51$, $\gamma = 0$, and $E/8\pi = 1.587$

(d) $s = 1$, $\gamma = \pi/4$, and $E/8\pi = 1.454$

Fig. 8.10. Charge-two skyrmions in the Skyrme case with $\kappa^2 = 0.03$ and $\rho = 2$: Contour plots of the charge densities for the hexagonal, square and other fundamental cells . As the captions of the individual subfigures indicate, the hexagonal configuration has the lowest energy.

The total energy of the skyrmions in the hexagonal setting turns out to be linearly proportional to the density of the skyrmions, reflecting the scale invariance of the model (Fig. 8.11). In particular, the global minimum of $E = 4\pi B = 8\pi$ is reached when $\rho \rightarrow 0$. This is expected since setting $\rho = 0$ is equivalent to setting the Skyrme parameter $\kappa$ to zero, in which case the model is effectively pure $O(3)$ and inequality (8.2.8) is saturated.

### 8.2.2.3. *The general case ($\kappa \neq 0$, $\mu \neq 0$)*

The hexagonal setting turns out to be the energetically favorable also in the general case. Moreover, since in this case the skyrmion has a definite size (as is demonstrated by the $\rho$ dependence in the energy functional), the skyrmion structure is different at low and at high densities and a phase transition occurs at a certain critical density. While at low densities the individual skyrmions are isolated from each other, at high densities they fuse together, forming the quarter-skyrmion crystal, as in the Skyrme case reported above. As the density $\rho$ decreases, or equivalently the value of $\mu$ increases, the size of the skyrmions becomes small compared to

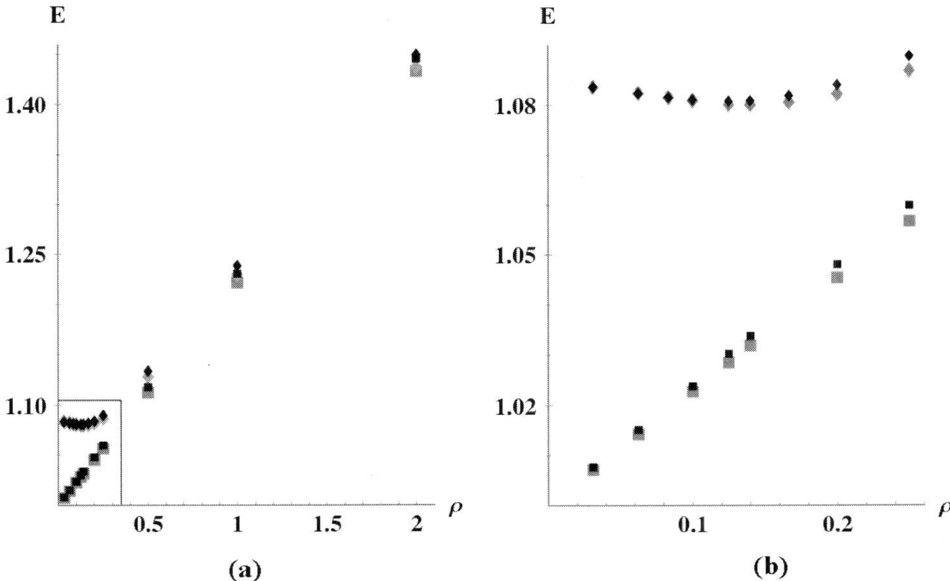

Fig. 8.11. Total energy $E$ (divided by $8\pi$) of the charge-two skyrmion in the hexagonal lattice (▨ — Skyrme case and ⬥ — general case) and in the square lattice (■ — Skyrme case and ◆ — general case) as function of the skyrmion density (in the Skyrme case, $\kappa^2 = 0.03$ and in the general case $\kappa^2 = 0.03$ and $\mu^2 = 0.1$). Note the existence of an optimal density (at $\rho \approx 0.14$) in the general case, for which the energy attains a global minimum. Figure (b) is an enlargement of the lower left corner of figure (a).

the cell size. The exact shape of the lattice loses its effects and the differences in energy among the various lattice types become very small. This is illustrated in Fig. 8.11.

Due to the finite size of the skyrmion, there is an optimal density for which the energy is minimal among all densities. Figure 8.12 shows the contour plots of the charge density of the charge-two skyrmion for several densities with $\kappa^2 = 0.03$ and $\mu^2 = 0.1$. The energy of the skyrmion is minimal for $\rho \approx 0.14$ (Fig. 8.11).

### 8.2.3. *Semi-analytical approach*

The energy functional (8.2.9) depends both on the Skyrme field $\phi$ and on the parallelogram parameters $\gamma$ and $s$. Formally, the minimal energy configuration over all parallelograms may be obtained by functional differentiation with respect to $\phi$ and regular differentiation with respect to $\gamma$ and $s$. However, since the resulting equations are very complicated, a direct numerical solution is quite hard. Nonetheless, some analytical progress may be achieved in the following way. As a first step, we

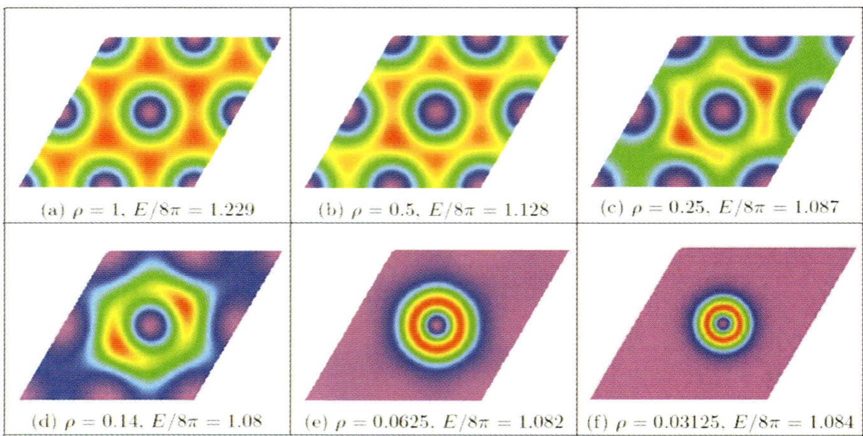

(a) $\rho = 1$, $E/8\pi = 1.229$    (b) $\rho = 0.5$, $E/8\pi = 1.128$    (c) $\rho = 0.25$, $E/8\pi = 1.087$

(d) $\rho = 0.14$, $E/8\pi = 1.08$    (e) $\rho = 0.0625$, $E/8\pi = 1.082$    (f) $\rho = 0.03125$, $E/8\pi = 1.084$

Fig. 8.12.  Charge-two skyrmions in the general case with $\mu^2 = 0.1$ and $\kappa^2 = 0.03$: Contour plots of the charge densities of the minimal-energy configurations in the hexagonal setting for different densities. Here, the energetically most favorable density is $\rho = 0.14$.

differentiate the energy functional (8.2.9) only with respect to $\gamma$ and $s$:

$$\frac{\partial E}{\partial \gamma} = \frac{1}{2s\cos^2\gamma}\left(\sin\gamma(\mathcal{E}^{yy} + s^2\mathcal{E}^{xx}) - 2s\mathcal{E}^{xy}\right) = 0\,,$$

$$\frac{\partial E}{\partial s} = \frac{1}{2s^2\cos\gamma}(\mathcal{E}^{yy} - s^2\mathcal{E}^{xx}) = 0\,, \tag{8.2.10}$$

where $\mathcal{E}^{ij} = \int_{\mathbb{T}_2} \mathrm{d}x\mathrm{d}y(\partial_i\phi\cdot\partial_j\phi)$ and $i,j \in \{x,y\}$. Solving these equations for $\gamma$ and $s$ yields

$$s = \sqrt{\frac{\mathcal{E}^{yy}}{\mathcal{E}^{xx}}}\,, \tag{8.2.11}$$

$$\sin\gamma = \frac{\mathcal{E}^{xy}}{\sqrt{\mathcal{E}^{xx}\mathcal{E}^{yy}}}\,.$$

Substituting these expressions into the energy functional (8.2.9), we arrive at a 'reduced' functional

$$E = \sqrt{\mathcal{E}^{xx}\mathcal{E}^{yy} - (\mathcal{E}^{xy})^2} + \frac{\kappa^2\rho}{2B}\mathcal{E}_{\mathrm{sk}} + \frac{\mu^2 B}{\rho}\mathcal{E}_{\mathrm{pot}}\,, \tag{8.2.12}$$

where $\mathcal{E}_{\mathrm{sk}} = \int_{\mathbb{T}_2} \mathrm{d}x\mathrm{d}y\,(\partial_x\phi\times\partial_y\phi)^2$ is the Skyrme energy and $\mathcal{E}_{\mathrm{pot}} = \int_{\mathbb{T}_2} \mathrm{d}x\mathrm{d}y\,(1-\phi_3)$ is the potential energy. Now that both $\gamma$ and $s$ are eliminated from the resultant expression, and the conditions for their optimization are built into the functional, the numerical minimization is carried out. We note here, however, that the procedure presented above should be treated with caution. This is since Eqs. (8.2.11) are only extremum conditions, and may correspond to a maximum or saddle-point. Hence, it is important to confirm any results obtained using this method by comparing them with corresponding results obtained from the method described in the previous section.

It is therefore reassuring that numerical minimization of the reduced functional (8.2.12) gives $\sin\gamma = 0.498$ ($\gamma \approx \pi/6$) and $s = 1$ (both for the Skyrme case and the general case), confirming the results presented in the previous section.

In the general ($\mu \neq 0$) case, the energy functional (8.2.12) may be further differentiated with respect to the skyrmion density $\rho$ to obtain the optimal density for which the skyrmion energy is minimal. Differentiating with respect to $\rho$, and substituting the obtained expression into the energy functional, results in the functional

$$E = \sqrt{\mathcal{E}^{xx}\mathcal{E}^{yy} - (\mathcal{E}^{xy})^2} + \kappa\mu\sqrt{2\mathcal{E}_{\text{sk}}\mathcal{E}_{\text{pot}}}\,. \tag{8.2.13}$$

Numerical minimization of the above expression for $\kappa^2 = 0.03$ and various $\mu$ values ($0.1 \leq \mu^2 \leq 10$) yields the hexagonal setting as in the Skyrme case. In particular, for $\mu^2 = 0.1$ the optimal density turns out to be $\rho \approx 0.14$, in accord with results presented in Sec. 8.2.2.3.

### 8.2.4. *Further remarks*

As pointed out earlier, the special role of the hexagonal lattice revealed here is not unique to the baby Skyrme model, but in fact arises in other solitonic models. In the context of Skyrme models, the existence of a hexagonal two dimensional structure of 3D skyrmions has also been found by Battye and Sutcliffe,[35] where it has already been noted that energetically, the optimum infinite planar structure of 3D skyrmions is the hexagonal lattice, which resembles the structure of a graphite sheet, the most stable form of carbon thermodynamically.[11] Other examples in which the hexagonal structure is revealed are Ginzburg-Landau vortices which are known to have lower energy in a hexagonal configuration than on a square lattice.[34] Thus, it should not come as a surprise that the hexagonal structure is found to be the most favorable in the baby Skyrme model.

As briefly noted in the opening paragraphs of this section, a certain type of baby skyrmions also arise in quantum Hall systems as low-energy excitations of the ground state near ferromagnetic filling factors (notably 1 and 1/3).[6] It has been pointed out that this state contains a finite density of skyrmions,[36] and in fact the hexagonal configuration has been suggested as a candidate for their lattice structure.[7] Our results may therefore serve as a supporting evidence in that direction, although a more detailed analysis is in order.

Our results also raise some interesting questions. The first is how the dynamical properties of baby skyrmions on the hexagonal lattice differ from their behavior in the usual rectangular lattice. Another question has to do with their behavior in non-zero-temperature.

One may also wonder whether and how these results can be generalized to the 3D case. Is the face-centered cubic lattice indeed the minimal energy crystalline structure of 3D skyrmions among all parallelepiped lattices? If not, what is the minimal energy structure, and how do these results depend on the presence of a mass term? These questions await a systematic study.

## 8.3. Baby Skyrmions on the Two-Sphere

Although skyrmions were originally introduced to describe baryons in three spatial dimensions,[1] they have been shown to exist for a very wide class of geometries,[37] specifically cylinders, two-spheres and three-spheres.[38–42]

Here, we consider a baby Skyrme model on the two-sphere.[a] We compute the full-field minimal energy solutions of the model up to charge 14 and show that they exhibit complex multi-skyrmion solutions closely related to the skyrmion solutions of the 3D model with the same topological charge. To obtain the minimum energy configurations, we apply two completely different methods. One is the full-field relaxation method, with which exact numerical solutions of the model are obtained, and the other is a rational map approximation scheme, which as we show yields very good approximate solutions.

In an exact analogy to the 3D Skyrme model, the charge-one skyrmion has a spherical energy distribution, the charge-two skyrmion is toroidal, and skyrmions with higher charge all have point symmetries which are subgroups of O(3). As we shall see, it is not a coincidence that the symmetries of these solutions are the same as those of the 3D skyrmions.

### 8.3.1. *The baby Skyrme model on the two-sphere*

The model in question is a baby Skyrme model in which both the domain and target are two-spheres. The Lagrangian density here is simply

$$\mathcal{L} = \frac{1}{2}\partial_\mu \boldsymbol{\phi} \cdot \partial^\mu \boldsymbol{\phi} + \frac{\kappa^2}{2}\left[(\partial_\mu \boldsymbol{\phi} \cdot \partial^\mu \boldsymbol{\phi})^2 - (\partial_\mu \boldsymbol{\phi} \cdot \partial_\nu \boldsymbol{\phi})(\partial^\mu \boldsymbol{\phi} \cdot \partial^\nu \boldsymbol{\phi})\right], \quad (8.3.14)$$

with metric $\mathrm{d}s^2 = dt^2 - \mathrm{d}\theta^2 - \sin^2\theta\,\mathrm{d}\varphi^2$, where $\theta$ is the polar angle $\in [0, \pi]$ and $\varphi$ is the azimuthal angle $\in [0, 2\pi)$. The Lagrangian of this model is invariant under rotations in both domain and the target spaces, possessing an $O(3)_{\mathrm{domain}} \times O(3)_{\mathrm{target}}$ symmetry. As noted earlier, in flat two-dimensional space an additional potential term is necessary to ensure the existence of stable finite-size solutions. Without it, the repulsive effect of the Skyrme term causes the skyrmions to expand indefinitely. In the present model, however, the finite geometry of the sphere acts as a stabilizer, so a potential term is not required. Furthermore, stable solutions exist even without the Skyrme term. In the latter case, we obtain the well known $O(3)$ (or $\mathbb{CP}^1$) nonlinear sigma model.[43]

---

[a]This type of model has been studied before,[39,40] although only rotationally-symmetric configurations have been considered.

As before, the field $\phi$ in this model is an $S^2 \to S^2$ mapping, so the relevant homotopy group is $\pi_2(S^2) = \mathbb{Z}$, implying that each field configuration is characterized by an integer topological charge $B$, the topological degree of the map $\phi$. In spherical coordinates $B$ is given by

$$B = \frac{1}{4\pi} \int d\Omega \frac{\phi \cdot (\partial_\theta \phi \times \partial_\varphi \phi)}{\sin \theta}, \tag{8.3.15}$$

where $d\Omega = \sin \theta \, d\theta \, d\varphi$.

Static solutions are obtained by minimizing the energy functional

$$E = \frac{1}{2} \int d\Omega \left( (\partial_\theta \phi)^2 + \frac{1}{\sin^2 \theta} (\partial_\varphi \phi)^2 \right) + \frac{\kappa^2}{2} \int d\Omega \left( \frac{(\partial_\theta \phi \times \partial_\varphi \phi)^2}{\sin^2 \theta} \right), \tag{8.3.16}$$

within each topological sector. Before proceeding, it is worthwhile to note that setting $\kappa = 0$ in Eq. (8.3.16) yields the energy functional of the $O(3)$ nonlinear sigma model. The latter has analytic minimal energy solutions within every topological sector, given by

$$\phi = (\sin f(\theta) \cos(B\varphi), \sin f(\theta) \sin(B\varphi), \cos f(\theta)), \tag{8.3.17}$$

where $f(\theta) = \cos^{-1}(1 - 2(1 + (\lambda \tan \theta/2)^{2B})^{-1})$ with $\lambda$ being some positive number.[43] These solutions are not unique, as other solutions with the same energy may be obtained by rotating (8.3.17) either in the target or in the domain spaces. The energy distributions of these solutions in each sector have total energy $E_B = 4\pi B$.

Analytic solutions also exist for the energy functional (8.3.16) with the Skyrme term only. They too have the rotationally symmetric form (8.3.17) with $f(\theta) = \theta$ and total energy $E_B = 4\pi B^2$. They can be shown to be the global minima by the following Cauchy-Schwartz inequality:

$$\left( \frac{1}{4\pi} \int d\Omega \frac{\phi \cdot (\partial_\theta \phi \times \partial_\varphi \phi)}{\sin \theta} \right)^2 \leq \left( \frac{1}{4\pi} \int d\Omega \phi^2 \right) \cdot \left( \frac{1}{4\pi} \int d\Omega (\frac{\partial_\theta \phi \times \partial_\varphi \phi}{\sin \theta})^2 \right).$$

$$\tag{8.3.18}$$

The left-hand side is simply $B^2$ and the first term in parenthesis on the right-hand side integrates to 1. Noting that the second term in the RHS is the Skyrme energy (without the $\kappa^2/2$ factor), the inequality reads $E \geq 4\pi B^2$, with an equality for the rotationally-symmetric solutions.

## 8.3.2. *Static solutions*

In general, if both the kinetic and Skyrme terms are present, static solutions of the model cannot be obtained analytically. This is with the exception of the $B = 1$ skyrmion which has an analytic "hedgehog" solution

$$\phi_{[B=1]} = (\sin \theta \cos \varphi, \sin \theta \sin \varphi, \cos \theta), \tag{8.3.19}$$

with total energy $\dfrac{E}{4\pi} = 1 + \dfrac{\kappa^2}{2}$.

For skyrmions with higher charge, we find the minimal energy configurations by utilizing the full-field relaxation method described earlier. In parallel, we also apply the rational map approximation method, originally developed for the 3D Skyrme model and compare the results with the relaxation method. Let us briefly discuss the rational map approximation method: computing the minimum energy configurations using the full nonlinear energy functional is a procedure which is both time-consuming and resource-hungry. To circumvent these problems, the rational map ansatz scheme has been devised. First introduced by Houghton, Manton and Sutcliffe,[44] this scheme has been used in obtaining approximate solutions to the 3D Skyrme model using rational maps between Riemann spheres. Although this representation is not exact, it drastically reduces the number of degrees of freedom in the problem, allowing computations to take place in a relatively short amount of time. The results in the case of 3D Skyrme model are known to be quite accurate.

Application of the approximation, begins with expressing points on the base sphere by the Riemann sphere coordinate $z = \tan\dfrac{\theta}{2}e^{i\varphi}$. The complex-valued function $R(z)$ is a rational map of degree $B$ between Riemann spheres

$$R(z) = \frac{p(z)}{q(z)}, \qquad (8.3.20)$$

where $p(z)$ and $q(z)$ are polynomials in $z$, such that $\max[\deg(p), \deg(q)] = B$, and $p$ and $q$ have no common factors. Given such a rational map, the ansatz for the field triplet is

$$\phi = (\frac{R + \bar{R}}{1 + |R|^2}, i\frac{R - \bar{R}}{1 + |R|^2}, \frac{1 - |R|^2}{1 + |R|^2}). \qquad (8.3.21)$$

It can be shown that rational maps of degree $B$ correspond to field configurations with charge $B$.[44] Substitution of the ansatz (8.3.21) into the energy functional (8.3.16) results in the simple expression

$$\frac{1}{4\pi}E = B + \frac{\kappa^2}{2}\mathcal{I}, \qquad (8.3.22)$$

with

$$\mathcal{I} = \frac{1}{4\pi}\int\left(\frac{1 + |z|^2}{1 + |R|^2}\left|\frac{dR}{dz}\right|\right)^4 \frac{2i\,dzd\bar{z}}{(1 + |z|^2)^2}. \qquad (8.3.23)$$

Note that in the $\kappa \to 0$ limit, where our model reduces to the $O(3)$ nonlinear sigma model, the rational maps become exact solutions and the minimal energy value $E = 4\pi B$ is attained. Furthermore, the minimal energy is reached independently of the specific details of the map (apart from its degree), i.e., all rational maps of a given degree are minimal energy configurations in the topological sector corresponding to this degree. This is a reflection of the scale- and the rotational invariance of the $O(3)$ model.

In the general case where $\kappa \neq 0$, the situation is different. Here, minimizing the energy (8.3.22) requires finding the rational map which minimizes the functional

$\mathcal{I}$. As we discuss in the next section, the expression for $\mathcal{I}$ given in Eq. (8.3.23) is encountered in the application of the rational map in the context of 3D skyrmions, where the procedure of minimizing $\mathcal{I}$ over all rational maps of the various degrees has been used.[44–46] Here we redo the calculations, utilizing a relaxation method: to obtain the rational map of degree $B$ that minimizes $\mathcal{I}$, we start off with a rational map of degree $B$, with the real and imaginary parts of the coefficients of $p(z)$ and $q(z)$ assigned random values from the segment $[-1, 1]$. Solutions are obtained by relaxing the map until a minimum of $\mathcal{I}$ is reached.

### 8.3.3. *Relation to the 3D Skyrme model*

In the 3D Skyrme model, the rational map ansatz can be thought of as taken in two steps. First, the radial coordinate is separated from the angular coordinates by taking the SU(2) Skyrme field $U(r, \theta, \varphi)$ to be of the form

$$U(r, \theta, \varphi) = \exp(if(r) \, \boldsymbol{\phi}(\theta, \varphi) \cdot \boldsymbol{\sigma}), \tag{8.3.24}$$

where $\boldsymbol{\sigma} = (\sigma_1, \sigma_2, \sigma_3)$ are Pauli matrices, $f(r)$ is the radial profile function subject to the boundary conditions $f(0) = \pi$ and $f(\infty) = 0$, and $\boldsymbol{\phi}(\theta, \varphi) : S^2 \mapsto S^2$ is a normalized vector which carries the angular dependence of the field. In terms of the ansatz (8.3.24), the energy of the Skyrme field is

$$E = \int 4\pi f'^2 r^2 \mathrm{d}r + \int 2(f'^2 + 1) \sin^2 f \mathrm{d}r \int \left( (\partial_\theta \boldsymbol{\phi})^2 + \frac{1}{\sin^2 \theta} (\partial_\varphi \boldsymbol{\phi})^2 \right) \mathrm{d}\Omega$$

$$+ \int \frac{\sin^4 f}{r^2} \mathrm{d}r \int \frac{(\partial_\theta \boldsymbol{\phi} \times \partial_\varphi \boldsymbol{\phi})^2}{\sin^2 \theta} \mathrm{d}\Omega. \tag{8.3.25}$$

Note that the energy functional (8.3.25) is actually the energy functional of our model (8.3.16) once the radial coordinate is integrated out. Thus, our 2D model can be thought of as a 3D Skyrme model with a 'frozen' radial coordinate.

The essence of the rational map approximation is the assumption that $\boldsymbol{\phi}(\theta, \varphi)$ takes the rational map form (8.3.21), which in turn leads to a simple expression for the energy

$$E = 4\pi \int \left( r^2 f'^2 + 2B(f'^2 + 1) \sin^2 f + \mathcal{I} \frac{\sin^4 f}{r^2} \right) \mathrm{d}r, \tag{8.3.26}$$

where $\mathcal{I}$ is given in Eq. (8.3.23). As in the baby model on the two-sphere, minimizing the energy functional requires minimizing $\mathcal{I}$ over all maps of degree $B$, which is then followed by finding the profile function $f(r)$.

Since the symmetries of the 3D skyrmions are determined solely by the angular dependence of the Skyrme field, it should not be too surprising that the solutions of the model discussed here share the symmetries of the corresponding solutions of the 3D Skyrme model.

### 8.3.4. *Results*

The configurations obtained from the full-field relaxation method are found to have the same symmetries as corresponding multi-skyrmions of the 3D model with the same charge. The $B = 2$ solution is axially symmetric, whereas higher-charge solutions were all found to have point symmetries which are subgroups of O(3). For $B = 3$ and $B = 12$, the skyrmions have a tetrahedral symmetry. The $B = 4$ and $B = 13$ skyrmions have a cubic symmetry, and the $B = 7$ is dodecahedral. The other skyrmion solutions have dihedral symmetries. For $B = 5$ and $B = 14$ a $D_{2d}$ symmetry, for $B = 6, 9$ and 10 a $D_{4d}$ symmetry, for $B = 8$ a $D_{6d}$ symmetry and for $B = 11$ a $D_{3h}$ symmetry. In Fig. 8.13 we show the energy distributions of the obtained solutions for $\kappa^2 = 0.05$.

While for solutions with $B < 8$ the energy density and the charge density are distributed in distinct peaks, for solutions with higher charge they are spread in a much more complicated manner. The total energies of the solutions (divided by $4\pi B$) are listed in Table 8.1, along with the symmetries of the solutions (again with $\kappa^2 = 0.05$).

Application of the rational map ansatz yields results with only slightly higher energies, only about 0.3% to 3% above the full-field results. The calculated values of $\mathcal{I}$ are in agreement with results obtained in the context of 3D skyrmions.[45] For $9 \le B \le 14$, the rational map approximation yielded slightly less symmetric solutions than the full-field ones. Considering the relatively small number of degrees of freedom, this method all-in-all yields very good approximations. The total energies of the solutions obtained with the rational map approximation is also listed in Table 8.1.

Table 8.1.   Total energies (divided by $4\pi B$) of the multi-solitons of the model for $\kappa^2 = 0.05$.

| Charge $B$ | Total energy Full-field | Total energy Rational maps | Difference in % | Symmetry of the solution |
|---|---|---|---|---|
| 2 | 1.071 | 1.073 | 0.177 | Toroidal |
| 3 | 1.105 | 1.113 | 0.750 | Tetrahedral |
| 4 | 1.125 | 1.129 | 0.359 | Cubic |
| 5 | 1.168 | 1.179 | 0.958 | $D_{2d}$ |
| 6 | 1.194 | 1.211 | 1.426 | $D_{4d}$ |
| 7 | 1.209 | 1.217 | 0.649 | Icosahedral |
| 8 | 1.250 | 1.268 | 1.406 | $D_{6d}$ |
| 9 | 1.281 | 1.304 | 1.771 | $D_{4d}$ |
| 10 | 1.306 | 1.332 | 1.991 | $D_{4d}$ |
| 11 | 1.337 | 1.366 | 2.224 | $D_{3h}$ |
| 12 | 1.360 | 1.388 | 2.072 | Tetrahedral |
| 13 | 1.386 | 1.415 | 2.137 | Cubic |
| 14 | 1.421 | 1.459 | 2.712 | $D_2$ |

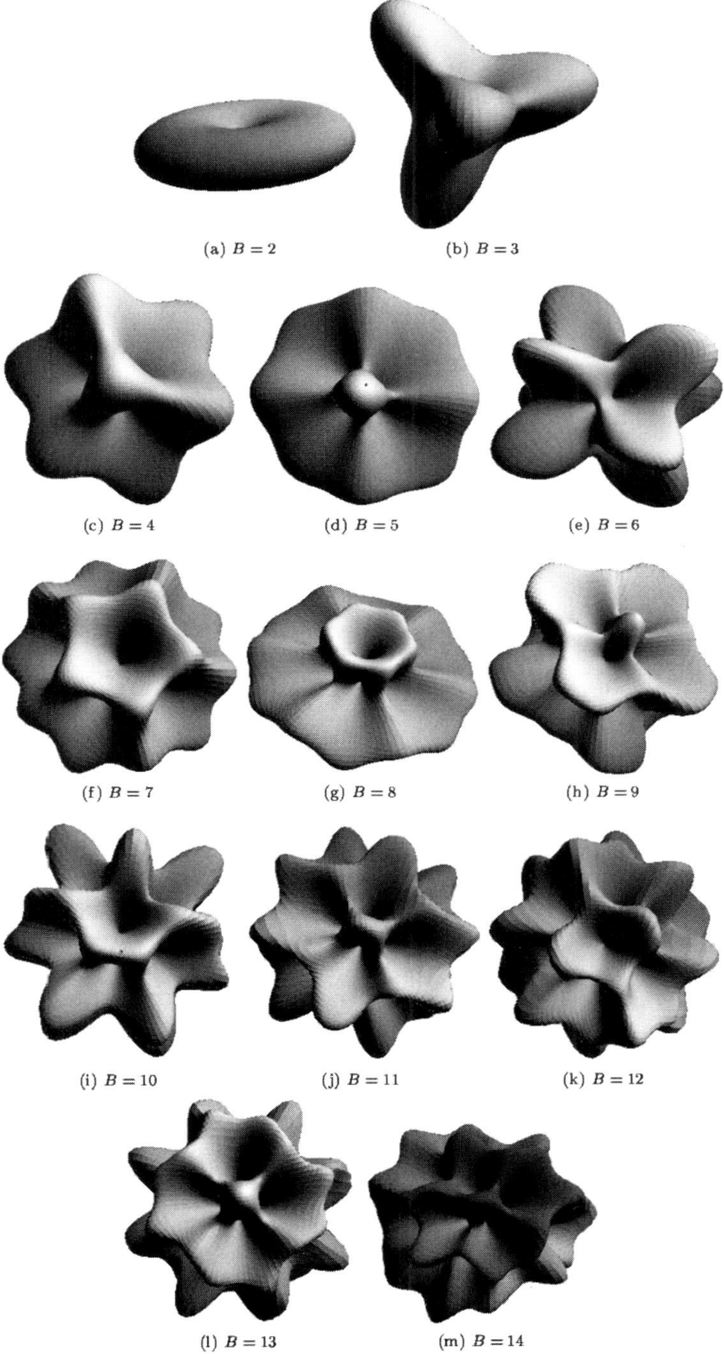

Fig. 8.13. The energy distributions of the multi-skyrmion solutions for charges $2 \leq B \leq 14$ ($\kappa^2 = 0.05$).

### 8.3.5.  *Further remarks*

As we have just seen, the baby Skyrme model on the two-sphere shares very significant similarities with the 3D model, especially in terms of multi-skyrmion symmetries. The fact that the model discussed here is two-dimensional makes it simpler to study and perform computations with, when compared with the 3D Skyrme model.

Some of the results presented above may, at least to some extent, also be linked to the baby skyrmions which appear in two-dimensional electron gas systems, exhibiting the quantum Hall effect. As briefly noted in the Introduction, baby skyrmions arise in quantum Hall systems as low-energy excitations of the ground state, near ferromagnetic filling factors (notably 1 and 1/3).[6,47] There, the skyrmion is a twisted two-dimensional configuration of spin, and its topological charge corresponds to the number of time the spin rotates by $2\pi$. While for the electron gas, the stability of the soliton arises from a balance between the electron-electron Coulomb energy and the Zeeman energy, in our model the repulsive Skyrme-term energy is balanced by the underlying geometry (i.e., the sphere). The connection between these two models suggests possible existence of very structured spin textures in quantum Hall systems, although a more detailed analysis of this analogy is in order.

### 8.4.  Rotating Baby Skyrmions

We now turn to analyze the phenomenon of spontaneous breaking of rotational symmetry (SBRS) as it appears in rotating baby skyrmions. In general, SBRS refers to cases where physical systems which rotate fast enough deform in a manner which breaks their rotational symmetry, a symmetry they posses when static or rotating slowly. The recognition that rotating physical systems can yield solutions with less symmetry than the governing equations is not new. One famous example which dates back to 1834 is that of the equilibrium configurations of a rotating fluid mass. It was Jacobi who was first to discover that if rotated fast enough, a self-gravitating fluid mass can have equilibrium configurations lacking rotational symmetry. In modern terminology, Jacobi's asymmetric equilibria appear through a symmetry breaking bifurcation from a family of symmetric equilibria as the angular momentum of the system increases above a critical value (a "bifurcation point").[48,49] Above this critical value, rotationally-symmetric configurations are no longer stable, and configurations with a broken rotational symmetry become energetically favorable.

By now it is widely recognized that symmetry-breaking bifurcations in rotating systems are of frequent occurrence and that this is in fact a very general phenomenon, appearing in a variety of physical settings, among which are fluid dynamics, star formation, heavy nuclei, chemical reactions, plasmas, and biological systems, to mention some diverse examples.

Recently, SBRS has also been observed in self-gravitating $N$-body systems,[50,51] where the equilibrium configurations of an $N$-body self-gravitating system enclosed in a finite 3 dimensional spherical volume have been investigated using a mean-field approach. It was shown that when the ratio of the angular momentum of the system to its energy is high, spontaneous breaking of rotational symmetry occurs, manifesting itself in the formation of double-cluster structures. These results have also been confirmed with direct numerical simulations.[52]

It is well-known that a large number of phenomena exhibited by many-body systems have their counterparts and parallels in field theory, which in some sense is a limiting case of $N$-body systems in the limit $N \to \infty$. Since the closest analogues of a lump of matter in field theories are solitons, the presence of SBRS in self-gravitating $N$-body systems has led us to expect that it may also be present in solitonic field theories.

Our main motivation towards studying SBRS in solitons is that in hadronic physics Skyrme-type solitons often provide a fairly good qualitative description of nucleon properties (see, *e.g.*, Refs. 53 and 54). In particular, it is interesting to ask what happens when such solitons rotate quickly, because this might shed some light on the non-spherical deformation of excited nucleons with high orbital angular momentum, a subject which is now of considerable interest.

We shall see that the baby Skyrme model on the two-sphere indeed exhibits SBRS, and we will try to understand why this is so.[55] First, we give a brief account for the occurrence of SBRS in physical systems in general, and then use the insights gained from this discussion to infer the conditions under which SBRS might appear in solitonic models and in that context we study its appearance in baby Skyrme models. Specifically, we shall show that SBRS emerges if the domain manifold of the model is a two-sphere, while if the domain is $\mathbb{R}^2$, SBRS does not occur.

### 8.4.1. *SBRS from a dynamical point of view*

The onset of SBRS may be qualitatively understood as resulting from a competition between the static energy of a system and its moment of inertia. To see this, let us consider a system described by a set of degrees of freedom $\phi$, and assume that the dynamics of the system is governed by a Lagrangian which is invariant under spatial rotations. When the system is static, its equilibrium configuration is obtained by minimizing its static energy $E_{\text{static}}$ with respect to its degrees of freedom $\phi$

$$\frac{\delta E}{\delta \phi} = 0 \quad \text{where} \quad E = E_{\text{static}}(\phi) \,. \tag{8.4.27}$$

Usually, if $E_{\text{static}}(\phi)$ does not include terms which manifestly break rotational symmetry, the solution to (8.4.27) is rotationally-symmetric (with the exception of degenerate spontaneously-broken vacua, which are not of our concern here). If the system rotates with a given angular momentum $\boldsymbol{J} = J\hat{z}$, its configuration is naturally deformed. Assuming that the Lagrangian of the system is quadratic in the time derivatives, stable rotating configurations (if such exist) are obtained by minimizing its total energy $E_J$

$$\frac{\delta E_J}{\delta \phi} = 0 \quad \text{where} \quad E_J = E_{\text{static}}(\phi) + \frac{J^2}{2I(\phi)}, \tag{8.4.28}$$

where $I(\phi)$ is the ratio between the angular momentum of the system and its angular velocity $\boldsymbol{\omega} = \omega \hat{z}$ (which for simplicity we assume is oriented in the direction of the angular momentum). $I(\phi)$ is the (scalar) moment of inertia of the system.

The energy functional (8.4.28) consists of two terms. The first, $E_{\text{static}}$, increases with the asymmetry. This is simply a manifestation of the minimal-energy configuration in the static case being rotationally-symmetric. The second term $J^2/2I$, having the moment of inertia in the denominator, decreases with the asymmetry. At low values of angular momentum, the $E_{\text{static}}$ term dominates, and thus asymmetry is not energetically favorable, but as the value of angular momentum increases, the second term becomes dominant, giving rise to a possible breaking of rotational symmetry.

### 8.4.2. *SBRS in baby Skyrme models*

In what follows, we show that the above mechanism of SBRS is present in certain types of baby Skyrme models.

As already discussed in previous sections, the static solutions of the baby Skyrme model (8.1.1) have rotationally-symmetric energy and charge distributions in the charge-one and charge-two sectors.[4] The charge-one skyrmion has an energy peak at its center which drops down exponentially. The energy distribution of the charge-two skyrmion has a ring-like peak around its center at some characteristic distance. The rotating solutions of the model in $\mathbb{R}^2$ are also known.[15,56] Rotation at low angular velocities slightly deforms the skyrmion but it remains rotationally-symmetric. For larger values of angular velocity, the rotationally-symmetric configuration becomes unstable but in this case the skyrmion does not undergo symmetry breaking. Its stability is restored through a different mechanism, namely that of radiation. The skyrmion radiates out the excessive energy and angular momentum, and as a result begins slowing down until it reaches equilibrium at some constant angular velocity, its core remaining rotationally-symmetric. Moreover, if the Skyrme fields are restricted to a rotationally-symmetric (hedgehog) form, the critical angular velocity above which the skyrmion radiates can be obtained analytically. It is simply the coefficient of the potential term $\omega_{\text{crit}} = \mu$.[15] Numerical full-field simulations also show that the skyrmion actually begins radiating well below $\omega_{\text{crit}}$, as radiation itself may be non-rotationally-symmetric. The skyrmion's core, however, remains rotationally-symmetric for every angular velocity.

The stabilizing effect of the radiation on the solutions of the model has lead us to believe that models in which radiation is somehow inhibited may turn out to be good candidates for the occurrence of SBRS. In what follows, we study the baby Skyrme model on the two-sphere, whose static solutions were presented in the previous section. Within this model, energy and angular momentum are not allowed to escape to infinity through radiation, and as a consequence, for high enough angular momentum the mechanism responsible for SBRS discussed in the previous section takes over, revealing solutions with spontaneously broken rotational symmetry.

### 8.4.3. *The baby Skyrme model on the two-sphere*

In order to find the stable rotating solutions of the model, we assume for simplicity that any stable solution would rotate around the axis of angular momentum (which is taken to be the $z$ direction) with some angular velocity $\omega$. The rotating solutions thus take the form $\phi(\theta, \varphi, t) = \phi(\theta, \varphi - \omega t)$. The energy functional to be minimized is

$$E = E_{\text{static}} + \frac{J^2}{2I} , \qquad (8.4.29)$$

where $I$ is the ratio of the angular momentum of the skyrmion to its angular velocity, or its "moment of inertia", given by

$$I = \frac{1}{4\pi B} \int d\Omega \left( (\partial_\varphi \phi)^2 + \kappa^2 (\partial_\theta \phi \times \partial_\varphi \phi)^2 \right) . \qquad (8.4.30)$$

### 8.4.4. *Results*

In what follows we present the results obtained by the minimization scheme applied to the rotating solutions of the model in the charge-one and charge-two sectors, which as mentioned above are rotationally-symmetric. For simplicity, we fix the parameter $\kappa$ at $\kappa^2 = 0.01$ although other $\kappa$ values were tested as well, yielding qualitatively similar solutions.

#### 8.4.4.1. *Rotating charge-one solutions*

The rotating charge-one skyrmion has spherically-symmetric energy and charge distributions in the static limit (Fig. 8.14(a)). When rotated slowly, its symmetry is reduced to $O(2)$, with the axis of symmetry coinciding with the axis of rotation (Fig. 8.14(b)). At some critical value of angular momentum (which in the current settings is $J_{\text{crit}} \approx 0.2$), the axial symmetry is further broken, yielding an ellipsoidal energy distribution with three unequal axes (Fig. 8.14(c)). Any further increase in angular momentum results in the elongation of the skyrmion in one horizontal direction and its shortening in the perpendicular one. The results are very similar to those of the rotating self-gravitating ellipsoid.

#### 8.4.4.2. *Rotating charge-two solutions*

SBRS is also observed in rotating charge-two skyrmions. The static charge-two skyrmion has only axial symmetry (Fig. 8.15(a)), with its symmetry axis having no

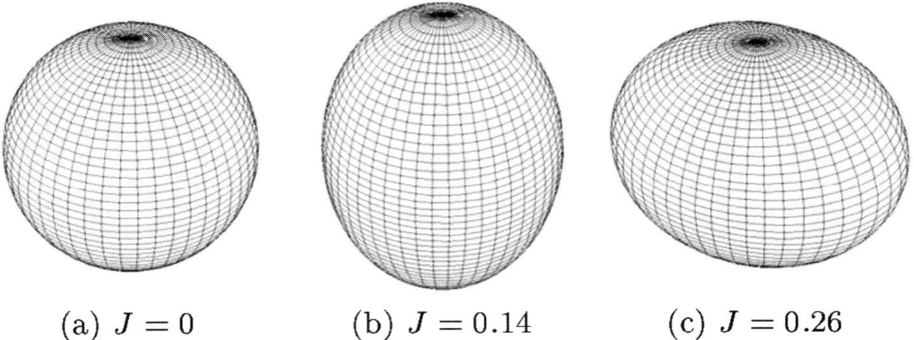

(a) $J = 0$            (b) $J = 0.14$            (c) $J = 0.26$

Fig. 8.14.    Baby skyrmions on the two-sphere ($\kappa^2 = 0.01$): The charge distribution $\mathcal{B}(\theta, \varphi)$ of the charge-one skyrmion for different angular momenta. In the figure, the vector $\mathcal{B}(\theta, \varphi)\hat{r}$ is plotted for the various $\theta$ and $\varphi$ values.

preferred direction. Nonzero angular momentum aligns the axis of symmetry with the axis of rotation. For small values of angular momentum, the skyrmion is slightly deformed but remains axially-symmetric (Fig. 8.15(b)). Above $J_{\mathrm{crit}} \approx 0.55$ however, its rotational symmetry is broken, and it starts splitting to its 'constituent' charge-one skyrmions (Fig. 8.15(c) and 8.15(d)). As the angular momentum is further increased, the splitting becomes more evident, and the skyrmion assumes a string-like shape. This is somewhat reminiscent of the well-known elongation, familiar from high-spin hadrons which are also known to assume a string-like shape with the constituent quarks taking position at the ends of the string.[58,59]

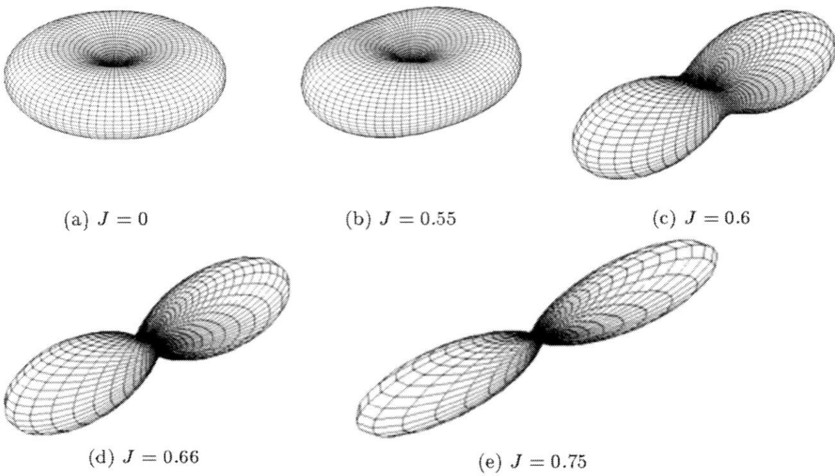

(a) $J = 0$            (b) $J = 0.55$            (c) $J = 0.6$

(d) $J = 0.66$            (e) $J = 0.75$

Fig. 8.15.    Baby skyrmions on the two-sphere ($\kappa^2 = 0.01$): The charge distribution $\mathcal{B}(\theta, \varphi)$ of the charge-two skyrmion for different angular momenta. In the figure, the vector $\mathcal{B}(\theta, \varphi)\hat{r}$ is plotted for the various $\theta$ and $\varphi$ values.

### 8.4.5. *The rational map ansatz*

A somewhat more analytical analysis of this system may be achieved by the use of the rational maps approximation scheme,[44] which as was shown earlier provides quite accurate results for the static solutions of the model.[57]

In its implementation here, we simplify matters even more and reduce the degrees of freedom of the maps by a restriction only to those maps which exhibit the symmetries observed in the rotating full-field solutions. This allows the isolation of those parameters which are the most critical for the minimization of the energy functional.

As shown in Fig. 8.14, the charge and energy densities of the charge-one skyrmion exhibit progressively lower symmetries as $J$ is increased. The static solution has an $O(3)$ symmetry, while the slowly-rotating solution has an $O(2)$ symmetry. Above a certain critical $J$, the $O(2)$ symmetry is further broken and only an ellipsoidal symmetry survives. Rational maps of degree one, however, cannot produce charge densities which have all the discrete symmetries of an ellipsoid with three unequal axes. Nonetheless, approximate solutions with only a reflection symmetry through the $xy$ plane (the plane perpendicular to the axis of rotation) and a reflection through one horizontal axis may be generated by the following one-parametric family of rational maps

$$R(z) = \frac{\cos\alpha}{z + \sin\alpha} \,, \tag{8.4.31}$$

which has the charge density

$$\mathcal{B}(\theta, \varphi) = \left( \frac{\cos\alpha}{1 + \sin\alpha \sin\theta \cos\varphi} \right)^2 . \tag{8.4.32}$$

Here, $\alpha \in [-\pi, \pi]$ is the parameter of the map, with $\alpha = 0$ corresponding to a spherically-symmetric solution and a non-zero value of $\alpha$ corresponding to a nonrotationally-symmetric solution. Results of a numerical minimization of the energy functional (8.4.29) for fields constructed from (8.4.31) for different values of angular momentum $J$ are shown in Fig. 8.16(a). While for angular momentum less than $J_{\text{crit}} \approx 0.1$, $\alpha = 0$ minimizes the energy functional (a spherically-symmetric solution), above this critical value bifurcation occurs and $\alpha = 0$ is no longer a minimum; the rotational symmetry of the charge-one skyrmion is broken and it becomes nonrotationally-symmetric.

A similar analysis of the charge-two rotating solution yields the one-parametric map

$$R(z) = \frac{\sin\alpha + z^2 \cos\alpha}{\cos\alpha + z^2 \sin\alpha} \,, \tag{8.4.33}$$

with corresponding charge density

$$\mathcal{B}(\theta, \varphi) = \left( \frac{2\cos 2\alpha \sin\theta}{2 + \sin^2\theta(\sin 2\alpha \cos 2\varphi - 1)} \right)^2 . \tag{8.4.34}$$

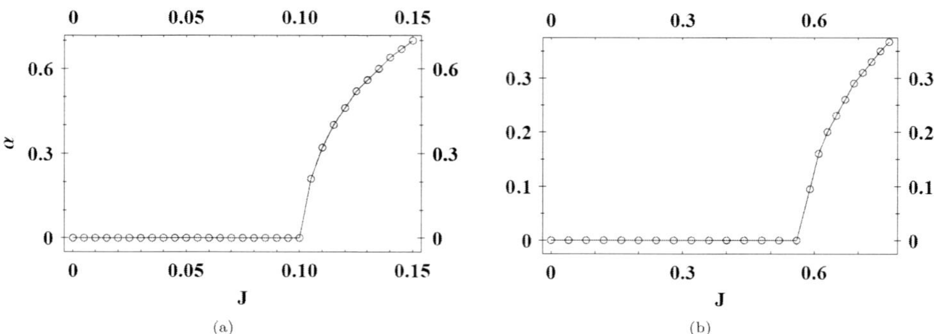

Fig. 8.16.  Spontaneous breaking of rotational symmetry in the restricted rational maps approximation for the baby skyrmions on the two-sphere: the parameter $\alpha$ as a function of the angular momentum $J$, for the charge-one (top) and the charge-two (bottom) solutions. The lines are to guide the eye.

In this case, $\alpha = 0$ corresponds to a torodial configuration, and a non-zero value of $\alpha$ yields solutions very similar to those shown in Fig. 8.15, having the proper discrete symmetries. The results in this case are summarized in Fig. 8.16(b), indicating that above $J_{\rm crit} \approx 0.57$ the minimal energy configuration is no longer axially-symmetric.

The discrepancies in the critical angular momenta $J_{\rm crit}$ between the full-field method (0.2 for charge-one and 0.55 for charge-two) and the rational maps scheme (0.1 for charge-one and 0.57 for charge-two) are of course expected, as in the latter method, the solutions have only one degree of freedom. Nonetheless, the qualitative similarity in the behavior of the solutions in both cases is strong.

### 8.4.6. *Further remarks*

We have seen that SBRS appears not only in rotating classical-mechanical systems but also in the baby Skyrme model on the two-sphere. We have argued that this is so because the phenomenon originates from general principles, and hence it is a universal one.

The results presented above may, at least to some extent, also be linked to recent advances in the understanding the non-sphericity of excited nucleons with of large orbital momentum. Non-spherical deformation of the nucleon shape is now a focus of considerable interest, both experimental[60,61] and theoretical.[62–64] As skyrmions are known to provide a good qualitative description of many nucleon properties, the results presented here may provide some corroboration to recent results on this subject (e.g., Ref. 64), although a more detailed analysis of this analogy is in order.

### Acknowledgments

This work was supported in part by a grant from the Israel Science Foundation administered by the Israel Academy of Sciences and Humanities.

## A.1. Obtaining Baby Skyrmion Solutions — The Relaxation Method

As a large part of the studies presented above is based on numerically obtaining the baby skyrmion configurations, in the following we describe the relaxation method that was used to obtain the solutions.

The multi-solitons of the baby Skyrme model are those field configurations which minimize the static energy functional within each topological sector. The energy functional is given by

$$E = \int d^2x \; \left( \frac{1}{2}(\partial_x \phi \cdot \partial_x \phi + \partial_y \phi \cdot \partial_y) + \frac{\kappa^2}{2}(\partial_x \phi \times \partial_y \phi)^2 + U(\phi_3) \right) . \quad (A.1)$$

As already noted, the baby Skyrme model is a nonintegrable system, so in general, explicit analytical solutions to its Euler-Lagrange equations are nearly impossible to find. Hence, one must resort to numerical techniques.

Generally speaking, there are two main approaches to finding the baby skyrmion solutions numerically. One approach is to employ standard techniques to numerically solve the Euler-Lagrange equations which follow from the energy functional (A.1). The other approach — the one taken here — is to utilize relaxation methods to minimize the energy of the skyrmion within any desired topological sector. In what follows, we describe in some detail the relaxation method we have used all throughout this research. This method is based on the work of Hale, Schwindt and Weidig.[67] We assume for simplicity that the base space is descretized to a rectangular grid. The implementation of this method in the case of curved spaces or for a polar grid is straightforward.

The relaxation method begins by defining a grid with $N^2$ points, where at each point a field triplet $\phi(x_m, y_n)$ is defined. All measurable quantities such as energy density or charge density are calculated at the centers of the grid squares, using the following expressions for the numerical derivatives, also evaluated at these points:

$$\left. \frac{\partial \phi}{\partial x} \right|_{(x_{m+\frac{1}{2}}, y_{n+\frac{1}{2}})} = \frac{1}{\Delta x} \left( \left\langle \frac{\phi(x_{m+1}, y_n) + \phi(x_{m+1}, y_{n+1})}{2} \right\rangle_{\text{normed}} \right.$$

$$\left. - \left\langle \frac{\phi(x_m, y_n) + \phi(x_m, y_{n+1})}{2} \right\rangle_{\text{normed}} \right), \quad (A.2)$$

with the $y$-derivatives analogously defined, and the "normed" subscript indicates that the averaged fields are normalized to one. If the field itself has to calculated at that center of a grid square, we use the prescription

$$\phi(x_{m+\frac{1}{2}}, y_{n+\frac{1}{2}}) \quad \quad (A.3)$$

$$= \left\langle \frac{1}{4} \left( \phi(x_m, y_n) + \phi(x_m, y_{n+1}) + \phi(x_{m+1}, y_n) + \phi(x_{m+1}, y_{n+1}) \right) \right\rangle_{\text{normed}} .$$

The basic updating mechanism of the relaxation process consists of the following two steps: A point $(x_m, y_n)$ on the grid is chosen at random, along with one of the three components of the field $\phi(x_m, y_n)$. The chosen component is then shifted by a value $\delta_\phi$ chosen uniformly from the segment $[-\Delta_\phi, \Delta_\phi]$ where $\Delta_\phi = 0.1$ initially. The field triplet is then scaled and the change in energy is calculated. If the energy decreases, the modification of the field is accepted and otherwise it is discarded.

The relaxation process, through which the energy of the baby skyrmion is minimized, is as follows:

(1) Initialize the field triplet $\phi$ to a rotationally-symmetric configuration

$$\phi_{\text{initial}} = (\sin f(r) \cos B\theta, \sin f(r) \sin B\theta, \cos f(r)). \tag{A.4}$$

In our setup, we have chosen the profile function $f(r)$ to be $f(r) = \pi \exp(-r)$, $r$ and $\theta$ being the usual polar coordinates.

(2) Perform the basic updating mechanism for $M \times N^2$ times (we took $M = 100$), and then calculate the average rate of acceptance. If it is smaller than 5%, decrease $\Delta_\phi$ by half.

(3) Repeat step (2) until $\Delta_\phi < 10^{-9}$, meaning no further decrease in energy is observed.

This procedure was found to work very well in practice, and its accuracy and validity were verified by comparison of our results to known ones. There is however one undesired feature to this minimization scheme, which we note here: it can get stuck at a local minimum. This problem can be resolved by using the "simulated annealing" algorithm,[65,66] which in fact has been successfully implemented before, in obtaining the minimal energy configurations of three dimensional skyrmions.[67] The algorithm is comprised of repeated applications of a Metropolis algorithm with a gradually decreasing temperature, based on the fact that when a physical system is slowly cooled down, reaching thermal equilibrium at each temperature, it will end up in its ground state. This algorithm, however, is much more expensive in terms of computer time. We therefore employed it only in part, just as a check on our method, which corresponds to a Metropolis algorithm algorithm at zero temperature. We found no apparent changes in the results.

## References


1. T.H.R. Skyrme, A non-linear field theory, *Proc. Roy. Soc. A* **260** (1961) 127.
2. T.H.R. Skyrme, A unified field theory of mesons and baryons, *Nucl. Phys.* **31** (1962) 556.
3. G.S. Adkins, C.R. Nappi and E. Witten, Static properties of nucleons in the Skyrme model, *Nucl. Phys. B* **228** (1983) 552.
4. B.M.A.G. Piette, B.J. Schoers and W.J. Zakrzewski, Multisolitons in a two-dimensional Skyrme model, *Z. Phys. C* **65** (1995) 165.
5. A.A. Belavin and A.M. Polyakov, Metastable states of two-dimensional isotropic ferromagnets, *JETP Lett.* **22** (1975) 245.
6. S.L. Sondhi, A. Karlhede, S.A. Kivelson and E.H. Rezayi, Skyrmions and the crossover

from the integer to fractional quantum Hall effect at small Zeeman energies, *Phys. Rev. B* **47** (1993) 16419.

7. N.R. Walet and T. Weidig, Full 2D numerical study of the quantum Hall Skyrme crystal, eprint:arXiv:cond-mat/0106157v2, (2001).

8. S.M. Girvin, *Topological Aspects of Low Dimensional Systems, Les Houches lectures, Vol. 29* edited by A. Comtet, T. Jolicoeur, S. Ouvry and F. David (Berlin, Springer Verlag, 1998).

9. Z.F. Ezawa, *Quantum Hall Effects: Field Theoretical Approach and Related Topics* (World Scientific, Singapore, 2000).

10. D.H. Lee and C.L. Kane, Boson-vortex-Skyrmion duality, spin-singlet fractional quantum Hall effect, and spin-1/2 anyon superconductivity, *Phys. Rev. Lett.* **64** (1990) 1313.

11. N.S. Manton and P.M. Sutcliffe, *Topological Solitons*, (Cambridge Univ. Press, Cambridge, 2004).

12. R.A. Leese, M. Peyrard and W.J. Zakrzewski, Soliton scatterings in some relativistic models in (2+1) dimensions, *Nonlinearity* **3** (1990) 773.

13. B.M.A.G. Piette and W.J. Zakrzewski, Skyrmion dynamics in (2+1) dimensions, *Chaos, Solitons and Fractals* **5** (1995) 2495.

14. P.M. Sutcliffe, The interaction of Skyrme-like lumps in (2+1) dimensions, *Nonlinearity* **4** (1991) 1109.

15. B.M.A.G. Piette, B.J. Schoers and W.J. Zakrzewski, Dynamics of baby skyrmions, *Nucl. Phys. B* **439** (1995) 205.

16. T. Weidig, The baby Skyrme models and their multi-skyrmions, *Nonlinearity* **12** (1999) 1489.

17. P. Eslami, M. Sarbishaei and W.J. Zakrzewski, Baby Skyrme models for a class of potentials, *Nonlinearity* **13** (2000) 1867.

18. I. Hen and M. Karliner, Rotational symmetry breaking in baby Skyrme models, *Nonlinearity* **21** (2008) 399.

19. E. Braaten, S. Townsend and L. Carson, Novel structure of static multisoliton solutions in the Skyrme model, *Phys. Lett. B* **235** (1990) 147.

20. R.A. Battye and P.M. Sutcliffe, Symmetric skyrmions, *Phys. Rev. Lett.* **79** (1997) 363.

21. N.R. Walet, Quantising the $B = 2$ and $B = 3$ skyrmion systems, *Nucl. Phys. A* **606** (1996) 429.

22. I. Klebanov, Nuclear matter in the skyrme model, *Nucl. Phys. B* **262** (1985) 133.

23. A.S. Goldhaber and N.S. Manton, Maximal symmetry of the Skyrme crystal, *Phys. Lett. B* **198** (1987) 231.

24. A.D. Jackson and J. Verbaarschot, Phase structure of the skyrme model, *Nucl. Phys. A* **484** (1988) 419.

25. M. Kugler and S. Shtrikman, A new skyrmion crystal, *Phys. Lett. B* **208** (1988) 491.

26. L. Castellejo, P. Jones, A.D. Jackson and J. Verbaarschot, Dense skyrmion systems, *Nucl. Phys. A* **501** (1989) 801.

27. O. Schwindt and N.R. Walet, Europhys. Lett. **55** (2001) 633.

28. R.S. Ward, Nonlinearity **17** (2004) 1033.

29. R.J. Cova and W.J. Zakrzewski, Soliton scattering in the $O(3)$ model on a torus, *Nonlinearity* **10** (1997) 1305.

30. R.J. Cova, Lump scattering on the torus, *Eur. Phys. J. B* **15** (2001) 673.

31. R.J. Cova and W.J. Zakrzewski, Scattering of periodic solitons, *Rev. Mex. Fis.* **50** (2004) 527

32. I. Hen and M. Karliner, Hexagonal structure of baby skyrmion lattices, *Phys. Rev. D* **77** (2008) 054009.

33. J. Bamberg, G. Cairns and D. Kilminster, The crystallographic restriction, permutations and Goldbach's conjecture, *Amer. Math. Mon.* **110** (2003) 202.

34. W.H. Kleiner, L.M. Roth and S.H. Antler, Bulk solution of Ginzburg-Landau equations for type II superconductors: upper critical field region, *Phys. Rev.* **133** A (1964) 1226.

35. R.A. Battye and P.M. Sutcliffe, A Skyrme lattice with hexagonal symmetry, *Phys. Lett. B* **416** (1998) 385.

36. L. Brey, H.A. Fertig, R. Côté and A.H. MacDonald, Skyrme crystal in a two-dimensional electron gas, *Phys. Rev. Lett.* **75** (1995) 2562.

37. N.S. Manton, Geometry of Skyrmions, *Commun. Math. Phys.* **111** (1987) 469.

38. R. Dandoloff and A. Saxena, Skyrmions on an elastic cylinder, *Eur. Phys. J. B* **29** (2002) 265.

39. M. de Innocentis and R.S. Ward, Skyrmions on the 2-sphere, *Nonlinearity* **14** (2001) 663.

40. N.N. Scoccola and D.R. Bes, Two-dimensional skyrmions on the sphere, *JHEP* **09** (1998) 012.

41. A. Wirzba and H. Bang, The mode spectrum and the stability analysis of skyrmions on a 3-sphere, *Nucl. Phys. A* **515** (1990) 571.

42. S. Krusch, $S^3$ skyrmions and the rational map ansatz, *Nonlinearity* **13** (2000) 2163.

43. A.A. Belavin and A.M. Polyakov, Metastable states of two-dimensional isotropic ferromagnets, *JETP Lett.* **22** (1975) 245.

44. C.J. Houghton, N.S. Manton and P.M. Sutcliffe, Rational maps, monopoles and skyrmions, *Nucl. Phys. B* **510** (1998) 507.

45. R.A. Battye and P.M. Sutcliffe, Skyrmions fullerenes and rational maps, *Rev. Math. Phys.* **14** (2002) 29.

46. N.S. Manton and B.M.A.G. Piette, Understanding skyrmions using rational maps, e-print hep-th/0008110.

47. J.G. Groshaus, I. Dujovne, Y. Gallais, C.F. Hirjibehedin, A. Pinczuk, Y. Tan, H. Stormer, B.S. Dennis, L.N. Pfeiffer and K.W. West, Spin texture and magnetoroton excitations at $\nu = 1/3$, *Phys. Rev. Lett.* **100** (2008) 046804.

48. R.A. Lyttleton, *The Stability of Rotating Liquid Masses*, (Cambridge Univ. Press., Cambridge, 1953).

49. S. Chandrasekhar, *Ellipsoidal Figures of Equilibrium*, (Yale Univ. Press, New Haven, 1969).

50. E.V. Votyakov, H.I. Hidmi, A. DeMartino and D.H.E. Gross, Microcanonical mean field thermodynamics of self-gravitating and rotating systems, *Phys. Rev. Lett.* **89** (2002) 031101.

51. E.V. Votyakov, A. DeMartino and D.H.E. Gross, Thermodynamics of rotating self-gravitating systems, *Eur. Phys. J. B* **29** (2002) 593.

52. M. Karliner, talk at London Mathematical Society Durham Symposium, `http://www.maths.dur.ac.uk/events/Meetings/LMS/2004/TSA/Movies/Karliner.wmv`, (2004).

53. I. Zahed and G.E. Brown, The Skyrme Model, *Phys. Rept.* **142** (1986) 1.

54. G. Holzwarth and B. Schwesinger, Baryons In The Skyrme Model, *Rept. Prog. Phys.* **49** (1986) 825.

55. I. Hen and M. Karliner, Spontaneous breaking of rotational symmetry in rotating solitons: a toy model of excited nucleons with high angular momentum, *Phys. Rev. D* **77** (2008) 116002.

56. M. Betz, H.B. Rodrigues and T. Kodama, Rotating Skyrmion in 2+1 dimensions, *Phys. Rev. D* **54** (1996) 1010.

57. I. Hen and M. Karliner, Baby skyrmions on the two-sphere, *Phys. Rev. E* **77** (2008) 036612.

58. Y. Nambu, Strings, monopoles, and gauge fields, *Phys. Rev. D* **10** (1974) 4262.

59. J.S. Kang and H.J. Schnitzer, Dynamics of light and heavy bound quarks, *Phys. Rev. D* **12** (1975) 841.

60. M.K. Jones *et al.* (The Jefferson Lab Hall A Collaboration), $G_{E_p}/G_{M_p}$ ratio by polarization transfer in $\vec{e}p \to \vec{e}p$, *Phys. Rev. Lett.* **84** (2000) 1398.

61. O. Gayou *et al.* (The Jefferson Lab Hall A Collaboration), Measurement of $G_{E_p}/G_{M_p}$ in $\vec{e}p \to \vec{e}p$ to $Q^2 = 5.6 GeV^2$, *Phys. Rev. Lett.* **88** (2002) 092301.

62. G.A. Miller, Shapes of the proton, *Phys. Rev. C* **68** (2003) 022201(R).

63. A. Kvinikhidze and G. A. Miller, Shapes of the nucleon, *Phys. Rev. C* **73**, 065203 (2006).

64. G.A. Miller, Densities, parton distributions, and measuring the nonspherical shape of the nucleon, *Phys. Rev. C* **76** (2007) 065209.

65. S. Kirkpatrick, C.D. Gellat and M.P. Vecchi, Optimization by simulated annealing, *Science* **220** (1983) 671.

66. S. Geman and D. Geman, Stochastic relaxation, Gibbs distributions, and the bayesian restoration of images, *IEEE Trans. Pattern Anal. Mach. Intell.* **6** (1984) 721.

67. M. Hale, C. Schwindt and T. Weidig, Simulated annealing for topological solitons, *Phys. Rev. E* **62** (2000) 4333.

# PART 2
# Condensed Matter

# Chapter 9

# Spin and Isospin: Exotic Order in Quantum Hall Ferromagnets

## Spin and a peculiar kind of isospin in two-dimensional electron gases can exhibit novel counterintuitive ordering phenomena.

Steven M. Girvin*

Quantum mechanics is a strange business, and the quantum physics of strongly correlated many-electron systems can be stranger still. Good examples are the various quantum Hall effects.[1-4] They are among the most remarkable many-body quantum phenomena discovered in the second half of the 20th century, comparable in intellectual import to superconductivity and superfluidity. The quantum Hall effects are an extremely rich set of phenomena with deep and truly fundamental theoretical implications.

The fractional quantum Hall effect has yielded fractional charge, with its attendant spin–statistics peculiarities, as well as phases with unprecedented order parameters. It has beautiful connections to a variety of different topological and conformal field theories more commonly studied as formal models in particle theory. But in the quantum Hall context, each of these theoretical constructs can be made manifest by the twist of an experimental knob. Where else but in condensed-matter physics can an experimenter change the number of flavors of relativistic chiral fermions in a sample, or create a system whose low energy description is a Chern–Simons gauge theory whose fundamental coupling constant (the $\theta$ angle) can be set by hand?

The first quantum Hall effect was discovered by Klaus von Klitzing 20 years ago, for which he won the 1985 Nobel Prize in physics. (See PHYSICS TODAY, December 1985, page 17.) Because of recent tremendous technological progress in molecular-beam epitaxy and the fabrication of artificial structures, quantum Hall experimentation continues to bring us striking new discoveries. The early experiments were limited to simple transport measurements that determined energy

---

*STEVEN GIRVIN *is a professor of physics at Indiana University in Bloomington.*

gaps for charged excitations. Recent advances, however, have given us many new probes — optical, acoustic, microwave, specific heat, tunneling spectroscopy, and NMR — that continue to pose intriguing new puzzles even as they advance our knowledge.


## Quantum Hall phenomena

The quantum Hall effect takes place in a two-dimensional electron gas formed in an artificial semiconductor quantum well and subjected to a high magnetic field normal to the plane. In essence, this macroscopic quantum effect is a result of commensuration between the number of electrons $N$ and the number of flux quanta $N_\Phi$ in the applied magnetic field. That is to say, the electron population undergoes a series of condensations into new states with highly non-trivial properties whenever the filling factor $\nu \equiv N/N_\Phi$ is an integer or a simple rational fraction.

Von Klitzing's original observation was, in effect, a sequence of energy gaps yielding (in the limit of zero temperature) electron transport without dissipation — much like a superconductor, but with radically different underlying physics.

The Hall conductivity $\sigma_{xy}$ in this dissipationless state turns out to be universal. It is given by $\nu e^2/h$ with great precision, irrespective of microscopic or macroscopic details. Therefore, one can exploit this remarkable phenomenon to make a very precise determination of the fine-structure constant and to realize a highly reproducible quantum-mechanical unit of electrical resistance. The quantum Hall effect is now used by standards laboratories around the world to maintain the ohm.

It is an amusing paradox that this ideal behavior occurs only in imperfect samples. That's because disorder produces Anderson localization of quasiparticles, preventing them from contributing to the transport properties. If the laboratory samples were ideal, the effect would go away!

The *integer* quantum Hall effect is due to an excitation gap associated with the filling of discrete kinetic-energy levels (Landau levels) of electrons executing quantized cyclotron orbits in the imposed magnetic field (see figure 1). Coulomb interactions between electrons would seem to be unimportant. When $\nu$ is an integer, the chemical potential lies in one of these kinetic energy gaps. The *fractional* quantum Hall effect occurs when one of the Landau levels is fractionally filled. Its physical origins — very different from those of the integer effect — are strong Coulomb corrections that produce a Mott-insulator-like excitation gap.

In some ways, this excitation gap is more like that in a superconductor, because it is not tied to a periodic lattice potential. That permits uniform charge flow of the incompressible electron liquid and hence a quantization of Hall conductivity. The electrons are strongly correlated because all the states in a given Landau level are completely degenerate in kinetic energy. Perturbation theory is therefore useless. But the novel correlation properties of this incompressible electron liquid are captured in a revolutionary wave function proposed by Robert Laughlin, for which

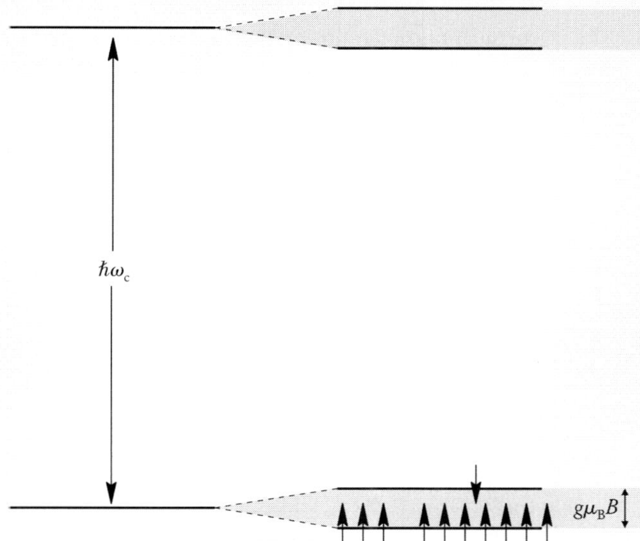

FIGURE 1. LANDAU LEVELS of uniformly spaced kinetic energy of a two-dimensional electron gas in a quantizing magnetic field $B$ whose cyclotron frequency is $\omega_c$. In free space, the Zeeman splitting $g\mu_B$ equals the Landau level splitting, but in GaAs heterostructures it is nearly two orders of magnitude smaller. At sufficiently low temperature, most of the electron spin orientations are in the lower Zeeman level.

he shared in 1998 Nobel Prize in physics with Horst Stormer and Daniel Tsui, who discovered the fractional quantum Hall effect in 1982. (See PHYSICS TODAY, December 1998, page 17.)

## Quantum Hall ferromagnetism

At $\nu = 1$ and certain other filling factors, quantum Hall systems exhibit spontaneous magnetic order. This constitutes a very peculiar kind of ferromagnetism: It is itinerant — the electrons are free to move around as in metals like iron — and yet it exhibits a charge excitation gap that manifests itself by precisely quantized Hall conductivity and the vanishing of the ordinary, dissipative longitudinal conductivity $\sigma_{xx}$.

My colleague Allan MacDonald refers to the $\nu = 1$ state as "the world's best understood ferromagnet." The lowest spin state of the lowest Landau level is completely filled and the exact ground state (neglecting small effects from Landau-level mixing) is very simple: It is a single Slater determinant precisely represented by Laughlin's wave function. (See the article by Jainendra Jain in PHYSICS TODAY, April 2000, page 39.) Unlike iron, this ferromagnet is 100% polarized, because the kinetic energy has been frozen into discrete Landau levels and polarizing the electron gas costs no kinetic energy.

For reasons peculiar to the electronic band structure of GaAs, the usual host semiconductor, the external magnetic field couples very strongly to the orbital motion (giving a large Landau level splitting) and very weakly to the spin degrees of freedom (giving an exceptionally small Zeeman gap, as shown in figure 1). Therefore, the spin orientation is not frozen in place, as one might naively expect. The low-energy spin degrees of freedom of this unusual ferromagnet have some rather novel properties that have recently been probed by specific-heat measurements, NMR, and other means.

The simplest excitations out of the ground state are spin waves (magnons), in which the spin orientation undergoes smooth fluctuations in space and time. Because of the unusual circumstance that the ground-state wave-function is a single, known Slater determinant, the single-magnon excited-state spectrum can also be computed exactly (see figure 2.) One can then use various approximate techniques to predict rather accurately the temperature dependence of the magnetization.[5–7]

One of the interesting features of the physics here is that two dimensions is the lowest dimensionality for which ordering is possible in magnets with Heisenberg ($SU_2$) symmetry. That is to say, the phase space for spin-wave excitations in two dimensions is large enough so that there is an infrared divergence in the number of excited magnons at any finite temperature. Hence the magnetization, which is 100% at zero temperature, crashes immediately to zero at any finite temperature. In the presence of a small Zeeman coupling, the magnetization begins to drop towards zero (as shown in figure 2b) at a temperature of a few K, characteristic of the Zeeman gap and the spin stiffness.

At filling factor $\nu = 1$, spin waves are the lowest energy excitations. But because they do not carry charge, they do not have a large impact on the electrical transport properties. Since the lowest spin state of the lowest Landau level is completely filled at $\nu = 1$, the Pauli exclusion principle tells us that we can add more charge, as illustrated in figure 1, only with reversed spin. In the absence of strong Coulomb interactions, the energy cost of this spin flip is simply the Zeeman energy, which is very small. So one might not expect to see a quantized Hall plateau near $\nu = 1$, because there would be a high density of thermally excited charges. However, the Coulomb interaction exacts a large exchange-energy penalty for having a reversed spin in a ferromagnetic state.[2,7] Thus magnetic order induced by Coulomb interactions turns out to be essential to the integer quantum Hall effect.

## Skyrmions

In 1993, Shivaji Sondhi and collaborators[8] made a notable discovery: Because the exchange energy is large and prefers locally parallel spins, the Zeeman energy being small, it is energetically cheaper to form a topogical spin texture by partially turning over some of the spins. (See the box on page 211.) Such a topological object is called a skyrmion, because of its provenance in the Skyrme model of nuclear physics.

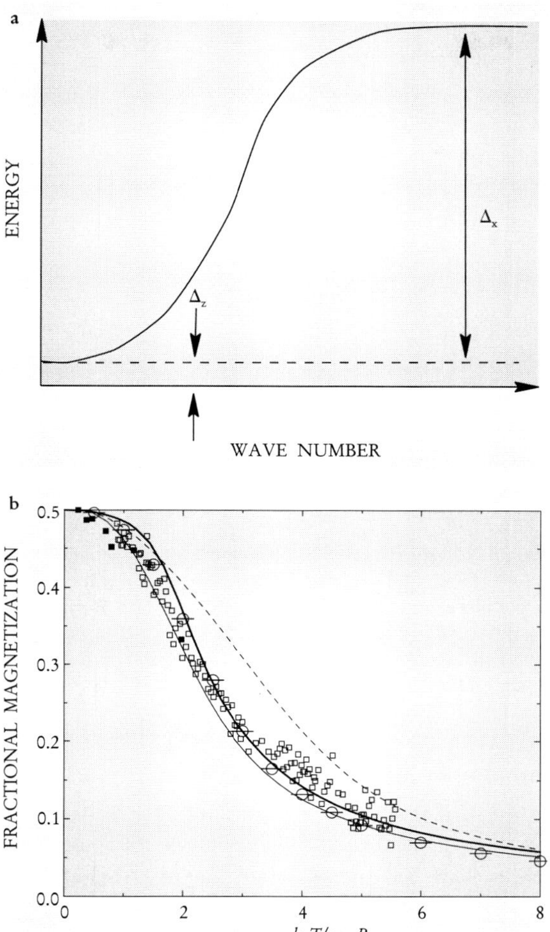


**FIGURE 2. SPIN WAVE EXCITATIONS** (magnons) from the quantum Hall ground state. **(a)** Dispersion relation for single magnons. At low frequency, the energy grows quadratically with wave number, starting from the Zeeman gap $\Delta_Z \sim$ 1K, and saturates at the Coulomb exchange energy $\Delta_X \sim$ 100K. **(b)** Temperature dependence of the magnetization at filling factor $\nu = 1$. Temperature is normalized to the Zeeman gap. Squares indicate experimental data.[5] Open circles and curves indicate various theoretical calculations.[6,7]

Since the system is an itinerant magnet with a quantized Hall conductivity, it turns out that the skyrmion texture accommodates precisely one extra unit of charge. NMR shifts and various optical and transport measurements have confirmed the prediction that each charge added to or removed from the state flips over a handful of spins. (See figure 3.)

In nuclear physics, the Skyrme model imagines the universe in a kind of ferromagnetic state, with a magnetization that is a four-component vector. Thus there

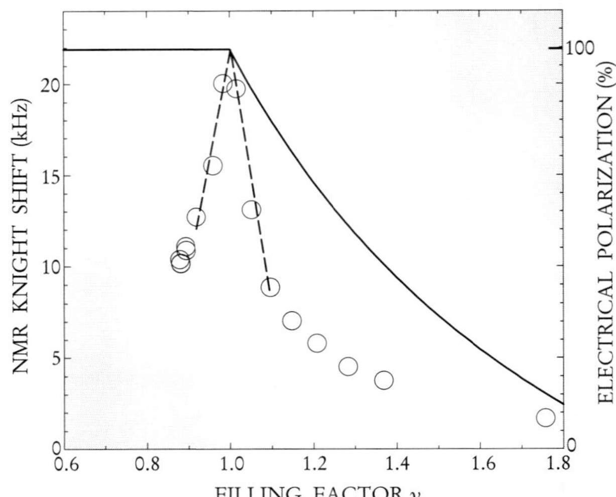



**Figure 3. Measured NMR shift** yields electron spin polarization as a function of filling factor near $\nu = 1$. This "Knight shift" is the change in nuclear precession frequency due to hyperfine coupling to the electron spin density. Circles are data from ref. 9. The steep fall-off on both sides of the 100% polarization peak at $\nu = 1$ indicates that typically 4 spins flip over for each charge added (or subtracted). The observed symmetry around the peak is due to the particle–hole symmetry between skyrmions and antiskyrmions. By contrast, the solid line is the prediction for non-interacting electrons.

are three directions in spin space for fluctuations around the (broken-symmetry) magnetization direction. So one has three different spin waves, representing the three light mesons $\pi^+$, $\pi^-$, and $\pi^0$. The nucleons (the protons, the neutron, and their antiparticles) are taken to be topological defects in this magnetization field. Through the magic of Berry-phase terms in the Lagrangian, these objects are fermions, even though they are excitations of a bosonic order-parameter field.

Essentially the same phenomenon occurs in quantum Hall ferromagnets, the only difference being that the spin waves have a non-relativistic (quadratic) dispersion relation, and the "nucleons" come in only one flavor: the electron and its antiparticle, the hole. Because the quantum Hall ferromagnetic order parameter is a three-component vector, there are only two directions in spin space for fluctuations around the broken-symmetry direction. One might think that this implies that there are two spin wave modes. But, in the nonrelativistic case, it turns out that the two coordinates are canonically conjugate and there is, in fact, only a single ferromagnetic spin wave.

Because it costs significant energy (about 30 K) to create a skyrmion or anti-skyrmion, they freeze out and disappear at low temperatures at $\nu = 1$. However, as one moves away from this filling factor, the cheapest way to add or subtract charge is through the formation of a finite density of skyrmions (proportional to $|\nu-1|$). Thus, away from $\nu = 1$, skyrmions do not freeze out, even at zero temperature. One might ask why skyrmions are not important in ordinary thin-film magnets. Skyrmions can exist there, in principle. But they always freeze out at low temperatures, because they do not carry charge and their density can not be controlled by varying the chemical potential.

Normally we think of manipulating spins by applying magnetic fields. A notable feature of quantum Hall ferromagnets is that, because skyrmions carry charge, one can move spins around by applying electrostatic potentials. For example, a random disorder potential can nucleate skyrmions.

In the presence of skyrmions, the ferromagnetic order is no longer colinear. The skyrmion configuration shown in the box on page 211 is only one of a continuous family of minimum-energy solutions. There exist two "zero modes," corresponding to translation of the skyrmion in real space and uniform rotation in spin space about the axis defined by the Zeeman field. In the presence of many skyrmions, these additional degrees of freedom lead to two totally new classes of low-energy collective excitations — "Goldstone modes" associated with the broken spin rotational and translational symmetry. Unlike ordinary spin waves, these Goldstone modes are not constrained by Larmor's theorem to have a minimum excitation gap given by the Zeeman energy. Indeed at long wavelengths, these excitations can go all the way down to zero frequency. That's because, in semiclassical terms, rotations about the Zeeman axis do not cost any Zeeman energy. In an ordinary ferromagnet, the ground state is invariant under rotations about the Zeeman axis. So the rotation produces no excitation. In a non-colinear system, however, states produced by different rotations are distinguishable from each other. Thus each skyrmion induces a new $xy$ quantumrotor degree of freedom.[10]

These low-frequency $xy$ spin fluctuations have been indirectly observed through a dramatic enhancement of the nuclear spin-relaxation rate $1/t_1$. Because nuclei precess at frequencies some three orders of magnitude below that of the Zeeman gap, they do not couple effectively to ordinary spin waves in the electron system. So the nuclear relaxation time $t_1$ can become many minutes, or even hours, at low temperature. But in the presence of skyrmions, $t_1$ becomes so short ($\sim$20 s) that the nuclei come into thermal equilibrium with the lattice through interactions with the electrons in the quantum well. This effect has recently been observed experimentally by Vincent Bayot, Mansour Shayegan and collaborators as a specific-heat enhancement of more than 5 orders of magnitude, due to the entropy of the nuclei[11] (see figure 4).

## Isospin Ordering in Bilayer Systems

Ordinary spin is not the only internal degree of freedom that can spontaneously become ordered. It is now possible to make a pair of identical electron gases in quantum wells separated by a distance ($\sim$10 nm) comparable to the electron spacing within a single quantum well. Under these conditions, one can expect strong interlayer correlations and new types of ordering phenomena associated with

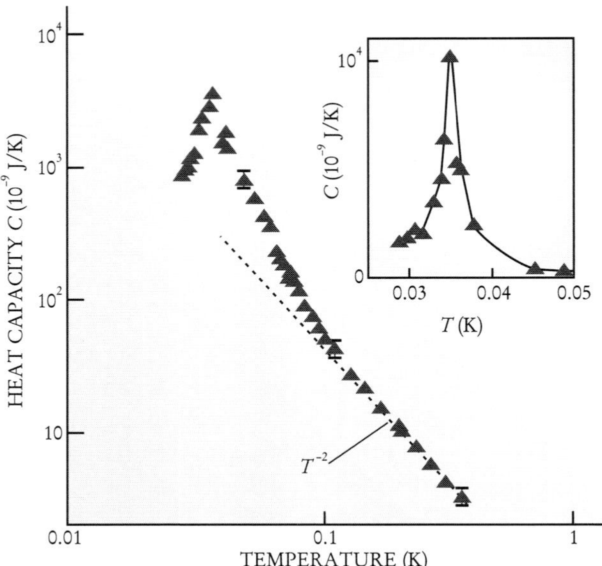

**FIGURE 4. SPECIFIC HEAT** is greatly enhanced by the presence of skyrmions. They dramatically shorten the nuclear spinlattice relaxation time, thus bringing the nuclei into thermal equilibrium. Dashed line is a calculation of the contribution of a model that assumes all nuclei in the quantum well contribute to the specific heat. At low temperatures, nuclei in the insulating barriers just outside the well raise the specific heat beyond this prediction.[11] The sharpness of this additional peak (inset linear plot) is not well understood.

the layer degree of freedom.[12] The many-body physics of two-layer systems can also be found in wide single-well systems with the two (nearly degenerate) lowest quantum-well subband states playing the role of a pseudospin degree of freedom.[13]

One of the peculiarities of quantum mechanics is that, even in the absence of tunneling between the layers, the electrons can be in a coherent state in which their layer index is uncertain. To understand the implications of this, we can define a pseudospin, which we also call "isospin," after the abstract spin Heisenberg introduced to distinguish neutrons from protons. In our case, the isospin is up if the electron is in the first layer and down if it is in the second. Spontaneous interlayer coherence corresponds to pseudospin magnetization lying in the $xy$ plane, corresponding to a coherent mixture of pseudospin up and down.

If the total filling factor for the two layers is $\nu = 1$, the Coulomb exchange energy will strongly favor this magnetic order, just as it does for real spins. That's because the spatial part of the fermionic wavefunction must vanish if two electrons with the same pseudospin orientation approach each other. (In contrast to the previous sections, we assume here that the real spins have been frozen into a ferromagnetic state and can be ignored.)

### Skyrmions and Topological Quantum Numbers

In this illustration of skyrmion spin texture in a quantum Hall ferromagnet, note that the spins are all up at infinity but down at the origin. At intermediate distances, they have a vortex-like configuration. Because of the quantized Hall conductivity, skyrmions carry extra charge. Although this extra charge is distributed throughout the core region, its total value is quantized. In fact, the skyrmion charge is directly proportional to the "topological charge" of the magnetization order-parameter field $m(r)$, and is given by the remarkable formula

$$Q = \frac{h}{e}\sigma_{xy} \int d^2r \frac{1}{8\pi} \varepsilon^{\mu\lambda} \varepsilon_{abc} m^a \partial_\mu m^b \partial_\lambda m^c .$$

where $\sigma_{xy}$ is the Hall conductivity. The epsilons are the fully antisymmetric tensors of second and third rank.

The physics behind this equation is the following: An electron traveling through a region will have its spin aligned with the local magnetization direction by the exchange field. Thus its spin direction will vary as the electron moves through the spin texture. If the spin direction is twisting in two directions at once (as required by the two spatial derivatives in the equation), the electron acquires a path-dependent Berry phase, much as if it were traveling through some additional magnetic flux. Adding flux draws in or expels charge proportional to the amount of this flux.

This same picture was used by Laughlin to derive the fractional charge of the quasiparticles in the case where the Hall conductivity $\sigma_{xy}$ is described by a fractional quantum number. At filling factor $\nu = 1$, the Hall conductivity $\sigma_{xy} = e^2/h$ and the skyrmion binds exactly one extra electron (or hole). Therefore it must be a fermion.

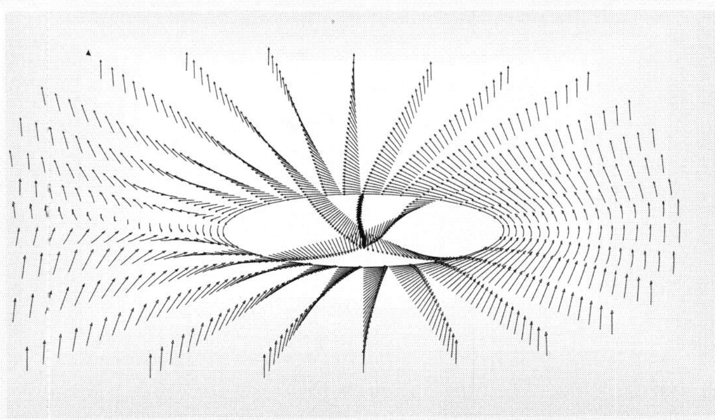

For real spins, the Coulomb interaction is spin invariant. For pseudospins, we must take into account the fact that intralayer repulsion is slightly stronger than interlayer repulsion. If the pseudospin were to become ordered in the $z$ direction, all of the electrons would be in one single layer, resulting in a large capacitive charging energy. That would lead to an "easy plane" anisotropy in which the pseudospin ferromagnetic order prefers to lie in the $xy$ plane.

When the charging energy is not severe, a good approximation to the $xy$ ordered state is

$$|\Psi\rangle = \prod_k (c_{k\uparrow}^\dagger + e^{i\varphi} c_{k\downarrow}^\dagger)|0\rangle\,, \tag{1}$$

where each $c^\dagger$ is the creation operator (acting on the vacuum state $|0\rangle$) for a given pseudospin in the $k$th single-particle spatial orbital. In this state, every single-particle orbital in the lowest Landau level is occupied by precisely one electron (hence $\nu = 1$). But each of these electrons is in a coherent superposition of the two pseudospin states. Much like the BCS wavefunction for a superconductor, this state has a definite phase $\varphi$, but an indefinite particle number. In our case, it is not the total particle number that is indefinite, but rather the particle-number difference between the two layers.[14] In contrast to the Cooper-pair field order parameter of a superconductor, the order parameter here

$$\Psi(r) \equiv \langle \psi_\uparrow^\dagger(r)\psi_\downarrow(r)\rangle \sim e^{i\varphi(r)} \tag{2}$$

is charge-neutral and thus able to condense despite the presence of the intense magnetic field. The order parameter at each point $\mathbf{r}$ is the expectation value of the spin-raising operator at that point. Because each electron is in a coherent superposition of states in different layers, one can destroy an electron in one layer and create an electron in the other, without leaving the ground state. In a certain sense, the coherent state is like an excitonic insulator with a particle and hole bound together — with the important difference that we do not know which layer each is in. This neutral object can travel through the magnetic field without suffering a classical Lorentz force or any Aharanov–Bohm phase shift.

In the absence of tunneling between the layers, the electrons have no way of determining the phase angle $\varphi$. Therefore, the energy must be independent of its global value. The exchange energy can, however, depend on spatial gradients of $\varphi$. The leading term in a gradient expansion is therefore

$$U = \frac{1}{2}\rho_{\rm s} \int d^2r |\nabla\varphi|^2\,, \tag{3}$$

where the pseudospin stiffness $\rho_{\rm s}$ has a typical value of about half a kelvin. (In general, spin stiffness is a measure of the energy cost of twisting spins out of perfect alignment.) Given the $xy$ symmetry of this model, we anticipate that the system will undergo a Kosterlitz–Thouless phase transition at a temperature on the order of $\rho_{\rm s}$.

This phase transition occurs when topological defects (vortices) in the phase field become unbound as a result of entropy gain, even though their interaction potential grows logarithmically with distance. In a superconducting film, such logarithmic interaction among vortices is due to the kinetic energy of supercurrents circulating around the vortices. But here there is no kinetic energy, and the energy cost is instead due to the loss of Coulomb exchange energy when there is a phase gradient. The "charge" conjugate to the order-parameter phase $\varphi$ is the $z$ component of the pseudospin, which is the charge difference between the layers. Therefore the supercurrent $J = \rho_{\mathrm{s}} \nabla \varphi$ corresponds to *oppositely* directed charge currents in the two layers.

One novel feature of the quantum Hall system is that vortices in the $\varphi$ field are "merons," carrying one half of the topological charge of skyrmions (see figure 5a). This implies that a meron carries *half* the fermion number of an ordinary fermion like an electron. The easy-plane anisotropy allows these "half skyrmions" to be topologically stable.

The onset of superfluidity below the Kosterlitz–Thouless temperature will manifest itself as an infinite antisymmetric conductivity between the two layers. One way to observe this would be to perform a drag experiment in which one sends current through one layer and then measures the voltage drop induced in the other layer. In ordinary fermi liquids, this drag is caused by collisions that transfer momentum between quasiparticles in different layers. Simple phase-space arguments show that this drag voltage should vanish like $T^2$ at low temperature. But in the superfluid phase, where the antisymmetric conductivity is infinite, the voltage drop must be exactly the same in both layers. That will lead to a very large drag that is not only opposite in sign to the usual drag effect, but actually increases in magnitude with decreasing temperature. Thus, as the temperature is lowered through the Kosterlitz–Thouless point, the drag should change sign and *increase* in magnitude, providing a very clear experimental signature.

This superfluid response of a phase-coherent inter-layer state has, in fact, not yet been directly observed. That's because it's hard to prevent tunneling between the layers when they are close enough to exhibit interlayer phase coherence. (A new generation of experiments is addressing this problem.) But long-range pseudospin $xy$ order has been observed experimentally through the strong response of the system to a weak magnetic field applied in the plane of the electron gases.

To understand this strong response, one has to consider the effects of weak tunneling. In the presence of tunneling, the particle-number difference between the two layers is no longer conserved and the global symmetry is lost. In addition to the exchange potential energy, there is now a tunneling energy term, which yields a preferred value $\varphi = 0$ for the order-parameter phase. We see from equation 1 that the vanishing of this phase represents the symmetric occupation of the two quantum-well states. In the presence of tunneling, this symmetric state is lower in energy than the antisymmetric combination.

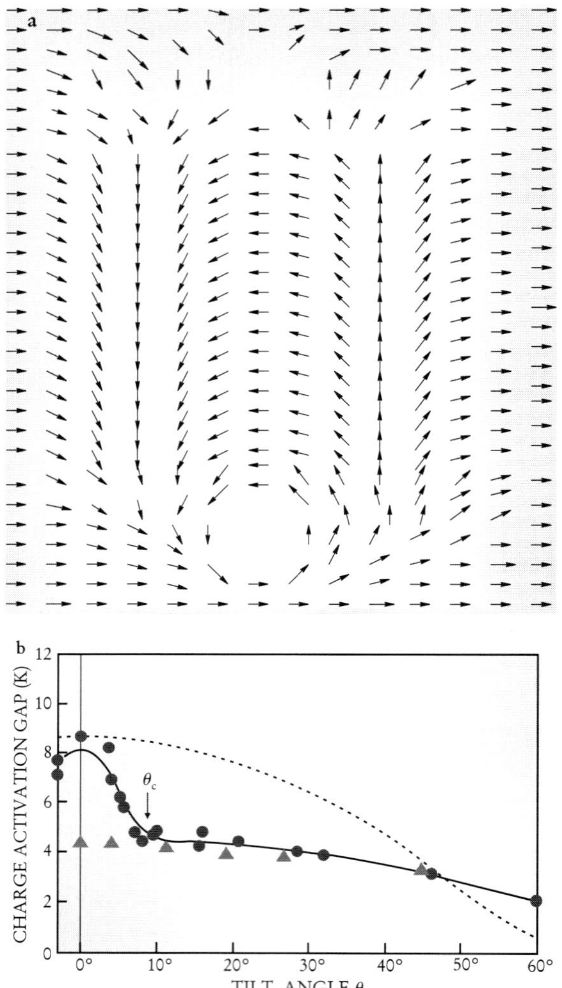



FIGURE 5. (a) MERON PAIR, formed by pseudospin orientation $\varphi$, is connected by a domain wall or "string." One half of an extra electron (or hole) resides in each defect.[12] (b) ENERGY GAP for charge activation, as a function of magnetic-field tilt angle in a weakly tunneling double-layer sample.[15] Red circles are for filling factor $\nu = 1$, blue triangles for $\nu = 2/3$. Arrow indicates critical angle $\theta_c$. Dashed line is an estimate of the renormalization (which we neglect) of the tunneling amplitude by the parallel magnetic-field component at nonzero tilt angle.

The tunneling term induces a _linear_ confining potential between vortices, thus destroying the Kosterlitz–Thouless phase transition. This comes about because pairs of right- and left-handed vortices are connected by a "string" or domain wall (see figure 5a). The energy of such a composite object of length $L$ is given by

$$E \approx WL + \frac{(e/2)^2}{L} + 2E_{\text{core}}, \tag{4}$$

where $W$ is the string tension (energy per unit length of the domain wall). The second term is the Coulomb repulsion between the half fermions bound to each vortex, and the third term is a constant governed by the ultraviolet details of the vortex cores.

The string tension for typical sample parameters is on the order of 0.1 kelvin per nanometer. That's 19 orders of magnitude weaker than the string tension that confines quarks inside nucleons and mesons! Furthermore, the string tension beween vortices, unlike that between quarks, is conveniently adjustable by simply tilting the magnetic field so that it has a component in the plane of electron gases (see figure 5b). This tilt causes tunneling particles to pick up a phase shift, making the order parameter prefer to tumble spatially. That, in turn, lowers the string tension and eventually drives it to zero, causing a phase transition to a deconfined phase in which domain walls proliferate.

In 1994, James Eisenstein and Sheena Murphy observed precisely this physics by exploiting the extreme sensitivity of the charge excitation gap to tilted magnetic fields.[12,15] As the string tension is lowered, the string stretches due to the Coulomb repulsion term in equation 4. That produces a readily observable rapid drop in the thermal activation energy needed to produce these charged objects.

The similarity between superconductivity and the physics of interlayer phase coherence has led to several suggestions of Josephson-like effects.[14] The equations of motion are indeed similar. But I believe that caution is required in their physical interpretation. For widely separated electron gas layers with no interlayer phase coherence, the tunneling current is extremely weak at small voltages. When an electron suddenly tunnels into an electron gas in a high magnetic field, it is very difficult for the other electrons to get out of the way of the newcomer, because the Lorentz force causes them to move in circular paths. Thus tunneling inevitably leaves the system in a highly excited state, with no ground-state overlap. Energy conservation then requires a finite voltage if there is to be any current.

By contrast, a system in a state with interlayer phase coherence has an indefinite number of particles in each layer, so that tunneling can leave the system in the ground state. Another way of saying this is to note that the tunneling operator that transfers an electron from one layer to the other is precisely the order parameter given by equation 2. Tunneling conductance is thus a spectroscopic probe of the order-parameter fluctuations. It should have a sharp peak at zero voltage in the broken-symmetry state, where the order parameter takes on a finite, nearly static value. This prediction, first made by Xiao-Gang Wen and Anthony Zee,[14] has recently received spectacular confirmation in some beautiful experiments carried out by Eisenstein's group at Caltech[16] (see figure 6).

## Other examples of pseudospin order

So far we have only discussed the case of pseudospin order at filling factor $\nu = 1$ under the assumption that the real spins are fully aligned. Another very inter-

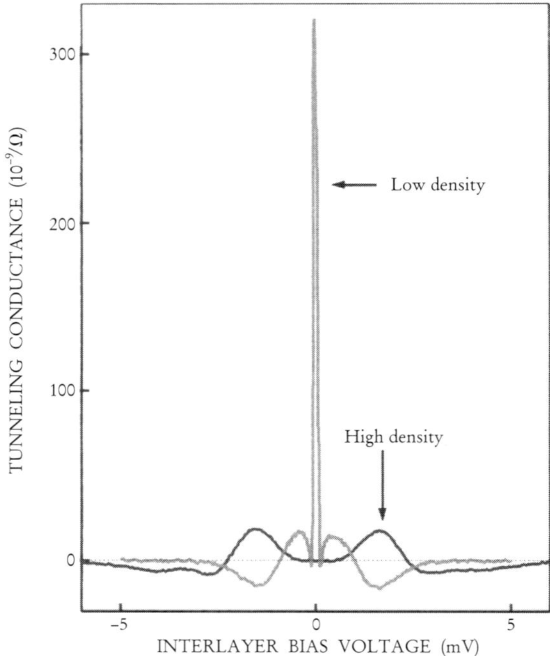


FIGURE 6. DIFFERENTIAL TUNNELING CONDUCTANCE between two adjacent two-dimensional electron gases. When the sample's electron density is high, the bilayer system is not in a phase-coherent state, and the tunneling shows a Coulomb pseudogap in the density of states. At lower electron density, the same sample goes into a phase-coherent state in which the electrons have strong interlayer correlations and the tunneling exhibits a huge anomaly at zero bias.[16]

esting situation at total filling factor $\nu = 2$, has recently been investigated theoretically by Sankar Das Sarma, Subir Sachdev and collaborators, and experimentally by Aron Pinczuk and his collaborators.[17] At $\nu = 2$, the situation is quite rich: There are four nearly degenerate levels (two spin and two isospin) producing a novel mixing of the pseudospin and real-spin order parameters that leads to a "canted anti-ferromagnetic" state for the real spins. The low-frequency fluctuations in the resulting $xy$ order parameter have been indirectly observed in light-scattering experiments.

In addition to the examples we have focused on here, there are several other examples where states of different Landau level, spin and/or electric-subband indices can be made degenerate by tuning tricks such as tilting the applied magnetic field. If the electron orbitals in question have little overlap, the pseudospin anisotropy tends to be of the easy-plane variety. But if the orbitals are fairly similar, the anisotropy tends to be of the Ising-like easy-axis type, leading to rather different physics, including the possibility of first-order phase transitions.[18]

*This article is based in part on lectures given in Les Houches.[4] The author's research is supported by a grant from the National Science Foundation. The work has been carried out in collaboration with Allan MacDonald, Herb Fertig, Patrik Henelius, Anders Sandvik, Ady Stern, Carsten Timm, Kun Yang, Kyungsun Moon, Jairo Sinova, and other friends and colleagues too numerous to list.*

# References


1. R. E. Prange, S. M. Girvin, eds., *The Quantum Hall Effect*, 2nd edition, Springer-Verlag, New York (1990).
2. T. Chakraborty, P. Pietiläinen, *The Fractional Quantum Hall Effect*, Springer-Verlag, New York (1988).
3. S. Das Sarma, A. Pinczuk, eds., *Perspectives in Quantum Hall Effects*, Wiley, New York (1997).
4. S. M. Girvin, in *1998 Les Houches Summer School Lecture Notes*, Springer-Verlag, New York and Editions de Physique, Les Ulis, France (2000). E-print: xxx.lanl.gox/abs/condmat/9907002.
5. M. J. Manfra, E. H. Aifer, B. B. Goldberg, D. A. Broido, L. Pfeiffer, K. West, *Phys. Rev. B* **54**, R17327 (1996).
6. N. Read, S. Sachdev, Phys. Rev. Lett. **75**, 3509 (1995). C. Timm, S. M. Girvin, P. Henelius, A. W. Sandvik, Phys. Rev. Lett. **58**, 1464 (1998). P. Henelius, A. W. Sandvik, C. Timm, S. M. Girvin, Phys. Rev. B **61**, 364 (2000).
7. B. I. Halperin, Helv. Phys. Acta **56**, 75 (1983).
8. S. L. Sondhi, A. Karlhede, S. A. Kivelson, E. H. Rezayi, Phys. Rev. B **47**, 16419 (1993).
9. S. E. Barrett, G. Dabbagh, L. N. Pfeiffer, K. W. West, R. Tycko, Phys. Rev. Lett. **74**, 5112 (1995). R. Tycko, S. E. Barrett, G. Dabbagh, L. N. Pfeiffer, and K. W. West, Science **268**, 1460 (1995).
10. S. Sachdev, T. Senthil, Annals of Physics **251**, 76 (1996). R. Côté, A. H. MacDonald, L. Brey, H. A. Fertig, S. M. Girvin, H. Stoof, Phys. Rev. Lett. **78**, 4825 (1997).
11. V. Bayot *et al.*, Phys. Rev. Lett. **76**, 4584 (1996); **79**, 1718 (1997).
12. J. P. Eisenstein, chap. 2, and S. M. Girvin, A. H. MacDonald, chap. 5, in ref. 3.
13. M. B. Santos, L. W. Engel, S. W. Hwang, M. Shayegan, Phys. Rev. B **44**, 5947 (1991). T. S. Lay, Y. W. Suen, H. C. Manoharan, X. Ying, M. B. Santos, M. Shayegan, Phys. Rev. B **50**, 17725 (1994).
14. X.-G. Wen, A. Zee, Phys. Rev. Lett. **69**, 1811 (1992); Phys. Rev. B **47**, 2265 (1993); Europhys. Lett. **5**, 22, (1996). Z. F. Ezawa, Phys. Rev. B **51**, 11152 (1995).
15. S. Q. Murphy, J. P. Eisenstein, G. S. Boebinger, L. N. Pfeiffer, K. W. West, Phys. Rev. Lett. **72**, 728 (1994).
16. I. B. Spielman, J. P. Eisenstein. L. N. Pfeifer, K. W. West, e-print: xxx.lanl.gov/abs/cond-mat./0002387, to be published in Phys. Rev. Lett.
17. S. Das Sarma, S. Sachdev, L. Zheng, Phys. Rev. B **58**, 4672 (1998). V. Pellegrini, A. Pinczuk, B. S. Dennis, A. S. Plaut, L. N. Pfeiffer, K. W. West, Phys. Rev. Lett. **78**, 310 (1997).
18. T. Jungwirth, S. P. Shukla, L. Smrčka, M. Shayegan, A. H. MacDonald, Phys. Rev. Lett. **81**, 2328 (1998). V. Piazza, V. Pellegrini, F. Beltram, W. Wegscheider, T. Jungwirth, A. H. MacDonald, Nature **402**, 638 (1999).

## Chapter 10

# Noncommutative Skyrmions in Quantum Hall Systems


Z.F. Ezawa[*] and G. Tsitsishvili[*,†]

*Department of Physics, Tohoku University,
Sendai, 980-8578 Japan
†Department of Theoretical Physics, A. Razmadze Mathematical Institute,
Tbilisi, 380093 Georgia



Charged excitations in quantum Hall (QH) systems are noncommutative skyrmions. QH systems present an ideal system equipped with noncommutative geometry. When an electron is confined within the lowest Landau level, its position is described solely by the guiding center, whose $X$ and $Y$ coordinates do not commute with one another. Topological excitations in such a noncommutative plane are noncommutative skyrmions flipping several spins coherently. We construct a microscopic skyrmion state by making a certain unitary transformation of a hole or electron state. A remarkable property is that a noncommutative skyrmion carries necessarily the electron number proportional to the topological charge. More remarkable is the bilayer QH system with the layer degree of freedom acting as the pseudospin, where the quasiparticle is a topological soliton to be identified with the pseudospin skyrmion. Such a skyrmion is deformed into a bimeron (a pair of merons) by the parallel magnetic field penetrated between the two layers. Each meron carries the electric charge $\pm e/2$.


## Contents


10.1 Introduction . . . . . . . . . . . . . . . . . . . . . . . . . . . . . . . . . . . . . . 234
10.2 Microscopic Theory of Quantum Hall Systems . . . . . . . . . . . . . . . . . 235
    10.2.1 Cyclotron motion . . . . . . . . . . . . . . . . . . . . . . . . . . . . . 235
    10.2.2 Noncommutative geometry . . . . . . . . . . . . . . . . . . . . . . . . 236
    10.2.3 Projected density operators . . . . . . . . . . . . . . . . . . . . . . . . 238
    10.2.4 Complex projective field . . . . . . . . . . . . . . . . . . . . . . . . . . 240
10.3 Skyrmion Excitations . . . . . . . . . . . . . . . . . . . . . . . . . . . . . . . 242
    10.3.1 Topological charge and electric charge . . . . . . . . . . . . . . . . . . 242
    10.3.2 Microscopic skyrmion states . . . . . . . . . . . . . . . . . . . . . . . 244
    10.3.3 Factorizable skyrmions . . . . . . . . . . . . . . . . . . . . . . . . . . 247
10.4 Coulomb Interactions . . . . . . . . . . . . . . . . . . . . . . . . . . . . . . . 248
    10.4.1 Projected Hamiltonians . . . . . . . . . . . . . . . . . . . . . . . . . . 248
    10.4.2 Exchange interaction . . . . . . . . . . . . . . . . . . . . . . . . . . . . 249
    10.4.3 Classical Hamiltonian . . . . . . . . . . . . . . . . . . . . . . . . . . . 250
    10.4.4 Spontaneous symmetry breaking . . . . . . . . . . . . . . . . . . . . . 251
    10.4.5 Skyrmion excitation energy . . . . . . . . . . . . . . . . . . . . . . . . 251

10.5 Bilayer QH Systems . . . . . . . . . . . . . . . . . . . . . . . . . . . . . . . 253
    10.5.1 Pseudospin QH ferromagnet . . . . . . . . . . . . . . . . . . . . . . 253
    10.5.2 Parallel magnetic field . . . . . . . . . . . . . . . . . . . . . . . . . 255
    10.5.3 Ground state . . . . . . . . . . . . . . . . . . . . . . . . . . . . . . 257
    10.5.4 Density operators . . . . . . . . . . . . . . . . . . . . . . . . . . . . 258
10.6 Skyrmion and Meron Excitations . . . . . . . . . . . . . . . . . . . . . . . 259
    10.6.1 Baby skyrmions . . . . . . . . . . . . . . . . . . . . . . . . . . . . . 259
    10.6.2 Coulomb energy . . . . . . . . . . . . . . . . . . . . . . . . . . . . . 260
    10.6.3 Meron states . . . . . . . . . . . . . . . . . . . . . . . . . . . . . . 262
    10.6.4 Large skyrmions . . . . . . . . . . . . . . . . . . . . . . . . . . . . 264
10.7 Conclusions . . . . . . . . . . . . . . . . . . . . . . . . . . . . . . . . . . . 265
References . . . . . . . . . . . . . . . . . . . . . . . . . . . . . . . . . . . . . . . 266

## 10.1. Introduction

According to Skyrme,[1] baryons are interpreted to be topological solitons in a non-linear field theory of mesons. Topological solitons are now called skyrmions. This idea has motivated an enormous amount of works in the nuclear and elementary particle physics. Recently, skyrmions have also been found to be relevant in the condensed matter physics. Indeed, charged excitations are skyrmions in quantum Hall (QH) systems.

The QH effect is one of the most remarkable phenomena discovered in the last century. There are already many textbooks on QH effects.[2-5] Nevertheless, theoretical and experimental developments are still being made in this sphere. Many novel ideas have been proposed to understand various novel experimental results, among which the concept of QH ferromagnet is prominent. Exchange Coulomb interactions play key roles in various strongly correlated electron systems. They are essential also in the QH system, where the long-range effective Hamiltonian is shown to be the nonlinear sigma model. In such a model, electron spins are spontaneously polarized even in the absence of the Zeeman effect, leading to quantum coherence and making the system into a ferromagnet. Furthermore, topological solitons are skyrmions.[6] What is remarkable is that a skyrmion carries the same charge as an electron or a hole at the filling factor $\nu = 1$.

Experimental evidence of skyrmion excitations is provided by observation of the number of flipped spins per unit charge.[7-9] Since a skyrmion is a coherent excitation of spins, its excitation changes the spin more than that of an electron or a hole excitation. A conventional way is to measure the activation energy by tilting a sample in a uniform external magnetic field, which increases more rapidly as the number of flipped spins increases.

Much more interesting phenomena associated with quantum coherence occur in the bilayer QH system. The bilayer system has the pseudospin degree of freedom, where the electron in the front (back) layer is assigned to carry the up (down) pseudospin. Provided the layer separation $d$ is reasonably small, the interlayer phase coherence[10,11] emerges due to the exchange interaction, and the system becomes a ferromagnet in pseudospins.

As we have stated, by tilting samples, the activation energy increases by the Zeeman effect in the monolayer QH system. On the contrary, an entirely opposite behavior has been observed in the bilayer QH system at the filling factor $\nu = 1$, where the activation energy decreases rapidly by tilting samples.[12–14] Note that we expect an increase since the $\nu = 1$ bilayer QH system is also a ferromagnet. This anomalous decrease occurs due to the loss of the exchange energy by a deformation of a skyrmion into a bimeron.

It is necessary to develop a microscopic theory of the QH system[15] to understand fully the mechanism how it becomes a ferromagnet and skyrmions arise as topological solitons. Electrons in a plane perform cyclotron motion under strong magnetic field $B_\perp$ and create Landau levels. Excitations across Landau levels are suppressed at sufficiently low temperature when the cyclotron energy is large enough. A self-consistent theory without these excitations is constructed by making the Landau-level projection.[16,17] An electron confined to a single Landau level is described by the guiding center $(X, Y)$ subject to the noncommutative relation, $[X, Y] = -i\ell_B^2$, with $\ell_B = \sqrt{\hbar/eB_\perp}$ the magnetic length. Thus the QH system provides us with a realistic world of noncommutative geometry[18] together with noncommutative topological solitons in a ferromagnet. A remarkable property is that the underlying symmetry is $W_\infty(N)$, which is an SU($N$) extension of $W_\infty$ familiar in the string theory, where $N = 2$ in the monolayer system and $N = 4$ in the bilayer system. This symmetry implies that the charge and the spin are entangled, as results in the fact that the topological charge and the electric charge are proportional one to another. We present a microscopic theory of skyrmions in the monolayer and bilayer QH ferromagnets, employing the framework of noncommutative geometry. We find that a skyrmion is constructed by dressing a cloud of spins (pseudospins) around an electron or a hole.

## 10.2. Microscopic Theory of Quantum Hall Systems

### 10.2.1. *Cyclotron motion*

Electrons perform cyclotron motion in an external magnetic field $B$. The Hamiltonian is given by

$$H = \frac{1}{2M}\left(P_x^2 + P_y^2\right) = \frac{1}{2M}(P_x - iP_y)(P_x + iP_y) + \frac{1}{2}\hbar\omega_c, \qquad (10.2.1)$$

where $\omega_c = eB_\perp/M$ is the cyclotron frequency, and $P_k$ is the covariant momentum,

$$P_x \equiv -i\hbar\partial_x + eA_x^{\text{ext}}, \qquad P_y \equiv -i\hbar\partial_y + eA_y^{\text{ext}}, \qquad (10.2.2)$$

with $A_k^{\text{ext}}$ the external electromagnetic potential describing the external magnetic field $B = (0, 0, -B_\perp)$,

$$B_\perp = -\varepsilon_{jk}\partial_j A_k^{\text{ext}} = \partial_y A_x^{\text{ext}} - \partial_x A_y^{\text{ext}} > 0. \qquad (10.2.3)$$

The electron charge is $-e$ with $e > 0$ in our convention.

We define the guiding-center coordinate by

$$X \equiv x + \frac{1}{eB_\perp}P_y, \qquad Y \equiv y - \frac{1}{eB_\perp}P_x, \tag{10.2.4}$$

where $\boldsymbol{R} \equiv (\frac{1}{eB_\perp}P_y, -\frac{1}{eB_\perp}P_x)$ is the relative coordinate. Since they satisfy

$$[X,Y] = -i\ell_B^2, \qquad [P_x, P_y] = i\frac{\hbar^2}{\ell_B^2},$$

$$[X, P_x] = [X, P_y] = [Y, P_x] = [Y, P_y] = 0, \tag{10.2.5}$$

the guiding center $(X,Y)$ and the covariant momentum $(P_x, P_y)$ are entirely independent variables.

To derive the energy spectrum, we construct two operators from these variables,

$$a \equiv \frac{\ell_B}{\sqrt{2}\hbar}(P_x + iP_y), \quad b \equiv \frac{1}{\sqrt{2}\ell_B}(X - iY), \tag{10.2.6}$$

obeying $[a, a^\dagger] = [b, b^\dagger] = 1$, $[a, b] = [a^\dagger, b] = 0$. The Fock vacuum is $a|0\rangle = 0$, $b|0\rangle = 0$, upon which Fock states are constructed,

$$|N, n\rangle = \sqrt{\frac{1}{N!n!}}(a^\dagger)^N (b^\dagger)^n |0\rangle. \tag{10.2.7}$$

The orthonormal completeness condition reads

$$\langle M, m|N, n\rangle = \delta_{MN}\delta_{mn}, \qquad \sum_{N,n} |N, n\rangle\langle N, n| = 1. \tag{10.2.8}$$

The Fock states present the Fock representations of the commutation relations (10.2.5).

The Hamiltonian (10.2.1) is rewritten as

$$H = (a^\dagger a + aa^\dagger)\frac{\hbar\omega_c}{2} = (a^\dagger a + \frac{1}{2})\hbar\omega_c. \tag{10.2.9}$$

The energy eigenvalue $E_N$ is that of the harmonic oscillator,

$$E_N = (N + \frac{1}{2})\hbar\omega_c, \tag{10.2.10}$$

with $|N, n\rangle$ the eigenstate. There exists a degeneracy in each Landau level, corresponding to the index $n$. We call $|N, n\rangle$ the $n$th Landau site in the $N$th Landau level. The degeneracy is proportional to the size of the system. We call $a$, $a^\dagger$ the Landau-level ladder operators, and $b$, $b^\dagger$ the Landau-site ladder operators.

## 10.2.2. *Noncommutative geometry*

We explore the physics of electrons confined to a single Landau level, where the electron position is specified solely by the guiding center $\boldsymbol{X} = (X, Y)$, whose $X$ and $Y$ components are noncommutative,

$$[X, Y] = -i\ell_B^2. \tag{10.2.11}$$

The QH system provides us with an ideal 2-dimensional world with the built-in noncommutative geometry.

We start with the plane wave $e^{-i\boldsymbol{p}\boldsymbol{x}}$, which we project to the $N$th Landau level. Since the coordinate $\boldsymbol{x}$ is decomposed into the guiding center $\boldsymbol{X}$ and the relative coordinate $\boldsymbol{R}$ as in (10.2.4), we find

$$\langle N|e^{-i\boldsymbol{p}\boldsymbol{x}}|N\rangle = \langle N|e^{-i\boldsymbol{p}\boldsymbol{R}}e^{-i\boldsymbol{p}\boldsymbol{X}}|N\rangle = F_N(\boldsymbol{p})e^{-i\boldsymbol{p}\boldsymbol{X}},$$

where

$$F_N(\boldsymbol{p}) = \langle N|e^{-i\boldsymbol{p}\boldsymbol{R}}|N\rangle \tag{10.2.12}$$

is called the Landau-level form factor. In particular it reads

$$F_0(\boldsymbol{p}) = e^{-\ell_B^2 \boldsymbol{p}^2/4} \tag{10.2.13}$$

for the lowest Landau level ($N = 0$). Apart from this factor, the projection maps the plane wave $e^{-i\boldsymbol{p}\boldsymbol{x}}$ to the operator $e^{-i\boldsymbol{p}\boldsymbol{X}}$ acting on the Fock space $\{|n\rangle; n = 0, 1, 2, \cdots\}$. Namely, it define a mapping,

$$W[e^{i\boldsymbol{p}\boldsymbol{x}}] = e^{i\boldsymbol{p}\boldsymbol{X}}, \tag{10.2.14}$$

from the plane wave $e^{i\boldsymbol{p}\boldsymbol{x}}$ to the operator $e^{i\boldsymbol{p}\boldsymbol{X}}$ in the noncommutative plane.

In general, based on the Fourier transformation,

$$f(\boldsymbol{x}) \equiv \int \frac{d^2p}{2\pi}\, e^{i\boldsymbol{p}\boldsymbol{x}} f(\boldsymbol{p}), \tag{10.2.15}$$

we find

$$W[f] \equiv \int \frac{d^2p}{2\pi}\, W[e^{i\boldsymbol{p}\boldsymbol{x}}] f(\boldsymbol{p}) = \frac{1}{(2\pi)^2}\int d^2q d^2x\, e^{i\boldsymbol{q}(\boldsymbol{X}-\boldsymbol{x})} f(\boldsymbol{x}), \tag{10.2.16}$$

where use was made of (10.2.14). Thus a classical function $f(\boldsymbol{x})$ is mapped to an operator $W[f]$ in the noncommutative plane. The operator $W[f]$ is known as the Weyl ordering of $f(\boldsymbol{x})$, while the function $f(\boldsymbol{x})$ is known as the symbol of $W[f]$. We also call $W[f]$ the Weyl operator. The inversion formula reads

$$f(\boldsymbol{x}) = \frac{\ell_B^2}{2\pi}\int d^2p\, e^{i\boldsymbol{p}\boldsymbol{x}}\mathrm{Tr}\left(e^{-i\boldsymbol{p}\boldsymbol{X}} W[f]\right), \tag{10.2.17}$$

where "Tr" is defined by $\mathrm{Tr}(O) = \sum_n \langle n|O|n\rangle$ for any operator $O$. Thus,

$$\mathrm{Tr}\left(W[f]\right) = \sum_n \langle n|W[f]|n\rangle = \frac{1}{2\pi\ell_B^2}\int d^2x\, f(\boldsymbol{x}). \tag{10.2.18}$$

There exists one to one correspondence between the symbol $f(\boldsymbol{x})$ and the Weyl operator $W[f]$

Any operator acting on the Fock space is expanded in terms of $|m\rangle\langle n|$,

$$W[f] = \sum_{mn} f_{mn}|m\rangle\langle n|, \tag{10.2.19}$$

where

$$f_{mn} = \langle m|W[f]|n\rangle = \frac{1}{2\pi}\int d^2q\,\langle m|e^{i\boldsymbol{q}\boldsymbol{X}}|n\rangle f(\boldsymbol{q}). \tag{10.2.20}$$

We can construct the symbol of the operator $|m\rangle\langle n|$ as[19]

$$\Xi_{mn}(\boldsymbol{x}) = \frac{\ell_B^2}{2\pi}\int d^2p\,e^{i\boldsymbol{p}\boldsymbol{x}}\mathrm{Tr}\left(e^{-i\boldsymbol{p}\boldsymbol{X}}|m\rangle\langle n|\right) = \frac{\ell_B^2}{2\pi}\int d^2p\,e^{i\boldsymbol{p}\boldsymbol{x}}\langle n|e^{-i\boldsymbol{p}\boldsymbol{X}}|m\rangle$$

$$= 2^{\frac{m-n}{2}+1}z^{m-n}e^{-|z|^2}\frac{(-1)^n\sqrt{n!}}{\sqrt{m!}}L_n^{m-n}(2\,|z|^2), \tag{10.2.21}$$

where $z = (x+iy)/\ell_B$. This formula is useful to calculate various physical quantities explicitly.

A product of two Weyl operators $W[f]W[g]$ is a Weyl operator, whose symbol we denote as $f(\boldsymbol{x})\star g(\boldsymbol{x})$,

$$W[f\star g] = W[f]W[g]. \tag{10.2.22}$$

It is called the star product[20–22] of $f(\boldsymbol{x})$ and $g(\boldsymbol{x})$.

The plane wave $e^{i\boldsymbol{p}\boldsymbol{x}}$ generates the translation in the ordinary space. It is easy to see that

$$e^{i\boldsymbol{p}\boldsymbol{X}}e^{i\boldsymbol{q}\boldsymbol{X}} = \exp\left[\frac{i}{2}\ell_B^2\boldsymbol{p}\wedge\boldsymbol{q}\right]e^{i(\boldsymbol{p}+\boldsymbol{q})\boldsymbol{X}}, \tag{10.2.23}$$

where $\boldsymbol{p}\wedge\boldsymbol{q} = p_xq_y - p_yq_x$. Thus the translation turns out to be non-Abelian within the lowest Landau level, and is called the magnetic translation.

Since $e^{i\boldsymbol{p}\boldsymbol{x}}\star e^{i\boldsymbol{q}\boldsymbol{x}}$ is the symbol of $e^{i\boldsymbol{p}\boldsymbol{X}}e^{i\boldsymbol{q}\boldsymbol{X}}$ by definition, (10.2.23) is equivalent to

$$e^{i\boldsymbol{p}\boldsymbol{x}}\star e^{i\boldsymbol{q}\boldsymbol{x}} = \exp\left(\frac{i}{2}\ell_B^2\boldsymbol{p}\wedge\boldsymbol{q}\right)e^{i(\boldsymbol{p}+\boldsymbol{q})\boldsymbol{x}}, \tag{10.2.24}$$

which implies

$$f(\boldsymbol{x})\star g(\boldsymbol{x}) = \frac{1}{(2\pi)^2}\lim_{\boldsymbol{y}\to\boldsymbol{x}}\int d^2p\,d^2q\,\exp\left(\frac{i}{2}\ell_B^2\boldsymbol{p}\wedge\boldsymbol{q}\right)e^{i(\boldsymbol{p}\boldsymbol{x}+\boldsymbol{q}\boldsymbol{y})}f(\boldsymbol{p})g(\boldsymbol{q})$$

$$= \lim_{\boldsymbol{y}\to\boldsymbol{x}}\exp\left(-\frac{i}{2}\ell_B^2\boldsymbol{\nabla}_{\boldsymbol{x}}\wedge\boldsymbol{\nabla}_{\boldsymbol{y}}\right)f(\boldsymbol{x})g(\boldsymbol{y}), \tag{10.2.25}$$

where $\boldsymbol{\nabla}_{\boldsymbol{x}}\wedge\boldsymbol{\nabla}_{\boldsymbol{y}} \equiv \varepsilon_{ij}\partial_i^x\partial_j^y$. This defines the star product $f\star g$ explicitly.

### 10.2.3. *Projected density operators*

We consider spinless electrons confined to the $N$th Landau level, where Fock states are given by $|N,n\rangle$ as in (10.2.7). The unique physical variable is the electron density $\rho(\boldsymbol{x})$ projected to the $N$th Landau level,

$$\rho(\boldsymbol{x}) = \psi^\dagger(\boldsymbol{x})\psi(\boldsymbol{x}), \tag{10.2.26}$$

where $\psi(\boldsymbol{x})$ is the field operator describing electrons in the $N$th Landau level, and given by

$$\psi(\boldsymbol{x}) = \sum_n \langle \boldsymbol{x}|N,n\rangle c(n), \qquad (10.2.27)$$

with $c(n)$ annihilating an electron in the Landau site $|N,n\rangle$,

$$\{c(n), c^\dagger(m)\} = \delta_{mn}. \qquad (10.2.28)$$

In the momentum space the projected density is given by

$$\rho(\boldsymbol{p}) \equiv \int \frac{d^2x}{2\pi} e^{-i\boldsymbol{px}} \rho(\boldsymbol{x}) = F_N(\boldsymbol{p})\hat{\rho}(\boldsymbol{p}), \qquad (10.2.29)$$

where $F_N(\boldsymbol{p})$ is the Landau-level form factor (10.2.12), and we have defined the bare density[15]

$$\hat{\rho}(\boldsymbol{p}) = \frac{1}{2\pi} \sum_{mn} \langle m|e^{-i\boldsymbol{pX}}|n\rangle c^\dagger(m)c(n). \qquad (10.2.30)$$

The inversion formula of (10.2.30) is

$$c^\dagger(m)c(n) = \ell_B^2 \int d^2q\, \langle n|e^{i\boldsymbol{qX}}|m\rangle \hat{\rho}(\boldsymbol{q}). \qquad (10.2.31)$$

It is convenient to construct a formalism based on the bare density $\hat{\rho}(\boldsymbol{p})$ rather than the projected density $\rho(\boldsymbol{q})$, though $\rho(\boldsymbol{x})$ is the physical density. This is because $\hat{\rho}(\boldsymbol{p})$ is more closely related to the Weyl operator. The difference between $\rho(\boldsymbol{p})$ and $\hat{\rho}(\boldsymbol{p})$ is negligible for sufficiently smooth field configurations since $F_N(\boldsymbol{p}) \to 1$ as $\boldsymbol{p} \to 0$.

It follows from (10.2.23) that the bare density satisfies the algebraic relation,[17]

$$[\hat{\rho}(\boldsymbol{p}), \hat{\rho}(\boldsymbol{q})] = \frac{i}{\pi}\hat{\rho}(\boldsymbol{p}+\boldsymbol{q})\sin\left(\ell_B^2 \frac{\boldsymbol{p}\wedge\boldsymbol{q}}{2}\right), \qquad (10.2.32)$$

which is isomorphic to the $W_\infty$ algebra.[23,24]

We may generalize the scheme to the $SU(N_{\mathrm{I}})$ theory, where the electron field $\Psi(\boldsymbol{x})$ has $N_{\mathrm{I}}$ isospin components and is given by

$$\psi_\mu(\boldsymbol{x}) = \sum_n \langle \boldsymbol{x}|N,n\rangle c_\mu(n) \qquad (10.2.33)$$

in the $N$th Landau level, with $\{c_\mu(n), c_\nu^\dagger(m)\} = \delta_{mn}\delta_{\mu\nu}$. The physical variables are the electron density $\rho$ and the isospin field $I_A$ projected to the $N$th Landau level,

$$\rho(\boldsymbol{x}) = \Psi^\dagger(\boldsymbol{x})\Psi(\boldsymbol{x}), \qquad I_A(\boldsymbol{x}) = \frac{1}{2}\Psi^\dagger(\boldsymbol{x})\lambda_A\Psi(\boldsymbol{x}), \qquad (10.2.34)$$

which are summarized into

$$D_{\mu\nu}(\boldsymbol{x}) \equiv \psi_\mu^\dagger(\boldsymbol{x})\psi_\nu(\boldsymbol{x}) = \frac{1}{N}\delta_{\mu\nu}\rho(\boldsymbol{x}) + (\lambda_A)_{\mu\nu}I_A(\boldsymbol{x}), \qquad (10.2.35)$$

where $\lambda_A$ are the generating matrices of $SU(N_{\mathrm{I}})$. They are the Pauli matrices $\tau_a$ for $SU(2)$.

The bare densities $\hat{\rho}(\boldsymbol{p})$ and $\hat{I}_A(\boldsymbol{p})$ are defined similarly as in (10.2.30) and summarized into the density matrix $\hat{D}_{\mu\nu}(\boldsymbol{p})$ as

$$\hat{D}_{\mu\nu}(\boldsymbol{p}) = \frac{1}{N}\delta_{\mu\nu}\hat{\rho}(\boldsymbol{p}) + (\lambda_A)_{\mu\nu}\hat{I}_A(\boldsymbol{p}) = \frac{1}{2\pi}\sum_{mn}\langle n|e^{-i\boldsymbol{p}\boldsymbol{X}}|m\rangle D_{\mu\nu}(m,n), \quad (10.2.36)$$

together with

$$D_{\mu\nu}(m,n) \equiv c_\nu^\dagger(n)c_\mu(m). \tag{10.2.37}$$

It is related to the physical density as

$$D_{\mu\nu}(\boldsymbol{p}) = F_N(\boldsymbol{p})\hat{D}_{\mu\nu}(\boldsymbol{p}), \tag{10.2.38}$$

where $F_N(\boldsymbol{p})$ is the Landau-level form factor (10.2.12).

It is straightforward to verify that

$$[D_{\mu\nu}(m,n), D_{\sigma\tau}(s,t)] = \delta_{\mu\tau}\delta_{mt}D_{\sigma\nu}(s,n) - \delta_{\sigma\nu}\delta_{sn}D_{\mu\tau}(m,t) \tag{10.2.39}$$

based on the anticommutation relation (10.2.28) of $c_\sigma(m)$. This is rewritten as[15]

$$2\pi[\hat{D}_{\mu\nu}(\boldsymbol{p}), \hat{D}_{\sigma\tau}(\boldsymbol{q})] = \delta_{\mu\tau}e^{+\frac{i}{2}\ell_B^2 \boldsymbol{p}\wedge\boldsymbol{q}}\hat{D}_{\sigma\nu}(\boldsymbol{p}+\boldsymbol{q}) - \delta_{\sigma\nu}e^{-\frac{i}{2}\ell_B^2 \boldsymbol{p}\wedge\boldsymbol{q}}\hat{D}_{\mu\tau}(\boldsymbol{p}+\boldsymbol{q}), \tag{10.2.40}$$

or[25]

$$[\hat{\rho}(\boldsymbol{p}), \hat{\rho}(\boldsymbol{q})] = \frac{i}{\pi}\hat{\rho}(\boldsymbol{p}+\boldsymbol{q})\sin\left(\ell_B^2\frac{\boldsymbol{p}\wedge\boldsymbol{q}}{2}\right), \tag{10.2.41a}$$

$$[\hat{I}_A(\boldsymbol{p}), \hat{\rho}(\boldsymbol{q})] = \frac{i}{\pi}\hat{I}_A(\boldsymbol{p}+\boldsymbol{q})\sin\left(\ell_B^2\frac{\boldsymbol{p}\wedge\boldsymbol{q}}{2}\right), \tag{10.2.41b}$$

$$[\hat{I}_A(\boldsymbol{p}), \hat{I}_B(\boldsymbol{q})] = \frac{i}{2\pi}f_{ABC}\hat{I}_C(\boldsymbol{p}+\boldsymbol{q})\cos\left(\ell_B^2\frac{\boldsymbol{p}\wedge\boldsymbol{q}}{2}\right)$$
$$+ \frac{i}{2\pi}d_{ABC}\hat{I}_C(\boldsymbol{p}+\boldsymbol{q})\sin\left(\ell_B^2\frac{\boldsymbol{p}\wedge\boldsymbol{q}}{2}\right)$$
$$+ \frac{i}{2\pi N}\delta_{AB}\hat{\rho}(\boldsymbol{p}+\boldsymbol{q})\sin\left(\ell_B^2\frac{\boldsymbol{p}\wedge\boldsymbol{q}}{2}\right), \tag{10.2.41c}$$

where $f_{ABC}$ is the structure constants characterizing the SU($N_{\mathrm{I}}$) algebra. It is referred to as the $W_\infty(N_{\mathrm{I}})$ algebra,[26] since it is an SU($N_{\mathrm{I}}$) extension of the $W_\infty$ algebra. The isospin field and the electron density become noncommutative within each Landau level. Consequently, the isospin rotation modulates the electron number density.

### 10.2.4. *Complex projective field*

What is observed experimentally is the classical field

$$D_{\mu\nu}^{\mathrm{cl}}(\boldsymbol{q}) = F_N(\boldsymbol{q})\hat{D}_{\mu\nu}^{\mathrm{cl}}(\boldsymbol{q}), \tag{10.2.42}$$

where $\hat{D}_{\mu\nu}^{\mathrm{cl}}(\boldsymbol{q})$ is the expectation value of $\hat{D}_{\mu\nu}(\boldsymbol{q})$ by a Fock state,

$$\hat{D}_{\mu\nu}^{\mathrm{cl}}(\boldsymbol{q}) = \langle\mathfrak{S}|\hat{D}_{\mu\nu}(\boldsymbol{q})|\mathfrak{S}\rangle = \frac{1}{2\pi}\sum_{mn}\langle n|e^{-i\boldsymbol{q}\boldsymbol{X}}|m\rangle D_{\mu\nu}^{\mathrm{cl}}(m,n), \tag{10.2.43}$$

with $D_{\mu\nu}^{\text{cl}}(m,n) = \langle\mathfrak{S}|D_{\mu\nu}(m,n)|\mathfrak{S}\rangle$. Here, as Fock states, we consider a wide class of states of the following form,[15,27]

$$|\mathfrak{S}\rangle = e^{iW}|\mathfrak{S}_0\rangle \quad \text{with} \quad |\mathfrak{S}_0\rangle = \prod_{\mu n} \left[c_\mu^\dagger(n)\right]^{\nu_\mu(n)}|0\rangle, \tag{10.2.44}$$

where $W$ is an arbitrary element of the $W_\infty(N_{\text{I}})$ algebra: $\nu_\mu(n)$ may take the value either 0 or 1 depending whether the isospin state $\mu$ at a site $n$ is occupied or not. The class of states (10.2.44) is quite general though it may not embrace all possible ones. Nevertheless all physically relevant states at integer filling factors seem to fall in this category. Indeed, as far as we know, perturbative excitations are spin waves and nonperturbative excitations are skyrmions in QH systems. They belong surely to this category. We call $\hat{D}_{\mu\nu}^{\text{cl}}$ the classical bare density.

Making the Fourier transformation of (10.2.43) we obtain the classical bare density

$$\hat{D}_{\mu\nu}^{\text{cl}}(\boldsymbol{x}) = \frac{1}{2\pi\ell_B^2} \sum_{mn} D_{\mu\nu}^{\text{cl}}(m,n)\Xi_{mn}(\boldsymbol{x}), \tag{10.2.45}$$

where $\Xi_{mn}(\boldsymbol{x})$, being given by (10.2.21), is the symbol of $|m\rangle\langle n|$. Hence the Weyl ordering of $\hat{D}_{\mu\nu}^{\text{cl}}$ is

$$W[\hat{D}_{\mu\nu}^{\text{cl}}] = \frac{1}{2\pi\ell_B^2} \sum_{mn} D_{\mu\nu}^{\text{cl}}(m,n)|m\rangle\langle n|. \tag{10.2.46}$$

We now use the definition (10.2.37) to derive a relation,[15]

$$\sum_k \sum_\kappa D_{\mu\kappa}^{\text{cl}}(m,k)D_{\kappa\nu}^{\text{cl}}(k,n) = D_{\mu\nu}^{\text{cl}}(m,n), \tag{10.2.47}$$

for the class of states (10.2.44), or $W[\hat{D}_{\mu\kappa}^{\text{cl}}]W[\hat{D}_{\kappa\nu}^{\text{cl}}] = W[\hat{D}_{\mu\nu}^{\text{cl}}]$. It yields the constraint condition

$$\sum_{\kappa=1}^N \hat{D}_{\mu\kappa}^{\text{cl}}(\boldsymbol{x}) \star \hat{D}_{\kappa\nu}^{\text{cl}}(\boldsymbol{x}) = \frac{1}{2\pi\ell_B^2} \hat{D}_{\mu\nu}^{\text{cl}}(\boldsymbol{x}) \tag{10.2.48}$$

on the classical bare density.

To resolve the constraint (10.2.48), we introduce an $N_{\text{I}}$-component complex field $n_\mu(\boldsymbol{x})$ and its complex conjugate $n_\mu^*(\boldsymbol{x})$ subject to the noncommutative normalization condition,

$$\sum_{\mu=1}^{N_{\text{I}}} n_\mu^*(\boldsymbol{x}) \star n_\mu(\boldsymbol{x}) = 1. \tag{10.2.49}$$

Indeed, when we set

$$\hat{D}_{\mu\nu}^{\text{cl}}(\boldsymbol{x}) = \frac{1}{2\pi\ell_B^2} n_\mu(\boldsymbol{x}) \star n_\nu^*(\boldsymbol{x}), \tag{10.2.50}$$

the constraint (10.2.48) is trivially satisfied. In terms of the Weyl operator $\mathfrak{n}_\mu = W[n_\mu]$, we may rewrite (10.2.50) as

$$\mathfrak{n}_\mu \mathfrak{n}_\nu^\dagger = \sum_{mn} |m\rangle\langle n| \cdot \langle \mathfrak{S}|c_\nu^\dagger(n)c_\mu(m)|\mathfrak{S}\rangle. \tag{10.2.51}$$

The field $n_\mu(\boldsymbol{x})$ is called the $\mathrm{CP}^{N_\mathrm{I}-1}$ field. For a given state $|\mathfrak{S}\rangle$, it is constructed by solving (10.2.51) together with the constraint $\sum \mathfrak{n}_\mu^\dagger \mathfrak{n}_\mu = 1$. Note that the $\mathrm{CP}^{N_\mathrm{I}-1}$ field is introduced as a classical field from the beginning.

The formula (10.2.50) is decomposed into

$$\hat{\rho}^{\mathrm{cl}}(\boldsymbol{x}) = \frac{1}{2\pi\ell_B^2} \sum_\mu n_\mu(\boldsymbol{x}) \star n_\mu^*(\boldsymbol{x}), \tag{10.2.52a}$$

$$\hat{I}_A^{\mathrm{cl}}(\boldsymbol{x}) = \frac{1}{2\pi\ell_B^2} \sum_{\mu\nu} \left(\frac{\lambda_A}{2}\right)_{\nu\mu} n_\mu(\boldsymbol{x}) \star n_\nu^*(\boldsymbol{x}). \tag{10.2.52b}$$

In the commutative limit and in the SU(2) theory, we have

$$\hat{\rho}^{\mathrm{cl}}(\boldsymbol{x}) = \frac{1}{2\pi\ell_B^2} n^\dagger(\boldsymbol{x})n(\boldsymbol{x}), \tag{10.2.53a}$$

$$\hat{S}_a^{\mathrm{cl}}(\boldsymbol{x}) = \frac{1}{2\pi\ell_B^2} n^\dagger(\boldsymbol{x})\frac{\tau_a}{2}n(\boldsymbol{x}), \tag{10.2.53b}$$

which become relevant to describe sufficiently smooth field configurations.

## 10.3. Skyrmion Excitations

### 10.3.1. *Topological charge and electric charge*

We employ the $\mathrm{CP}^{N_\mathrm{I}-1}$ field to discuss topological issues. We define the topological charge density by the formula[15]

$$J_0(\boldsymbol{x}) = \frac{1}{2\pi\ell_B^2} \sum_{\mu=1}^{N_\mathrm{I}} [n_\mu^*(\boldsymbol{x}) \star n_\mu(\boldsymbol{x}) - n_\mu(\boldsymbol{x}) \star n_\mu^*(\boldsymbol{x})], \tag{10.3.54}$$

since it is reduced to

$$J_0(\boldsymbol{x}) = \frac{1}{2\pi i}\epsilon_{ij} \sum_{\mu=1}^{N_\mathrm{I}} \partial_i n_\mu^*(\boldsymbol{x})\partial_j n_\mu \tag{10.3.55}$$

in the commutative limit ($\ell_B \to 0$), which is the standard formula for the topological charge density in the commutative $\mathrm{CP}^{N_\mathrm{I}-1}$ theory.

Here we recall that the electron density excitation is given by (10.2.52a). Hence we conclude that

$$\Delta\hat{\rho}^{\mathrm{cl}}(\boldsymbol{x}) \equiv \hat{\rho}^{\mathrm{cl}}(\boldsymbol{x}) - \rho_0 = -J_0(\boldsymbol{x}). \tag{10.3.56}$$

Namely, the density moduration around a topological soliton is equal to the topological charge density, as implies that a soliton carries necessarily the electron number,

$$\Delta N_{\mathrm{e}}^{\mathrm{cl}} = -Q_{\mathrm{sky}}, \tag{10.3.57}$$

with $\Delta N_{\mathrm{e}}^{\mathrm{cl}} = \int d^2x \, \Delta\rho^{\mathrm{cl}}(\boldsymbol{x})$ and

$$Q_{\mathrm{sky}} = \int d^2x \, J_0(\boldsymbol{x}) = \mathrm{Tr}([\mathfrak{n}_\mu^\dagger, \mathfrak{n}_\mu]), \tag{10.3.58}$$

where $\mathfrak{n}_\mu = W \, n_\mu]$ is the Weyl operator.

A topological soliton with $Q_{\mathrm{sky}} \neq 0$ is called a noncommutative $\mathrm{CP}^{N_1-1}$ skyrmion. According to the formula (10.3.57) such a soliton carries the electric charge $-eN_{\mathrm{e}}^{\mathrm{cl}} = eQ_{\mathrm{sky}}$. (Note that we take a convention that a skyrmion has $Q_{\mathrm{sky}} > 0$ while an antiskyrmion has $Q_{\mathrm{sky}} < 0$.)

The formula (10.3.57) implies also that an excitation possessing the electric charge necessarily carries the topological charge. It follows that an electron and a hole are topological solitons in the noncommutative plane, though it sounds odd.

We investigate this problem in order to understand the difference of the noncommutative theory from the commutative one. We study the electron state $(+)$ and the hole $(-)$ state explicitly in the spin $\mathrm{SU}(2)$ theory,

$$|+\rangle = c_\downarrow^\dagger(0)|g\rangle, \qquad |-\rangle = c_\uparrow(0)|g\rangle, \tag{10.3.59}$$

where $|g\rangle = \prod_{n=0}^{\infty} c_\uparrow^\dagger(n)|0\rangle$ is the ground state with all up-spin states filled up to minimize the Zeeman energy. It is easy to see

$$\langle g|c_\mu^\dagger(m)c_\nu(n)|g\rangle = \delta_{\mu\uparrow}\delta_{\nu\uparrow}\delta_{mn}, \tag{10.3.60a}$$

$$\langle +|c_\mu^\dagger(m)c_\nu(n)|+\rangle = \delta_{\mu\uparrow}\delta_{\nu\uparrow}\delta_{mn} + \delta_{\mu\downarrow}\delta_{\nu\downarrow}\delta_{m0}\delta_{n0}, \tag{10.3.60b}$$

$$\langle -|c_\mu^\dagger(m)c_\nu(n)|-\rangle = \delta_{\mu\uparrow}\delta_{\nu\uparrow}\delta_{mn} - \delta_{\mu\uparrow}\delta_{\nu\uparrow}\delta_{m0}\delta_{n0}. \tag{10.3.60c}$$

The bare densities are

$$\hat{\rho}_\pm^{\mathrm{cl}}(\boldsymbol{x}) = \rho_0 \left( 1 \pm 2e^{-r^2/\ell_B^2} \right), \tag{10.3.61a}$$

$$\hat{S}_x^{\mathrm{cl}}(\boldsymbol{x}) = \hat{S}_y^{\mathrm{cl}}(\boldsymbol{x}) = 0, \quad \hat{S}_z^{\mathrm{cl}}(\boldsymbol{x}) = \frac{1}{2}\hat{\rho}_\pm^{\mathrm{cl}}(\boldsymbol{x}). \tag{10.3.61b}$$

Thus the spin texture is trivial.

In the commutative theory the topological number is given by (10.3.55), or equivalently by the Pontryagin number,

$$Q_{\mathrm{P}} = \frac{1}{\pi} \int d^2x \, \varepsilon_{abc}\varepsilon_{ij} S_a \partial_i S_b \partial_j S_c, \tag{10.3.62}$$

where $S_a$ is the normalized spin density, $S_a \equiv \frac{1}{2}n_\mu^*(\tau_a)_{\mu\nu}n_\nu$. We would obviously conclude $Q_{\mathrm{P}} = 0$ for the trivial spin texture such as (10.3.61b). On the other hand, the $\mathrm{CP}^1$ field giving the trivial spin texture (10.3.61b) is

$$n_\uparrow(\boldsymbol{x}) = e^{i\vartheta}, \qquad n_\downarrow(\boldsymbol{x}) = 0. \tag{10.3.63}$$

Though it carries a winding number, since it is ill-defined at the origin, we cannot make a naive calculation of the topological number based on the formula (10.3.55). A careful examination shows that $Q_{\text{sky}} = 0$, as is consistent with $Q_{\text{P}} = 0$.

However, this argument is not applicable to the noncommutative theory. To construct the noncommutative $\text{CP}^1$ field, we write down (10.2.51) explicitly with the use of (10.3.60), which we solve. The result reads

$$\mathfrak{n}_\uparrow^+ = \sum_{n=0}^\infty |n\rangle\langle n+1|, \qquad \mathfrak{n}_\downarrow^+ = \sum_{n=0}^\infty |0\rangle\langle 0|, \tag{10.3.64a}$$

$$\mathfrak{n}_\uparrow^- = \sum_{n=0}^\infty |n+1\rangle\langle n|, \qquad \mathfrak{n}_\downarrow^- = 0, \tag{10.3.64b}$$

for electron $(+)$ and hole $(-)$. According to the formula (10.3.58), the topological charge is given by

$$Q_{\text{sky}}^\pm = \text{Tr}([\mathfrak{n}_\mu^{\pm\dagger}, \mathfrak{n}_\mu^\pm]) = \mp 1. \tag{10.3.65}$$

Let us explain this by calculating the topological number explicitly in the real space.

The symbol of the Weyl operator (10.3.64) is the noncommutative $\text{CP}^1$ field. The $\text{CP}^1$ field which gives the trivial spin texture (10.3.61b) is highly nontrivial. It reads

$$n_\uparrow^+(\boldsymbol{x}) = f(r)e^{-i\vartheta}, \qquad n_\downarrow^+(\boldsymbol{x}) = 2e^{-r^2/\ell_B^2}, \tag{10.3.66a}$$

for electron, and

$$n_\uparrow^-(\boldsymbol{x}) = f(r)e^{+i\vartheta}, \qquad n_\downarrow^-(\boldsymbol{x}) = 0, \tag{10.3.66b}$$

for hole, with

$$f(r) = 2\sqrt{2}\frac{r}{\ell_B}e^{-r^2/\ell_B^2}\sum_{n=0}^\infty \frac{(-1)^n}{\sqrt{n+1}}L_n^1(2r^2/\ell_B^2). \tag{10.3.67}$$

It is well-defined everywhere: In particular, $f(0) = 0$ at the origin, and $f(\theta) \to 1$ asymptotically. Hence, the topological number is clearly $Q_{\text{sky}}^\pm = \mp 1$, which agrees with (10.3.65). Consequently, there exist no conceptual differences between a hole (electron) and a skyrmion (antiskyrmion). It is reasonable to regard a hole and an electron as a baby skyrmion and a baby antiskyrmion, respectively. Indeed, as we shall see in the succeeding subsection, a skyrmion (antiskyrmion) state is constructed as a continuous unitary transformation of a hole (electron) state: See (10.3.68).

### 10.3.2. *Microscopic skyrmion states*

The skyrmion is a classical solution to the nonlinear sigma model.[28] Indeed, the concept of skyrmion was introduced[6] into QH ferromagnets first in this context.

Subsequently a microscopic skyrmion state was considered to carry out a Hartree-Fock approximation.[29–32] This idea can be elaborated to construct a microscopic theory of skyrmions[15,27] in the framework of noncommutative geometry.

We study the spin SU(2) system. We introduce the state $|\mathfrak{S}_{\text{sky}}^{\mp}\rangle$ as a $W_\infty(2)$-rotated state of the hole state $|h\rangle$ or the electron state $|e\rangle$,

$$|\mathfrak{S}_{\text{sky}}^{-}\rangle = e^{iW^-}|h\rangle = e^{iW^-}c_\uparrow(0)|g\rangle, \tag{10.3.68a}$$

$$|\mathfrak{S}_{\text{sky}}^{+}\rangle = e^{iW^+}|e\rangle = e^{iW^+}c_\downarrow^\dagger(0)|g\rangle, \tag{10.3.68b}$$

where $W^\mp$ is an arbitrary element of the $W_\infty(2)$ algebra.[27] An important property of the $W_\infty(2)$-rotated state $|\mathfrak{S}\rangle$ is that the electron number is the same as that of the state $|\mathfrak{S}_0\rangle$,

$$\langle\mathfrak{S}|N_e|\mathfrak{S}\rangle = \langle\mathfrak{S}_0|e^{-iW}N_e e^{+iW}|\mathfrak{S}_0\rangle = \langle\mathfrak{S}_0|N_e|\mathfrak{S}_0\rangle, \tag{10.3.69}$$

since the total electron number

$$N_e = \int d^2x\, \rho(\boldsymbol{x}) = \int d^2x\, \hat{\rho}(\boldsymbol{x}) = \sum_n \sum_\mu c_\mu^\dagger(n)c_\mu(n) \tag{10.3.70}$$

is a Casimir operator. According to the properties (10.3.69) and (10.3.57), the states $|\mathfrak{S}_{\text{sky}}^{-}\rangle$ and $|\mathfrak{S}_{\text{sky}}^{+}\rangle$ have the same electron numbers as $|h\rangle$ and $|e\rangle$, respectively, and hence the topological number $Q_{\text{sky}} = \pm 1$. Hence it a skyrmion or antiskyrmion state. We shall later show that it yields the familiar expression of the $CP^1$ skyrmion in the commutative limit provided a certain choice is made of the $W_\infty(2)$ rotation $W^\mp$: See (10.3.86).

The simplest $W_\infty(2)$ rotation mixes only the nearest neighboring sites, and is given by $W^\mp = \sum_{n=0}^{\infty} W_n^\mp$ with

$$iW_n^\mp = \alpha_n^\mp\left[c_\downarrow^\dagger(n)c_\uparrow(n+1) - c_\uparrow^\dagger(n+1)c_\downarrow(n)\right], \tag{10.3.71}$$

where $\alpha_n^\mp$ is a real parameter. After a straightforward calculation we find

$$|\mathfrak{S}_{\text{sky}}^{\mp}\rangle = \prod_{n=0}^{\infty} \xi_{\mp}^\dagger(n)|0\rangle, \tag{10.3.72}$$

where

$$\xi_{\mp}^\dagger(n) = u_{\mp}(n)c_\downarrow^\dagger(n)_n + v_{\mp}(n)c_\uparrow^\dagger(n+1) \tag{10.3.73}$$

with $u_{\mp}(n) = \sin\alpha_n^\mp$, and $v_{\mp}(n) = \cos\alpha_n^\mp$.

We calculate the classical density explicitly. The physical density (10.2.38) reads

$$\frac{\rho^{\mathrm{cl}\pm}(\boldsymbol{x})}{\rho_0} = e^{-|z|^2/2} \sum_{n=0} \frac{u_\pm^2(n) + v_\pm^2(n-1)}{n!} \left(\frac{|z|^2}{2}\right)^n, \tag{10.3.74a}$$

$$\frac{S_z^{\mathrm{cl}\pm}(\boldsymbol{x})}{\rho_0} = \pm\frac{1}{2} e^{-|z|^2/2} \sum_{n=0} \frac{u_\pm^2(n) - v_\pm^2(n-1)}{n!} \left(\frac{|z|^2}{2}\right)^n, \tag{10.3.74b}$$

$$\frac{S_x^{\mathrm{cl}\pm}(\boldsymbol{x})}{\rho_0} = \frac{x}{\sqrt{2}\ell_B} e^{-|z|^2/2} \sum_{n=0} \frac{u_\pm(n)v_\pm(n)}{n!\sqrt{n+1}} \left(\frac{|z|^2}{2}\right)^n, \tag{10.3.74c}$$

$$\frac{S_y^{\mathrm{cl}\pm}(\boldsymbol{x})}{\rho_0} = \pm\frac{y}{\sqrt{2}\ell_B} e^{-|z|^2/2} \sum_{n=0} \frac{u_\pm(n)v_\pm(n)}{n!\sqrt{n+1}} \left(\frac{|z|^2}{2}\right)^n. \tag{10.3.74d}$$

The classical densities of the skyrmion and antiskyrmion states are characterized by infinitely many variables $u_\pm(n)$ and $v_\pm(n)$. By minimizing the energies of these states, we can determine them and hence the microscopic state (10.3.72). We do this later in subsection 10.4.5.

We proceed to reformulate the skyrmion state $|\mathfrak{S}_{\mathrm{sky}}^-\rangle$ in terms of the noncommutative $\mathrm{CP}^1$ field. For notational simplicity, we set $u(n) \equiv u_-(n)$, $v(n) \equiv v_-(n)$ and

$$\begin{pmatrix} u_\uparrow \\ u_\downarrow \end{pmatrix} = \begin{pmatrix} 0 \\ u \end{pmatrix}, \qquad \begin{pmatrix} v_\uparrow \\ v_\downarrow \end{pmatrix} = \begin{pmatrix} v \\ 0 \end{pmatrix}. \tag{10.3.75}$$

The classical density matrix $D_{\mu\nu}^{\mathrm{cl}}(m,n) \equiv \langle\mathfrak{S}_{\mathrm{sky}}^-|c_\nu^\dagger(n)c_\mu(m)|\mathfrak{S}_{\mathrm{sky}}^-\rangle$ is calculated as

$$D_{\mu\nu}^{\mathrm{cl}}(n,n) = u_\mu(n)u_\nu^*(n) + v_\mu(n-1)v_\nu^*(n-1), \tag{10.3.76a}$$

$$D_{\mu\nu}^{\mathrm{cl}}(n,n+1) = u_\mu(n)v_\nu^*(n), \tag{10.3.76b}$$

$$D_{\mu\nu}^{\mathrm{cl}}(n+1,n) = v_\mu(n)u_\nu^*(n). \tag{10.3.76c}$$

All other matrix elements vanish. It follows from (10.2.51) that

$$\begin{aligned}
\mathfrak{n}_\mu\mathfrak{n}_\nu^\dagger &= u_\mu(n)u_\nu^*(n)|n\rangle\langle n| + v_\mu(n-1)v_\nu^*(n-1)|n\rangle\langle n| \\
&\quad + u_\mu(n)v_\nu^*(n)|n\rangle\langle n+1| + v_\mu(n)u_\nu^*(n)|n+1\rangle\langle n|.
\end{aligned} \tag{10.3.77}$$

This is uniquely solved as

$$\mathfrak{n}_\mu = u_\mu(n)|n\rangle\langle n| + v_\mu(n)|n+1\rangle\langle n|, \tag{10.3.78}$$

or

$$\mathfrak{n}_\uparrow = \sum_n v(n)|n+1\rangle\langle n|, \qquad \mathfrak{n}_\downarrow = \sum_n u(n)|n\rangle\langle n|. \tag{10.3.79}$$

The symbol reads

$$n_\uparrow(\boldsymbol{x}) = \sum_n v(n)\Xi_{n+1,n}(\boldsymbol{x}), \qquad n_\downarrow(\boldsymbol{x}) = \sum_n u(n)\Xi_{n,n}(\boldsymbol{x}), \tag{10.3.80}$$

where $\Xi_{m,n}(\boldsymbol{x})$ is given by (10.2.21). The noncommutative $\mathrm{CP}^1$ skyrmion is described by this set of $n_\uparrow(\boldsymbol{x})$ and $n_\downarrow(\boldsymbol{x})$.

The noncommutative $CP^1$ skyrmion has quite a complicated expression. We now argue when it is reduced to the familiar expression of the $CP^1$ skyrmion in the commutative limit. From (10.3.79) we find

$$b^\dagger \mathfrak{n}_\downarrow = \sum_n u(n) b^\dagger |n\rangle\langle n| = \sum_n u(n)\sqrt{n+1}|n+1\rangle\langle n|, \tag{10.3.81}$$

which has the same operator structure as $\mathfrak{n}_\uparrow$. Hence, if we require

$$u(n)\sqrt{n+1} = \frac{\lambda}{\sqrt{2}} v(n) \tag{10.3.82}$$

with $\lambda$ being a real constant, we obtain

$$b^\dagger \mathfrak{n}_\downarrow = \frac{\lambda}{\sqrt{2}} \mathfrak{n}_\uparrow, \tag{10.3.83}$$

or

$$z \star n_\downarrow(\boldsymbol{x}) = \lambda n_\uparrow(\boldsymbol{x}). \tag{10.3.84}$$

Thus,

$$\begin{pmatrix} n_\uparrow(\boldsymbol{x}) \\ n_\downarrow(\boldsymbol{x}) \end{pmatrix} = \frac{1}{\lambda}\begin{pmatrix} z \\ \lambda \end{pmatrix} \star n_\downarrow(\boldsymbol{x}). \tag{10.3.85}$$

This is reduced to the familiar expression of the $CP^1$ skyrmion,

$$\begin{pmatrix} n_\uparrow(\boldsymbol{x}) \\ n_\downarrow(\boldsymbol{x}) \end{pmatrix} = \frac{1}{\sqrt{|z|^2 + \lambda^2}}\begin{pmatrix} z \\ \lambda \end{pmatrix}, \tag{10.3.86}$$

in the commutative limit, since the constraint condition (10.2.49) yields $n_\downarrow(\boldsymbol{x}) = \lambda/\sqrt{|z|^2 + \lambda^2}$.

### 10.3.3. *Factorizable skyrmions*

We make a further study of the skyrmion satisfying the condition (10.3.82), which is solved as

$$u^2(n) = \frac{\omega^2}{n+1+\omega^2}, \quad v^2(n) = \frac{n+1}{n+1+\omega^2}, \tag{10.3.87}$$

with $\lambda = \sqrt{2}\omega$. Substituting them into (10.3.74), after somewhat tedious calculations, we obtain

$$S_a(\boldsymbol{x}) = \rho(\boldsymbol{x}) \mathcal{S}_a(\boldsymbol{x}) \tag{10.3.88}$$

with

$$\mathcal{S}_x(\boldsymbol{x}) = \frac{\lambda x}{r^2 + \lambda^2}, \quad \mathcal{S}_y(\boldsymbol{x}) = \frac{-\lambda y}{r^2 + \lambda^2}, \quad \mathcal{S}_z(\boldsymbol{x}) = \frac{1}{2}\frac{r^2 - \lambda^2}{r^2 + \lambda^2} \tag{10.3.89}$$

and

$$\rho(\boldsymbol{x}) = -\rho_0 \frac{r^2 + \lambda^2}{2\ell_B^2 + \lambda^2} M(1; \omega^2 + 2, -|z|^2/2), \tag{10.3.90}$$

where $M(a; b; x)$ is the Kummer function. Thus the spin density of the skyrmion is factorized into the electron density $\rho(\boldsymbol{x})$ and the normalized spin field $\mathcal{S}_a(\boldsymbol{x})$ describing the familiar skyrmion. Let us call it a factorizable skyrmion.

Using the property of the Kummer function we can derive the asymptotic behavior of the density,

$$\frac{\rho(\boldsymbol{x})}{\rho_0} = 1 - \frac{2\lambda^2}{(|z|^2 + \lambda^2)^2} + \cdots, \tag{10.3.91}$$

as $|\boldsymbol{x}| \to \infty$. The density $\rho(\boldsymbol{x})$ approaches the ground-state value $\rho_0$ only polynomially unless $\lambda = 0$.

We estimate the number of spins flipped around a skyrmion,

$$N_{\text{spin}} = \int d^2x \left\{ S_z(\boldsymbol{x}) - \frac{1}{2}\rho_0 \right\}. \tag{10.3.92}$$

The asymptotic behavior of $S_z(\boldsymbol{x})$ is

$$S_z(\boldsymbol{x}) = \frac{\rho_0}{2} \left( 1 - 2\frac{\lambda^2}{r^2} \right) + \cdots. \tag{10.3.93}$$

Unless $\lambda = 0$, we find $N_{\text{spin}}$ to diverge logarithmically. This is because the density $\rho(\boldsymbol{x})$ approaches the ground-state value $\rho_0$ only polynomially. The Zeeman energy is given by

$$H_{\text{Z}} = -\Delta_{\text{Z}} N_{\text{spin}} \tag{10.3.94}$$

with the Zeeman gap

$$\Delta_{\text{Z}} \equiv |g^*\mu_B B|. \tag{10.3.95}$$

It is divergent however small the Zeeman gap is. The factorizable skyrmion (10.3.88) has an infinitely large Zeeman energy, and hence it cannot be physical in the QH system. We shall present a skyrmion state having a finite energy in subsection 10.4.5.

## 10.4. Coulomb Interactions

### 10.4.1. *Projected Hamiltonians*

We proceed to analyze Coulomb interactions among electrons confined within the lowest Landau level ($N = 0$). The Coulomb Hamiltonian is given in terms of the physical density $\rho(\boldsymbol{q})$,

$$H_{\text{C}} = \pi \int d^2q \, V(\boldsymbol{q})\rho(-\boldsymbol{q})\rho(\boldsymbol{q}). \tag{10.4.96}$$

We rewrite this in terms of the bare density $\hat{\rho}(\boldsymbol{q})$, for which we introduce a new notation,

$$H_{\text{D}} = \pi \int d^2q \, V_{\text{D}}(\boldsymbol{q})\hat{\rho}(-\boldsymbol{q})\hat{\rho}(\boldsymbol{q}), \tag{10.4.97}$$

though $H_{\mathrm{D}} \equiv \bar{H}_{\mathrm{C}}$. Here, $V_{\mathrm{D}}(\boldsymbol{q})$ is the effective potential in the lowest Landau level,

$$V_{\mathrm{D}}(\boldsymbol{q}) = V(\boldsymbol{q})F_0(-\boldsymbol{q})F_0(\boldsymbol{q}), \qquad (10.4.98)$$

with $F_0(\boldsymbol{q})$ the Landau-level form factor (10.2.12) for the lowest Landau level. It is given by

$$V_{\mathrm{D}}(\boldsymbol{q}) = \frac{e^2}{4\pi\varepsilon|\boldsymbol{q}|}e^{-\ell_B^2\boldsymbol{q}^2/2}, \qquad (10.4.99)$$

or

$$V_{\mathrm{D}}(\boldsymbol{x}) = \frac{e^2\sqrt{2\pi}}{8\pi\varepsilon\ell_B}I_0(\boldsymbol{x}^2/4\ell_B^2)e^{-\boldsymbol{x}^2/4\ell_B^2}, \qquad (10.4.100)$$

where $I_0(x)$ is the modified Bessel function. It approaches the ordinary Coulomb potential at large distance, as expected,

$$V_{\mathrm{D}}(\boldsymbol{x}) \to V(\boldsymbol{x}) = \frac{e^2}{4\pi\varepsilon|\boldsymbol{x}|} \quad \text{as} \quad |\boldsymbol{x}| \to \infty, \qquad (10.4.101)$$

but at short distance it does not diverge in contrast to the ordinary Coulomb potential,

$$V_{\mathrm{D}}(\boldsymbol{x}) \to \frac{e^2\sqrt{2\pi}}{8\pi\varepsilon\ell_B} \quad \text{as} \quad |\boldsymbol{x}| \to 0. \qquad (10.4.102)$$

This is physically reasonable because an electron cannot be localized to a point within the lowest Landau level.

### 10.4.2.  *Exchange interaction*

It is necessary to rewrite the Coulomb Hamiltonian (10.4.97) into another equivalent form to reveal the intrinsic feature of the system such as a spontaneous symmetry breaking together with development of quantum coherence. The bare density operator (10.2.30) reads

$$\hat{\rho}(\boldsymbol{q}) = \frac{1}{2\pi}\sum_{mn}\langle m|e^{-i\boldsymbol{q}\boldsymbol{X}}|n\rangle c_\mu^\dagger(m)_\mu c(n) \qquad (10.4.103)$$

with the isospin index included. Substituting this into (10.4.97), we find

$$H_{\mathrm{D}} = \sum_{mnij} V_{mnij}\rho(n,m)\rho(j,i), \qquad (10.4.104)$$

where

$$\rho(n,m) = \sum_\sigma c_\sigma^\dagger(m)c_\sigma(n). \qquad (10.4.105)$$

Due to the algebraic relation

$$\delta_{\alpha\beta}\delta_{\sigma\tau} = \frac{1}{2}\sum_A^{N_I^2-1}\lambda_{\alpha\tau}^A\lambda_{\sigma\beta}^A + \frac{1}{N_I}\delta_{\alpha\tau}\delta_{\sigma\beta}, \qquad (10.4.106)$$

there holds the relation

$$\rho(n,m)\rho(j,i) = -2[I_A(j,m)I_A(n,i) + \frac{1}{2N_I}\rho(j,m)\rho(n,i)]$$
$$+ \frac{2N_I^2 - 1}{N_I}\rho(n,m)\delta_{ij} + \rho(j,m)\delta_{in}, \tag{10.4.107}$$

with

$$I_A(n,m) = \frac{1}{2}\sum_{\sigma\tau} c_\sigma^\dagger(m)\lambda_{\sigma\tau}^A c_\tau(n). \tag{10.4.108}$$

With the aid of this relation we can rewrite the Hamiltonian (10.4.104) as

$$H_{\mathrm{X}} = -2\sum_{mnij} V_{mnij}[I_A(j,m)I_A(n,i) + \frac{1}{2N}\rho(j,m)\rho(n,i)], \tag{10.4.109}$$

where we have introduced a new notation $H_{\mathrm{X}}$, though $H_{\mathrm{X}} \equiv H_{\mathrm{D}}$.

We use the inversion formula (10.2.31) to convert this in the momentum space,

$$H_{\mathrm{X}} = -\pi\int d^2p\, V_{\mathrm{X}}(\boldsymbol{p})\left[\sum_{A=1}^{N_I^2-1}\hat{I}_A(-\boldsymbol{p})\hat{I}_A(\boldsymbol{p}) + \frac{1}{2N}\hat{\rho}(-\boldsymbol{p})\hat{\rho}(\boldsymbol{p})\right], \tag{10.4.110}$$

where the exchange potential $V_{\mathrm{X}}(\boldsymbol{p})$ is defined by

$$V_{\mathrm{X}}(\boldsymbol{p}) = \frac{\ell_B^2}{\pi}\int d^2k\, e^{-i\ell_B^2\boldsymbol{p}\wedge\boldsymbol{k}}V_{\mathrm{D}}(\boldsymbol{k}) = 4\epsilon_{\mathrm{X}}e^{-\frac{1}{4}\ell_B^2 p^2}I_0\left(\frac{\ell_B^2 p^2}{4}\right), \tag{10.4.111}$$

where

$$\epsilon_{\mathrm{X}} \equiv \sum_j V_{njjn} = \frac{1}{4\pi}\int d^2k\, V_{\mathrm{D}}(\boldsymbol{k}) = \frac{1}{4\ell_B^2}V_{\mathrm{X}}(\boldsymbol{p}=0) = \frac{1}{2}\sqrt{\frac{\pi}{2}}\frac{e^2}{4\pi\varepsilon\ell_B}. \tag{10.4.112}$$

It has a typical property of the exchange interaction with a short-range potential $V_{\mathrm{X}}(\boldsymbol{x})$. Two Hamiltonians (10.4.97) and (10.4.110) are equivalent as the microscopic Hamiltonian.

### 10.4.3. *Classical Hamiltonian*

Though these two Hamiltonians are equivalent as quantum mechanical ones, $H_{\mathrm{D}} = H_{\mathrm{X}}$, they are not when they are regarded as the corresponding classical Hamiltonians. For instance, let us consider two well-separated charged excitations. There is a long-range Coulomb interaction $V_{\mathrm{D}}(\boldsymbol{x})$ between them, but not a short-range exchange interaction $V_{\mathrm{X}}(\boldsymbol{x})$. Hence, $H_{\mathrm{D}}^{\mathrm{cl}} \neq H_{\mathrm{X}}^{\mathrm{cl}}$, where $H_{\mathrm{D}}^{\mathrm{cl}}$ and $H_{\mathrm{X}}^{\mathrm{cl}}$ are the Hamiltonians of the direct-interaction form (10.4.97) and of the exchange-interaction form (10.4.110) with the density operators $\hat{\rho}(\boldsymbol{x})$ and $\hat{I}_A(\boldsymbol{x})$ being replaced by the classical ones $\hat{\rho}^{\mathrm{cl}}(\boldsymbol{x})$ and $\hat{I}_A^{\mathrm{cl}}(\boldsymbol{x})$,

$$H_{\mathrm{D}}^{\mathrm{cl}} = \pi\int d^2k\, V_{\mathrm{D}}(\boldsymbol{k})\hat{\rho}^{\mathrm{cl}}(-\boldsymbol{k})\hat{\rho}^{\mathrm{cl}}(\boldsymbol{k}), \tag{10.4.113a}$$

$$H_{\mathrm{X}}^{\mathrm{cl}} = -\pi\int d^2k\, V_{\mathrm{X}}(\boldsymbol{k})\left[\hat{\boldsymbol{I}}^{\mathrm{cl}}(-\boldsymbol{k})\hat{\boldsymbol{I}}^{\mathrm{cl}}(\boldsymbol{k}) + \frac{1}{2N}\hat{\rho}^{\mathrm{cl}}(-\boldsymbol{k})\hat{\rho}^{\mathrm{cl}}(\boldsymbol{k})\right]. \tag{10.4.113b}$$

We can demonstrate[15] that, for the class of the state (10.2.44), the total energy is simply the sum of them,

$$H_{\mathrm{C}}^{\mathrm{cl}} = \langle H_{\mathrm{C}} \rangle = H_{\mathrm{D}}^{\mathrm{cl}} + H_{\mathrm{X}}^{\mathrm{cl}}. \tag{10.4.114}$$

We call this the decomposition formula.

### 10.4.4. *Spontaneous symmetry breaking*

Since the Coulomb Hamiltonian (10.4.97) does not involve isospin variables, it seems that the energy of a state is independent of isospin orientations. This is not the case because of the decomposition formula (10.4.114), according to which the energy consists of the direct and exchange ones. Though the direct energy $H_{\mathrm{D}}^{\mathrm{cl}}$ does not involves isospins, the exchange energy $H_{\mathrm{X}}^{\mathrm{cl}}$ does. The exchange energy is minimized when all isospins are polarized into one arbitrary direction, leading to a spontaneous breaking of the $\mathrm{SU}(N_I)$ symmetry. Accordingly the resulting system is called the QH ferromagnet.

We should note that the direction of polarization is chosen externally in actual QH systems. Indeed, there exists the Zeeman effect, which determines the spin direction. Nevertheless, provided that the Zeeman energy is much smaller than the exchange energy, it is still proper to regard the QH system to be a ferromagnet.

### 10.4.5. *Skyrmion excitation energy*

We now calculate the excitation energy of the skyrmion (10.3.73) in the spin SU(2) system as a function of the Zeeman gap $\Delta_{\mathrm{Z}}$. We propose an ansatz,[27]

$$u_-^2(n) = v_+^2(n) = \frac{\omega^2 t^{2n+2}}{n+1+\omega^2}, \tag{10.4.115}$$

where $\omega$ and $t$ are parameters to be fixed to minimize the skyrmion energy. After somewhat tedious calculations with use of the ansatz (10.4.115), we express the bare densities as[27]

$$\Delta\hat{\rho}_\pm(\boldsymbol{k}) = \pm \frac{1}{2\pi}\varpi(k), \tag{10.4.116a}$$

$$\hat{S}_{x,y}^\pm(\boldsymbol{k}) = \frac{1}{4\pi i}\left\{\frac{k_x}{k}, \pm\frac{k_y}{k}\right\}\xi(k), \qquad \hat{S}_z^\pm(\boldsymbol{k}) = \frac{1}{2}\delta(\ell_B \boldsymbol{k}) - \frac{1}{4\pi}\sigma(k), \tag{10.4.116b}$$

where we have introduced the notations

$$\varpi(k) = e^{-\frac{1}{4}\ell_B^2 k^2} \sum_{n=0}^{\infty}\left[\frac{\omega^2 t^{2n}}{n+\omega^2} - \frac{\omega^2 t^{2n+2}}{n+1+\omega^2}\right] L_n\left(\frac{\ell_B^2 k^2}{2}\right), \tag{10.4.117a}$$

$$\xi(k) = \sqrt{2}\omega t k e^{-\frac{1}{4}\ell_B^2 k^2} \sum_{n=0}^{\infty}\frac{\vartheta_n(\omega,t)t^n}{n+1+\omega^2} L_n^{(1)}\left(\frac{\ell_B^2 k^2}{2}\right), \tag{10.4.117b}$$

$$\sigma(k) = e^{-\frac{1}{4}\ell_B^2 k^2} \sum_{n=0}^{\infty}\left[\frac{\omega^2 t^{2n}}{n+\omega^2} + \frac{\omega^2 t^{2n+2}}{n+1+\omega^2}\right] L_n\left(\frac{\ell_B^2 k^2}{2}\right), \tag{10.4.117c}$$

with

$$\vartheta_n(\omega, t) = \sqrt{1 + \frac{1 - t^{2n+2}}{n + 1}\omega^2}.$$
(10.4.118)

Substituting these into (10.4.114) we get

$$\langle H_{\mathrm{C}}\rangle_{\mathrm{sky}} = \frac{1}{2}E_{\mathrm{C}}^0 \int_0^\infty dk\, e^{-\frac{1}{2}\ell_B^2 k^2}\, \varpi^2(k)$$
$$+ \frac{1}{2}\varepsilon_{\mathrm{X}} \int_0^\infty kdk\, \left[1 - e^{-\frac{1}{4}\ell_B^2 k^2} I_0\left(\frac{\ell_B^2 k^2}{4}\right)\right] [\xi^2(k) + \sigma^2(k) + \varpi^2(k)].$$
(10.4.119)

The convergence of these integrals is easily checked. The Zeeman energy (10.3.94) is expressed as

$$\langle H_{\mathrm{Z}}\rangle_{\mathrm{sky}} = -\Delta_{\mathrm{Z}} N_{\mathrm{spin}} = -\Delta_{\mathrm{Z}}\sigma(k = 0),$$
(10.4.120)

which is finite.

We are able to determine the parameters $\omega$ and $t$ by minimizing the sum of the Coulomb and Zeeman energies, $\langle H\rangle_{\mathrm{sky}} = \langle H_{\mathrm{C}}\rangle_{\mathrm{sky}} + \langle H_{\mathrm{Z}}\rangle_{\mathrm{sky}}$, as a function of the Zeeman gap $\Delta_{\mathrm{Z}}$. For this purpose we calculate $\langle H\rangle_{\mathrm{sky}}$ numerically as a function of $\omega$ and $t$ for a given value of $\Delta_{\mathrm{Z}}$, and we determine the values of $\omega$ and $t$ which minimizes $\langle H\rangle_{\mathrm{sky}}$ at each $\Delta_{\mathrm{Z}}$. In this way we obtain the skyrmion excitation energy $\langle H\rangle_{\mathrm{sky}}$ as a function of $\Delta_{\mathrm{Z}}$. A comment is in order with respect to fitting experimental data by a theoretical result [Fig. 10.1]. First, we have so far assumed an ideal two-dimensional space for electrons. This is not the case. Electrons are confined within

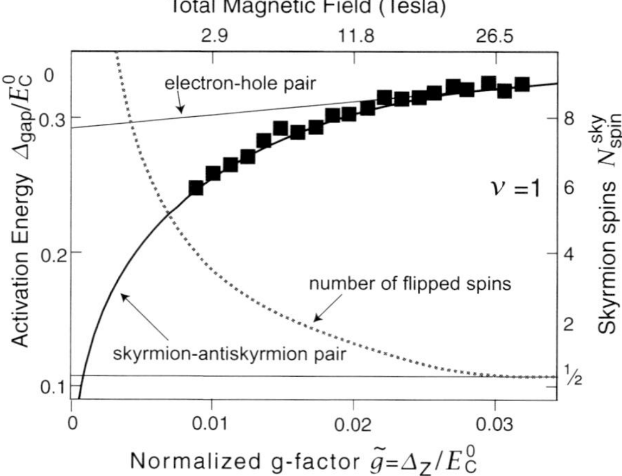


Fig. 10.1. The skrmion-antiskyrmion pair excitation energy is plotted as a function of the normalized Zeeman gap $\tilde{g} = \Delta_{\mathrm{Z}}/E_{\mathrm{C}}^0$. The thin line represents the electron-hole pair excitation energy. The heavy solid curve is obtained by the numerical analysis based on (10.4.119) and (10.4.120). Experimental data are taken from Schmeller et al.[8]

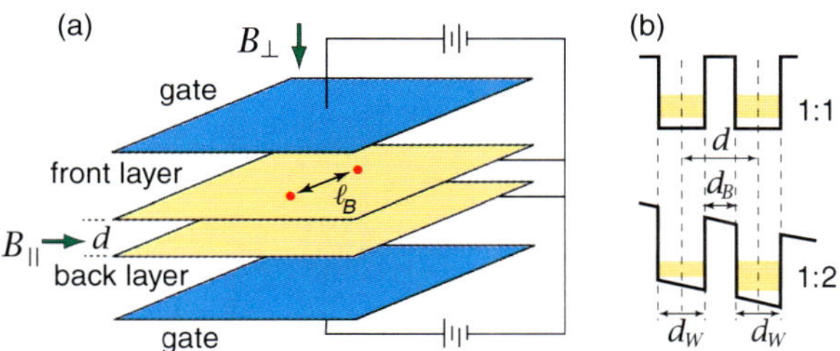



Fig. 10.2.  (a) The bilayer system has four scale parameters, the magnetic length $\ell_B$, the Zeeman gap $\Delta_Z$, the interlayer distance $d$ and the tunneling gap $\Delta_{\text{SAS}}$. It is customary to take $d = d_B + d_W$ with the width $d_W$ and the separation $d_B$ of the two quantum wells. Here, $d_W \simeq 200$ Å and $d_B \simeq 31$ Å in typical samples. The number density $\rho^\alpha$ in each quantum well is controlled by applying gate bias voltages. A parallel magnetic field $B_\parallel$ may be additionally applied to the system. (b) Two typical examples are given where $\rho^f : \rho^b = 1 : 1$ (balanced) and $1 : 2$ (imbalanced).

a quantum well of a finite width of order 200 Å. This will reduce the Coulomb energy considerably. It is quite difficult to make a rigorous analysis of the Coulomb energy in an actual quantum well. We simulate the effect by including the reduction factor $\gamma$ to the Coulomb energy, $E_{\text{sky}} = \gamma(E_X + E_D) + E_Z$, where $0 < \gamma < 1$. Next, the skyrmion excitation takes place in the presence of charged impurities. The existence of charged impurities reduces the activation energy considerably. We include an offset parameter $\Gamma_{\text{offset}}$ to treat this effect phenomenologically as in

$$\Delta_{\text{gap}} = E_{\text{sky}} - \Gamma_{\text{offset}}. \tag{10.4.121}$$

The best fit is obtained by choosing $\gamma \simeq 0.56$ and $\Gamma_{\text{offset}} = 0.41 E_C^0$. We have plotted the excitation energy $\langle H \rangle_{\text{sky}}$ as a function of the normalized Zeeman gap $\widetilde{g} = \Delta_Z / E_C^0$ in Fig. 10.1.

## 10.5.  Bilayer QH Systems

### 10.5.1.  *Pseudospin QH ferromagnet*

A bilayer system is made by trapping electrons in two thin layers at the interface of semiconductors, where the structure introduces an additional degree of freedom in the third direction. We label the two layers by the index $\alpha = $ f, b, and call the $\alpha = $ f layer the front layer and the $\alpha = $ b layer the back layer. The bilayer QH system possesses the SU(2) pseudospin structure, where the SU(2) index corresponds to the index $\alpha = $ f, b. The pseudospin rotates when electrons are transferred from one layer to the other. When the number density is balanced between the two quantum wells, it is referred to as the balanced configuration and otherwise as the imbalanced configuration. We solely analyze the balanced configuration in this article.

There are four types of electrons associated with the field operators $\psi_{f\uparrow}$, $\psi_{f\downarrow}$, $\psi_{b\uparrow}$ and $\psi_{b\downarrow}$, constituting the SU(4) algebra. After the projection to a single Landau level, the field operator $\psi = (\psi_{f\uparrow}, \psi_{f\downarrow}, \psi_{b\uparrow}, \psi_{b\downarrow})$ is given by (10.2.33). The density matrix satisfies the $W_\infty(4)$ density algebra (10.2.40). A topological soliton in the bilayer system is a noncommutative $CP^3$ skyrmion. There are some experimental indications[14] supporting $CP^3$ skyrmion excitations.[33] However, a microscopic formulation is technically quite difficult. We present only a subset, that is the pseudospin SU(2) sector with the spin degree of freedom frozen.

The Coulomb interaction operates electrons in the front and back layers. The Hamiltonian is

$$H_C = \frac{1}{2} \sum_{\alpha,\beta} \int d^2x d^2y \, V^{\alpha\beta}(\boldsymbol{x} - \boldsymbol{y}) \rho^\alpha(\boldsymbol{x}) \rho^\beta(\boldsymbol{y}), \qquad (10.5.122)$$

where

$$V^{\alpha\beta}(\boldsymbol{x}) = \frac{e^2}{4\pi\varepsilon} \frac{1}{\sqrt{|\boldsymbol{x}|^2 + d_{\alpha\beta}^2}}, \qquad (10.5.123)$$

with $d_{\mathrm{ff}} = d_{\mathrm{bb}} = 0$ and $d_{\mathrm{fb}} = d_{\mathrm{bf}} = d$. We decompose (10.5.122) into the SU(2)-invariant Coulomb term $H_C^+$, the capacitance term $H_C^-$. By substituting the projected density operators into (10.5.122) we rewrite the Hamiltonian, for which we introduce a new notation,

$$H_D = \pi \int d^2p \, V_D^+(\boldsymbol{p}) \hat{\rho}(-\boldsymbol{p}) \hat{\rho}(\boldsymbol{p}) + 4\pi \int d^2p \, V_D^-(\boldsymbol{p}) \hat{P}_z(-\boldsymbol{p}) \hat{P}_z(\boldsymbol{p}), \qquad (10.5.124)$$

with

$$V_D^\pm(\boldsymbol{q}) = \frac{e^2}{8\pi\varepsilon|\boldsymbol{q}|} \left(1 \pm e^{-|\boldsymbol{q}|d}\right) e^{-\ell_B^2 q^2/2}, \qquad (10.5.125)$$

where $\hat{\rho}(\boldsymbol{p})$ and $\hat{P}_z(\boldsymbol{p})$ are the bare densities.

The Coulomb Hamiltonian (10.5.124) can be rewritten into the exchange interaction form,

$$H_X = -\frac{\pi}{2} \int d^2p \left[ V_X^d(\boldsymbol{p}) \hat{P}_a(-\boldsymbol{p}) \hat{P}_a(\boldsymbol{p}) + 2V_X^-(\boldsymbol{p}) \hat{P}_z(-\boldsymbol{p}) \hat{P}_z(\boldsymbol{p}) + \frac{1}{4} V_X(\boldsymbol{p}) \hat{\rho}(-\boldsymbol{p}) \hat{\rho}(\boldsymbol{p}) \right], \qquad (10.5.126)$$

where $V_X(\boldsymbol{p})$ is given by (10.4.112),

$$V_X^d(\boldsymbol{p}) = \frac{e^2 \ell_B^2}{2\pi\varepsilon} \int_0^\infty dk \, e^{-\frac{1}{2}\ell_B^2 k^2 - dk} J_0(\ell_B^2 |\boldsymbol{p}|k), \qquad (10.5.127)$$

and $V_X^-(\boldsymbol{p}) = (V_X - V_X^d)/2$. Two Hamiltonians (10.5.124) and (10.5.126) are equivalent as the microscopic Hamiltonian, $H_X \equiv H_D$.

However, there holds the decomposition formula (10.4.114) for the classical Hamiltonian,

$$H_C^{\mathrm{cl}} = \langle H_C \rangle = H_D^{\mathrm{cl}} + H_X^{\mathrm{cl}}, \qquad (10.5.128)$$

where $H_{\mathrm{D}}^{\mathrm{cl}}$ and $H_{\mathrm{X}}^{\mathrm{cl}}$ are the Hamiltonians of the direct-interaction form (10.5.124) and of the exchange-interaction form (10.5.126) with various density operators being replaced by the corresponding classical ones. It is found that the isospin is spontaneously polarized to minimize the exchange energy $H_{\mathrm{X}}^{\mathrm{cl}}$, leading to the pseudospin QH ferromagnet.

### 10.5.2. *Parallel magnetic field*

Due to the similarity between the spin QH ferromagnet and the pseudospin QH ferromagnet, charged excitations are noncommutative skyrmions in the bilayer system as well. However there exists a new feature to pseudospin skyrmions because of the layer structure. Namely, we may introduce the parallel magnetic field between the two layers in addition to the perpendicular magnetic field.

This is carried out simply by tilting a sample by angle $\Theta$ in the external magnetic field, as creates the parallel magnetic field such that $B_{\parallel} = B_{\perp}\tan\Theta$. We choose the gauge $\boldsymbol{A}_{\parallel} = (B_{\parallel}z, 0, 0)$. The kinetic Hamiltonian reads

$$H_{\mathrm{K}} = \frac{1}{2M}\sum_{\alpha=\mathrm{f,b}}\int d^2x\,\psi_{\alpha}^{\dagger}(\boldsymbol{x}; B_{\parallel})(P_x^{\alpha} - iP_y^{\alpha})(P_x^{\alpha} + iP_y^{\alpha})\psi_{\alpha}(\boldsymbol{x}; B_{\parallel}), \qquad (10.5.129)$$

with the covariant momentum

$$P_x^{\mathrm{f}} \equiv -i\hbar\partial_x + eA_x^{\mathrm{ext}} + \frac{1}{2}\hbar\delta_{\mathrm{m}}, \qquad P_y^{\mathrm{f}} \equiv -i\hbar\partial_y + eA_y^{\mathrm{ext}}, \qquad (10.5.130\mathrm{a})$$

$$P_x^{\mathrm{b}} \equiv -i\hbar\partial_x + eA_x^{\mathrm{ext}} - \frac{1}{2}\hbar\delta_{\mathrm{m}}, \qquad P_y^{\mathrm{b}} \equiv -i\hbar\partial_y + eA_y^{\mathrm{ext}}, \qquad (10.5.130\mathrm{b})$$

and $\delta_{\mathrm{m}} = edB_{\parallel}/\hbar$. Accordingly, the guiding-center coordinates are shifted in the opposite directions along the parallel magnetic field in the front and back layers,

$$X_{\alpha} \equiv x + \frac{1}{eB_{\perp}}P_y^{\alpha} = X, \qquad Y_{\alpha} \equiv y - \frac{1}{eB_{\perp}}P_x^{\alpha} = Y \mp \frac{1}{2}d\tan\Theta, \qquad (10.5.131)$$

where $(X, Y)$ is the guiding-center coordinate at $B_{\parallel} = 0$. This leads to a deformation of a skyrmion into a pair of excitations located at $(0, -\frac{1}{2}d\tan\Theta)$ in the front layer and $(0, \frac{1}{2}d\tan\Theta)$ in the back layer [Fig. 10.3], as we shall see. The deformed skyrmion is called a bimeron (a pair of merons).

We evaluate the classical densities at $B_{\parallel} \neq 0$ by relating them to those at $B = 0$. For this purpose we introduce an auxiliary field operator $\psi_{\alpha}(\boldsymbol{x})$ by

$$\psi_{\mathrm{f}}(\boldsymbol{x}; B_{\parallel}) = e^{-i\delta_{\mathrm{m}}x/2}\psi_{\mathrm{f}}(\boldsymbol{x}), \qquad \psi_{\mathrm{b}}(\boldsymbol{x}; B_{\parallel}) = e^{+i\delta_{\mathrm{m}}x/2}\psi_{\mathrm{b}}(\boldsymbol{x}). \qquad (10.5.132)$$

The kinetic Hamiltonian is reduced to

$$H_{\mathrm{K}} = \frac{1}{2M}\sum_{\alpha=\mathrm{f,b}}\int d^2x\,\psi_{\alpha}^{\dagger}(\boldsymbol{x})(P_x - iP_y)(P_x + iP_y)\psi_{\alpha}(\boldsymbol{x}), \qquad (10.5.133)$$

where the parallel magnetic field has been removed. Hence, it is identified with the field operator at $B_{\parallel} = 0$.

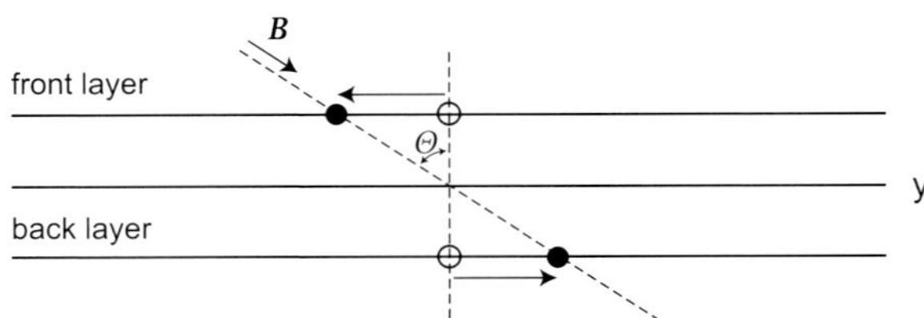


Fig. 10.3. The center of a skyrmion is shifted to the opposite direction along the magnetic field in the front and back layers.

We expand the field operators $\psi_\alpha(\boldsymbol{x}; B_\parallel)$ and $\psi_\alpha(\boldsymbol{x})$ as

$$\psi_\alpha(\boldsymbol{x}; B_\parallel) = \sum_n \langle \boldsymbol{x}|n\rangle_\alpha c_\alpha(n), \qquad (10.5.134\text{a})$$

$$\psi_\alpha(\boldsymbol{x}) = \sum_n \langle \boldsymbol{x}|n\rangle c_\alpha(n), \qquad (10.5.134\text{b})$$

in terms of the wave functions $\langle \boldsymbol{x}|n\rangle_\alpha$ and $\langle \boldsymbol{x}|n\rangle$, where $c_\alpha(n)$ obeys the anticommutation relation, $\{c_\alpha(m), c_\beta^\dagger(n)\} = \delta_{\alpha\beta}\delta_{mn}$, together with the vacuum $|0\rangle$ defined by $c_\alpha(n)|0\rangle = 0$. Any physical states are created by applying $c_\alpha^\dagger(n)$ to the vacuum $|0\rangle$.

It follows from (10.5.132) and (10.5.134) that their wave functions are related as

$$\langle \boldsymbol{x}|n\rangle_\alpha = e^{\mp i\delta_{\rm m} x/2}\langle \boldsymbol{x}|n\rangle = \langle \boldsymbol{x}|e^{\mp i\delta_{\rm m}(X+R_x)/2}|n\rangle, \qquad (10.5.135)$$

where we have moved the c-number factor $e^{\mp i\delta_{\rm m} x/2}$ within the scalar product and changed the c-number $x$ to the sum of the operators, $x = X + R_x$, since it act on the state $\langle \boldsymbol{x}|$. Hence we have

$$|n\rangle_\alpha = e^{\mp i\delta_{\rm m}(X+R_x)/2}|n\rangle, \qquad (10.5.136)$$

or

$$_{\rm f}\langle m|n\rangle_{\rm b} = \langle m|e^{+i\delta_{\rm m}(X+R_x)}|n\rangle = e^{-\delta_{\rm m}^2 \ell_B^2/4}\langle m|e^{+i\delta_{\rm m} X}|n\rangle, \qquad (10.5.137\text{a})$$

$$_{\rm b}\langle m|n\rangle_{\rm f} = \langle m|e^{-i\delta_{\rm m}(X+R_x)}|n\rangle = e^{-\delta_{\rm m}^2 \ell_B^2/4}\langle m|e^{-i\delta_{\rm m} X}|n\rangle. \qquad (10.5.137\text{b})$$

The Fock states $|n\rangle_{\rm f}$ and $|n\rangle_{\rm b}$ for the front and back layers do not belong to the same Landau level due to the existence of the Landau-level mixing operator $e^{\mp i\delta_{\rm m} R_x}$. The suppression factor $e^{-\delta_{\rm m}^2 \ell_B^2/4}$ has arisen due to the Landau-level mismatch between the front and back layers.

### 10.5.3. *Ground state*

The ground state is the one that minimizes the tunneling energy. The tunneling Hamiltonian is

$$H_{\mathrm{T}} = -\frac{1}{2}\Delta_{\mathrm{SAS}}\int d^2x\,[\psi_{\mathrm{f}}^{\dagger}(\boldsymbol{x};B_{\parallel})\psi_{\mathrm{b}}(\boldsymbol{x};B_{\parallel}) + \psi_{\mathrm{b}}^{\dagger}(\boldsymbol{x};B_{\parallel})\psi_{\mathrm{f}}(\boldsymbol{x};B_{\parallel})]. \qquad (10.5.138)$$

Substituting (10.5.134a) into it and using (10.5.137), we find

$$H_{\mathrm{T}} = -\frac{1}{2}e^{-\delta_{\mathrm{m}}^2\ell_B^2/4}\Delta_{\mathrm{SAS}}\sum_n[\langle m|e^{+i\delta_{\mathrm{m}}X}|n\rangle c_{\mathrm{f}}^{\dagger}(m)c_{\mathrm{b}}(n) + \langle m|e^{-i\delta_{\mathrm{m}}X}|n\rangle c_{\mathrm{b}}^{\dagger}(m)c_{\mathrm{f}}(n)]. \qquad (10.5.139)$$

We may rewrite this as

$$H_{\mathrm{T}} = -\frac{1}{2}e^{-\delta_{\mathrm{m}}^2\ell_B^2/4}\Delta_{\mathrm{SAS}}\sum_n[C_{\mathrm{f}}^{\dagger}(n;B_{\parallel})C_{\mathrm{b}}(n;B_{\parallel}) + C_{\mathrm{b}}^{\dagger}(n;B_{\parallel})C_{\mathrm{f}}(n;B_{\parallel})], \qquad (10.5.140)$$

by defining a new operator,

$$C_\alpha(m;B_{\parallel}) = \sum_n\langle m|e^{\mp i\delta_{\mathrm{m}}X/2}|n\rangle c_\alpha(n). \qquad (10.5.141)$$

It satisfies the canonical anticommutation relation, $\{C_\alpha(m;B_{\parallel}),C_\beta^{\dagger}(n;B_{\parallel})\} = \delta_{\alpha\beta}\delta_{mn}$, and $C_\alpha(n;B_{\parallel})$ annihilates the vacuum, $C_\alpha(n;B_{\parallel})|0\rangle = 0$.

We then introduce the "symmetric" and "antisymmetric" operators,

$$t_{\uparrow}(n;B_{\parallel}) = \frac{1}{\sqrt{2}}[C_{\mathrm{f}}(n;B_{\parallel}) + C_{\mathrm{b}}(n;B_{\parallel})], \qquad (10.5.142a)$$

$$t_{\downarrow}(n;B_{\parallel}) = \frac{1}{\sqrt{2}}[C_{\mathrm{f}}(n;B_{\parallel}) - C_{\mathrm{b}}(n;B_{\parallel})]. \qquad (10.5.142b)$$

They satisfy the canonical anticommutation relation,

$$\{t_\mu(m;B_{\parallel}),t_\nu^{\dagger}(n;B_{\parallel})\} = \delta_{\mu\nu}\delta_{mn}, \qquad (10.5.143)$$

and $t_\mu(n;B_{\parallel})$ annihilates the vacuum, $t_\mu(n;B_{\parallel})|0\rangle = 0$. We may now rewrite the Hamiltonian (10.5.139) as

$$H_{\mathrm{T}} = \frac{1}{2}e^{-\delta_{\mathrm{m}}^2\ell_B^2/4}\Delta_{\mathrm{SAS}}\sum_n[t_{\downarrow}^{\dagger}(n;B_{\parallel})t_{\downarrow}(n;B_{\parallel}) - t_{\uparrow}^{\dagger}(n;B_{\parallel})t_{\uparrow}(n;B_{\parallel})], \qquad (10.5.144)$$

which has the diagonalized expression.

The ground state is given by filling up the "symmetric" state at $\nu = 1$,

$$|\mathrm{g};B_{\parallel}\rangle = \prod_{n\geq 0}t_{\uparrow}^{\dagger}(n;B_{\parallel})|0\rangle. \qquad (10.5.145)$$

The hole and electron states located at $n = 0$ are

$$|-;B_{\parallel}\rangle \equiv |\mathrm{h};B_{\parallel}\rangle = \prod_{n\geq 1}t_{\uparrow}^{\dagger}(n;B_{\parallel})|0\rangle, \qquad (10.5.146a)$$

$$|+;B_{\parallel}\rangle \equiv |\mathrm{e};B_{\parallel}\rangle = t_{\downarrow}^{\dagger}(0;B_{\parallel})|\mathrm{g};B_{\parallel}\rangle. \qquad (10.5.146b)$$

In general, we consider a wide class of states of the following form,

$$|\mathfrak{S}; B_{\parallel}\rangle = e^{iW}|\mathfrak{S}_0; B_{\parallel}\rangle \quad \text{with} \quad |\mathfrak{S}_0; B_{\parallel}\rangle = \prod_{\mu n} \left[t_{\mu}^{\dagger}\left(n; B_{\parallel}\right)\right]^{\nu_{\mu}(n)} |0\rangle, \quad (10.5.147)$$

where $W$ is an arbitrary element of the $W_{\infty}(2)$ algebra: $\nu_{\mu}(n)$ may take the value either 0 or 1 depending whether the pseudospin state $\mu = \uparrow, \downarrow$ at the site $n$ is occupied or not.

An important property is that the matrix element

$$\langle \mathfrak{S}; B_{\parallel}|t_{\mu_1}^{\dagger}\left(m_1; B_{\parallel}\right) t_{\mu_2}^{\dagger}\left(m_2; B_{\parallel}\right) \cdots t_{\nu_1}\left(n_1; B_{\parallel}\right) t_{\nu_2}\left(n_2; B_{\parallel}\right) \cdots |\mathfrak{S}; B_{\parallel}\rangle \quad (10.5.148)$$

is independent of $B_{\parallel}$. The matrix element

$$\langle \mathfrak{S}; B_{\parallel}|C_{\alpha_1}^{\dagger}\left(m_1; B_{\parallel}\right) C_{\alpha_2}^{\dagger}\left(m_2; B_{\parallel}\right) \cdots C_{\beta_1}\left(n_1; B_{\parallel}\right) C_{\beta_2}\left(n_2; B_{\parallel}\right) \cdots |\mathfrak{S}; B_{\parallel}\rangle$$

$$(10.5.149)$$

is calculable by rewriting $C_{\alpha}\left(n; B_{\parallel}\right)$ in terms of $t_{\mu}\left(n; B_{\parallel}\right)$, and also independent of $B_{\parallel}$. Consequently we may evaluate them at $B_{\parallel} = 0$. In this way we can relate the classical densities at $B_{\parallel} \neq 0$ to those at $B = 0$.

### 10.5.4. Density operators

We analyze the density operators. First, using (10.5.134a), (10.5.136) and (10.5.141), we find

$$\rho_{\alpha}(\boldsymbol{q}; B_{\parallel}) = e^{-\boldsymbol{q}^2 \ell_B^2/4} \sum_{mn} \langle m|e^{-i\boldsymbol{q}(\boldsymbol{X} \mp \boldsymbol{\delta}_{\mathrm{m}} \ell_B^2/2)}|n\rangle C_{\alpha}^{\dagger}(m; B_{\parallel})C_{\alpha}(n; B_{\parallel}), \quad (10.5.150)$$

with $\boldsymbol{\delta}_{\mathrm{m}} = (\delta_{\mathrm{m}}, 0)$. Hence, the guiding center $\boldsymbol{X}$ is shifted in the front and back layers into the opposite direction as in (10.5.131), where $d \tan \Theta = \ell_B^2 \delta_{\mathrm{m}}$. Sandwiching this with a generic Fock state $|\mathfrak{S}; B_{\parallel}\rangle$, and taking into account the comment given below (10.5.149), we have

$$\rho_{\alpha}^{\mathrm{cl}}(\boldsymbol{q}; B_{\parallel}) = e^{-\boldsymbol{q}^2 \ell_B^2/4} e^{\pm i q_y \delta_{\mathrm{m}} \ell_B^2/2} \hat{\rho}_{\alpha}^{\mathrm{cl}}(\boldsymbol{q}), \quad (10.5.151)$$

where $\hat{\rho}_{\alpha}^{\mathrm{cl}}(\boldsymbol{q})$ is the classical bare density at $B_{\parallel} = 0$, or

$$\hat{\rho}_{\alpha}^{\mathrm{cl}}(\boldsymbol{q}) = \frac{1}{2\pi} \sum_{mn} \langle m|e^{-i\boldsymbol{q}\boldsymbol{X}}|n\rangle \langle \mathfrak{S}|c_{\alpha}^{\dagger}(m)c_{\alpha}(n)|\mathfrak{S}\rangle. \quad (10.5.152)$$

Here, $|\mathfrak{S}\rangle = \lim_{B_{\parallel} \to 0} |\mathfrak{S}; B_{\parallel}\rangle$. It follows from (10.5.151) that

$$\hat{\rho}_{\alpha}^{\mathrm{cl}}(\boldsymbol{x}; B_{\parallel}) = \hat{\rho}_{\alpha}^{\mathrm{cl}}(x, y \pm \ell_B^2 \delta_{\mathrm{m}}/2). \quad (10.5.153)$$

The electron densities are shifted in the front and back layers by the amount of $\mp \frac{1}{2}\boldsymbol{\delta}_{\mathrm{m}}\ell_B^2$ in the front $(-)$ and back $(+)$ layers, as illustrated in Fig. 10.3.

Next, we study

$$P_a(\boldsymbol{x}; B_{\parallel}) = \frac{1}{2}\psi_{\alpha}(\boldsymbol{x}; B_{\parallel})(\tau_a)_{\alpha\beta}\psi_{\beta}(\boldsymbol{x}; B_{\parallel}). \quad (10.5.154)$$

We set

$$P_{\pm}(\boldsymbol{x}; B_{\parallel}) = P_x(\boldsymbol{x}; B_{\parallel}) \pm iP_y(\boldsymbol{x}; B_{\parallel}). \tag{10.5.155}$$

Using the expansion (10.5.134a), (10.5.136) and (10.5.141), we find

$$P_+(\boldsymbol{q}; B_{\parallel}) = \frac{1}{2\pi} \sum_{mn} e^{-\frac{1}{4}\ell_B^2(\boldsymbol{q}-\boldsymbol{\delta}_{\mathrm{m}})^2} \langle m|e^{-i\boldsymbol{q}\boldsymbol{X}}|n\rangle C_{\mathrm{f}}^{\dagger}(m; B_{\parallel})C_{\mathrm{b}}(n; B_{\parallel}), \tag{10.5.156a}$$

$$P_-(\boldsymbol{q}; B_{\parallel}) = \frac{1}{2\pi} \sum_{mn} e^{\frac{1}{4}\ell_B^2(\boldsymbol{q}+\boldsymbol{\delta}_{\mathrm{m}})^2} \langle m|e^{-i\boldsymbol{q}\boldsymbol{X}}|n\rangle C_{\mathrm{b}}^{\dagger}(m; B_{\parallel})C_{\mathrm{f}}(n; B_{\parallel}) \tag{10.5.156b}$$

Sandwiching them with a generic Fock state $|\mathfrak{S}; B_{\parallel}\rangle$, we have

$$P_{\pm}^{\mathrm{cl}}(\boldsymbol{q}; B_{\parallel}) = e^{-\frac{1}{4}\ell_B^2(\boldsymbol{q}\mp\boldsymbol{\delta}_{\mathrm{m}})^2} \hat{P}_{\pm}^{\mathrm{cl}}(\boldsymbol{q}), \tag{10.5.157}$$

where $\hat{P}_{\pm}^{\mathrm{cl}}(\boldsymbol{q})$ is the classical bare density at $B_{\parallel} = 0$. Hence, the densities at $B_{\parallel}$ are obtainable once they are given at $B_{\parallel} = 0$.

## 10.6. Skyrmion and Meron Excitations

### 10.6.1. *Baby skyrmions*

We calculate explicitly the classical bare densities $D_{\mu\nu}^{\mathrm{cl}\pm}(\boldsymbol{p}) = \langle\pm|D_{\mu\nu}(\boldsymbol{p})|\pm\rangle$ for baby skyrmions, that is, an electron $(+)$ and a hole $(-)$, placed at the Landau site $|0\rangle$. First we calculate them at $B_{\parallel} = 0$. It is easy to see

$$\langle g|t_{\mu}^{\dagger}(m)t_{\nu}(n)|g\rangle = \delta_{\mu\uparrow}\delta_{\nu\uparrow}\delta_{mn}, \tag{10.6.158a}$$

$$\langle+ t_{\mu}^{\dagger}(m)t_{\nu}(n)|+\rangle = \delta_{\mu\uparrow}\delta_{\nu\uparrow}\delta_{mn} + \delta_{\mu\downarrow}\delta_{\nu\downarrow}\delta_{m0}\delta_{n0}, \tag{10.6.158b}$$

$$\langle- t_{\mu}^{\dagger}(m)t_{\nu}(n)|-\rangle = \delta_{\mu\uparrow}\delta_{\nu\uparrow}\delta_{mn} - \delta_{\mu\uparrow}\delta_{\nu\uparrow}\delta_{m0}\delta_{n0}. \tag{10.6.158c}$$

The density modulation at $B_{\parallel} = 0$ reads

$$\rho_{\mathrm{f}}^{\mathrm{cl}\pm}(\boldsymbol{q}) = \rho_{\mathrm{b}}^{\mathrm{cl}\pm}(\boldsymbol{q}) = \frac{1}{2}\delta(\ell_B\boldsymbol{q}) \pm \frac{1}{2}\rho_0 e^{-\ell_B^2 q^2/2}. \tag{10.6.159a}$$

$$P_x^{\mathrm{cl}\pm}(\boldsymbol{q}) = \frac{1}{2}\delta(\ell_B\boldsymbol{q}) - \frac{1}{2}\rho_0 e^{-\ell_B^2 q^2/2}, \qquad P_y^{\mathrm{cl}\pm}(\boldsymbol{q}) = 0. \tag{10.6.159b}$$

We substitute these into (10.5.151) and (10.5.157) to construct the physical densities at $B_{\parallel}$. After making their Fourier transformation, we find

$$\rho_{\mathrm{f}}^{\mathrm{cl}\pm}(\boldsymbol{x}; B_{\parallel}) = \frac{1}{2}\rho_0 \pm \frac{1}{2}\rho_0 e^{-x^2/2\ell_B^2} e^{-(y+a)^2/2\ell_B^2}, \tag{10.6.160a}$$

$$\rho_{\mathrm{b}}^{\mathrm{cl}\pm}(\boldsymbol{x}; B_{\parallel}) = \frac{1}{2}\rho_0 \pm \frac{1}{2}\rho_0 e^{-x^2/2\ell_B^2} e^{-(y-a)^2/2\ell_B^2} \tag{10.6.160b}$$

$$P_x^{\mathrm{cl}\pm}(\boldsymbol{x}; B_{\parallel}) = \frac{1}{2}\rho_0 e^{-a^2/\ell_B^2} - \frac{1}{2}\rho_0 e^{-a^2/2\ell_B^2} e^{-r^2/2\ell_B^2} \cos(x\delta_{\mathrm{m}}/2), \tag{10.6.160c}$$

$$P_y^{\mathrm{cl}\pm}(\boldsymbol{x}; B_{\parallel}) = -\frac{1}{2}\rho_0 e^{-a^2/2\ell_B^2} e^{-r^2/2\ell_B^2} \sin(x\delta_{\mathrm{m}}/2), \tag{10.6.160d}$$

with $a = \frac{1}{2}\ell_B^2\delta_{\mathrm{m}}$.

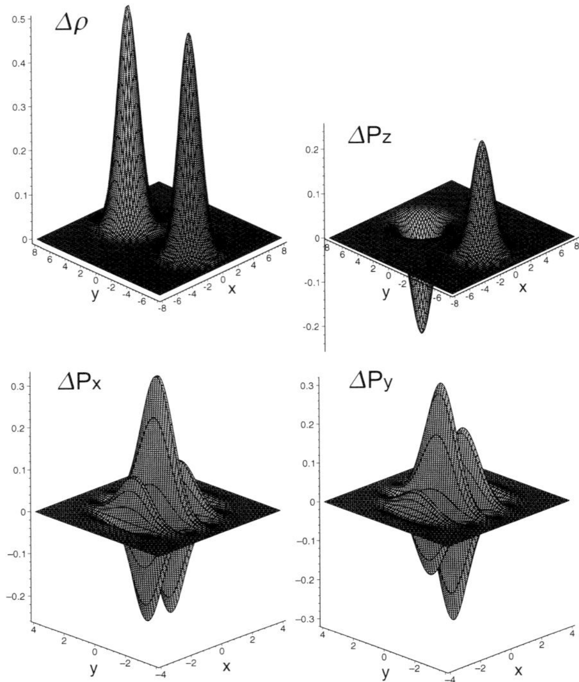

Fig. 10.4. Various physical densities of an electron excitation in the presence of the parallel magnetic field $B_\parallel$. An electron is made of two identical pieces whose separation increases as $B_\parallel$ increases. We have taken $\delta_{\mathrm{m}} = 8/\ell_B$ to emphasize the separation. There is no separation at $B_\parallel = 0$.

In Fig. 10.4, we demonstrate the spatial structures of $\Delta\rho^{\mathrm{cl}}(\boldsymbol{x})$ and $\Delta P_a^{\mathrm{cl}}(\boldsymbol{x})$, removing the ground-state contribution. A charged excitation consists of two identical pieces (merons) located at $(0, -a)$ in the front layer and $(0, +a)$ in the back layer [Fig. 10.3]. By increasing the parallel magnetic field, the separation between two merons also increases.

### 10.6.2. *Coulomb energy*

The Coulomb Hamiltonian is

$$H_{\mathrm{C}} = \pi \int d^2p\, V^+(\boldsymbol{p})\rho(-\boldsymbol{p}; B_\parallel)\rho(\boldsymbol{p}; B_\parallel) + 4\pi \int d^2p\, V^-(\boldsymbol{p})P_z(-\boldsymbol{p}; B_\parallel)P_z(\boldsymbol{p}; B_\parallel),$$

$$(10.6.161)$$

which reads

$$H_{\mathrm{C}} = \pi \int d^2p\, V^+(\boldsymbol{p})\rho(-\boldsymbol{p})\rho(\boldsymbol{p}) + 4\pi \int d^2p\, V^-(\boldsymbol{p})P_z(-\boldsymbol{p})P_z(\boldsymbol{p}) \qquad (10.6.162)$$

in terms of the auxiliary field (10.5.132), where the parallel magnetic field has been removed. As in the system with $B_\parallel = 0$, we can rewrite $H_C$ as

$$H_D = \pi \int d^2p\, V_D^+(\boldsymbol{p})\hat{\rho}(-\boldsymbol{p})\hat{\rho}(\boldsymbol{p}) + 4\pi \int d^2p\, V_D^-(\boldsymbol{p})\hat{P}_z(-\boldsymbol{p})\hat{P}_z(\boldsymbol{p}), \qquad (10.6.163)$$

and also into the exchange interaction form,

$$H_X = -\pi \int d^2p \left[ V_X^d(\boldsymbol{p})\hat{P}_a(-\boldsymbol{p})\hat{P}_a(\boldsymbol{p}) + 2V_X^-(\boldsymbol{p})\hat{P}_z(-\boldsymbol{p})\hat{P}_z(\boldsymbol{p}) + \frac{1}{4}V_X(\boldsymbol{p})\hat{\rho}(-\boldsymbol{p})\hat{\rho}(\boldsymbol{p}) \right].$$
$$(10.6.164)$$

Note that $H_X \equiv H_D \equiv H_C$.

For the type of states (10.5.147) the decomposition formula holds, and the classical Coulomb energy reads $H_C^{cl} = H_D^{cl} + H_X^{cl}$ with

$$H_D^{cl} = \pi \int d^2p\, V_D^+(\boldsymbol{p})\hat{\rho}^{cl}(-\boldsymbol{p}; B_\parallel)\hat{\rho}^{cl}(\boldsymbol{p}; B_\parallel) + 4\pi \int d^2p\, V_D^-(\boldsymbol{p})\hat{\mathbb{P}}_z^{cl}(-\boldsymbol{p}; B_\parallel)\hat{\mathbb{P}}_z^{cl}(\boldsymbol{p}; B_\parallel),$$
$$(10.6.165a)$$

$$H_X^{cl} = -\pi \int d^2p [V_X^d(\boldsymbol{p})\hat{\mathbb{P}}_a^{cl}(-\boldsymbol{p}; B_\parallel)\hat{\mathbb{P}}_a^{cl}(\boldsymbol{p}; B_\parallel) + 2V_X^-(\boldsymbol{p})\hat{\mathbb{P}}_z^{cl}(-\boldsymbol{p}; B_\parallel)\hat{\mathbb{P}}_z^{cl}(\boldsymbol{p}; B_\parallel)$$
$$+ \frac{1}{4}V_X(\boldsymbol{p})\hat{\rho}^{cl}(-\boldsymbol{p}; B_\parallel)\hat{\rho}^{cl}(\boldsymbol{p}; B_\parallel)], \qquad (10.6.165b)$$

where we have defined the classical fields,

$$\hat{\rho}_\alpha^{cl}(\boldsymbol{q}; B_\parallel) = \langle \mathfrak{S}; B_\parallel | \hat{\rho}_\alpha(\boldsymbol{q}) | \mathfrak{S}; B_\parallel \rangle, \qquad (10.6.166a)$$

$$\hat{\mathbb{P}}_a^{cl}(\boldsymbol{q}; B_\parallel) = \langle \mathfrak{S}; B_\parallel | \hat{P}_a(\boldsymbol{p}) | \mathfrak{S}; B_\parallel \rangle. \qquad (10.6.166b)$$

Note that (10.6.166a) is equal to (10.5.151). On the other hand, while (10.6.166b) is obtained from

$$\hat{\mathbb{P}}_a^{cl}(\boldsymbol{q}; B_\parallel) = \frac{1}{2\pi} \sum_{mn} \langle m | e^{-i\boldsymbol{q}\boldsymbol{X}} | n \rangle \langle \mathfrak{S}; B_\parallel | c_\alpha^\dagger(m) \left( \frac{\sigma_a}{2} \right)_{\alpha\beta} c_\beta(n) | \mathfrak{S}; B_\parallel \rangle. \qquad (10.6.167)$$

We are able to show

$$\hat{\mathbb{P}}_\pm^{cl}(\boldsymbol{q}; B_\parallel) = \hat{\mathbb{P}}_x^{cl}(\boldsymbol{q}; B_\parallel) \pm i\hat{\mathbb{P}}_y^{cl}(\boldsymbol{q}; B_\parallel) = \hat{P}_\pm^{cl}(\boldsymbol{q} \pm \boldsymbol{\delta}_m). \qquad (10.6.168)$$

Hence the classical Coulomb energy at $B_\parallel$ is written in terms of the classical bare densities at $B_\parallel = 0$.

We consider the electron and hole states. Using (10.6.158) we find

$$\hat{\rho}_f^{cl\pm}(\boldsymbol{q}; B_\parallel) = \frac{1}{2}\delta(\ell_B\boldsymbol{q}) \pm \frac{1}{2}\rho_0 e^{+i\ell_B^2 k_y \delta_m} e^{-\ell_B^2 \boldsymbol{q}^2/4}, \qquad (10.6.169a)$$

$$\hat{\rho}_b^{cl\pm}(\boldsymbol{q}; B_\parallel) = \frac{1}{2}\delta(\ell_B\boldsymbol{q}) \pm \frac{1}{2}\rho_0 e^{-i\ell_B^2 k_y \delta_m} e^{-\ell_B^2 \boldsymbol{q}^2/4}, \qquad (10.6.169b)$$

$$\hat{\mathbb{P}}_+^{cl\pm}(\boldsymbol{q}; B_\parallel) = \frac{1}{2}\delta(\ell_B\boldsymbol{q} + \ell_B\boldsymbol{\delta}_m) - \frac{1}{2}\rho_0 e^{-\ell_B^2(\boldsymbol{q}+\delta_m)^2/4}, \qquad (10.6.169c)$$

$$\hat{\mathbb{P}}_-^{cl\pm}(\boldsymbol{q}; B_\parallel) = \frac{1}{2}\delta(\ell_B\boldsymbol{q} - \ell_B\boldsymbol{\delta}_m) - \frac{1}{2}\rho_0 e^{-\ell_B^2(\boldsymbol{q}-\delta_m)^2/4}. \qquad (10.6.169d)$$

Substituting these into the Coulomb energy (10.6.165) we perform some integrations and come to

$$E_{\mathrm{D}}^{\pm} = E_{\mathrm{D}}^{\mathrm{g}} + \frac{1}{4}E_{\mathrm{C}}^0 \int_0^{\infty} \left[1 + e^{-(d/\ell_B)z} J_0(\ell_B \delta_{\mathrm{m}} z)\right] e^{-z^2} dz, \qquad (10.6.170a)$$

$$E_{\mathrm{X}}^{\pm} = E_{\mathrm{X}}^{\mathrm{g}} - \frac{1}{4}E_{\mathrm{C}}^0 \int_0^{\infty} \left[1 + e^{-(d/\ell_B)z} J_0(\ell_B \delta_{\mathrm{m}} z)\right] e^{-z^2} dz$$

$$+ \frac{1}{2}E_{\mathrm{C}}^0 \int_0^{\infty} e^{-\frac{1}{2}z^2 - (d/\ell_B)z} J_0(\ell_B \delta_{\mathrm{m}} z) dz, \qquad (10.6.170b)$$

where the ground-state values are

$$E_{\mathrm{D}}^{\mathrm{g}} = 0,$$

$$E_{\mathrm{X}}^{\mathrm{g}} = \frac{1}{2}\epsilon_{\mathrm{X}} \left[1 - \sqrt{\frac{2}{\pi}} \int_0^{\infty} e^{-\frac{1}{2}z^2 - (d/\ell_B)z} J_0(\ell_B \delta_{\mathrm{m}} z) dz\right] N_{\Phi}, \qquad (10.6.171a)$$

where $N_{\Phi}$ is the number of Landau sites. Adding the tunneling energy, the pair excitation energy is

$$\Delta E_{\mathrm{e\text{-}h}}(\delta_{\mathrm{m}}) = e^{-\frac{1}{4}\ell_B^2 \delta_{\mathrm{m}}^2} \Delta_{\mathrm{SAS}} + E_{\mathrm{C}}^0 \int_0^{\infty} e^{-\frac{1}{2}t^2 - (d/\ell_B)t} J_0(\ell_B \delta_{\mathrm{m}} t) dt. \qquad (10.6.172)$$

The origin of the second term is the exchange interaction. As the parallel magnetic field increases, the distance between two merons increases, as results in the decrease of the exchange energy.

### 10.6.3. *Meron states*

A hole or an electron, which is a baby skyrmion or a baby antiskyrmion, consists of two merons located at $y = \pm a$. There are four types of merons, a hole-like meron in the front or back layer and an electron-like meron in the front or back layer. To construct them we consider a configuration with an excitation only in one of the layers. At $B_{\parallel} = 0$ they read

$$|\mathrm{f}-\rangle = \prod_{n=0}^{\infty} \frac{1}{\sqrt{2}}[c_{\mathrm{f}}(n+1) + c_{\mathrm{b}}(n)]|0\rangle, \qquad (10.6.173a)$$

$$|\mathrm{b}-\rangle = \prod_{n=0}^{\infty} \frac{1}{\sqrt{2}}[c_{\mathrm{f}}(n) + c_{\mathrm{b}}(n+1)]|0\rangle, \qquad (10.6.173b)$$

$$|\mathrm{f}+\rangle = c_{\mathrm{f}}^{\dagger}(0)|\mathrm{f}-\rangle, \qquad (10.6.173c)$$

$$|\mathrm{b}+\rangle = c_{\mathrm{b}}^{\dagger}(0)|\mathrm{b}-\rangle, \qquad (10.6.173d)$$

where $|\mathrm{f}-\rangle$ denotes a hole-like meron in the front layer, and so on. Their electric charges are clearly $\pm\frac{1}{2}e$.

Their classical densities are calculated precisely as in the case of an electron or a hole. They read

$$\langle \text{f}\pm|c_\text{f}^\dagger(m)c_\text{f}(n)|\text{f}\pm\rangle = \frac{1}{2}(\delta_{mn}\pm\delta_{m0}\delta_{n0}), \tag{10.6.174a}$$

$$\langle \text{f}\pm|c_\text{b}^\dagger(m)c_\text{b}(n)|\text{f}\pm\rangle = \frac{1}{2}\delta_{mn}, \tag{10.6.174b}$$

$$\langle \text{f}\pm|c_\text{f}^\dagger(m)c_\text{b}(n)|\text{f}\pm\rangle = \frac{1}{2}\delta_{m-1,n}, \tag{10.6.174c}$$

$$\langle \text{f}\pm|c_\text{b}^\dagger(m)c_\text{f}(n)|\text{f}\pm\rangle = \frac{1}{2}\delta_{m,n-1}, \tag{10.6.174d}$$

and similar ones for $|\text{b}\pm\rangle$. Substituting these into (10.5.151) and (10.5.157), and making the Fourier transformation, we evaluate the physical density for these states at $B_\| \neq 0$,

$$\rho_\text{f}^\text{cl}(\boldsymbol{x};B_\|;\text{f}\pm) = \frac{1}{2}\rho_0 \pm \frac{1}{2}\rho_0 e^{-x^2/2\ell_B^2}e^{-(y+a)^2/2\ell_B^2}, \quad \rho_\text{b}^\text{cl}(\boldsymbol{x};B_\|;\text{f}\pm) = \frac{1}{2}\rho_0 \tag{10.6.175a}$$

$$\rho_\text{f}^\text{cl}(\boldsymbol{x};B_\|;\text{b}\pm) = \frac{1}{2}\rho_0, \quad \rho_\text{b}^\text{cl}(\boldsymbol{x};B_\|;\text{b}\pm) = \frac{1}{2}\rho_0 \pm \frac{1}{2}\rho_0 e^{-x^2/2\ell_B^2}e^{-(y-a)^2/2\ell_B^2}. \tag{10.6.175b}$$

They describe each part of a hole or electron in (10.6.160), as expected.

We may calculate the noncommutative $\text{CP}^1$ field for merons by solving (10.2.51) with (10.6.174).

$$\mathfrak{n}_{\text{f}-} = \frac{1}{\sqrt{2}}\sum_{n=0}^{\infty}\begin{pmatrix}|n+1\rangle\langle n|\\|n\rangle\langle n|\end{pmatrix}, \tag{10.6.176a}$$

$$\mathfrak{n}_{\text{f}+} = \frac{1}{\sqrt{2}}\sum_{n=0}^{\infty}\begin{pmatrix}|n+1\rangle\langle n+1|\\|n\rangle\langle n+1|\end{pmatrix} + \begin{pmatrix}|0\rangle\langle 0|\\0\end{pmatrix}, \tag{10.6.176b}$$

and similar ones for $\mathfrak{n}_{\text{b}\pm}$, where the upper (lower) component is for the front (back) layer. The noncommutative $\text{CP}^1$ fields are the symbols of these Weyl operators. Their asymptotic behaviors are

$$\mathfrak{n}_{\text{f}-} \rightarrow \frac{1}{\sqrt{2}}\begin{pmatrix}e^{i\vartheta}\\1\end{pmatrix}, \quad \mathfrak{n}_{\text{b}-} \rightarrow \frac{1}{\sqrt{2}}\begin{pmatrix}1\\e^{i\vartheta}\end{pmatrix}, \tag{10.6.177a}$$

$$\mathfrak{n}_{\text{f}+} \rightarrow \frac{1}{\sqrt{2}}\begin{pmatrix}e^{-i\vartheta}\\1\end{pmatrix}, \quad \mathfrak{n}_{\text{b}+} \rightarrow \frac{1}{\sqrt{2}}\begin{pmatrix}1\\e^{-i\vartheta}\end{pmatrix}. \tag{10.6.177b}$$

Note that those of the hole and electrons states are given by (10.3.66), which yield

$$n_-(\boldsymbol{x}) \rightarrow \frac{e^{i\vartheta}}{\sqrt{2}}\begin{pmatrix}1\\1\end{pmatrix}, \quad n_+(\boldsymbol{x}) \rightarrow \frac{e^{-i\vartheta}}{\sqrt{2}}\begin{pmatrix}1\\1\end{pmatrix} \tag{10.6.178}$$

in terms of the front-back layer basis.

### 10.6.4. *Large skyrmions*

Microscopic skyrmion states are constructed by making $W_\infty(2)$ transformations of the electron and hole states (10.5.146) as in the case of the monolayer spin system. However, since the analysis is too complicated, here we are satisfied with an effective approach by assuming sufficiently large factorizable skyrmions. Namely we use the classical skyrmion configuration (10.3.89) for the densities. To reproduce the ground-state configuration $\mathcal{P}_x = \frac{1}{2}, \quad \mathcal{P}_y = \mathcal{P}_z = 0$ at $r \to \infty$, we take the skyrmion configuration

$$\mathcal{P}_x^\mp(\boldsymbol{x}) = \frac{1}{2}\frac{r^2 - \lambda^2}{r^2 + \lambda^2}, \quad \mathcal{P}_y^\mp(\boldsymbol{x}) = \pm\frac{\lambda y}{r^2 + \lambda^2}, \quad \mathcal{P}_z^\mp(\boldsymbol{x}) = \frac{\lambda x}{r^2 + \lambda^2},$$

$$\Delta\rho^\mp(\boldsymbol{x}) = \mp\frac{1}{\pi}\frac{\lambda^2}{(r^2 + \lambda^2)^2}, \tag{10.6.179}$$

at $B_\parallel = 0$ for a skyrmion $(-)$ and an antiskyrmion $(+)$, with $\lambda$ an arbitrary dimensional constant. The corresponding $CP^1$ field reads

$$n^-(\boldsymbol{x}) = \frac{1}{\sqrt{(|z|^2\ell_B^2 + \lambda^2)}}\begin{pmatrix} z\ell_B + \lambda \\ z\ell_B - \lambda \end{pmatrix} \tag{10.6.180}$$

for the skyrmion. The antiskyrmion is given by replacing $z$ with $z^*$.

The density configuration at $B_\parallel \neq 0$ is calculated by substituting this for $\hat{\rho}_\alpha^{cl}(\boldsymbol{x})$ and $\hat{P}_\pm^{cl}(\boldsymbol{x})$ in (10.5.151) and (10.5.157), and illustrated in Fig. 10.5. A skyrmion excitation consists of two merons separated by the distance $2a = \ell_B^2\delta_m$. The electron density of the meron located at $(x, y) = (0, a)$ is given by

$$\Delta\rho_{+a}^\mp(\boldsymbol{x}) = \mp\frac{1}{2\pi}\left[\frac{\lambda^2}{(x^2 + (y - a)^2)^2} + \frac{\lambda x}{x^2 + (y - a)^2}\right], \tag{10.6.181a}$$

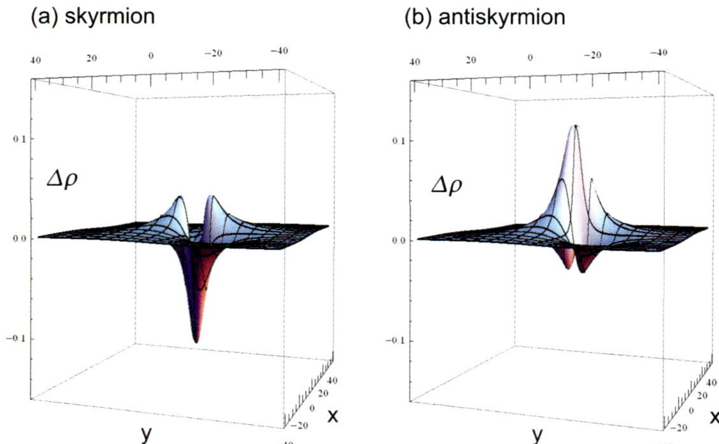


Fig. 10.5. The density of a skyrmion (quasihole) and an antiskyrmion (quasielectron) in the presence of the parallel magnetic field $B_\parallel$. We have taken $\delta_m = 1/\ell_B$ and $\lambda = 4\ell_B$.

while that of the meron located at $(x, y) = (0, -a)$ is given by

$$\Delta \rho_{-a}^{\mp}(\boldsymbol{x}) = \mp \frac{1}{2\pi} \left[ \frac{\lambda^2}{(x^2 + (y+a)^2)^2} - \frac{\lambda x}{x^2 + (y+a)^2} \right]. \qquad (10.6.181b)$$

They merge into a simple skyrmion at $B_\parallel = 0$. By increasing the parallel magnetic field, the separation also increases. Each meron has the electron number $\mp \frac{1}{2}$, and hence the topological number $Q_{\text{sky}} = \pm \frac{1}{2}$ according to the formula (10.3.57).

We may estimate the excitation energy of a skyrmion-antiskyrmion pair as a function of $B_\parallel$. As $B_\parallel$ increases, the distance between two merons increases, as results in the decrease of the exchange energy. This accounts for the anomalous decrease of the activation energy observed experimentally.[12]

In passing we comment on the bimeron picture of a skyrmion excitation. A bimeron has the same quantum numbers as a skyrmion, and it is a deformed skyrmion with two meron excitations. It is found that the meron core exists at $(x, y) = (0, -a)$ in the front layer and at $(0, a)$ in the back layer: See (10.6.160) and (10.6.181) for an explicit example. Originally it was proposed[11,34,35] to explain the activation energy anomaly. According to their mechanism, the bimeron excitation energy consists of the core energy, the string energy and the Coulomb repulsive energy between the two cores. It was argued that the parallel magnetic field decreases the string tension and hence the bimeron activation energy. It is clear that the present skyrmion picture has a close resemblance to but also differences from their bimeron picture. First, a deformation of a skyrmion into two parts occurs only for $B_\parallel \neq 0$ in the present picture but also for $B_\parallel = 0$ in their bimeron picture. Second, the separation between the two parts is determined to be $2a = \ell_B^2 \delta_{\text{m}}$ in the present picture but assumed to be much longer in their bimeron picture. (The origin of this separation is the LLL projection.) The maximum value of the separation $2a$ is given at the phase transition point, and hence

$$2a \leq \ell_B^2 \delta_{\text{m}}^* = \frac{4\ell_B^2}{\pi \lambda_J} \approx \ell_B \qquad (10.6.182)$$

with the use of typical samples parameters.[12] Namely the separation is of the order of the core size. It is hard to say that a bimeron consists of two merons with a string between them. Furthermore, the dominant contribution to the exchange-energy loss comes from the halo region and not from the string region of a bimeron.


## 10.7. Conclusions

We have presented a microscopic theory of skyrmions in QH systems at the filling factor $\nu = 1$. They are topological solitons in the noncommutative plane, where the $x$ and $y$ coordinates are noncommutative. The underlying symmetry is an SU(2) extension of $W_\infty$, which entangles the charge density and the spin (pseudospin) density in the monolayer spin (bilayer pseudospin) system. We have shown that a skyrmion is constructed by dressing a cloud of spins (pseudospin) around an electron or a hole. Consequently, a skyrmion carries the same electric charge as that of an electron or a hole, which implies that its excitation is observed by magnetotransport experiments. Indeed, a charge carrier is a skyrmion rather than an electron or a hole in QH systems. An experimental evidence is a rapid increase (decrease) of the activation energy in the monolayer spin (bilayer pseudospin) system as a function of the tilting angle of the sample against the external magnetic field.

## References


1. T.H.R. Skyrme, *Proc. Roy. Soc. (London)* A **260** (1961) 1271.
2. R.E. Prange and S.M. Girvin (eds), *The Quantum Hall Effect* (Springer, 1990) 2nd edn.
3. M. Stone (ed), *Quantum Hall Effect* (World Scientific, 1992) page 289.
4. T. Chakraborty and P. Pietiläinen, *The Quantum Hall Effects: Fractional and Integral* (Springer, 1995) 2nd edn.
5. Z.F. Ezawa, *Quantum Hall Effects: Field-Theoretical Approach and Related Topics* (World Scientific, 2000; 2008, 2nd edn.)
6. S.L. Sondhi, A. Karlhede, S.A. Kivelson and E.H. Rezayi, *Phys. Rev. B* **47** (1993) 16419.
7. S.E. Barrett, G. Dabbagh, L.N. Pfeiffer, K.W. West and R. Tycko, *Phys. Rev. Lett.* **74** (1995) 5112.
8. A. Schmeller, J.P. Eisenstein, L.N. Pfeiffer and K.W. West, *Phys. Rev. Lett.* **75** (1995) 4290.
9. E.H. Aifer, B.B. Goldberg and D.A. Broido, *Phys. Rev. Lett.* **76** (1996) 680.
10. Z.F. Ezawa and A. Iwazaki, *Int. J. Mod. Phys. B* **6** (1992) 3205.
11. K. Moon, H. Mori, K. Yang, S.M. Girvin, A.H. MacDonald, L. Zheng, D. Yoshioka and S.-C. Zhang, *Phys. Rev. B* **51** (1995) 5138.
12. S.Q. Murphy, J.P. Eisenstein, G.S. Boebinger, L.N. Pfeiffer and K.W. West, *Phys. Rev. Lett.* **72** (1994) 728.
13. A. Sawada, D. Terasawa, N. Kumada, M. Morino, K. Tagashira, Z.F. Ezawa, K. Muraki, T. Saku and Y. Hirayama, *Physica E* **18** (2003) 118.
14. D. Terasawa, M. Morino, K. Nakada, S. Kozumi, A. Sawada, Z.F. Ezawa, N. Kumada, K. Muraki, T. Saku and Y. Hirayama, *Physica E* **22** (2004) 52.
15. Z.F. Ezawa and G. Tsitsishvili, *Phys. Rev. D* **72** (2005) 85002.
16. S.M. Girvin and T. Jach, *Phys. Rev. B* **29** (1984) 5617.
17. S.M. Girvin, A.H. MacDonald and P.M. Platzman, *Phys. Rev. B* **33** (1986) 2481.
18. A. Connes, Noncommutative Geometry (Academic Press, 1994).
19. J.A. Harvey, *Komaba Lectures on Noncommutative Solitons and D-Branes*, hep-th/0102076.
20. J. Moyal, *Proc. Camb. Phil. Soc.* **45** (1949) 99.
21. P. Fletcher, *Phys. Lett. B* **248** (1990) 323.
22. I.A. Strachan, *Phys. Lett. B* **283** (1992) 63.
23. S. Iso, D. Karabali and B. Sakita, *Phys. Lett. B* **296** (1992) 143.
24. A. Cappelli, C. Trugenberger and G. Zemba, *Nucl. Phys. B* **396** (1993) 465.
25. Z.F. Ezawa, *Phys. Lett. A* **229** (1997) 392; *Phys. Rev. B* **55** (1997) 7771.
26. Z.F. Ezawa, G. Tsitsishvili and K. Hasebe, *Phys. Rev. B* **67** (2003) 125314.
27. G. Tsitsishvili and Z.F. Ezawa, *Phys. Rev. B* **72** (2005) 115306.
28. R. Rajaraman, *Solitons and Instantons* (North-Holland, 1982).

29. H.A. Fertig, L. Brey, R. Cote and A.H. MacDonald, *Phys. Rev. B* **50** (1994) 11018.
30. A.H. MacDonald, H.A. Fertig and L. Brey, *Phys. Rev. Lett.* 76 (1996) 2153.
31. M. Abolfath, J.J. Palacios, H.A. Fertig, S.M. Girvin and A.H. MacDonald, *Phys. Rev. B* **56** (1997) 6795.
32. H.A. Fertig, L. Brey, R. Cote, A.H. MacDonald, A. Karlhede and S.L. Sondhi, *Phys. Rev. B* **55** (1997) 10671.
33. Z.F. Ezawa and G. Tsitsishvili, *Phys. Rev. B* **70** (2004) 125304.
34. K. Yang and A.H. MacDonald, *Phys. Rev. B* **51** (1995) 17247.
35. N. Read, *Phys. Rev. B* **52** (1995) 1926.

Chapter 11

# Skyrmions and Merons in Bilayer Quantum Hall System


Kyungsun Moon

*Department of Physics and IPAP, Yonsei University,*
*134 Shinchon, Seoul 120-749 Korea,*
*kmoon@yonsei.ac.kr*



The bilayer two-dimensional electron gas systems can form an unusual broken symmetry state with spontaneous inter-layer phase coherence at certain filling factors. At total filling factor $\nu_T = 1$, the lowest energy charged excitation of the system is proposed to be a meron-pair, which is topologically identical to a single skyrmion. We will review how this remarkable excitation arises theoretically and can help unravel the novel experimental results of bilayer quantum Hall system.


## Contents


11.1 Introduction . . . . . . . . . . . . . . . . . . . . . . . . . . . . . . . . . . . . . . 269
11.2 Lowest Landau Level Systems and Spin Charge Entanglement . . . . . . . . . . . . . . . 271
11.3 Effective Action . . . . . . . . . . . . . . . . . . . . . . . . . . . . . . . . . . . . . 273
    11.3.1 SU(2)-invariant interactions . . . . . . . . . . . . . . . . . . . . . . . . . . . . 274
    11.3.2 Skyrmion excitation . . . . . . . . . . . . . . . . . . . . . . . . . . . . . . . . 277
    11.3.3 Symmetry-breaking interactions . . . . . . . . . . . . . . . . . . . . . . . . . . . 278
    11.3.4 Merons . . . . . . . . . . . . . . . . . . . . . . . . . . . . . . . . . . . . . . . 280
    11.3.5 Inter-layer tunneling . . . . . . . . . . . . . . . . . . . . . . . . . . . . . . . . 282
    11.3.6 Linearly confined meron-pair excitations . . . . . . . . . . . . . . . . . . . . . . 283
11.4 Parallel Magnetic Field . . . . . . . . . . . . . . . . . . . . . . . . . . . . . . . . . 284
    11.4.1 Unbinding of linearly confined meron-pair excitation . . . . . . . . . . . . . . . . 287
11.5 Summary . . . . . . . . . . . . . . . . . . . . . . . . . . . . . . . . . . . . . . . . . 287
References . . . . . . . . . . . . . . . . . . . . . . . . . . . . . . . . . . . . . . . . . 288


## 11.1. Introduction

The study of the low-dimensional strongly correlated quantum and statistical systems has been one of the most formidable and challenging subject in the field of condensed matter physics.[1-5] Since these systems defy a naive perturbative treatment, we often express that the whole is greater than the sum of its parts and look into their novel emergent properties. For recent decades, the two most extensively studied examples of the low-dimensional strongly correlated electron systems are probably high temperature superconductors and quantum Hall effect.

For high  temperature superconductors, one needs to study the strong correlation effect among electrons living in a 2D lattice. As for quantum Hall effect, the correlations among electrons living in a Landau level are investigated. Recently, graphene, a two-dimensional sheet of graphite monolayer, has attracted a lot of attention by providing Dirac fermions living in a 2D plane.

In the strong magnetic field regime, correlations among electrons are especially important, since all electrons can be accommodated within the lowest Landau level and hence the kinetic energy is quenched. The quantum Hall effect occurs when the system has a charge excitation gap.[1,2] The integer quantum Hall effect occurs due to one body gap (Landau level splitting or Zeeman gap) formed by filling up the integer number of Landau levels, while the fractional quantum Hall effect is thought to occur due to the strong correlations among electrons in certain partially filled Landau levels leading to a correlated insulator. The remarkable property of the correlated insulator is the possible emergence of the topological defects upon doping, often carrying a fractional charge. It is generally believed that the Coulomb interaction is not crucial to realize the integer quantum Hall effect. Here we will review the interesting physics of bilayer quantum Hall system with a special focus on the emergence of topological excitations such as skyrmions and merons.[6–10] The bilayer system is composed of two laterally placed quantum wells producing a pair of 2D electron gases as shown schematically in Fig. 11.1.[6,7] The 2D electron gases are separated by a distance $d$ small enough ($d \sim 100$Å) to be comparable to the typical spacing between electrons within the same layer. There have been several theoretical and experimental evidences, which led to apparent charged excitation gaps in bilayer systems at certain Landau level filling factors for sufficiently strong inter-layer interactions.[11–13] For some cases, it has also been suggested that the bilayer system can form an unusual broken symmetry state with spontaneous phase coherence between layers in the absence of inter-layer electron tunneling.[8,9,14–20]

We will concentrate on the bilayer system with total filling factor $\nu_T = 1$ with ($\nu_T \equiv N/N_\phi$ where $N$ is the number of electrons and $N_\phi$ is the number of single-particle levels per Landau level.). At strong magnetic field regime, one can take the Landau level spacing and the Zeeman gap to be large enough to be infinity. Since the kinetic energy of electrons is quenched and electron spins are completely

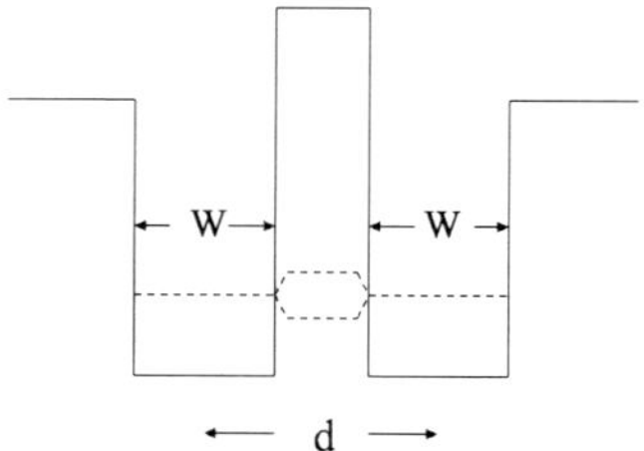


Fig. 11.1.   Schematic diagram of the bilayer quantum Hall system.

frozen, electron-electron interaction $E_c$ and inter-layer tunneling $\Delta_{SAS}$ are the two important factors to control the system. When the layers are widely separated, there will be no correlations between them and quantum Hall effect will not occur, since each layer has $\nu = 1/2$, which we believe to be a Fermi liquid.[21] For smaller separations, it is observed experimentally that there is a large excitation gap on the scale of 20K even when $\Delta_{SAS} \sim 1$K and a quantized Hall plateau.[6,7,22] This indicates that the excitation gap is highly collective in nature.

By using a 'pseudospin' magnetic language in which pseudospin 'up' ('down') refers to an electron in the 'upper' ('lower') layer, we will demonstrate that the bilayer system at total filling factor $\nu_T = 1$ can be viewed as an easy-plane quantum itinerant ferromagnet.[9] The lowest charged excitation of our system is shown to be a linearly confined meron-pair excitation, which carries a charge of $\pm e$. Based on our theory, we will unravel the novel experimental results of the bilayer quantum Hall system. In section 1.2, we review the commutation relations between operators projected to the lowest landau level (LLL) and demonstrate the entanglement of the spin and charge at the LLL. Throughout section 1.3, we will derive the low energy effective action for the bilayer system at total filling factor $\nu_T = 1$. In section 1.3.1, we consider a SU(2) invariant interaction and derive the effective action of a smooth spin texture. In section 1.3.2, a skyrmion is shown to be the lowest charged excitation of the system. In section 1.3.3, we study the pseudospin dependent interaction, which lowers SU(2) symmetry to U(1) symmetry. In section 1.3.4, we show that a skyrmion is fractionalized into a meron-pair bound logarithmically, which is topologically identical to a single skyrmion. In section 1.3.5, we study the effect of electron tunneling between layers, which explicitly breaks U(1) symmetry. In section 1.3.6, we show that the logarithmic interaction between merons changes to the linear confinement. In section 1.4, the role of an additional parallel magnetic field is studied. We demonstrate that parallel magnetic field induces a commensurate-incommensurate transition. In section 1.4.1, the parallel magnetic field is shown to reduce the string tension and the bound meron-pair becomes unbound above $B_{\parallel}^{*}$. Finally, section 1.5 is the summary.

## 11.2. Lowest Landau Level Systems and Spin Charge Entanglement

We will begin with a brief review of single particle states in a strong magnetic field $B$ along z-direction in terms of creation and annihilation operators. In the symmetric gauge, the magnetic field is given by the vector potential $A = (-By/2, Bx/2, 0)$. The magnetic length of the system $\ell = (\hbar c / eB)^{1/2}$ is taken to be unity in case of no confusion. In terms of the complex coordinates $z = x + iy$ and $\bar{z} = x - iy$, one can define the following two sets of oscillators:

$$a = \sqrt{2}(\bar{\partial} + z/4)\,, \ \ a^+ = \sqrt{2}(-\partial + \bar{z}/4)$$
$$b = \sqrt{2}(\partial + \bar{z}/4)\,, \ \ b^+ = \sqrt{2}(-\bar{\partial} + z/4)\,, \tag{11.2.1}$$

where $\partial$ $(\bar{\partial})$ is a holomorphic (anti-holomorphic) derivative satisfying $\partial z = 1$ $(\bar{\partial}\bar{z} = 1)$. The operators satisfy the following commutation relations, $[a, a^\dagger] = 1$, $[b, b^\dagger] = 1$ and otherwise mutually commuting. The Hamiltonian can be described in terms of the first set of oscillators, $a$ and $a^+$,

$$H = \hbar\omega_c \left( a^+ a + \frac{1}{2} \right) \tag{11.2.2}$$

where $\omega_c = (m^* c / eB)$ is the cyclotron frequency. The energy eigenvalues of the Landau levels (LL) are $E_n = \hbar\omega_c(n + 1/2)$ with $n$ being a non-negative integer.

Each LL is degenerate and the degenerate states can be distinguished in terms of the second set of oscillators, $b$ and $b^+$, which describe the guiding center coordinates. Using the angular momentum operator $L = 2(b^+ b - a^+ a)$ which commutes with the Hamiltonian $[L, H] = 0$, one may assign quantum numbers $m$ to the degenerate states: $|n, m\rangle$. Since $L$ satisfies the following algebra

$$[L, b^+] = b^+ , \quad [L, b] = -b \tag{11.2.3}$$

one can raise (lower) the angular momentum using $b^+$ $(b)$ within each LL. The LLL is the set of states with $n = 0$, which is annihilated by applying $a$. The wave functions in the LLL are of the form of $\psi(z) = f(z)e^{-z\bar{z}/4}$ with $f(z)$ a complex analytic function of $z$ alone. The angular momentum eigenstate in the LLL as $|m\rangle \equiv |0, m\rangle = \frac{1}{\sqrt{m!}} b^{\dagger m} |0, 0\rangle$ is given by

$$\langle z|m\rangle \equiv \Phi_m(z) = \frac{z^m}{\sqrt{2\pi 2^m m!}} e^{-z\bar{z}/4}. \tag{11.2.4}$$

This state forms thin shells of radius $\sqrt{2(m + 1)}$ occupying an area of $2\pi\ell^2$ since it is normalized as $\langle z|z\rangle = 1/2\pi$. The orbital degeneracy of a given Landau level is thus $N_\phi = A/(2\pi\ell^2)$, where $A$ is the total area of the system.

The LLL is the projection of the Hilbert space into a subspace. Any operator $\mathcal{O}$ acting on the larger Hilbert space of LL can be similarly projected out so that the projected operator denoted as $\overline{\mathcal{O}}$ acts only on the LLL states. We will put the overbar to represent the projection onto the LLL. The projection onto the LLL can be effectively done by taking a normal ordering of $a$-oscillators and then putting it to be zero. We will take an example of the one-body density operator in momentum space given by

$$\rho_q = \frac{1}{\sqrt{A}} e^{-iq\cdot r} = \frac{1}{\sqrt{A}} e^{-\frac{i}{2}(q^* z + q z^*)} = \frac{1}{\sqrt{A}} e^{-\frac{i}{\sqrt{2}}(q^* b + q b^+)} e^{-\frac{i}{\sqrt{2}}(q^* a + q a^+)} \tag{11.2.5}$$

where $A$ is the total area of the system, and $q = q_x + iq_y$. Hence

$$\bar{\rho}_q = \frac{1}{\sqrt{A}} e^{-\frac{i}{\sqrt{2}}(q^* b + q b^+)} : e^{-\frac{i}{\sqrt{2}}(q^* a + q a^+)} := \frac{1}{\sqrt{A}} e^{-|q|^2/4} \tau_q \tag{11.2.6}$$

where

$$\tau_q = e^{-\frac{i}{\sqrt{2}}(q^* b + q b^+)} \tag{11.2.7}$$

is a magnetic translation operator satisfying the closed Lie algebra

$$[\tau_q, \tau_k] = 2i\,\tau_{q+k}\,\sin\frac{q \wedge k}{2}, \tag{11.2.8}$$

where $q \wedge k \equiv \ell^2(q \times k) \cdot \hat{z}$.

This formalism is readily generalized to the case of many particles with spin. In a system with area $A$ and $N$ particles the projected charge and spin density operators are

$$\bar{\rho}_q = \frac{1}{\sqrt{A}} \sum_{i=1}^{N} \overline{e^{-iq\cdot r_i}} = \frac{1}{\sqrt{A}} \sum_{i=1}^{N} e^{-\frac{|q|^2}{4}}\,\tau_q(i) \tag{11.2.9}$$

$$\bar{S}_q^\mu = \frac{1}{\sqrt{A}} \sum_{i=1}^{N} \overline{e^{-iq\cdot r_i}}\,S_i^\mu = \frac{1}{\sqrt{A}} \sum_{i=1}^{N} e^{-\frac{|q|^2}{4}}\,\tau_q(i)\,S_i^\mu, \tag{11.2.10}$$

where $\tau_q(i)$ is the magnetic translation operator for the $i$th particle and $S_i^\mu$ is the $\mu$th component of the spin operator for the $i$th particle. We immediately find that unlike the unprojected operators, the projected spin and charge density operators do not commute:

$$[\bar{\rho}_k, \bar{S}_q^\mu] = \frac{2i}{\sqrt{A}}\,e^{\frac{|k+q|^2 - |k|^2 - |q|^2}{4}}\,\bar{S}_{k+q}^\mu\,\sin\left(\frac{k \wedge q}{2}\right) \neq 0. \tag{11.2.11}$$

This implies that within the LLL, the dynamics of spin and charge are entangled, i.e., when you rotate spin, charge gets moved.[9]

## 11.3. Effective Action

In this section, we want to show that the bilayer system at total filling factor $\nu_{\rm T} = 1$ can be viewed as an easy-plane quantum itinerant ferromagnet. We will begin with a brief review of the single layer system at $\nu = 1$ in the limit of zero Zeeman gap for the real spins.[23,24] The system can be viewed as a giant atom with electrons occupying exactly a half of the available spin degenerate states in a single orbital of energy $\hbar\omega_c/2$. In the presence of Coulomb repulsion between the particles, Hund's rule would suggest that the system could lower its interaction energy by maximizing its total spin since states with maximum total spin are symmetric under spin exchange and hence the spatial wave function is necessarily fully antisymmetric. In an ordinary ferromagnet the Hund's rule tendency to maximize the total spin is partially counteracted by the increase in kinetic energy (due to the Pauli principle) that accompanies spin polarization. In the lowest Landau level however, the kinetic energy has been quenched by the magnetic field and the system will spontaneously develop 100% polarization. An explicit microscopic wave function believed to exactly describe the ground state of $N$ electrons at $\nu = 1$ is[25]

$$\Psi = \Psi_V |\uparrow\uparrow\uparrow\uparrow\uparrow\uparrow\uparrow\uparrow\uparrow\uparrow\uparrow\uparrow \cdots \uparrow\rangle, \tag{11.3.12}$$

where $\Psi_V$ is a Vandermonde determinant wave function[11] of the form

$$\Psi_V \equiv \prod_{i<j} (z_i - z_j) \prod_k \exp(-|z_k|^2/4). \qquad (11.3.13)$$

The first term in Eq. (11.3.12) is simply the Laughlin spatial wave function for the filled Landau level and the second term indicates that every spin is up. This state has total spin $S = N/2$ and $S^z\Psi = (N/2)\Psi$. Because Coulomb interactions are spin independent, $[H, S^\mu] = 0$ and $\Psi$ is simply one of a total of $2S+1$ degenerate states, all with $S = N/2$. The other states are simply created using the total spin lowering operator $S^- \equiv \sum_{j=1}^{N} s_j^-$ which is itself fully symmetric under spin exchange.

### 11.3.1. *SU(2)-invariant interactions*

We will use a 'pseudospin' magnetic language in which pseudospin 'up' ('down') refers to an electron in the 'upper' ('lower') layer.[26] When $d/\ell = 0$, the Coulomb interaction between a pair of electrons is pseudospin independent. Although this limit of $d/\ell = 0$ can not be realized experimentally, the real bilayer systems of relatively small $d/\ell$ are adiabatically connected to the SU(2) invariant case of $d/\ell = 0$ with only the symmetry being reduced. From now on we will ignore 'real' spin, since it is frozen out by the Zeeman gap at strong magnetic fields. For bilayer system, the 'real' spin degrees of freedom are replaced with pseudospins. Following an analogy to the single layer system at $\nu = 1$ with zero Zeeman gap, the ground state at $\nu_{\rm T} = 1$ without inter-layer tunneling has its pseudospin fully polarized spontaneously, which minimizes the interaction energy. We expect that the low-lying excited states can be described by spin textures in which the local spin alignment varies slowly with position. To be explicit, we define the following as a spin texture state:

$$|\tilde{\psi}[m(r)]\rangle = e^{-i\bar{O}} |\psi_0\rangle. \qquad (11.3.14)$$

Here $|\psi_0\rangle$ is the $S^z = N/2$ member of the ground state spin-multiplet given in Eq. (11.3.12) and the operator $O$ is a non-uniform spin rotation which reorients the local spin direction from $\hat{z}$ to $m(r)$ ($m$ is a unit vector). We limit ourselves to small tilts away from the $\hat{z}$ direction so that

$$O = \sum_{j=1}^{N} \Omega(r_j) \cdot S_j \equiv \sum_q e^{\frac{|q|^2}{4}} \Omega_q^\mu S_{-q}^\mu, \qquad (11.3.15)$$

where $\Omega(r) = \hat{z} \times m(r)$ is the angle over which a spin is rotated. [Note that $\Omega^z(r) \equiv 0$, $\Omega^x(r) = -m^y(r)$, and $\Omega^y(r) = m^x(r)$]. Our final result requires only that $\Omega$ is slowly varying in space and not that $\Omega$ is small. Projecting $O$ onto the LLL ensures that $|\tilde{\psi}\rangle$ has no projection on higher Landau levels as required in the strong perpendicular magnetic field limit.

We want to calculate the effective action of a smooth spin texture for SU(2)-invariant electron-electron interactions. The considerations in this section apply to a single-layer with zero Zeeman energy as well. We assume, for the sake of convenience, that the spins are almost aligned in the $\hat{z}$ direction, and that they vary slowly in space, i.e., $\Omega_q$ is negligible when $q\ell \geq 1$. The interaction, after projection onto the LLL, is

$$\overline{V} = \frac{1}{2} \sum_q V_q \left( \overline{\rho}_q \, \overline{\rho}_{-q} - N e^{-\frac{q^2}{2}} \right), \tag{11.3.16}$$

where $V_k = \int d^2r \, V(r) e^{-ik \cdot r}$. The expectation value of the energy of spin texture subtracting the ground state energy can be obtained by the following formula

$$\delta E = \langle \psi_0 | e^{i\overline{O}} \, [\overline{V}, e^{-i\overline{O}}] | \psi_0 \rangle. \tag{11.3.17}$$

By expanding $\delta E$ in powers of $\overline{O}$, one can obtain the leading non-vanishing contribution to the energy from the second-order term

$$\delta E = \frac{\rho_s^0}{2} \int d^2r \, [(\nabla \Omega^x)^2 + (\nabla \Omega^y)^2] = \frac{\rho_s^0}{2} \int d^2r (\nabla m)^2 . \tag{11.3.18}$$

where we used the relation $\Omega(r) = \hat{z} \times m(r)$ and assumed $q\ell \ll 1$. The spin stiffness $\rho_s^0$, implicitly defined above, is related to the pair correlation function of $|\psi_0\rangle$ by

$$\rho_s^0 = \frac{-1}{32\pi^2} \int dk k^3 V_k h(k) \tag{11.3.19}$$

where the pair-correlation function $h(k)$ is known analytically for $\nu = 1$: $h(k) = -\exp(-|k|^2/2)$. The physical origin of the stiffness is the loss of exchange and correlation energy when the spin orientation varies with position. For the Coulomb interaction, $\rho_s^0 = e^2/(16\sqrt{2\pi}\epsilon\ell) \sim (e^2/\epsilon\ell)2.49 \times 10^{-2}$.[23] The classical model defined by Eq. (11.3.18) is called the O(3) non-linear sigma model and has been studied in great detail.[27] We note in passing that for the SU(2) invariant case, the spin stiffness $\rho_s$ found here is exact.

Following a detailed analysis of the higher order terms, one can see that the next important contribution comes from the fourth order term for the case of Coulomb interaction

$$E^{(4)} = \frac{1}{2} \iint d^2r d^2r' \, V(r - r') \delta\rho(r) \delta\rho(r'), \tag{11.3.20}$$

where $\delta\rho = \frac{1}{8\pi} \epsilon_{ij} (\partial_i m \times \partial_j m) \cdot m$ is the topological charge density. We will demonstrate below that $\delta\rho$ represents the excess charge density in a given spin texture and hence $E^{(4)}$ stands for the direct Coulomb energy for density fluctuations. Since the projection to the LLL entangles spin and charge, one can easily imagine that spin texture can induce a local charge density. The excess charge density $\delta\rho_k$ in a spin texture can be explicitly calculated by

$$\delta\rho_k = \langle \psi_0 | e^{i\overline{O}} \left[ \overline{\rho}_k , e^{-i\overline{O}} \right] | \psi_0 \rangle. \tag{11.3.21}$$

By expanding in powers of $\bar{O}$ for a smooth spin texture, we obtain the local density of the spin texture

$$\delta\rho(r) = -\frac{1}{8\pi}\,\epsilon_{\mu\nu}\,m(r)\cdot[\partial_\mu m(r)\times\partial_\nu m(r)] \qquad (11.3.22)$$

which is exactly the Pontryagin index density, or topological charge density.[23,28] Thus we have shown that for spin-states with $S = N/2$, the physical charge density is the topological charge density in the long wavelength limit. This remarkable result was first obtained by Sondhi *et al.* within the context of a Chern-Simons effective field theory description of spin textures.[23] The present derivation gives a microscopic proof of their result. The total extra charge carried by the spin texture is exactly the Pontryagian index:

$$\triangle N = -\frac{1}{8\pi}\int d^2r\,\epsilon_{\mu\nu}\,m(r)\cdot[\partial_\mu m(r)\times\partial_\nu m(r)]. \qquad (11.3.23)$$

For the case of homotopy group of $S^2 \to S^2$, the Pontryagian index can only take an integer value because it is the number of times a unit sphere is wrapped around by the order parameter, i.e., it is the winding number of the spin texture.[28]

So far we have calculated the most relevant terms of the static energy of a spin texture. The dynamics can be obtained by studying the equation of motion. The quantum equation of motion is

$$\frac{dm_q^\mu}{dt} = 4\pi\left\langle\frac{dS_q^\mu}{dt}\right\rangle = \frac{4\pi}{\hbar}\frac{\delta}{\delta\Omega_{-q}^\mu}E[m], \qquad (11.3.24)$$

where $E[m]$ is the energy functional of the spin texture,

$$E[m] = \frac{\rho_s^0}{2}\int d^2r(\nabla m)^2 + \frac{1}{2}\iint d^2rd^2r'\,V(r-r')\delta\rho(r)\delta\rho(r'). \qquad (11.3.25)$$

If we include only the leading gradient term in the energy functional, an approximation which is always valid for sufficiently slowly varying spin-textures, we obtain

$$\frac{dm_q}{dt} = \frac{4\pi\rho_s^0 q^2}{\hbar}\hat{z}\times m_q \qquad (11.3.26)$$

The equation of motion has spin-wave solutions in which the magnetization precesses around the $\hat{z}$ direction with wavevector $q$ and frequency $\hbar\omega = 4\pi\rho_s^0 q^2$. This is precisely the energy of the long-wavelength spin-waves of the system.[29,30] This equation of motion immediately leads to the following effective Lagrangian:

$$L = \frac{1}{4\pi}\int d^2r A[m(r)]\cdot\partial_t m(r) - E[m], \qquad (11.3.27)$$

where $A$ is the vector potential of a unit magnetic monopole[28,31] in the spin space; i.e., $\nabla_m\times A = m$. The first term simply contributes to the action a geometric phase proportional to the solid angle traced out by the spin vector during its motion. This is exactly the Berry's phase for the spin and appears at the adiabatic level as expected.[32]

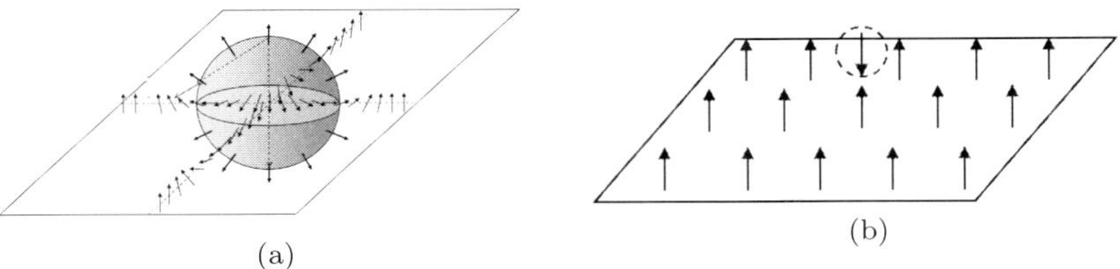

Fig. 11.2.   Charged excitation for SU(2) invariant case. (a) Spin profile of skyrmion excitation and stereographic projection onto $S^2$ sphere. (b) Single particle excitation.

### 11.3.2. *Skyrmion excitation*

The effective gradient energy of bilayer system at $d/\ell = 0$ is given by the O(3) non-linear sigma model. It is well known that the model possesses a topological excitation called a skyrmion. For a single layer system at $\nu = 1$, Sondhi *et. al.* have shown that skyrmions are the lowest energy charged excitations of the system carrying a charge $\pm e$.[23,33,34] Hence for bilayer system in a SU(2) invariant case, the skyrmions made of pseudospins are the lowest charged excitations. This is not surprising because for the skyrmion spin configuration, the spins are nearly parallel locally as shown in Fig. 11.2(a), so the exchange energy is only slightly reduced. In contrast, for ordinary single particle excitations [see Fig. 11.2(b)], an added electron has its spin opposite to the others and has no exchange energy. As pointed out earlier, in the SU(2) invariant limit we know the *exact* spin stiffness. Hence the exact energy of a single (large scale) skyrmion can be obtained[23,27]:

$$E_s = 4\pi\rho_s. \tag{11.3.28}$$

For the case of a system with Coulomb interactions at $\nu = 1$, we obtain from the non-linear sigma model energy expression

$$E_s = \frac{1}{4}\sqrt{\frac{\pi}{2}\frac{e^2}{\epsilon\ell}}, \tag{11.3.29}$$

where $2E_s$ is only a half of the ordinary single particle-hole pair excitation energy.[23] We can write down simple microscopic variational wave functions for the skyrmion. Consider the following state in the plane

$$\psi_\lambda = \prod_m \begin{pmatrix} z_m \\ \lambda \end{pmatrix}_m \Psi_V, \tag{11.3.30}$$

where $\Psi_V$ is defined in Eq. (11.3.13), $()_m$ refers to the spinor for the $m$th particle, and $\lambda$ is a fixed length scale. This is a skyrmion because it has its spin purely down at the origin (where $z_m = 0$) and has spin purely up at infinity (where $z_m \gg \lambda$). The parameter $\lambda$ is simply the size scale of the skyrmion.[23,27] Notice that in the limit $\lambda \longrightarrow 0$ (where the continuum effective action is invalid, but this microscopic wave function is still sensible) we recover a fully spin polarized filled Landau level with a charge-1 Laughlin quasihole at the origin. Hence the number of flipped spins interpolates continuously from zero to infinity as $\lambda$ increases.

In order to analyze the skyrmion wave function in Eq. (11.3.30), we use the Laughlin plasma analogy. In this analogy the norm of $\psi_\lambda$, $Tr_{\{\sigma\}} \int D[z] |\Psi[z]|^2$ is viewed as the partition function of a Coulomb gas. In order to compute the density distribution we simply need to take a trace over the spin

$$Z = \int \mathcal{D}[z]\, e^{2\left\{\sum_{i>j} \log|z_i-z_j| + \frac{1}{2}\sum_k \log(|z_k|^2+\lambda^2) - \frac{1}{4}\sum_k |z_k|^2\right\}}. \tag{11.3.31}$$

This partition function describes the usual logarithmically interacting Coulomb gas with uniform background charge plus a spatially varying impurity background charge $\Delta\rho_b(r)$,

$$\Delta\rho_b(r) \equiv -\frac{1}{2\pi}\nabla^2 V(r) = -\frac{\lambda^2}{\pi(r^2+\lambda^2)^2}, \tag{11.3.32}$$

$$V(r) = \frac{1}{2}\log(r^2+\lambda^2). \tag{11.3.33}$$

For large enough scale size $\lambda \gg \ell$, local neutrality of the plasma[19] implies that the excess electron number density is precisely $\Delta\rho_b(r)$, so that Eq. (11.3.33) is in agreement with the standard result for the topological density.[27] Finally, we note that by replacing $\left(\begin{smallmatrix} z \\ \lambda \end{smallmatrix}\right)$ with $\left(\begin{smallmatrix} z^n \\ \lambda^n \end{smallmatrix}\right)$, we can generate a skyrmion with a Pontryagin index $n$.

### 11.3.3. *Symmetry-breaking interactions*

For bilayer system with finite layer separation $d/\ell \neq 0$, the electron-electron interaction strengths will depend on whether two electrons belong to the same layer or to opposite layers. We define

$$V^0 \equiv \frac{1}{2}\left(V_k^A + V_k^E\right) \tag{11.3.34}$$

$$V_k^z \equiv \frac{1}{2}\left(V_k^A - V_k^E\right) \tag{11.3.35}$$

where $V_k^A$ is the Fourier transform with respect to the planar coordinate of the (intra-layer) interaction potential between a pair of electrons in the same layer and $V_k^E$ is Fourier transform of the (inter-layer) interaction potential between a pair of electrons in opposite layers. If we neglect the finite thickness of the layers, $V_k^A = 2\pi e^2/k$ and $V_k^E = \exp(-kd)V_k^A$. The interaction Hamiltonian can then be separated into a pseudospin-independent part with interaction $V^0$ and a pseudospin-dependent part. The pseudospin dependent term in the Hamiltonian is

$$\overline{V}_{\text{sb}} = 2\sum_k V_k^z \overline{S}_k^z \overline{S}_{-k}^z. \tag{11.3.36}$$

Since $V_k^A > V_k^E$, this term produces an easy-plane rather than an Ising anisotropy. The pseudospin symmetry of the Hamiltonian is reduced from $SU(2)$ to $U(1)$ by this term.

In order to calculate the expectation value of the pseudospin-dependent interaction, it is convenient to take the ground state $|\psi_0\rangle$ to be spin polarized along the $\hat{x}$ direction. In the limit of slowly varying spin-textures we obtain the following result for the contribution of the symmetry-breaking term to the energy of the spin-texture

$$E_{\text{sb}}[m] \simeq \int d^2r \left\{ \beta_m (m^z)^2 + \frac{\rho_s^z}{2}(\nabla m^z)^2 - \frac{\rho_s^z}{2}\left[(\nabla m^x)^2 + (\nabla m^y)^2\right] \right\}, \quad (11.3.37)$$

where

$$\rho_s^z = \frac{-1}{32\pi^2} \int_0^\infty dk V^z(k) h(k) k^3, \quad (11.3.38)$$

and the total mass term $\beta_m$ is given by

$$\beta_m = \frac{-1}{8\pi^2} \int_0^\infty dk \left[V^z(0) - V^z(k)\right] k\, h(k), \quad (11.3.39)$$

which represents the capacitive charging energy. We see immediately from this term that the symmetry-breaking interactions favor equal population of the two layers, or in pseudospin language they favor spin-textures where the pseudospin-orientation is in the $\hat{x} - \hat{y}$ plane.

Including both SU(2) invariant contribution defined in Eq. (11.3.18) and the symmetry breaking contributions, the total energy-functional for a spin-texture is given by

$$E_{\text{tot}}[m] \simeq \int d^2r \left\{ \beta_m (m^z)^2 + \frac{\rho_A}{2}(\nabla m^z)^2 + \frac{\rho_E}{2}\left[(\nabla m^x)^2 + (\nabla m^y)^2\right] \right\} \quad (11.3.40)$$

where

$$\rho_A = \frac{-1}{32\pi^2} \int_0^\infty dk V_k^A h(k) k^3, \quad (11.3.41)$$

and

$$\rho_E = \frac{-1}{32\pi^2} \int_0^\infty dk V_k^E h(k) k^3. \quad (11.3.42)$$

The term proportional to $(\nabla m^z)^2$ in the energy density captures the reduction of the exchange-correlation energy from within each layer when the density in the layer is not constant and therefore $\rho_A = \rho_s^0$ is dependent only on the intra-layer interaction. On the other hand, pseudospin-order in the $\hat{x} - \hat{y}$ plane represents interlayer phase coherence. An inter-layer phase relationship which changes as a function of position results in a loss of inter-layer correlation energy so that $\rho_E$ depends only on inter-layer interactions.

For the following discussion, we will neglect the tunneling of electrons between layers. The effective energy functional in Eq. (11.3.40) can be minimized by forcing pseudospins to lie in the $\hat{x} - \hat{y}$ spin space ($m_z = 0$). Neglecting the quantum spin fluctuations, the corresponding spin texture can be well represented by taking $m = (\cos\varphi, \sin\varphi, 0)$. Then the gradient energy of the spin texture can be reduced to

$$E(\varphi) = \frac{1}{2}\rho_s \int d^2r |\nabla\varphi|^2. \tag{11.3.43}$$

This Hamiltonian defines an effective 2D XY model which will contain vortex excitations interacting logarithmically. In a thin film of superfluid $^4$He, vortices interact logarithmically because of the energy cost of supercurrents circulating around the the vortex centers. (In superconducting thin films the same logarithmic interaction appears but is cut off on length scales exceeding the penetration depth.) Here the same logarithmic interaction appears. Microscopically this interaction is due to the potential energy cost (loss of exchange) associated with the phase gradients (circulating pseudospin currents). Hartree-Fock estimates[9] indicate that $\rho_s$ and hence the Kosterlitz-Thouless (KT) critical temperature $T_{\mathrm{KT}}$ are on the scale of 1K in typical samples.

### 11.3.4. *Merons*

We notice that the $\hat{z}$ component of the order parameter becomes massive and the system has $U(1)$ symmetry for $d/\ell \neq 0$. In this case, there is another class of topologically stable charged objects, merons.[35,36] A meron carries an electronic charge of $\pm e/2$. As shown in Fig. 11.3(a), far away from the core of a meron the order parameter lies in the (massless) XY plane and forms a vortex configuration with $\pm$ vorticity, while inside the core region the order parameter smoothly rotates either up or down out of the XY plane to avoid singularity. In contrast, for a vortex in superconductor, the singularity is avoided by making a normal core region and so the magnitude of the order parameter vanishes at the core center. There exist four flavors of merons. The energy of a single meron diverges logarithmically with the system size with a coefficient proportional to the inter-layer spin stiffness. The interaction between merons has a contribution from the stiffness energy which is also logarithmic, attractive for opposite vorticity pairs and repulsive for same vorticity

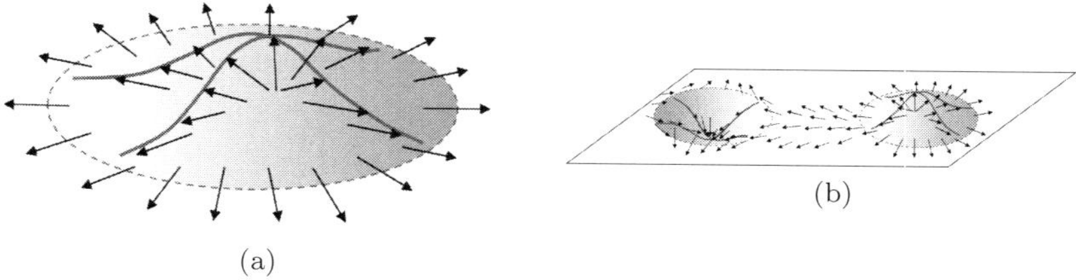

(a)

(b)

Fig. 11.3.   Charged excitation for U(1) symmetry.   (a) Meron excitation.   (b) Meron-pair excitation.

pairs. These properties are exactly the same as the vortices in the classical XY model. In order to determine the sign of the charge carried by a meron, one has to specify both its vorticity and the spin configuration in the core region. Merons will also have a long range $1/r$ interaction due to their charges which is attractive for oppositely charged merons and repulsive for like-charged merons.

The fact that merons carry topological charge one half can be seen by the following argument. Imagine a vortex in the spin system. If an electron circles the vortex at a large distance, its spin rotates through $2\pi$. This induces a Berry's phase of $\exp(i2\pi S) = -1$ which is equivalent to that induced by a charge moving around one-half of a flux quantum. Since $\sigma_{xy} = e^2/h$, the vortex picks up charge $1/2$. The topological charge of a meron can also be understood by considering a variational function for the meron spin texture:

$$m = \left\{ \sqrt{1 - (m^z(r))^2} \cos\varphi , \sqrt{1 - (m^z(r))^2} \sin\varphi , m^z(r) \right\}. \tag{11.3.44}$$

The topological charge $Q$ of a meron calculated from $\delta\rho = -\frac{1}{8\pi} \int d^2r \, \epsilon_{ij}(\partial_i m \times \partial_j m) \cdot m$ is given by

$$Q = \frac{1}{2}\left[m^z(\infty) - m^z(0)\right]. \tag{11.3.45}$$

For a meron, the spin points up or down at the core center and tilts away from the $\hat{z}$ direction as the distance from the core center increases. Asymptotically it points purely radially in the $\hat{x} - \hat{y}$ plane. Thus the topological charge is $\pm\frac{1}{2}$ depending on the polarity of core spin. The variational function mentioned above corresponds to a vortex with positive vorticity. In order to make a vortex with negative vorticity (anti-vortex), we need to apply the space-inversion operation to the vortex solution. Since topological charge is a pseudo-scalar quantity, it is odd with respect to parity. Hence the general result for the topological charge of the four meron flavors may be summarized by the following formula:

$$Q = \frac{1}{2}\left[m^z(\infty) - m^z(0)\right] n_{\mathrm{v}}, \tag{11.3.46}$$

where $n_{\mathrm{v}}$ is the vortex winding number.

It is well known that 2D XY model exhibits the KT phase transition at critical temperature $T_{\mathrm{KT}} = (\pi/2)\rho_s$. The KT transition can be understood as the vortex-anti-vortex unbinding transition. It seems likely that under appropriate circumstances the lowest energy charged excitations of the system will consist of a bound pair of merons. (A skyrmion can be viewed as a closely bound pair of merons with the same charge and opposite vorticity and a meron can be viewed as half a skyrmion as demonstrated in Fig. 11.3(b).) They are somewhat analogous to Laughlin quasiparticles, however they differ considerably in that, below $T_{\mathrm{KT}}$, they are confined together in vorticity neutral pairs by their logarithmic interaction. The cheapest object with a net charge is then a vortex-antivortex pair, with each vortex carrying charge $+\frac{1}{2}$ (or $-\frac{1}{2}$) for a total charge of $+1$ (or $-1$). The charge excitation

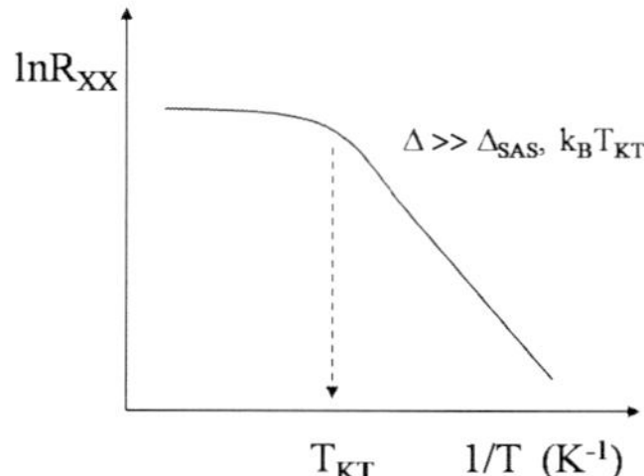


Fig. 11.4.  Typical Arrhenius plot of longitudinal resistance $R_{\mathrm{xx}}$: The slope indicates $\Delta/2k_B$ of the lowest charged excitation.

cost can be estimated by minimizing

$$E_{\mathrm{pair}} = 2E_{\mathrm{mc}} + \frac{e^2}{4\epsilon R} + 2\pi\rho_s \ln\left[\frac{R}{R_{\mathrm{mc}}}\right], \tag{11.3.47}$$

where $E_{\mathrm{mc}}$ is the meron core energy[37] , and $R_{\mathrm{mc}}$ is the meron core size. The optimal separation is given by[9] $R_0 = e^2/(8\pi\epsilon\rho_s)$. The typical value of $E_{\mathrm{pair}}$ is estimated to be on the order of $10K$.

In Fig. 11.4, typical experimental behaviors of the Arrhenius plots of longitudinal resistance $R_{\mathrm{xx}}$ are schematically shown, which exhibit a thermally activated dissipation.[6] The low temperature activation energy $\Delta$ is, as already noted, much larger than $\Delta_{SAS}$. If $\Delta$ were nevertheless somehow a single-particle gap, one would expect the Arrhenius law to be valid up to temperatures of order $\Delta$. Instead one observes a rather abrupt leveling off in the dissipation as the temperature increases past values as low as $\sim 0.1\Delta$. In our theoretical picture, the low temperature activation energy $\Delta$ can be interpreted as a creation energy of a charge neutral excitation composed of two meron-pairs $2E_{\mathrm{pair}}$ due to the charge conservation. Above $T_{\mathrm{KT}}$, the gap collapses due to the proliferation of free merons. We want to emphasize that a small but finite inter-layer tunneling induces a rapid cross-overs rather than true phase transitions because the phase $\varphi$ is compact.

### 11.3.5. *Inter-layer tunneling*

In this section, we will investigate the effect of electron tunneling between layers.[10] Since the pseudospins are restricted to lie in the $\hat{x} - \hat{y}$ spin space to minimize the mass term, it is quite convenient to use the following variational wave function

$$|\psi\rangle = \prod_X \left\{c^\dagger_{X\uparrow} + e^{i\varphi}c^\dagger_{X\downarrow}\right\}|0\rangle, \tag{11.3.48}$$

where $X$ is a state label (for instance, the Landau gauge orbital guiding center[9]). The interpretation of this wave function is that every Landau orbital $X$ is occupied (hence $\nu_{\mathrm{T}} = 1$), but the system is in a coherent linear combination of pseudospin up and down states determined by the phase angle $\varphi$. This means that the system has a definite total number of particles ($\nu_{\rm T} = 1$ exactly) but an indefinite number of particles in each layer. In the absence of inter-layer tunneling, the particle number in each layer is a good quantum number. Hence this state has a spontaneously broken symmetry[9,14,15,17] in the same sense that the BCS state for a superconductor has indefinite (total) particle number but a definite phase relationship between states of different particle number. The tunneling Hamiltonian can be written by

$$H_{\rm T} = -t \int d^2r \left\{ \psi_\uparrow^\dagger(r)\psi_\downarrow(r) + \psi_\downarrow^\dagger(r)\psi_\uparrow(r) \right\} \tag{11.3.49}$$

which can be written in the spin representation as

$$H_{\rm T} = -2t \int d^2r\, S_x(r). \tag{11.3.50}$$

(Recall that the eigenstates of $S_x$ are symmetric and antisymmetric combinations of up and down.) A finite tunneling amplitude $t$ between the layers explicitly breaks the U(1) symmetry and the expectation value of the tunneling energy is given by

$$H_{\rm eff} = \langle \psi | H_{\rm T} | \psi \rangle = -\frac{t}{2\pi\ell^2} \int d^2r \cos\varphi \tag{11.3.51}$$

by giving a preference to symmetric tunneling states. By adding the gradient energy, the total Hamiltonian of the system is given by

$$H_{\rm eff} = \int d^2r \left[ \frac{1}{2}\rho_s |\nabla\varphi|^2 - \frac{t}{2\pi\ell^2} \cos\varphi \right]. \tag{11.3.52}$$

### 11.3.6. *Linearly confined meron-pair excitations*

The introduction of finite tunneling amplitude explicitly destroys the U(1) symmetry and makes the simple vortex-pair configuration extremely expensive. To lower the energy the system distorts the spin deviations into a domain wall or 'string' connecting the vortex cores as shown in Fig. 11.5. The spins are oriented in the $\hat{x}$ direction everywhere except in the domain line region where they tumble rapidly through $2\pi$. The domain line has a fixed energy per unit length and so the vortices are now confined by a linear potential corresponding to a fixed 'string tension'

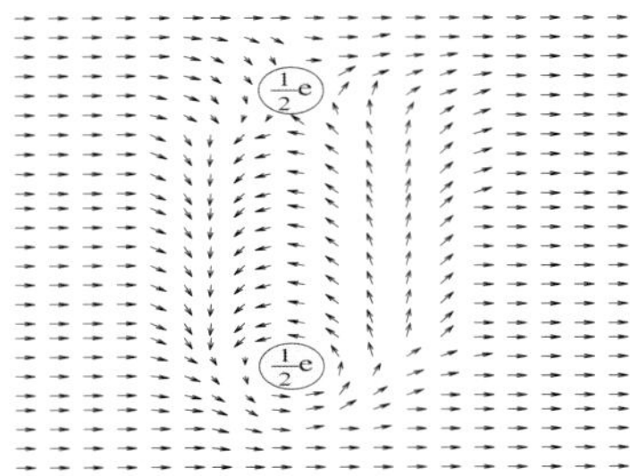

Fig. 11.5.   Meron-pair excitation bound by linear string tension.

rather than being confined only logarithmically. We can estimate the string tension by examining the energy of a domain line of infinite length. The optimal form for a domain line lying along the $y$ axis is given by

$$\varphi(r) = \pm 4 \arctan[\exp(x/\xi)], \tag{11.3.53}$$

where the characteristic width of the string is

$$\xi = \left[\frac{2\pi \ell^2 \rho_s}{t}\right]^{\frac{1}{2}}. \tag{11.3.54}$$

The resulting string tension is[10]

$$T_0 = 8 \left[\frac{t\rho_s}{2\pi \ell^2}\right]^{\frac{1}{2}} = \frac{8\rho_s}{\xi}, \tag{11.3.55}$$

which is independent of the sign of $\varphi(r)$. Provided the string is long enough ($R \gg \xi$), the total energy of a segment of length $R$ will be well-approximated by the expression

$$E'_{\text{pair}} = 2E'_{\text{mc}} + \frac{e^2}{4\epsilon R} + T_0 R. \tag{11.3.56}$$

The prime on $E_{mc}$ in Eq. (11.3.56) indicates that the meron core energy can depend on $\Delta_{SAS}$. $E'_{\text{pair}}$ is minimized at $R = R'_0 \equiv \sqrt{e^2/4\epsilon T_0}$ where it has the value

$$E^*_{pair} = 2E'_{\text{mc}} + \sqrt{e^2 T_0/\epsilon}. \tag{11.3.57}$$

Note that apart from the core energies, the charge gap at fixed layer separation (and hence fixed $\rho_s$) is $\propto T_0^{1/2} \propto t^{1/4} \sim \Delta_{SAS}^{1/4}$, which contrasts with the case of free electrons, for which the charge gap is linearly proportional to $\Delta_{SAS}$. Note that because the exponent $1/4$ is so small, there is an extremely rapid initial increase in the charge gap as tunneling is turned on. Here we want to reiterate that the Hamiltonian in Eq. (11.3.52) is qualitatively different from the sine-Gordon Hamiltonian, since the phase $\varphi$ is a compact variable. Thus no KT transition will occur.

## 11.4. Parallel Magnetic Field

It has been experimentally demonstrated[6,7] that the charge gap in bilayer systems is remarkably sensitive to the application of relatively weak magnetic fields $B_\parallel$, oriented in the plane of the 2D electron gas, as shown schematically in Fig. 11.6(a). Experimentally this field component is generated by slightly tilting the sample relative to the magnetic field orientation. Tilting the field (or sample) has traditionally been an effective method for identifying effects due to (real) spins because orbital motion in a single-layer 2DEG system is primarily[38] sensitive to $B_\perp$, while the (real) spin Zeeman splitting is proportional to the full magnitude of $B$. Adding a parallel field component will tend to favor more strongly spin-polarized states.

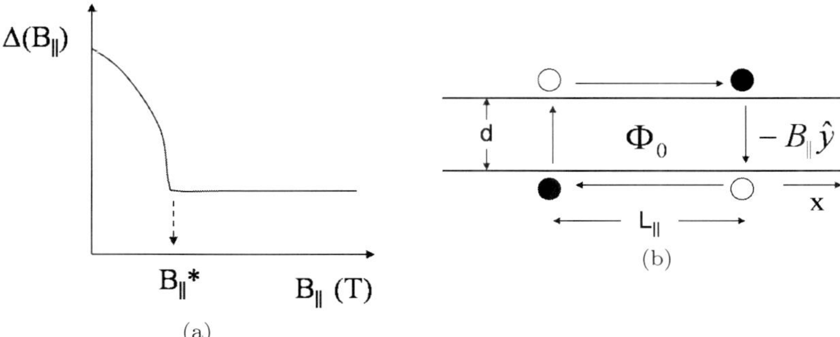


Fig. 11.6.   Inter-layer electron tunneling in the presence of $B_{||}$.   (a) The $B_{||}$ dependence of activation energy. (b) Schematic diagram for inter-layer tunneling.

For the case of the bilayer $\nu_T = 1$ systems,[6] we have assumed that the real spins are fully polarized, which is quite consistent with the experiments. Hence the addition of a parallel field would not, at first glance, be expected to influence the low energy states since they are already fully spin-polarized. (At a fixed Landau level filling factor $B_\perp$ is fixed and so both the total $B$ and the corresponding Zeeman energy increase with tilt). Nevertheless experiments[6] have shown that these systems are very sensitive to $B_{||}$. The activation energy drops rapidly (by factors varying from two up to an order-of-magnitude in different samples) with increasing $B_{||}$. At $B_{||} = B_{||}^*$ there appears to be a phase transition to a new state whose activation gap is approximately independent of further increases in $B_{||}$. We will study the effect of $B_{||}$ on the pseudospin system using the following gauge in which $B_{||} = \nabla \times A_{||}$ where $A_{||} = B_{||}(0, 0, x)$. In this gauge the vector potential points in the $\hat{z}$ direction (perpendicular to the layers) and varies with position $x$ as one moves parallel to the layers. As an electron tunnels from one layer to the other it moves along the direction in which the vector potential points and so the tunneling matrix element acquires a position-dependent phase $t \to t\, e^{iQx}$ where $Q = 2\pi/L_{||}$ and $L_{||} = \Phi_0/B_{||}d$ is the length associated with one flux quantum $\Phi_0$ between the layers [defined in Fig. 11.6(b)]. This modifies the tunneling Hamiltonian to $H_T = -\int d^2r\, h(r) \cdot S(r)$ where $h(r)$ 'tumbles': i.e., $h(r) = 2t\,(\cos Qx, \sin Qx, 0)$. The effective XY model now becomes

$$H = \int d^2r \left\{ \frac{1}{2}\, \rho_s |\nabla \varphi|^2 - \frac{t}{2\pi \ell^2}\, \cos\left[\varphi(r) - Qx\right] \right\}, \qquad (11.4.58)$$

which is precisely the Pokrovsky-Talapov (P-T) model[39] and has a very rich phase diagram. For small $Q$ and/or small $\rho_s$ the phase obeys (at low temperatures) $\varphi(r) \equiv Qx$; the moment rotates commensurately with the pseudospin Zeeman field. However, as $B_{|}$ is increased, the local field tumbles too rapidly and a continuous phase transition to an incommensurate state with broken translation symmetry occurs. This is because at large $B_\parallel$ it costs too much exchange energy to remain commensurate and the system rapidly gives up the tunneling energy in order to return to a uniform state $\nabla\varphi \approx 0$ which becomes independent of $B_\parallel$.

The effect of $B_\parallel$ is most easily studied by changing variables to

$$\theta(r) \equiv \varphi(r) - Qx. \tag{11.4.59}$$

In terms of this new variable, the P-T model energy is

$$H = \int d^2r \left\{ \frac{1}{2} \rho_s[(\partial_x\theta + Q)^2 + (\partial_y\theta)^2] - \frac{t}{2\pi\ell^2} \cos\theta \right\}. \tag{11.4.60}$$

Since the extra term induced by $Q$ represents a total derivative, the optimal form of the soliton solution is unchanged. The differential equation for the extremal solution of Eq. (11.4.60) is given by

$$\frac{\partial^2\theta}{\partial x^2} - \frac{1}{\xi^2} \sin\theta = 0. \tag{11.4.61}$$

The trivial solution of Eq. (11.4.61) is that of $\theta = 0$, which corresponds to the commensurate phase. Previously we argued that for large $B_\parallel$, the system will move to a uniform phase to minimize the exchange energy by giving up the tunneling energy. In reality, the system wisely selects to make a nontrivial soliton solution (domain wall) given by $\theta(x) = \pm 4\arctan[\exp(x/\xi)]$, which takes advantage of both tunneling and exchange energy. We calculate the energy difference $\delta E$ of the commensurate state and the domain wall solution

$$\delta E = E_{\text{dom}} - E_{\text{com}} = T_0 \left[ 1 - \frac{B_\parallel}{B_\parallel^*} \right] L_y, \tag{11.4.62}$$

where $T_0$ is the tension in the absence of parallel B field given by Eq. (11.3.55) and $L_y$ is the dimension of the system along the $\hat{y}$ direction. We see that $B_\parallel$ defines a preferred direction in the problem. The phase transition occurs at zero temperature for

$$B_\parallel^* = B_\perp \, (2\ell/\pi d)(2t/\pi\rho_s)^{1/2}, \tag{11.4.63}$$

which is the critical parallel field at which the string tension goes to zero.[10] One can notice that it is energetically favorable to have a domain wall condensation above $B_\parallel^*$.

Using the experimental parameters of the relevant samples and neglecting quantum fluctuation[40] renormalizations of both $t$ and $\rho_s$ we find that the critical field for the transition is $\approx 1.6$T which is within a factor of two of the observed value.[6] Note that the observed value $B_\parallel^* = 0.8$T corresponds in these samples to a large value for $L_\parallel$: $L_\parallel/\ell \sim 20$ indicating that the transition is highly collective in nature. We emphasize again that these very large length scales are possible in a magnetic field only because of the inter-layer phase coherence in the system associated with condensation of a neutral object.

### 11.4.1. *Unbinding of linearly confined meron-pair excitation*

Having argued for the existence of the commensurate-incommensurate transition, we must now connect it to the experimentally observed transport properties. In the commensurate phase, the order parameter tumbles more and more rapidly as $B_\parallel$ increases. As we shall see below, it is this tumbling which causes the charge gap to drop rapidly. In the incommensurate phase the state of the system is approximately independent of $B_\parallel$ and this causes the charge excitation gap to saturate at a fixed value. Recall that in the presence of tunneling, the cheapest charged excitation was found to be a pair of vortices of opposite vorticity and like charge (each having charge $\pm 1/2$) connected by a domain line with a constant string tension. In the absence of $B_\parallel$ the energy is independent of the orientation of the string. Following Eq. (11.4.62), the energy per unit length of the soliton, which is the domain line string tension, decreases linearly with $Q$ and hence $B_\parallel$[9,41]:

$$T = T_0 \left[ 1 - \frac{B_\parallel}{B_\parallel^*} \right], \qquad (11.4.64)$$

Recall that the charge excitation gap is given by the energy of a vortex pair separated by the optimal distance $R_0 = \sqrt{e^2/4\epsilon T}$. From Eq. (11.3.56) we have that the energy gap for the commensurate state of the phase transition is given by

$$\Delta = 2E'_{\mathrm{mc}} + \left[ e^2 T/\epsilon \right]^{\frac{1}{2}}$$
$$= \Delta_0 + \sqrt{e^2 T_0/\epsilon} \left[ 1 - \left( \frac{B_\parallel}{B_\parallel^*} \right) \right]^{\frac{1}{2}}. \qquad (11.4.65)$$

As $B_\parallel$ increases, the reduced string tension allows the Coulomb repulsion of the two vortices to stretch the string and lower the energy. Far on the incommensurate side of the phase transition the possibility of inter-layer tunneling becomes irrelevant. From the discussion of the previous section it follows that the ratio of the charge gap at $B_\parallel = 0$ to the charge gap at $B_\parallel \to \infty$ should be given approximately by

$$\frac{\Delta_0}{\Delta_\infty} = (t/t_{\mathrm{cr}})^{1/4} \approx (e^2/\epsilon\ell)^{1/2} t^{1/4} 8\rho_s^{3/4}. \qquad (11.4.66)$$

Putting in typical values of $t$ and $\rho_s$ gives gap ratios $\sim 1.5 - 7$ in agreement with experiment.

### 11.5. **Summary**

We have studied the bilayer quantum Hall system at total filling factor $\nu_{\mathrm{T}} = 1$ with the layer separation $d$ comparable to the mean particle spacing within the same layer. Several experiments have suggested that the system exhibits a novel quantum Hall effect due to a strong inter-layer electron-electron correlations. By using a pseudospin language based on layer indices, we have derived the low energy effective action of a smooth spin texture state. We have demonstrated that the bilayer system can be viewed as an easy-plane quantum itinerant ferromagnet. Based on the above action, we have proposed that the lowest charged excitation of our system is the linearly confined meron-pair with opposite vorticity, which is topologically identical to a single skyrmion. It is quite amusing to notice that spin textures actually carry true electronic charges due to the spin charge entanglement at the LLL. A meron is known to carry a half of electron charge $\pm e/2$. Hence a meron-pair can be charge neutral or carry a unit of electron charge. Upon applying an additional magnetic field parallel to the 2D plane, the system exhibits a sharp decrease of an activation energy with increasing $B_\parallel$, which gets flattened above $B_\parallel^*$. We have explained the experimental result by analyzing the dependence of the activation energy of meron-pair on $B_\parallel$. As $B_\parallel$ increases, the string tension between meron-pair becomes reduced linearly and vanishes above $B_\parallel^*$. The decrease of string tension makes the meron-pair further apart and thus dramatically reduces the activation energy of a meron-pair. Above $B_\parallel^*$, a meron-pair is deconfined and the activation energy becomes insensitive to the parallel magneic field.

## Acknowledgments

In the work, I have given a review of the results of a valuable collaboration with my colleagues, S. Girvin, A. MacDonald, H. Mori, Kun Yang, Lotfi Belkhir, L. Zheng, D. Yoshioka, and Shou-Cheng Zhang. This work was supported by the Korea Research Foundation Grant funded by the Korean Government(MOEHRD) through Grant No. KRF-2008-313-C00243.

## References

1. *The Quantum Hall Effect* ed. by R.E. Prange and S.M. Girvin (Springer, New York, 1990).
2. *Perspectives in Quantum Hall Effects* ed. by S. Das Sarma and A. Pinczuk (Wiley, New York, 1997).
3. K.S. Novoselov, A.K. Geim, S.V. Morozov, D. Jiang, M.I. Katsnelson, I.V. Grigorieva, S.V. Dubonos and A.A. Firsov, *Nature* **438** (2005) 197.
4. Yuanbo Zhang, Yan-Wen Tan, Horst L. Stormer and Philip Kim, *Nature* **438** (2005) 201.
5. K.S. Novoselov, E. McCann, S.V. Morozov, V.I. Falko, M.I. Katsnelson, U. Zeitler, D. Jiang, F. Schedin and A.K. Geim, *Nature Physics* **2** (2006) 177.
6. S.Q. Murphy, J.P. Eisenstein, G.S. Boebinger, L.N. Pfeiffer and K.W. West, *Phys. Rev. Lett.* **72** (1994) 728.
7. M.B. Santos, L.W. Engel, S.W. Hwang and M. Shayegan, *Phys. Rev. B* **44** (1991) 5947; T.S. Lay, Y.W. Suen, H.C. Manoharan, X. Ying, M.B. Santos and M. Shayegan, *Phys. Rev. B* **50** (1994) 17725.

8. Kun Yang, K. Moon, L. Zheng, A.H. MacDonald, S.M. Girvin, D. Yoshioka and Shou-Cheng Zhang, *Phys. Rev. Lett.* **72** (1994) 732.

9. K. Moon, H. Mori, Kun Yang, S.M. Girvin, A.H. MacDonald, L. Zheng, D. Yoshioka and Shou-Cheng Zhang, *Phys. Rev. B* **51** (1995) 5138.

10. Kun Yang, K. Moon, Lofti Belkhir, H. Mori, S.M. Girvin, A.H. MacDonald, L. Zheng and D. Yoshioka, *Phys. Rev. B* **54** (1996) 11644.

11. B. I. Halperin, *Helv. Phys. Acta* **56** (1983) 75.

12. T. Chakraborty and P. Pietiläinen, *Phys. Rev. Lett.* **59** (1987) 2784; E.H. Rezayi and F.D.M. Haldane, *Bull. Am. Phys. Soc.* **32** (1987) 892; Song He, S. Das Sarma and X.C. Xie, *Phys. Rev. B* **47** (1993) 4394; D. Yoshioka, A.H. MacDonald and S.M. Girvin, *Phys. Rev. B* **39** (1989) 1932.

13. Y.W. Suen *et al.*, *Phys. Rev. Lett.* **68** (1992) 1379; J.P. Eisenstein *et al.*, *Phys. Rev. Lett.* **68** (1992) 1383.

14. X.G. Wen and A. Zee, *Phys. Rev. Lett.* **69** (1992) 1811; X.G. Wen and A. Zee, *Phys. Rev. B* **47** (1993) 2265.

15. Z.F. Ezawa and A. Iwazaki, *Int. J. of Mod. Phys. B* **19** (1992) 3205; Z.F. Ezawa and A. Iwazaki, *Phys. Rev. B* **47** (1993) 7295; Z.F. Ezawa and A. Iwazaki, *Phys. Rev. B* **48** (1993) 15189.

16. A.H. MacDonald, P.M. Platzman and G.S. Boebinger, *Phys. Rev. Lett.* **65** (1990) 775.

17. Luis Brey, *Phys. Rev. Lett.* **65** (1990) 903; H.A. Fertig, *Phys. Rev. B* **40** (1989) 1087; O. Narikiyo and D. Yoshioka, *J. Phys. Soc. Jpn.* **62** (1993) 1612.

18. R. Côté, L. Brey and A.H. MacDonald, *Phys. Rev. B* **46** (1992) 10239; X.M. Chen and J.J. Quinn, *Phys. Rev. B* **45** (1992) 11054.

19. Tin-Lun Ho, *Phys. Rev. Lett.* **73** (1994) 874.

20. Ana Lopez and Eduardo Fradkin, *Phys. Rev. B* **51** (1995) 4347.

21. B.I. Halperin, Patrick A. Lee and Nicholas Read, *Phys. Rev. B* **47** (1993) 7312 and work cited therein; V. Kalmeyer and S.C. Zhang, *Phys. Rev. B* **46** (1992) 9889.

22. G.S. Boebinger, H.W. Jiang, L.N. Pfeiffer and K.W. West, *Phys. Rev. Lett.* **64** (1990) 1793; G.S. Boebinger, L.N. Pfeiffer and K.W. West, *Phys. Rev. B* **45** (1992) 11391.

23. S.L. Sondhi, A. Karlhede, S.A. Kivelson and E.H. Rezayi, *Phys. Rev. B* **47** (1993) 16419.

24. D.-H. Lee and C.L. Kane, *Phys. Rev. Lett.* **64** (1990) 1313.

25. See for example, D. Yoshioka, A.H. MacDonald and S.M. Girvin, *Phys. Rev. B* **38** (1988) 3636.

26. S. Datta, *Phys. Lett.* **103**A (1984) 381.

27. R. Rajaraman, *Solitons and Instantons*, North Holland, Amsterdam (1982).

28. Eduardo Fradkin, *Field Theories of Condensed Matter Systems*, Addison-Wesley (1990).

29. Mark Rasolt, F. Perrot and A.H. MacDonald, *Phys. Rev. Lett.* **55** (1985) 433; Mark Rasolt and A.H. MacDonald, *Phys. Rev. B* **34** (1986) 5530; M. Rasolt, B.I. Halperin and D. Vanderbilt, *Phys. Rev. Lett.* **57** (1986) 126.

30. C. Kallin and B.I. Halperin, *Phys. Rev. B* **31** (1985) 3635.

31. F.D.M. Haldane, *Phys. Lett.* **93**A (1983) 464; *Phys. Rev. Lett.* **50** (1983) 1153.

32. Michael Stone, *Phys. Rev. D* **33** (1986) 1191.

33. A.H. MacDonald, H.A. Fertig and L. Brey, *Phys. Rev. Lett.* **76** (1996) 2153.

34. K. Moon and K. Mullen, *Phys. Rev. B* **57** (1998) 14833.

35. D.J. Gross, *Nucl. Phys. B* **132** (1978) 439.

36. Ian Affleck, *Phys. Rev. Lett.* **56** (1986) 408.

37. Kun Yang and A.H. MacDonald, *Phys. Rev. B* **51** (1995) 17247.

38. J.D. Nickila, Ph.D. thesis, Indiana University, 1991.

39. Per Bak, *Rep. Prog. Phys.* **45** (1982) 587; *Marcel den Nijs in Phase Transitions and Critical Phenomena* **12**, ed. by C. Domb and J.L. Lebowitz (Academic Press, New York, 1988), pp. 219–333.
40. K. Moon, *Phys. Rev. Lett.* **78** (1997) 3741.
41. K. Moon and K. Mullen, *Phys. Rev. B* **57** (1998) 1378.

# Chapter 12

# Spin and Pseudospin Textures in Quantum Hall Systems

H.A. Fertig[*] and L. Brey[†]

*Department of Physics, Indiana University, Bloomington,
Indiana 47405, USA
†Instituto de Ciencia de Materiales de Madrid (CSIC), Cantoblanco,
Madrid 28049, Spain

## Contents

12.1 Introduction . . . . . . . . . . . . . . . . . . . . . . . . . . . . . . . . . . . . . . 291
12.2 Microscopic Theory for Skyrmions . . . . . . . . . . . . . . . . . . . . . . . 293
12.3 The Skyrme Crystal . . . . . . . . . . . . . . . . . . . . . . . . . . . . . . . . . 294
12.4 Collective Modes and Quantum Fluctuations . . . . . . . . . . . . . . . . . 296
12.5 The Bilayer Quantum Hall System . . . . . . . . . . . . . . . . . . . . . . . 299
    12.5.1 Two-dimensional superfluidity and Josephson physics . . . . . . . . . . 302
    12.5.2 The coherence network model . . . . . . . . . . . . . . . . . . . . . . . . 303
    12.5.3 Effect of interlayer bias . . . . . . . . . . . . . . . . . . . . . . . . . . . . 306
12.6 Conclusion . . . . . . . . . . . . . . . . . . . . . . . . . . . . . . . . . . . . . . . 308
References . . . . . . . . . . . . . . . . . . . . . . . . . . . . . . . . . . . . . . . . . . 308

## 12.1. Introduction

In strong perpendicular magnetic fields, a two-dimensional electron system (2DES) may exhibit the quantum Hall effect (QHE).[1] This occurs when the filling factor, defined as $\nu = N/N_\phi$, where $N$ is the total number of electrons in the 2DES and $N_\phi$ the total number of magnetic flux quanta penetrating the plane, is either close to an integer or a rational fraction (usually with odd denominator) $\nu_0$. In the vicinity of $\nu_0$, the Hall conductivity of the system is quantized at $\sigma_{xy} = \nu_0 \frac{e^2}{\hbar}$, and the diagonal conductance $\sigma_{xx}$ vanishes in the same range of filling factors. The explanation of this remarkable phenomenon ultimately resides in the presence of an energy gap between excited states and the groundstate of the system. For the integer quantum Hall effect, this gap is associated with Landau quantization of the kinetic energy, whereas for the fractional quantum Hall effect, it arises due to many-body correlations in the groundstate induced by electron-electron interactions.

Another important energy scale in this problem arises because of the electron spin. In spite of the large magnetic fields required to induce the QHE, the Zeeman splitting of the electrons is surprisingly small compared both to the kinetic and interaction energy scales of the electrons. This occurs because the effective $g$ factors for the electrons in their semiconductor environments is remarkably small. The inclusion of the spin degree of freedom in the dynamics of the electrons introduces a rich set of phenomena. In the QHE the incompressible ground state at filling factor $\nu=1$ can be a strong ferromagnet, and its total spin quantum number $S$ can equal $N/2$, so that the electronic spins are completely aligned by a small Zeeman coupling. Because of this behavior, this system has come to be known as a quantum Hall ferromagnet (QHF). In a first approximation one may describe the spin density in terms of a Heisenberg ferromagnet, and employ a non-linear sigma model (NLSM) to describe its low-energy physics.[2,3] An immediate consequence of this is that the system should support skrymions as excitations from the groundstate.

As first noticed in numerical exact diagonalization[4,5] and demonstrated experimentally,[6,7] the spin polarization in these systems is strongly reduced away from filling factor unity, where the ground state must incorporate the charge excitations of the $\nu=1$ QHF. This turns out to reflect two unique features of skyrmions in the quantum Hall system. First, the skyrmions carry electrical charge as a consequence of their topological charge, and hence they have a stable finite size for small but nonzero Zeeman coupling.[3,8,9] Quantitative estimates of the skyrmions quasiparticle energies[3,8,9] indicate, for filling factor unity, that are lower in energy than spin-polarized qusiparticles. Second, skyrmions are present in the ground state near $\nu=1$, and have an obvious influence on observable properties. Skyrmions may be injected or removed from the ground state of the 2DES, by adjusting the filling factor slightly away from $\nu=1$. This explains the reduction of the magnetization in the QHF when injecting these quasiparticles into the ground state.[6,7]

Closely related effects to those occurring in the QHF also occur in double layer systems, in which the layer index is analogous to spin.[10,11] Here we assume that the Zeeman energy is large enough that fluctuations of the real spin can be ignored, and we consider only the lowest electric subband in each quantum well. Thus we have a two state system that can be labeled by a pseudospin 1/2 degree of freedom — for example, pseudospin up can be identified with locating the electron in the top layer, and pseudospin down places the electron in the bottom layer. Theory has predicted[10,12] that at some filling factors, energy gaps which are needed to support the quantum Hall effect occur in double layer systems only if the interlayer interaction is sufficiently strong. When this is the case, the interlayer interaction can also lead to unusual broken symmetry states with a novel kind of spontaneous phase coherence between the layers, even when there is no tunneling between the layers. As we will discuss later in this chapter, the spontaneous phase coherence is responsible for a variety of novel features seen experimentally.

## 12.2. Microscopic Theory for Skyrmions

Near filling factor $\nu=1$, it is possible to quantitatively compute properties of the skyrmions using a Hartree-Fock approximation.[8,9] Because of the symmetry of the Skyrmion charged excitations, it is convenient to work in the symmetric gauge where single-particle wave functions in the lowest Landau level have the form

$$\phi_m(z) = \frac{z^m \exp(-|z|^2/4\ell^2)}{(2^{m+1}\pi\ell^2 m!)^{1/2}}. \tag{12.2.1}$$

Here $m=0,1,...$ is the angular momentum, $z = x + iy$ expresses the two dimensional coordinate as a complex number and $\ell_0 = \sqrt{\hbar c/eB}$ is the magnetic length.

A Hartree-Fock version of the quasihole skyrmion state may be written down in second quantized form as

$$|\Psi> = \prod_{m=0}^{\infty} (u_m a_m^+ + v_m b_{m+1}^+)|0>, \tag{12.2.2}$$

where $|0>$ is the particle vacuum and $|u_m|^2 + |v_m|^2=1$, so that the wave function is normalized. Here $a_m^+$ creates a down-spin electron and $b_m^+$ creates an up-spin electron in the $m$th angular momentum state. [The quasielectron (anti-)skyrmion may be generated from this state using particle-hole symmetry.[9]]

The form of this wave function is essentially dictated by the symmetry of the classical skyrmion solutions which are invariant under the action of $L_z \pm S_z$ for the skyrmion. It is also easy to demonstrate that the expectation value of the total spin operator in this state describes a spin texture with unit topological charge, provided that $u_m$ varies slowly with $m$ from $u_{m=0} = 1$ to $u_{m\to\infty} = 0$. Far from origin, this state is locally identical to a ferromagnetic ground state, and all spins are aligned with the Zeeman magnetic field which is assumed to point in the "up" direction. Near the origin the projection of the total spin along the field direction becomes negative.

The parameters $u_m$ and $v_m$ are essentially variational parameters, and as usual one chooses them to minimize the energy of the state.[8] The size of the skyrmion is determined by a competition[3] between the Coulomb interaction (which favors large skyrmions, to spread out the excess charge of the quasiparticle) and the Zeeman coupling (which favors small skyrmions, to minimize the number of spins pointing in the minority direction.) The skyrmion size may be characterized by the number of overturned spins (relative to the spin-polarized quasiparticle),

$$K = \sum_{m=0}^{\infty} |u_m|^2. \tag{12.2.3}$$

Figure 12.1 illustrates the values of $K$ computed using the Hartree-Fock approach as a function of the Zeeman splitting $\tilde{g} = g\mu_B B\kappa\ell_0/e^2$, where $\mu_B$ is the Bohr magneton, $g$ the Landé g-factor, $B$ the magnetic field and $\kappa$ the dielectric constant of the host semiconductor. For typical experiments, $\tilde{g} \sim 0.015$, which yields $K \sim 3-5$.

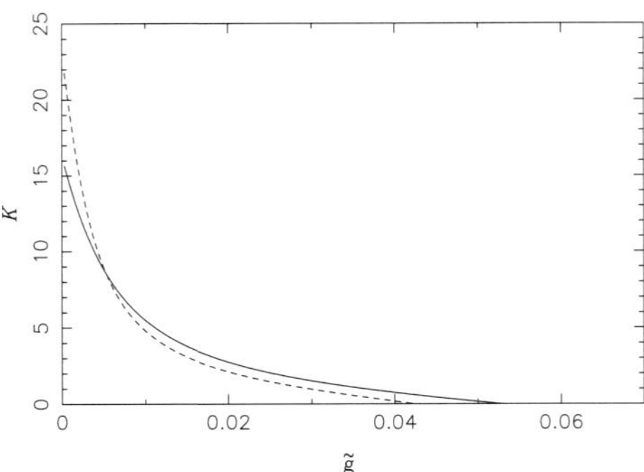

Fig. 12.1.   Number of flipped spins $K$ in a skyrmion as a function of Zeeman splitting $\tilde{g}$. Solid line is for a purely two-dimensional electron system; dotted line is includes a finite thickness correction.

Remarkably, this is very close to the value found experimentally[7,13] for the number of overturned spins per quasiparticle when real quantum Hall systems are doped away from $\nu = 1$.

## 12.3.  The Skyrme Crystal

That $K$ excess spins are flipped whenever a quasielectron or quasihole is injected into the groundstate by varying $\nu$ manifests itself as a rapid decay in the spin polarization of the 2DES as one moves away from $\nu = 1$. This is illustrated in Fig. 12.2, which reproduces the experimentally measured spin polarization $P$ (open and closed circles) of a 2DES using an NMR approach.[7] While the value of $K$ for single skyrmions reproduces the initial slope of $P$ as one moves away from $\nu = 1$ (as opposed to the expectations for spin-polarized quasiparticles, shown as the dashed-dotted line), to understand how $P$ develops as $|\nu - 1|$ increases one needs to understand how skyrmions interact with one another.

At filling factors near but not precisely at $\nu=1$, a finite density of skyrmions exists in the 2DEG. Since the skyrmions are localized charged objects, at low densities they form a crystal, with lattice parameter proportional to $|1 - \nu|^{-1/2}$. At very low densities and finite Zeeman coupling the resulting lattice is triangular, in order to minimize the Coulomb repulsive energy. However, because the charge density of a single skyrmion is tied to the spin density, this will be spread out about the skyrmion center. The Coulomb repulsion among the skyrmions thus tends to shrink the individual skyrmions, and for high enough density one expects them to collapse into spin-polarized quasiparticles. The spin polarization of such a periodic state may also be computed using a Hartree-Fock approach,[14] and the result is illustrated as the dotted line in Fig. 12.2. Surprisingly this collapse occurs quite close to $\nu = 1$, much closer than is seen in experiment. To understand why this happens, it is helpful to examine the spin texture associated with the Skyrme lattice state.

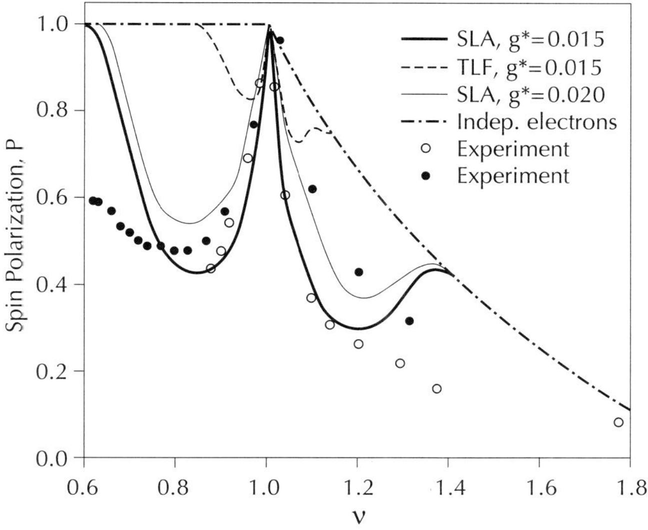


Fig. 12.2. Spin polarization of the 2DES as a function of filling factor. Filled and open circles represent experimental data for two different values of Zeeman coupling, $\tilde{g} = g^* \approx 0.015$ and 0.020. Dashed-dotted line represents the expected spin polarization for non-interacting electrons (equivalent to spin-polarized quasiparticles), dashed line is for a triangular lattice of skyrmions, solid lines are for square lattices of skyrmions. Reproduced from L. Brey *et al.* Phys. Rev. Lett. **75**, 2562 (1995).

Figure 12.3(a) illustrates this for the triangular lattice, where the in-plane $(x - y)$ component of the spin density is shown. An apparent property of the spin texture for this lattice symmetry is that the spins must rotate rapidly along nearest neighbor bonds. This property is unavoidable for the triangular lattice: if each skyrmion is identical, then the spin density must rotate in this way. From the NLSM point of view this might be expected to be a state with high energy, since large gradients in the spin are associated with large energy densities.

However, the triangular lattice structure with one skyrmion per unit cell does not allow the system to take advantage of an important degree of freedom available to the skyrmions. Skyrmion energies are unaffected by global rotations around the $z$ axis of the electron spin, as should be expected for a Heisenberg ferromagnet. From the Hartree-Fock point of view, this is manifested in Eq. (12.2.2) by the fact that the energy of the state is independent of the relative phase of the $u_m$'s and the $v_m$'s (provided this phase is the same for all values of $m$.) Indeed, if $u_m v_m^* = |u_m v_m| e^{i\theta}$, then the angle between the in-plane spin density and a radial vector from the center of the skyrmion is just $\theta$. It is thus possible to relieve the large spin gradient between nearest neighbors if they are not identical, but rather have a relative phase of $\pi$ between them. On the triangular lattice this is not possible due to the inherent frustration of the lattice. However, a square lattice with two skyrmions per unit

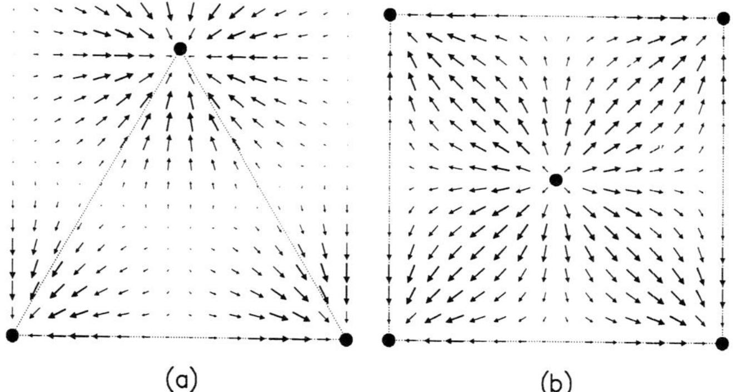

Fig. 12.3.   Two-dimensional vector representation of the $x-y$ components of the spin density of a crystal of skyrmions as obtained using the Hartree-Fock approximation. (a) Triangular lattice with one skyrmion per unit cell and $\tilde{g}$=0.015.  (b) Square lattice with two skyrmions mutually rotated per unit cell and $\tilde{g}$=0.015.  Reproduced from L. Brey *et al.* Phys. Rev. Lett. **75**, 2562 (1995).

cell, differing by $\pi$ in their relative phase, can have a smooth spin density between nearest neighbors. This is illustrated in Fig. 12.3(b). Remarkably, it is found that the square lattice state is lower in energy over a very broad range of filling factors away from $\nu = 1$. We find that that the square skyrmion lattice has lower energy than the triangular skyrmion lattice except at very small $|\nu - 1|$. As may be seen in Fig. 12.2, the square lattice structure gives a spin polarization that agrees with experiment over a broad range of filling factors.

## 12.4.  Collective Modes and Quantum Fluctuations

The structure illustrated in Fig. 12.3(b) is highly reminiscent of a two-dimensional *XY* anti*ferromagnet, where the phase angle $\theta$ between the in-plane spin density and the radial vectors from each skyrmion center play the roles of the effective *XY* spin degree of freedom for each site. Since the system has a new broken symmetry (spin rotations in the $x-y$ plane), we expect the system to support a new Goldstone mode, analogous to spin waves in the *XY* antiferromagnet. This can be demonstrated by using the Hartree-Fock state as a basis for a time-dependent Hartree-Fock analysis of the system.[15] This allows the computation of various response functions, whose poles appear at the collective modes of the system. The results of such a calculation are illustrated in Fig. 12.4, which shows the values of $\omega$ where poles of the density-density response function $\chi_{nn}(\vec{k},\vec{k};\omega)$ and the spin response functions $\chi_{zz}(\vec{k},\vec{k};\omega)$ and $\chi_{+-}(\vec{k},\vec{k};\omega)$ appear. (Here the subscript $z$ refers to the response of the spin density operator $S_z$, and $\pm$ refers to $S_x \pm iS_y$.) For small values of $k$, one may see two modes dispersing from $\omega = 0$. One is the usual phonon mode of a crystal of charged particles in two-dimensions, dispersing as $k^{3/2}$. The second disperses

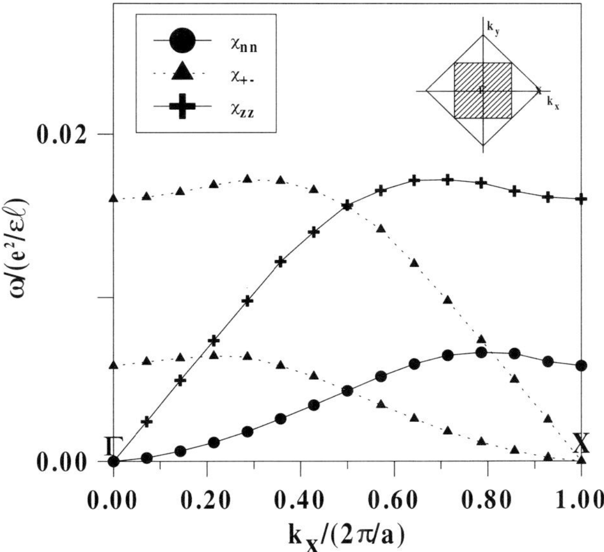



Fig. 12.4. Collective mode energies in $e^2/(\varepsilon\ell)$ units ($\varepsilon \equiv \kappa$ is the dielectric constant) for $\tilde{g} = 0.016$ and $\nu = 1.10$. The wavevector runs from the origin to a primitive reciprocal vector of the magnetic lattice with the mid-point on the edge of the magnetic BZ. At each $k$ a mode is labeled by the response function component with the largest in $\chi(k,k)$ residue. Reproduced from R. Côté *et al.* Phys. Rev. Lett. **78**, 4825 (1997).

linearly, and is the direct analog of spin waves in an $XY$ antiferromagnet. As indicated in the figure, the poles appear principally in the density response function for the phonons and in the spin response functions for the other modes for small values of $k$. Away from $k = 0$, all the poles appear in all the response functions due to the inherent spin-charge coupling of the skyrmion state; interestingly, as the Brillouin zone boundary is approached, the gapless modes previously corresponding to $\chi_{nn}$ and $\chi_{zz}$ now appear in the $\chi_{+-}$ response functions. It should be noted that due to an exact symmetry of the groundstate ,consisting of translation by the nearest neighbor distance and a global $\pi$ spin rotation in $x - y$ plane, the modes on the right hand side of the Brillouin zone may be folded back to overlap with those on the left.

One of the important consequences of the presence of the gapless collective spin mode is that it can couple to nuclear spins, allowing a rapid nuclear spin relaxation. Normally at low temperatures such relaxation is very slow, because for spin-polarized systems there is a gap for spin wave excitations, so that very few of these are available for the nuclear spins to scatter. The new gapless modes provide such spin waves even at very low temperature, and it seems quite reasonable that these could be responsible for a rapid increase in the nuclear spin relaxation rate[6] observed as one moves away from $\nu = 1$.

Another aspect of the collective mode spectrum is that it may be used as a basis for understanding the effects of quantum fluctuations on the state of the system. A convenient language for doing so is the Bose-Hubbard model, a tight-binding model where each site accommodates bosons, whose number is analogous to the $\hat{z}$ component of a skyrmion spin, and the phase of the bosonic wavefunction is analogous to the phase angle associated with the in-plane spins. Since the skyrmions have a preferred size due to the electron-electron interaction, one in principle may minimize the energy of the Hartree-Fock state (Eq. (12.2.2)) subject to the constraint that $K$ is fixed at some value, and thereby find a form for $E(K)$ that should have a minimum with $K > 0$.[9] Near the minimum of $E(K)$, one may form a quadratic approximation for the skyrmion energy: $E(K_i) \approx U(K_i - K_0)^2$. In the boson-Hubbard model, $U$ then plays the role of an on-site interaction energy, $K_i$ is the number of bosons on site $i$, and $2K_0$ is the effective chemical potential for the bosons. The effective Hamiltonian then takes the form

$$H^{BH} = U \sum_i (\hat{K}_i - K^0)^2 + J \sum_{\langle ij \rangle} \cos\left(\varphi_i - \varphi_j\right). \tag{12.4.4}$$

The Josephson coupling amplitude $J > 0$ encourages the phase angles of nearest neighbors $< ij >$ to be rotated by $\pi$, as we saw is the most favored situation for the square skyrmion lattice.

The boson-Hubbard model defined above has been studied extensively,[16] and much is known about its phase diagram in the $K_0 - U/J$ plane at zero temperature. In particular, for a fixed value of $U/J$, as $K_0$ is increased the system oscillates between a superconducting state (with well-defined $< \theta_i >$ on each site and a gapless Goldstone mode) and a Mott insulator state with well-defined $< K_i >$ on each site. The transitions between these states are driven by quantum fluctuations. For Eq. (12.4.4), the Hartree-Fock calculations allow *quantitative* estimates of all the parameters entering into $H^{BH}$: as described above, $U$ and $K_0$ may be found by computing $E(K)$ for a given filling factor $\nu$ and Zeeman coupling $\tilde{g}$, and $J$ may be inferred by matching the slope of the "antiferromagnetic" spin wave (Fig. 12.4) to the expected dispersion of the Goldstone mode in the superconducting state of the boson-Hubbard model. In this way, it is possible to find a phase diagram for the Skyrme lattice showing where skyrmion coupling leads to a "superconducting" state supporting a gapless mode (and presumably allowing anomalously fast nuclear spin relaxation at low temperatures) and a state in which the skyrmions essentially have an integral spin $< K_i >$, the analog of the Mott insulator state.

Figure 12.5 illustrates the phase diagram of the Skyrme lattice system. The most prominent feature is the heavy line separating the triangular from the square lattice. This structural transition occurs because when the skyrmions are very dilute (small $|\nu - 1|$) or the skyrmions are very small (large $g$), the coupling between them is too small to outweigh the Madelung energy, which favors a triangular rather than square symmetry. The signal for this transition in the time-dependent Hartree-Fock approximation is $\omega^2(k) < 0$ for the phonon near $k = 0$, indicating a lattice

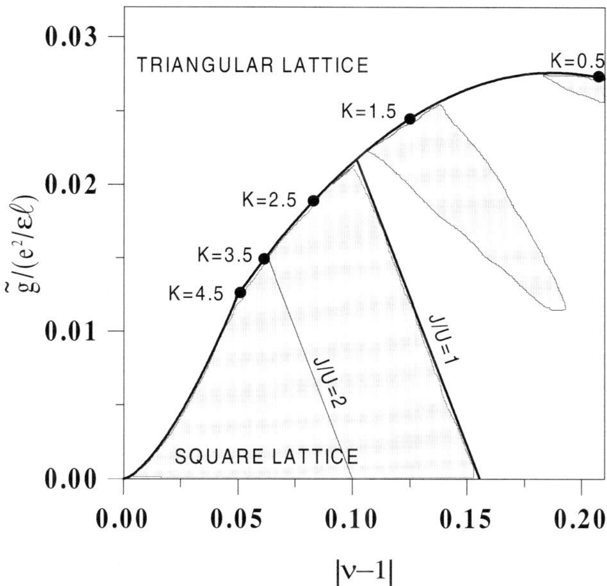


Fig. 12.5. $T = 0$ phase diagram for Skyrme crystal states. $\varepsilon \equiv \kappa$ is the dielectric constant. See text. Reproduced from R. Côté *et al.* Phys. Rev. Lett. **78**, 4825 (1997).

instability. Inside the square lattice phase, quantum fluctuations can drive the system out of the gapless "superconducting" state as described above; the parameters for which this state survives is indicated by the shaded regions. Figure 12.5 also indicates where along the triangular-square phase boundary the skyrmion sizes $K$ are equal to half-integral values (presumably the region of greatest stability for the superconducting state[16]), and the lines at which $J/U = 1, 2$. We note for $J/U > 1$ the superconducting state is always stable with respect to the Mott insulating state, so that the entire lower left corner of the phase diagram is shaded.

## 12.5. The Bilayer Quantum Hall System

A completely different experimental realization of an effective spin-1/2 quantum Hall system that has been studied for a number of years is the bilayer two dimensional electron gas.[17] These systems can be fabricated as double quantum well structures, resulting in two layers of electron gas very close to one another. The electrons then have a discrete degree of freedom in the two choices of layer. If we label one layer as "up" and the other "down", it is clear that the Hilbert space for single electron states is essentially the same as that of the spinful electrons discussed above. This layer index is often called a pseudospin, and many of the ideas discussed above can be applied to this system. One can even consider the situation in which both real spin and pseudospin are active degrees of freedom.[18] In this review we will focus on the limit in which the Zeeman coupling polarizes the real spin, and focus on the unique aspects of the bilayer pseudospin.

An important difference between the spin and the (bilayer) pseudospin degrees of freedom is that interactions are not SU(2) invariant in the latter as they are in the former. This is because with finite layer separation $d$ the Coulomb repulsion is larger for a pair of electrons in the same layer than it is for a pair in different layers with the same in-plane separation $\mathbf{r}$. Nevertheless, in real samples $d$ can be made of order or smaller than the average distance between electrons so that interlayer interactions and correlations are important. Indeed it is useful to consider the limit $d \to 0$ as a starting point for understanding this system. In this case the Hamiltonian for the bilayer quantum Hall system at filling factor $\nu = 1$ is identical to the single layer system with spin. The Zeeman term in the Hamiltonian maps onto a tunneling term for the bilayer system, which energetically favors single particle states that are symmetric linear combinations of states in the two wells over antisymmetric combinations. Real samples may be grown such that there is wide range of possibilities for the scale of this term, from rather large so that all electrons are firmly in the symmetric state — essentially removing the layer degree of freedom from the problem — to very small, orders of magnitude below accessible temperatures. This latter situation has resulted in some of the most interesting and puzzling experimental observations on this system.

Armed with this mapping, we expect that if the tunneling term is sufficiently small then the charged excitations will be skyrmions. However if we identify the "top" layer with spin up and the bottom with spin down, then the tunneling term has the form $H_T = -t \sum_X \{ c_{T,X}^\dagger c_{B,X} + c_{B,X}^\dagger c_{T,X} \}$, where $c_{B(T),X}$ annihilates an electron in the bottom (top) layer and $X$ is the guiding center coordinate quantum number. Written in terms of a Pauli matrix this has the form

$$H_t = -t \sum_X \left\{ \left( c_{T,X}^\dagger \, c_{B,X}^\dagger \right) \sigma_x \begin{pmatrix} c_{T,X} \\ c_{B,X} \end{pmatrix} \right\}, \qquad (12.5.5)$$

so that we identify the spin quantization axis with the $\hat{x}$ direction. Thus in representing a skyrmion one should execute a spin rotation, and the resulting pseudospin texture has an interesting structure, as illustrated in Fig. 12.6.

It is apparent in this representation that the skyrmion contains a vortex-antivortex structure. The primary difference between these and vortices that arise in thin film superconductors and superfluids[19] is in the core. For superfluids, there is a scaler complex order parameter which vanishes at the center of the core, eliminating the ordered phase in this region. For merons, the analog of the order parameter is the magnetization, which does *not* vanish at the center; rather it tilts out of the $x - y$ pseudospin plane, into either the positive or negative $\hat{z}$ direction. This seemingly minor difference turns out to have very interesting physical consequences, as we describe below.

When a finite layer separation $d$ is included, inter- and intra-plane interactions are no longer the same. If we describe the ferromagnet by a unit vector $\mathbf{m}$, an

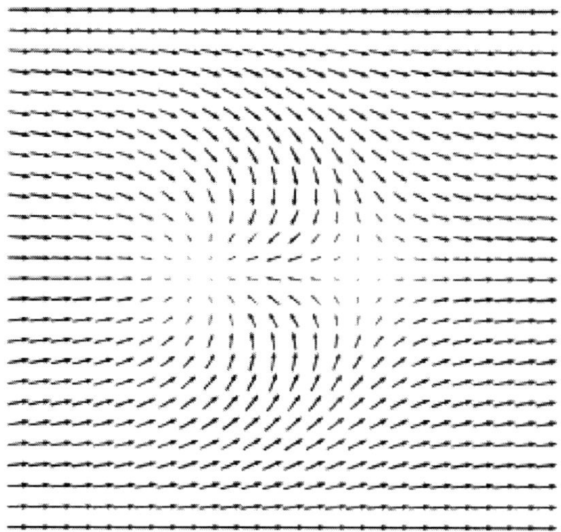

Fig. 12.6.   In plane pseudospin of a bimeron. Reproduced from L. Brey *et al.*, Phys. Rev. B **54**, 16888 (1996).

effective energy functional for the system has the form[20]

$$E[\mathbf{m}] = \frac{\rho_E}{2} \int d\mathbf{r}(\nabla m^\mu)^2 + \frac{1}{2} \int d\mathbf{r} d\mathbf{r}' q(\mathbf{r}) V(\mathbf{r} - \mathbf{r}') q(\mathbf{r})$$
$$- \frac{\Delta_{SAS}}{4\pi \ell_0^2} \int d\mathbf{r}[m^x(d\mathbf{r}) - 1] + \beta \int d\mathbf{r}(m^z)^2$$
$$- \frac{e^2 d^2}{16\pi\kappa} \int \frac{d\mathbf{q}}{4\pi^2} q m^z_{-\mathbf{q}} m^z_{\mathbf{q}} + \frac{\rho_A - \rho_E}{2} \int d\mathbf{r}(\nabla m^z)^2. \qquad (12.5.6)$$

The first two terms of the energy are SU(2) invariant contributions. The leading gradient term is the only one that appears in the nonlinear $\sigma$ model for Heisenberg ferromagnets, and $\rho_E$ is the spin stiffness in the $\hat{x} - \hat{y}$ plane. The second term describes the SU(2) invariant Hartree energy corresponding to the charge density associated with spin textures in quantum Hall ferromagnets. $V(r)$ is the Coulomb interaction screened by the dielectric constant $\kappa$ of the host semiconductor. The third term describes the loss in tunneling energy when electrons are promoted from symmetric to antisymmetric states; here $\Delta_{SAS} = 2t$ is the single-particle splitting between symmetric and antisymmetric states. The last three terms are the leading interaction anisotropy terms at long wavelengths. The $(\nabla m^z)^2$ term accounts for the anisotropy of the spin stiffness. Pseudospin order in the $\hat{x} - \hat{y}$ plane physically corresponds to interlayer phase coherence so that $\rho_A - \rho_E$ will become larger with increasing $d$. The sum of the first and sixth terms in Eq. (12.5.6) gives an $XY$-like anisotropic nonlinear $\sigma$ model. However, this gradient term is not the most important source of anisotropy at long wavelengths. The fourth term produces the leading anisotropy, and is basically the capacitive energy of the double-layer system.

The fifth term appears due to the long-range nature of the Coulomb interaction; its presence demonstrates that a naive gradient expansion of the anisotropic terms is not valid ($m_q$ is the Fourier transform of the unit vector field **m**). Equation (12.5.6) can be rigorously derived from the Hartree-Fock approximation in the limit of slowly varying spin textures,[11] and explicit expressions are obtained for $\rho_E$ (which is due in this approximation entirely to interlayer interactions), $\rho_A$ (due to intralayer interactions), and $\beta$. Quantum fluctuations will alter the values of these parameters from those implied by the Hartree-Fock theory.

Equation (12.5.6) is an energy functional for an easy-plane ferromagnet. The effect of the anisotropy on the structure of bimeron states such as illustrated in Fig. 12.6 is to further separate the vortex-antivortex pair relative to the skyrmion state. This effect appears to be relatively small in Hartree-Fock calculations,[20] but is expected to become larger when thermal and quantum fluctuations are introduced. For $\Delta_{SAS} = 0$, it is clear that above the Kosterlitz-Thouless temperature the meron pairs will unbind. Renormalization group calculations and simulation studies suggest that such unbinding can still occur if $\Delta_{SAS}$ is sufficiently small, either from thermal fluctuations[21] or disorder.[22,23] The presence of unbound merons in the system qualitatively explains a number of remarkable phenomena that are observed in experiments on this system.

### 12.5.1. *Two-dimensional superfluidity and Josephson physics*

The analogy with easy-plane ferromagnetism suggests a different way to interpret the energy functional in Eq. (12.5.6). If $\beta$ is sufficiently large then out-of-plane fluctuations will be strongly suppressed, and in a first approximation one may ignore $m_z$ as a dynamical degree of freedom. Writing $m_x + i m_y = e^{i\theta}$, to lowest order in gradients the energy functional may be written in the simple form

$$E_{SF} = \frac{\rho_s}{2} \int \left[ d\mathbf{r} (\nabla \theta)^2 - \frac{\Delta_{SAS}}{4\pi\ell_0^2} \cos\theta \right]. \tag{12.5.7}$$

For $\Delta_{SAS} = 0$ (i.e., negligible tunneling), this has exactly the form expected for a two-dimensional thin film superfluid, with $\theta$ the condensate wavefunction phase, and $\rho_s$ an effective two-dimensional "superfluid stiffness." In this case one expects the system to have a linearly dispersing "superfluid mode" which is analogous to the spin wave of an easy-plane ferromagnet. The presence of such a mode has been verified in microscopic calculations using the underlying electron degrees of freedom.[10] This suggests the possibility that one might observe some form of superfluidity in this system. To see exactly what this means, it is convenient to consider momentarily a wavefunction for the groundstate of the system in terms of the electron degrees of freedom,

$$|\Psi_{ex}> = \Pi_X \left[ u_X + v_X c_{T,X}^\dagger c_{B,X} \right] |Bot> \tag{12.5.8}$$

where $|Bot>$ represents the state in which all the single particle states in the lowest Landau level have been filled. For a state with uniform density and equal populations in each well, $u_X = v_X = 1/\sqrt{2}$. More generally, one can represent an imbalanced state, obtained physically with an electric field applied perpendicular to the bilayer, by taking $u_X = \sqrt{\nu_T}$ and $v_X = \sqrt{\nu_B}$, with $\nu_T + \nu_B = 1$. The constants $\nu_T$ and $\nu_B$ represent the filling fractions in each of the layers, and the situation where $\nu_T \neq \nu_B$ turns out to be quite interesting, as we will discuss below.

Equation (12.5.8) turns out to be an excellent trial wavefunction, provided the layer separation $d$ is not too large.[24] It shows that the condensed objects in the groundstate are *excitons*, particle-hole pairs with each residing in a different layer. This immediately implies that the superfluidity in this system will be in *counterflow*, when electron current in each layer runs in opposite directions.

Remarkably something much like this has been observed in experiments where electrical contact is made separately with each layer.[25,26] Current may be made to flow in opposite directions in each layer through the $\nu = 1$ quantum Hall state of the bilayer. By measuring the voltage drop in a single layer along the direction of current, one finds that the dissipation extrapolates to zero in the zero temperature limit.

Another type of experiment takes advantage of the fact $\Delta_{SAS}$, while very small (typically several tens of microKelvin), is not zero. When the last term in Eq. (12.5.7) is included, the energy functional has a form very similar to that of a Josephson junction, so that one may posit that this system supports a Josephson effect.[27] In tunneling experiments, where one separately contacts to each layer such that current must tunnel between layers, the tunneling $I - V$ is nearly vertical near zero interlayer bias,[28] which appears very similar to a Josepshon $I - V$ characteristic.

While these results look quite similar to what one might expect for exciton superfluidity, it is important to recognize that these results clearly are *not* genuine superfluid behavior. If the condensate could truly flow without dissipation, one would expect zero dissipation at any finite temperature below the Kosterlitz-Thouless transition, where vortex-antivortex pairs unbind. In experiment this truly dissipationless flow appears to emerge, if at all, only in the zero temperature limit. Similarly, the Josepshon effect should be truly dissipationless, whereas in experiment there is always a measurable tunneling resistance at zero bias. The superfluidity in this system is *imperfect*. What kind of state can be nearly superfluid in this way? The answer likely involves disorder, which as mentioned above can cause the meron-antimeron pairs to unbind at arbitrarily low temperature. We next discuss a model which seems to capture much of the physics found in experiment.

### 12.5.2. *The coherence network model*

One important way in which skyrmions and merons of the $\nu = 1$ quantum Hall system are different than those of more standard ferromagnets is that they carry charge. This means that they couple to electric potential fluctuations due to disorder. In these systems, disorder is ubiquitous because electrons are provided to the layers by dopants, which leave behind charged centers when they donate electrons. The resulting potential fluctuations are extremely strong, creating large puddles of positive and negative charge, separated by narrow strips of incompressible Hall fluid at with local filling factor near $\nu = 1$.[29,30] For the bilayer system, the charge flooding the puddles should take the form of merons and antimerons, whose high density spoils the interlayer coherence. The coherence however will remain strongest in the regions separating the puddles, even though some meron-antimerons pairs will likely straddle them. Thus one forms a network structure for the regions where the coherence is strong, and these should dominate the "superfluid" properties of the system. A schematic picture of the system is illustrated in Fig. 12.7.

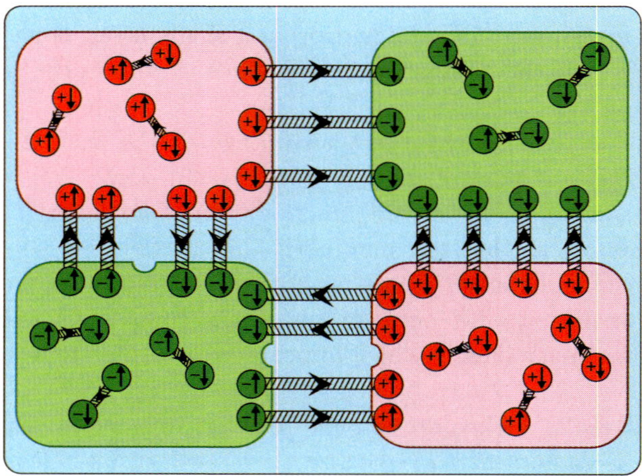

Fig. 12.7. Representation of coherence network. Links and nodes separate puddles of merons (circles). Meron charge and electric dipole moments indicated inside circles, as are strings of overturned phase connecting meron-antimeron pairs. Reproduced from Ref. 23.

The key assumption in this model is that with such dense puddles, merons are able to diffuse independently through the system. This is supported by a renormalization group analysis, which suggests there exists a state in which disorder enters as an effective temperature, so that one would likely be above any meron-antimeron unbinding transition for such strong disorder.[23] Motion of the merons is then limited by energy barriers for them to cross the coherent links between puddles. The tendency for dissipationless counterflow to emerge only at zero temperature now becomes very natural. When condensed excitons flow down the system, these produce a force on the merons perpendicular to that current.[19] The resulting meron current is limited by the activation energy to hop over the coherent links, and vanishes rapidly but only completely when the temperature drops to zero. This meron current induces a voltage drop in the direction of the exciton current via the Josephson relation, rendering the counterflow current dissipative. True superfluid response in this system can only occur at zero temperature.

Dissipation in the tunneling geometry also emerges naturally in this model.[23] Since the current flows into (say) the top layer on the left and leaves via the bottom layer on the right, the current in the system must be decomposed into a sum of symmetric "co-flow" and antisymmetric counterflow (CF). The former is likely carried by edge currents which are essentially dissipationless in the quantum Hall state. To obtain the correct current geometry, the CF current must point in opposite directions at the two ends of the sample. Thinking of the network as a Josephson array, the current of excitons — i.e., CF current — is proportional to $\nabla\theta$. In order to inject CF currents in opposite directions at each end of the sample, the phase angle at the sample edges should be rotated in the *same* direction. This means the phase angle throughout the system will tend to rotate at a uniform rate, which is limited by the $\frac{\Delta_{SAS}}{4\pi\ell_0^2}\cos\theta$ term in Eq. (12.5.7). This is most effective at the nodes of the network, where the coherence is least compromised by the disorder-induced merons.

The dynamics of a typical node with phase angle $\theta$ may be described by a Langevin equation

$$\Gamma\frac{d^2\theta}{dt^2} = \sum_{links} F_{link} - \gamma_0\frac{d\theta}{dt} - h\sin\theta + \xi(t). \qquad (12.5.9)$$

The quantities $F_{link}$ represent the torque on an individual rotor due its neighbors, transmitted through the links. $\Gamma$ is the effective moment of inertia of a rotor, proportional to the capacitance of the node, $h = \frac{\Delta_{SAS}}{4\pi\ell_0^2}$, $\xi$ is a random (thermal) force, and $\gamma_0$ is the viscosity due to dissipation from the other node rotors in the system. For a small driving force, the node responds viscously, and the resulting rotation rate has the form $\gamma\dot\theta = \sum F_{link}$. The Josephson relation $V = \frac{\hbar}{e}\frac{d\theta}{dt}$ then implies that the viscosity $\gamma$ is proportional to the tunneling *conductance* $\sigma_T$ of the system. For $k_BT \gg h$ one may show the viscosity for an individual node to be[31]

$$\gamma = \gamma_0 + \Delta\gamma = \gamma_0 + \sqrt{\frac{\pi}{2\Gamma}}\frac{h^2}{(k_BT)^{3/2}}. \qquad (12.5.10)$$

As each node contributes the same amount to the total viscosity, the *total* response of the system to the injected CF current obeys

$$I_{CF} \propto N_{nodes}\Delta\gamma\frac{e^2V_{int}}{\hbar} = \sigma_T V_{int} \qquad (12.5.11)$$

Note that because the nodes respond viscously, the tunneling conductance is proportional to the area of the bilayer. This is a non-trivial prediction of the model discussed here, which has recently been confirmed in experiment.[32] The proportionality of the tunneling conductance to $\Delta_{SAS}^2$ is another non-trivial prediction which appears to be consistent with experimental data, and which contrasts with the result one expects in the absence of disorder, for which $\sigma_T \propto \Delta_{SAS}$.

### 12.5.3. *Effect of interlayer bias*

When an electric field is applied perpendicular to the layers, the density in the two layers becomes imbalanced. The effect of this can be incorporated into the model, Eq. (12.5.6), by replacing $\beta \int d\mathbf{r}(m^z)^2$ with $\beta \int d\mathbf{r}(m^z - m_0)^2$, with $m_0 = \nu_T - \nu_B$. The imbalance has interesting consequences for merons: due to the connection between charge density and (pseudo-)spin texture in quantum Hall ferromagnets, the four types of merons now have four different charges. These charges are specifically given by $q_{s,T(B)} = -s\,\sigma\,\nu_{B(T)}$, where $s = \pm$ is the vorticity of the meron, and the $T(B)$ subscript reflects the layer in which the magnetization at the core of the meron – its polarization – resides. The index $\sigma$ indicates a sign associated with the polarization: $\sigma = 1$ for polarization in the top layer, $\sigma = -1$ for the bottom layer.

The connection between polarization and charge has very interesting consequences for another type of transport experiment specific to bilayers, known as drag. In these experiments, one drives a current through only a single layer, and measures voltage drops either in the drive layer or the drag layer. Within the coherence network model, the activation barrier for merons to hop across incompressible strips will clearly depend on the relative orientation of the meron polarization and the applied bias. Naively one would think that at low temperature, transport will be dominated by only the smallest activation energy, so that a measurement of resistance will reveal an activation energy that is symmetric around zero bias, which drops as the bias increases.

But this is not what is seen in experiment. The activation energy as measured in the drive layer is highest when the density is biased into the drive layer, and decreases monotonically as the imbalance is changed so that more density is transferred to the drag layer. In the drag layer, measured voltage drops turn out to be much smaller than in the drive layer, and are symmetric, but *increase* as the layer is imbalanced.[33]

A careful analysis of the situation requires a method for determining voltage drops in individual layers, not just the interlayer voltage difference, which is what the Josephson relation applied above actually reveals. This can be accomplished[34] by adopting a "composite boson" description of the $\nu = 1$ quantum Hall state. The idea is to model electrons as *bosons*, each carrying a single magnetic flux quantum in an infinitesimally thin solenoid. The Aharonov-Bohm effect then implements the correct phase (minus sign) when two of these objects are interchanged.[35] By orienting the flux quanta opposite to the direction of the applied magnetic field, on average the field is canceled, and in mean-field theory the system may be modeled as a collection of bosons in zero field. The quantum Hall state is then equivalent to a Bose condensate of these composite bosons. For the coherent bilayer state, there is an additional sense in which the bosons are condensed: they carry a pseudospin with an in-plane ferromagnetic alignment.

Because merons carry physical charge, they will carry a quantity of magnetic flux proportional to this charge. In analogy with a thin-film superconductor,[19] this means that a net current in the bilayer (i.e., a coflow) creates a force on the meron perpendicular to the current. This has to be added to the force due to any counterflow component. Together, these yield a net force which may be shown to be[34]

$$F_T = \frac{es}{2}\Phi_0[(1+\sigma)J_B - (1-\sigma)J_T] \times \hat{z}, \tag{12.5.12}$$

where $\mathbf{J}_{T(B)}$ is the current density in the top (bottom) layer. As is clear from this expression, only one polarization of meron is subject to a force in a drag experiment, since one of the two current densities vanishes.

The force $\vec{F}_{s,\sigma}$ on merons of vorticity $s$ and polarization $\sigma$ will cause them to flow with a velocity $u_{s,\sigma} = \mu_{s,\sigma}F_{s,\sigma}$ where $\mu_{s,\sigma}$ is an effective mobility, which we expect to be thermally activated, with a bias dependence of the activation energy as discussed above. The resulting motion of the vortices induces voltages in two ways. The first is through the Josepshon relation for the interlayer phase, yielding the relation[34]

$$\Delta V = \Delta V_T - \Delta V_B = -\frac{2\pi h}{e}y_0 \sum_{s,\sigma} n_{s\sigma}su_{s\sigma} \tag{12.5.13}$$

for the voltage drops between $\Delta V_\sigma$ between two points a distance $y_0$ apart along the direction of electron current, in layer $\sigma$, where $n_{s,\sigma}$ is the meron density. The second is due to the effective magnetic flux moving with the merons, which induces a voltage drop between electrons at different points along the current flow that is independent of the layer in which they reside. This contribution is given by[34]

$$(\nu_U\Delta V_U + \nu_L\Delta V_L) = -\frac{h}{e}y_0 \sum_{s,\sigma} n_{s\sigma}q_{s\sigma}u_{s\sigma}. \tag{12.5.14}$$

In a drag geometry we have, for example, $J_B = 0$ and $J_T = \frac{I}{W}\hat{y}$, with $I$ the total current and $W$ the sample width. Combining Eqs. (12.5.13) and (12.5.14), we obtain $\Delta V_B = 0$ and

$$\frac{\Delta V_T}{I} = \frac{y_0}{W}h\Phi_0(n_{1,-1}\mu_{1,-1} + n_{-1,-1}\mu_{-1,-1}). \tag{12.5.15}$$

Notice the final result depends on the mobility of *only* merons with polarization $\sigma = -1$. It immediately follows that the voltage drop in the drive layer is asymmetric with respect to bias, precisely as observed in experiment.

In order to explain the voltage drop in the drag layer ($\Delta V_L \neq 0$) we must identify how forces on the $\sigma = +1$ merons might arise. A natural candidate for this is the attractive interaction between merons with opposite vorticities, which in the absence of disorder binds them into pairs at low meron densities. Assuming that driven merons crossing incompressible strips will occasionally be a component of these bimerons, a voltage drop in the drag layer will result. The mobility of such bimerons is limited by the energy barrier to cross an incompressible strip. These strips are likely to be narrow compared to the size scale of the constituents of the bimeron,[34] so we expect the activation energy to be given approximately by the maximum of the activation energies for merons of the two polarizations $\sigma = \pm 1$. This leads to a drag resistance much smaller than that of the drive layer, with an activation energy that is *symmetric* with respect to and increases with bias. These are the behaviors observed in experiment.[33]

We see this result followed from the precise cancelation between the counter-flow current force on the vorticity of merons of a particular polarization, and the Lorentz force associated with meron charge and its associated effective flux. The experiments thus provide indirect evidence that the meron charges vary in precisely the way one expects from the connection between spin textures and physical charge density, verifying the spin-charge relation that is so special in the quantum Hall context.

## 12.6. Conclusion

In summary, we have outlined some of the properties of skyrmions in the quantum Hall effect, including their sizes, coupling effects, collective modes, and phase diagram. Their effect of measurements of electron spin polarization and coupling to nuclei was discussed, and a surprisingly rich phase diagram was shown to emerge for filling factors close to unity due to their presence. The same physics may be applied to the bilayer quantum Hall system, which shows properties highly reminiscent of superfluidity. The existence of dissipation in this system may be understood if the "skyrmions" – bimerons in this case – break up into their constituent merons due to disorder. The results of drag experiments in biased bilayers yield indirect evidence of the close connection between charge and spin textures in the quantum Hall system.

## Acknowledgements

The authors have benefited from discussions and collaborations with many colleagues in the course of the research described here. We would like in particular to thank René Côté, Jim Eisenstein, Allan MacDonald, Kieran Mullen, Ganpathy Murthy, Bahman Roostaei, Steve Simon, Joseph Straley. LB acknowledges the support of MCyT of Spain trough frant No. MAT2006-03741. HAF acknowledges the support of the NSF through grant No. DMR-0704033. The authors would like to thank the Kavli Institute for Theoretical Physics where some of this work was performed. While at the KITP this research was supported in part by the National Science Foundation under Grant No. PHY05-51164.

## References

1. R.E. Prange and S.M. Girvin, *The Quantum Hall Effect* (Springer-Verlag, New-York, 1987).

2. D. Lee and C. Kane, *Phys. Rev. Lett.* **64** (1990) 1313.
3. S. Sondhi, A. Karlhede, S.A. Kivelson and E.H. Rezayi, *Phys. Rev. B* **47** (1993) 16419.
4. E.H. Rezayi, *Phys. Rev. B* **36** (1987) 5454.
5. E.H. Rezayi, *Phys. Rev. B* **43** (1991) 5944.
6. G.L.R. Tycko, S.E. Barret and K.W. West, *Science* **268** (1995) 1460.
7. S.E. Barret, G. Dabbagh, L.N. Pfeiffer, K.W. West and R. Tycko, *Phys. Rev. Lett.* **74** (1995) 5112.
8. H.A. Fertig, L. Brey, R. Côté and A.H. MacDonald, *Phys. Rev. B* **50** (1994) 11018.
9. H.A. Fertig, L. Brey, R. Côté, A.H. MacDonald, A. Karlhede and S.L. Sondhi, *Phys. Rev. B* **55** (1997) 10671.
10. H.A. Fertig, *Phys. Rev. B* **40** (1989) 1087.
11. K. Moon, H. Mori, K. Yang, S. M. Girvin, A. H. MacDonald, L. Zheng, D. Yoshioka and S.-C. Zhang, *Phys. Rev. B* **51** (1995) 5138.
12. L. Brey, *Phys. Rev. Lett.* **65** (1990) 903.
13. E.H. Aifer, B.B. Goldberg and D.A. Broido, *Phys. Rev. B* **76** (1996) 680.
14. L. Brey, H. Fertig, R. Côté and A.H. MacDonald, *Phys. Rev. Lett,* **75** (1995) 2562.
15. R. Côté, A.H. MacDonald, L. Brey, H.A. Fertig, S.M. Girvin and H.T.C. Stoof, *Phys. Rev. Lett.* **78** (1997) 4825.
16. M. Fisher, P.B. Weichman, G. Grinstein and D.S. Fisher, *Phys. Rev. B* **40** (1989) 546.
17. A. MacDonald and S. Girvin, *Perspectives in Quantum Hall Effects* (Wiley, 1996).
18. J. Bourassa, B. Roostaei, R. Côté, H. Fertig and K. Mullen, *Phys. Rev. B* **74** (2006) 195320.
19. A.J. Leggett, *Quantum Liquids* (Oxford University Press, New York, 2006).
20. L. Brey, H. Fertig, R. Côté and A. MacDonald, *Phys. Rev. B* **54** (1996) 16888.
21. H.A. Fertig, *Phys. Rev. Lett.* **89** (2002) 035703.
22. H.A. Fertig and J.P. Straley, *Phys. Rev. Lett.* **91** (2003) 046806.
23. H.A. Fertig and G. Murthy, *Phys. Rev. Lett.* **95** (2005) 156802.
24. H. Fertig, A.H. MacDonald and L. Brey, *Phys. Rev. Lett.* **76** (1996) 2153.
25. M. Kellogg, J.P. Eisenstein, L.N. Pfeiffer and K.W. West, *Phys. Rev. Lett.* **93** (2004) 036801.
26. E. Tutuc, M. Shayegan and D.A. Huse, *Phys. Rev. Lett.* **93** (2004) 036802.
27. X. Wen and A. Zee, *Phys. Rev. B* **47** (1993) 2265.
28. I.B. Spielman, J.P. Eisenstein, L.N. Pfeiffer and K.W. West, *Phys. Rev. Lett.* **84** (2000) 5808.
29. A. Efros, *Solid State Commun.* **65** (1988) 1281.
30. F.A. Efros and V. Burnett, *Phys. Rev. B* **47** (1993) 2233.
31. I.P.W. Dieterich and W. Schneider, *Z. fur Physik B* **27** (1977) 177.
32. A. Finck, A. Champagne, J. Eisenstein, L. Pfeiffer and K. West, *Phys. Rev. B* **78** (2008) 075302.
33. R.D. Wiersma, J.G.S. Lok, S. Kraus, W. Dietsche, K. von Klitzing, D. Schuh, M. Bichler, H.-P. Tranitz and W. Wegscheider, *Phys. Rev. B* **51** (1995) 5138.
34. H.F.B. Roostaei, K. Mullen and S. Simon, *Phys. Rev. Lett.* **101** (2008) 046804.
35. S.C. Zhang, T.H. Hansson and S. Kivelson, *Phys. Rev. Lett.* **62** (1989) 82.

**Chapter 13**

# Half-Skyrmion Theory for High-Temperature Superconductivity

Takao Morinari

*Yukawa Institute for Theoretical Physics, Kyoto University,*
*Kyoto 606-8502, Japan*

We review the half-Skyrmion theory for copper-oxide high-temperature superconductivity. In the theory, doped holes create a half-Skyrmion spin texture which is characterized by a topological charge. The formation of the half-Skyrmion is described in the single hole doped system, and then the half-Skyrmion excitation spectrum is compared with the angle-resolved photoemission spectroscopy results in the undoped system. Multi-half-Skyrmion configurations are studied by numerical simulations. We show that half-Skyrmions carry non-vanishing topological charge density below a critical hole doping concentration $\sim 30\%$ even in the absence of antiferromagnetic long-range order. The magnetic structure factor exhibits incommensurate peaks in stripe ordered configurations of half-Skyrmions and anti-half-Skyrmions. The interaction mediated by half-Skyrmions leads to $d_{x^2-y^2}$-wave superconductivity. We also describe pseudogap behavior arising from the excitation spectrum of a composite particle of a half-Skyrmion and doped hole.


## Contents


13.1 Introduction . . . . . . . . . . . . . . . . . . . . . . . . . . . . . . . . . . . . . 311
13.2 Review of High-Temperature Superconductivity . . . . . . . . . . . . . . . . . . . 312
13.3 Single Hole Doped System . . . . . . . . . . . . . . . . . . . . . . . . . . . . . . 314
13.4 Multi Half-Skyrmion Configurations . . . . . . . . . . . . . . . . . . . . . . . . 318
13.5 Mechanism of d-Wave Superconductivity . . . . . . . . . . . . . . . . . . . . . 321
13.6 Pseudogap . . . . . . . . . . . . . . . . . . . . . . . . . . . . . . . . . . . . . . 325
13.7 Summary . . . . . . . . . . . . . . . . . . . . . . . . . . . . . . . . . . . . . . . 328
References . . . . . . . . . . . . . . . . . . . . . . . . . . . . . . . . . . . . . . . . 329


## 13.1. Introduction

One of the most challenging problems in condensed matter physics is to unveil the mechanism of high-temperature superconductivity in the copper oxides. Although it has past more than two decades since its discovery,[1] no established theory exists. The most difficult aspect is to cope with strong electron correlations: The undoped system of high-temperature superconductors is an insulator. Contrary to conventional band insulators, strong Coulomb repulsion makes the system insulating. High-temperature superconductivity occurs by doping holes in such a Mott insulator.[2] The pairing symmetry is not conventional $s$-wave but $d_{x^2-y^2}$-wave.[3] It is believed that electron-phonon couplings do not play an essential role in the mechanism of high-temperature superconductivity because of the strong Coulomb repulsion. Searching for a mechanism based on the strong electron correlation is necessary. In this chapter, as a candidate providing such a mechanism the half-Skyrmion theory is reviewed.

The plan of the review is as follows. In Sec. 13.2, we review the structure, electronic properties, and the phase diagram of high-temperature superconductors. Then, we describe the half-Skyrmion spin texture in a single hole doped system in Sec. 13.3. The half-Skyrmion excitation spectrum is compared with the angle-resolved photoemission spectroscopy results in the undoped system. Topological character and magnetic properties of multi-half-Skyrmion configurations are described in Sec. 13.4. In Sec. 13.5, we describe a mechanism of $d_{x^2-y^2}$-wave superconductivity based on half-Skyrmions. In Sec. 13.6, we describe a pseudogap behavior in the half-Skyrmion system. It is shown that the energy dispersion of a composite particle of a half-Skyrmion and doped hole leads to an arc-like Fermi surface.

## 13.2. Review of High-Temperature Superconductivity

Although there are a number of high-temperature superconductors, the essential structure is the $CuO_2$ plane. Material differences arise from an insulating layer sandwiched by $CuO_2$ planes.[2] In the parent compound, nine electrons occupy $3d$ orbitals at each copper ion. In the hole picture, there is one hole at each copper site. The hole band is half-filled but the system is an insulator because of a strong Coulomb repulsion. The system is well described by the spin $S = 1/2$ antiferromagnetic Heisenberg model on the square lattice with the superexchange interaction $J \simeq 1500K$.[4] Experimentally and theoretically it is established that the ground state is an antiferromagnetic long-range ordered state.[4] The structure of the $CuO_2$ plane and the arrangement of spins at copper sites in the undoped system are schematically shown in Fig. 13.1(a).

This antiferromagnetic long-range order is rapidly suppressed by hole doping. In fact only $2 - 3\%$ doping concentration is enough to kill antiferromagnetic long-range order. This critical hole concentration is much lower than the percolation limit of $\sim 40\%$. High-temperature superconductivity occurs by introducing about 0.05 holes per copper ion. A schematic phase diagram is shown in Fig. 13.1(b). (In this review we focus on the hole doped system and do not discuss the electron doped system.)

In the high-temperature superconductors anomalous behaviors are observed in physical quantities for temperatures above the superconducting transition temperature, $T_c$.[5] The phenomenon is called pseudogap. The Fermi surface observed by

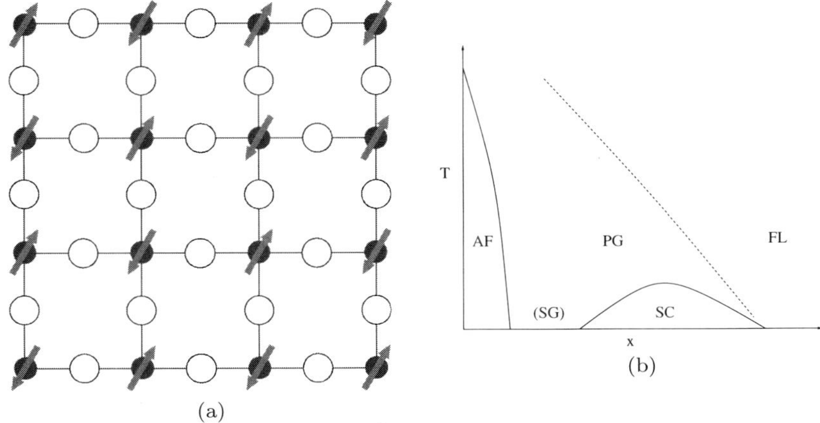

Fig. 13.1. (a) Two-dimensional $CuO_2$ plane. Filled circles represent copper ions. Open circles represent oxygen ions. Arrows are localized spin moments at each copper site. In the ground state, those spins have anti-ferromagnetic long-range order. (b) A schematic phase diagram of the high-temperature superconductors. The horizontal axis represents the doped hole concentration and the vertical axis represents temperature. AF indicates antiferromagnetic long-range order and SC indicates superconductivity. Below a characteristic temperature curve denoted by the dashed line, the system shows a pseudogap behavior (denoted by PG). SG and FL indicate spin-glass-like state and Fermi liquid state, respectively.

angle-resolved photoemission spectroscopy (ARPES) in the underdoped regime is a truncated, arc-like Fermi surface.[6] (See for a review, Ref. 7.) In scanning tunneling spectroscopy, a gap like feature appears below the pseudogap temperature $T^*$ which is higher than $T_c$.[8] For temperatures below $T^*$ gap-like behaviors are observed in NMR, transport coefficients, and optical conductivity. (See for a review, Ref. 5.)

In the doped system, because of the strong Coulomb repulsion at each copper site[2] doped holes occupy oxygen $p$-orbitals. Hole spins interact with copper site spins with strong antiferromagnetic Kondo interaction. Because Kondo interaction coupling, $J_K$, is much larger than $J$ and hole hopping matrix elements, there is correlation of forming a singlet pair called the Zhang-Rice singlet.[9]

The strong $J_K$ limit leads to the $t$-$J$ model.[9] The $t$-$J$ model has been studied extensively. (See for a review, Ref. 10.) In the $t$-$J$ model, double occupancy is projected out. One way to deal with this constraint is to use slave-particle formulations. Based on the resonating valence bond picture proposed by Anderson,[11] a spin-charge separation scenario has been applied to the physics of high-temperature superconductors.[10] From various physical view points different theories have been proposed. There is a view in which incommensurate spin correlations observed in neutron scattering are associated with stripe order. (See for a reivew, Ref. 12.) Chakravarty *et al.* proposed d-density wave order as competing order against superconductivity to explain the pseudogap phenomenon.[13] The half-Skyrmion theory has some connection with these theories which will be discussed later.

## 13.3. Single Hole Doped System

The high-temperature superconductors are characterized by a rich phase diagram shown in Fig. 13.1(b). Remarkably this phase diagram is essentially controlled by a single parameter $x$, the doped hole concentration. Therefore, to understand the physics of high-temperature superconductivity it is necessary to establish how to describe doped holes. Here we consider a half-Skyrmion spin texture created by a doped hole in an antiferromagnetically correlated spins.

As the simplest model we consider the single hole doped system. As stated in the previous section, the undoped system is described by the antiferromagnetic Heisenberg model on the square lattice,

$$H = J \sum_{\langle i,j \rangle} S_i \cdot S_j, \qquad (13.3.1)$$

where the summation is taken over the nearest neighbor sites and the vector $S_i$ describes a spin $S = 1/2$ at site $i$. Theoretically and experimentally it has been established that the ground state is the antiferromagnetic long-range ordered state.[4] A convenient description of the state is obtained by introducing Schwinger bosons[14] and then describing the long-range ordered state in terms of a Bose-Einstein condensate of those Schwinger bosons. In the Schwinger boson theory the spin $S_i$ is represented by

$$S_i = \frac{1}{2} \begin{pmatrix} \zeta_{i\uparrow}^{\dagger} & \zeta_{i\downarrow}^{\dagger} \end{pmatrix} \boldsymbol{\sigma} \begin{pmatrix} \zeta_{i\uparrow} \\ \zeta_{i\downarrow} \end{pmatrix},$$

where the components of the vector $\boldsymbol{\sigma} = (\sigma_x, \sigma_y, \sigma_z)$ are Pauli spin matrices. To describe the spin $S = 1/2$, the Schwinger bosons must satisfy the constraint, $\sum_{\sigma=\uparrow,\downarrow} \zeta_{j\sigma}^{\dagger} \zeta_{j\sigma} = 1$. We introduce a mean field $A_{ij} = \langle \zeta_{i\uparrow}\zeta_{j\downarrow} - \zeta_{i\downarrow}\zeta_{j\uparrow} \rangle$ and introduce a Lagrange multiplier $\lambda_j$ to impose the constraint. In the Schwinger boson mean field theory,[14] we assume uniform values for these quantities as $A_{ij} = A$ and $\lambda_j = \lambda$. The energy dispersion of Schwinger bosons is given by $\omega_k = \sqrt{\lambda^2 - 4J^2 A^2 \gamma_k^2}$ with $\gamma_k = (\sin k_x + \sin k_y)/2$. In the ground state, $\lambda = 2JA$. Bose-Einstein condensation occurs[15–17] at $k = (\pm\pi/2, \pm\pi/2)$.

Now we consider a hole introduced in the system. The strong interaction between the doped hole spin and copper site spins leads to correlation of forming a Zhang-Rice spin singlet[9] as mentioned in the previous section. If the singlet is formed, then the Bose-Einstein condensate of the Schwinger bosons is suppressed around the doped hole position. Generally if the condensate is suppressed at some point in two-dimensional space, then a vortex is formed around that point. The vortex solution is found by solving the Gross-Pitaevskii equation.[18] For the Schwinger bosons, the vortex turns out to be a half-Skyrmion as shown below.

For the description of the half-Skyrmion, it is convenient to use the non-linear sigma model.[19] Low-energy physics of the antiferromagnetic Heisenberg model is well described by the non-linear sigma model,[19]

$$S = \frac{\rho_s}{2} \int_0^{(k_B T)^{-1}} d\tau \int d^2 r \left[ \frac{1}{c_{sw}^2} \left( \frac{\partial n}{\partial \tau} \right)^2 + (\nabla n)^2 \right], \qquad (13.3.2)$$

where $\rho_s$ is the spin stiffness and $c_{sw}$ is the antiferromagnetic spin-wave velocity. (Hereafter we use units in which $\hbar = 1$.) The unit vector $n$ represents the staggered moment and $\tau$ is the imaginary time.

In order to describe the correlation of forming a Zhang-Rice singlet pair between doped hole spins and copper site spins, one has to be careful about its description. Obviously forming a static singlet state which is realized in the $J_K \to \infty$ limit does not work. Because such a simple singlet state contradicts with the rapid suppression of antiferromagnetic long-range order by hole doping. If static singlet states are formed, then sites occupied by singlets do not interact with the other spins at all. The situation is similar to site dilution, and suppression of magnetic long-range order is described by the percolation theory. In other words, considering a strongly localized wave function of a doped hole at a copper site is not realistic. We need to consider a hole wave function extending over some area so that the doped hole spin interacts with the other spins. In fact, numerical diagonalization studies of the $t$-$J$ model show a Skyrmion-like spin texture[20] when a hole motion is restricted to one plaquette. A similar situation may be realized in Li-doped system as discussed in Ref. 21.

To include the effect of the interaction with the other spins, we formulate the correlation of forming a Zhang-Rice singlet in the following way. The spin singlet wave function of a copper site spin and a hole spin is described by

$$\frac{1}{\sqrt{2}} \left( |\uparrow\rangle_h |\downarrow\rangle_{Cu} - |\downarrow\rangle_h |\uparrow\rangle_{Cu} \right).$$

This wave function has the form of superposition of the hole spin-up and copper spin-down state, $|\uparrow\rangle_h |\downarrow\rangle_{Cu}$ and the hole spin-down and copper spin-up state, $|\downarrow\rangle_h |\uparrow\rangle_{Cu}$. In order to include the interaction effect, we consider these states separately and construct superposition of them. We assume that the spin state at site $j$ is spin-up before the introduction of a doped hole. Under this assumption, the spin-up state does not change directions of the neighboring spins. So the system is uniform for the staggered spin $n$. By contrast, the spin-down state at site $j$ creates a Skyrmion spin texture characterized by a topological charge,

$$Q = \frac{1}{8\pi} \int d^2 r \varepsilon_{\alpha\beta} \, n(r) \cdot [\partial_\alpha n(r) \times \partial_\beta n(r)],$$

where $\varepsilon_{xx} = \varepsilon_{yy} = 0$ and $\varepsilon_{xy} = -\varepsilon_{yx} = 1$. Following Ref. 22, the Skyrmion solution is found by making use of an inequality

$$\int d^2 r \left[ \partial_\alpha n \pm \varepsilon_{\alpha\beta} (n \times \partial_\beta n) \right]^2 \geq 0.$$

The classical energy satisfies,

$$E = \frac{\rho_s}{2} \int d^2r \, (\nabla n)^2 \geq 4\pi\rho_s Q.$$

The equality holds if and only if

$$\partial_\alpha n \pm \varepsilon_{\alpha\beta} \left( n \times \partial_\beta n \right) = 0. \tag{13.3.3}$$

This equation is rewritten in a simple form. If we introduce

$$w = \frac{n_x + i n_y}{1 - n_z},$$

then Eq. (13.3.3) is rewritten as

$$\left( \partial_x \mp i\partial_y \right) w = 0.$$

This equation is the Cauchy-Riemann equation. Noting

$$n = \left( \frac{2\mathrm{Re}w}{|w|^2 + 1}, \frac{2\mathrm{Im}w}{|w|^2 + 1}, \frac{|w|^2 - 1}{|w|^2 + 1} \right),$$

the solutions satisfying the boundary condition $n(r_j) = -\hat{z}$ and $n(r \to \infty) = \hat{z}$ with $\hat{z}$ the unit vector along the $z$-axis are the Skyrmion spin texture,

$$n \left( r \right) = \left( \frac{2\eta x}{r^2 + \eta^2}, \frac{2\eta y}{r^2 + \eta^2}, \frac{r^2 - \eta^2}{r^2 + \eta^2} \right),$$

with $Q = 1$ and the anti-Skyrmion spin texture,

$$n \left( r \right) = \left( \frac{2\eta x}{r^2 + \eta^2}, -\frac{2\eta y}{r^2 + \eta^2}, \frac{r^2 - \eta^2}{r^2 + \eta^2} \right),$$

with $Q = -1$.

Now we consider superposition of the uniform state and the Skyrmion state. Although superposition of the two spin configurations is not the solution of the field equation, these solutions suggest that the resulting spin configuration is characterized by a topological charge $Q$ with $0 < |Q| < 1$. The value of $Q$ is determined by making use of the fact that the antiferromagnetic long-range ordered state is described by Bose-Einstein condensation of Schwinger bosons. In the $\mathrm{CP}^1$ representation of the non-linear sigma model,[23] the U(1) gauge field is introduced by

$$\alpha_\mu = -i \sum_{\sigma=\uparrow,\downarrow} z_\sigma^* \partial_\mu z_\sigma \tag{13.3.4}$$

where the complex field $z_\sigma$ is defined through

$$n = \left( z_\uparrow^* \; z_\downarrow^* \right) \sigma \begin{pmatrix} z_\uparrow \\ z_\downarrow \end{pmatrix}. \tag{13.3.5}$$

In terms of the gauge field $\alpha_\mu$, the topological charge $Q$ is rewritten as

$$Q = \int \frac{d^2r}{2\pi} \left( \partial_x \alpha_y - \partial_y \alpha_x \right).$$

From this expression, we see that a spin configuration with $Q$ corresponds to the flux $2\pi Q$ in the condensate of the Schwinger bosons. On the other hand, each Schwinger boson carries the spin $S = 1/2$. However, there are no $S = 1/2$ excitations in the low-energy excitation spectrum. Low-lying excitations are antiferromagnetic spin waves which carry the spin one. Therefore, all Schwinger bosons are paired and the flux quantum is $\pi$ similarly to conventional BCS superconductors.[24] The flux value is not arbitrary and $Q$ must be in the form of $Q = n/2$, with $n$ an integer. Taking into account the constraint $0 < |Q| < 1$, we may conclude $|Q| = 1/2$.[25] The spin texture with $|Q| = 1/2$ is called half-Skrymion spin texture because the topological charge is half of the Skyrmion spin texture. The half-Skyrmion spin texture and the anti-half-Skyrmion spin texture are schematically shown in Fig. 13.2.

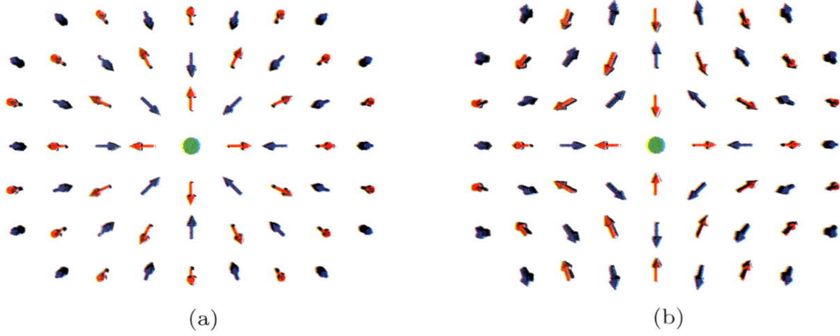

Fig. 13.2. (a) Half-Skyrmion spin texture. Arrows represent the directions of the spin at the copper sites. Neighboring spins are almost anti-parallel because of the antiferromagnetic correlations. Filled circle at the center denotes the core of the half-Skyrmion. (b) Anti-half-Skyrmion spin texture.

Moving half-Skyrmion spin texture is obtained by applying Lorentz boost on the static solution above by making use of the Lorentz invariance of the non-linear sigma model.[23] The energy dispersion is

$$E_k = \sqrt{c_{\text{sw}}^2 k^2 + E_0^2}, \tag{13.3.6}$$

where $E_0 = 2\pi\rho_{\text{s}}$ is the half-Skyrmion creation energy. On the square lattice the dispersion is transformed into

$$E_k = \sqrt{c_{\text{sw}}^2 \left(\cos^2 k_x + \cos^2 k_y\right) + E_0^2}. \tag{13.3.7}$$

Note that the lowest energy states are at $(\pm\pi/2, \pm\pi/2)$ because the Schwinger bosons are gapless at those points in the antiferromagnetic long-range ordered state. The half-Skyrmion spin texture are mainly formed by Schwinger bosons around those points.

Now we compare the half-Skyrmion excitation spectrum with the ARPES result in the undoped system. The excitation spectrum Eq. (13.3.7) is qualitatively in good agreement with excitation spectrum obtained by Wells *et al.*[26] The parameters $c_{\text{sw}}$

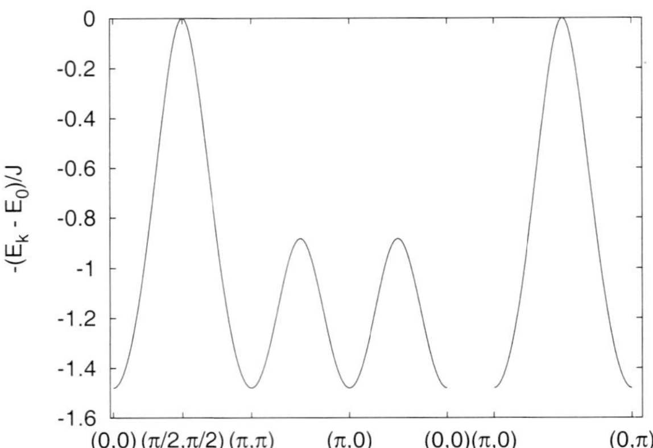

Fig. 13.3. The half-Skyrmion dispersion. Horizontal axis represents positions in the Brillouin zone.

and $E_0$ are determined from the values for the Heisenberg antiferromagnet. We use the renormalized factors $Zc = 1.17$ and $Z_\rho = 0.72$, which are estimated from quantum Monte Carlo simulations[27,28] and a series expansion technique.[29] Using these values, we find that the bandwidth is $1.5J$ and $E_0 = 1.1J$. The experimentally estimated bandwidth by Wells et al. is $2.2J$. From the fitting of the experimental data assuming Eq. (13.3.7), we find $E_0 \sim J$.

In the undoped system, anomalously broad line shapes are observed by ARPES. Line shape broadening is associated with scattering of excitations by fluctuation modes. In the half-Skyrmion theory, half-Skymions couple to spin-wave excitations. Describing those spin wave excitations in terms of the gauge field fluctuations line shape broadening is studied by applying a strong coupling analysis.[30] The width of the broadening is in good agreement with the experiment.

## 13.4. Multi Half-Skyrmion Configurations

In the previous section, the single half-Skyrmion has been considered. The most important physical quantity carried by the half-Skyrmion is the topological charge. The topological charge density, which is defined in the continuum as

$$q_c\left(r\right) = \frac{1}{4\pi} n\left(r\right) \cdot \left[\partial_x n\left(r\right) \times \partial_y n\left(r\right)\right],$$

has the following form on the lattice,

$$q_c\left(x_j, y_j\right) = \frac{1}{16\pi} n\left(x_j, y_j\right) \cdot \left[n\left(x_j + 1, y_j\right) - n\left(x_j - 1, y_j\right)\right]$$
$$\times \left[n\left(x_j, y_j + 1\right) - n\left(x_j, y_j - 1\right)\right].$$

In the single half-Skyrmion state, $q_c(x_j, y_j)$ has a peak around the half-Skyrmion position, and vanishes at infinity. The integration of $q_c(r)$ leads to the quantized value $Q = \pm 1/2$.

If there are many half-Skyrmions, do half-Skyrmions keep topological charge? In order to answer this question, we carry out a simple numerical simulation. First, we put either an XY-spin-vortex or an anti-XY-spin-vortex randomly. (A similar numerical simulation is discussed in Ref. 31.) A multi-XY-spin-vortex configuration is defined by

$$n_x(r) = \sum_j q_j \frac{x - x_j}{\left(x - x_j\right)^2 + \left(y - y_j\right)^2}, \qquad (13.4.8)$$

$$n_y(r) = \sum_j q_j \frac{y - y_j}{\left(x - x_j\right)^2 + \left(y - y_j\right)^2}, \qquad (13.4.9)$$

where $q_j = +1$ for XY-spin-vortices and $q_j = -1$ for anti-XY-spin-vortices. A doped hole is sitting at each spin-vortex position $(x_j, y_j)$, and $n(x_j, y_j) = 0$. Then, a random number which ranges from $-0.1$ to $0.1$ is assigned to the $z$-component of the vector $n(x_i, y_i)$ except for the nearest neighbor sites $(x_i \pm 1, y_i)$ and $(x_i, y_i \pm 1)$. After that, the equilibrium configuration of the vectors $n(x_i, y_i)$ is obtained by the relaxation method. At site $(x_\ell, y_\ell)$, $n(x_\ell, y_\ell)$ is updated by

$$n\left(x_\ell, y_\ell\right) = \frac{1}{4}\left[n\left(x_\ell + 1, y_\ell\right) + n\left(x_\ell - 1, y_\ell\right) + n\left(x_\ell, y_\ell + 1\right) + n\left(x_\ell, y_\ell - 1\right)\right].$$

The constraint $|n(x_\ell, y_\ell)| = 1$ is imposed by taking the normalization after the update. This update procedure is carried out over all lattice sites except for the hole positions $(x_j, y_j)$ and its nearest neighbor sites, $(x_j \pm 1, y_j)$ and $(x_j, y_j \pm 1)$. The resulting converged state is an approximate state for a multi-half-Skyrmion-anti-half-Skyrmion configuration. As an example, Fig. 13.4 shows the topological charge density distribution at the doping concentration $x = 0.107$. There are regions where topological charge density is non-zero. Positive (negative) topological charge density region is associated with half-Skyrmions (anti-half-Skyrmions). Non-vanishing distribution patterns are observed for $x < x_c$ with $x_c \sim 0.30$. For $x > x_c$, half-Skyrmions and anti-half-Skyrmions are heavily overlapped. As a result topological charges are canceled out. Therefore, above $x_c$ the topological nature of half-Skyrmions is lost. (For related discussions about the effect of thermally excited skyrmions and hole induced skyrmions, see Refs. 32 and 33.)

Now we discuss magnetic properties of multi-half-Skyrmion configurations. The magnetic correlation is investigated by the static magnetic structure factor,

$$S\left(q\right) = \sum_{\alpha,\beta} \left(\delta_{\alpha\beta} - \frac{q_\alpha q_\beta}{q^2}\right) S_{\alpha\beta}\left(q\right).$$

Here

$$S_{\alpha\beta}\left(q\right) = \frac{1}{N} \sum_{i,j} e^{iq\cdot(R_i - R_j)} S_{i\alpha} S_{j\beta}.$$

$S(q)$ is measured by neutron scattering experiments. Introducing,

$$S\left(q\right) = \sum_j e^{iq\cdot R_j} S_j,$$

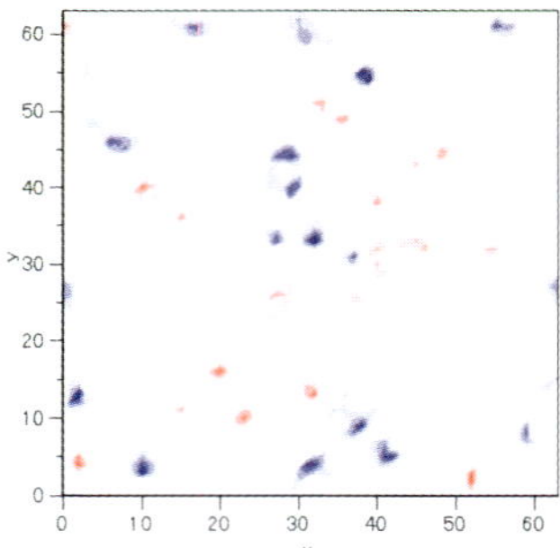

Fig. 13.4. Topological charge density distribution in real space at the doping concentration $x = 0.107$ on a $64 \times 64$ lattice. Positive values are shown in red and negative values are shown in blue.

$S(q)$ is rewritten as

$$S(q) = \frac{1}{N} \left[ |S(q)|^2 - \frac{1}{q^2} |q \cdot S(q)|^2 \right].$$

If there is antiferromagnetic long-range order, then $S(q)$ has a peak at $q = (\pi, \pi) \equiv Q$, and the peak height is proportional to the number of lattice sites. From numerical simulations above, we find that $S(q)$ shows incommensurate peaks at positions shifted from $q = Q$. Furthermore, we find that around $x = 0.10$ the maximum peak height is on the order of the square root of the number of lattice sites. Therefore, the magnetic long-range order disappears around that doping concentration.

The physical origin of the incommensurate peaks is found by studying a regular configuration of half-Skyrmions. Taking a vortex-anti-vortex configuration given by $q_{(x_j, y_j)} = (-1)^{x_j + y_j}$, an approximate "antiferromagnetic" configuration of half-Skyrmions and anti-half-Skyrmions is obtained by the numerical simulation. Figure 13.5(a) shows the magnetic structure factor of the resulting state at $x = 0.0625$ on a $64 \times 64$ lattice. The incommensulate peaks are found at $(\pi(1 \pm 2\delta), \pi(1 \pm 2\delta))$ with $\delta = 0.125$. These peaks are associated with the superlattice formed by half-Skyrmions and anti-half-Skyrmions. A stripe order case is shown in Fig. 13.5(b) which is obtained by taking a vortex-anti-vortex configuration with $q_{(x_j, y_j)} = (-1)^{x_j}$. The dominant incommensulate peaks are found at $(\pi(1 \pm 2\delta), \pi)$ with $\delta = 0.125$.

Experimentally neutron scattering experiments show incommensurate peaks at $q = (\pi(1 \pm 2\delta), \pi(1 \pm 2\delta))$ for $x < 0.05$ and $q = (\pi(1 \pm 2\delta), \pi)$ and $q = (\pi, \pi(1 \pm 2\delta))$ for $x > 0.05$.[34-36] As shown above, such incommensurate peaks are found in some configurations of half-Skyrmions and anti-half-Skyrmions. However, there is a quantitative difference. Experimentally it is found that $\delta \simeq x$.[35,36] In order to explain this experimental result, it is necessary to consider stripe-like configurations of half-Skyrmions and anti-half-Skyrmions. To determine the stable configuration of half-Skyrmions and anti-half-Skyrmions, we need to take into account the inter-

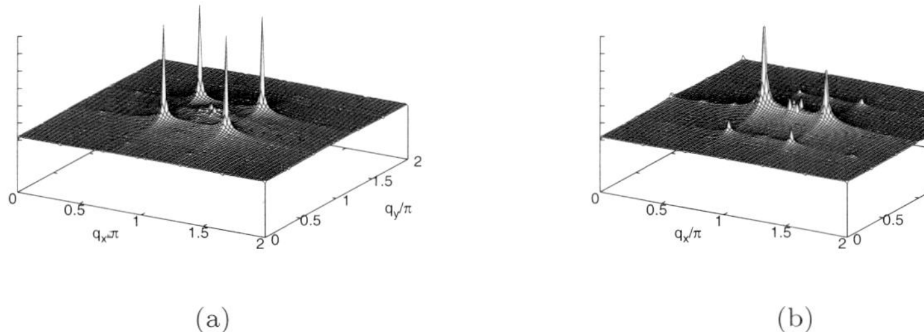

Fig. 13.5.   (a) Magnetic structure factor in a roughly antiferromagnetically ordered half-Skyrmions and anti-half-Skyrmions.   (b) Magnetic structure factor in a stripe-like configuration of half-Skyrmions and anti-half-Skyrmions.

action between half-Skyrmions, which is not included in the numerical simulation above. Determination of the stable half-Skyrmion configuration is left for future work.

## 13.5.  Mechanism of d-Wave Superconductivity

In the half-Skyrmion theory, doped holes create either a half-Skyrmion or an anti-half-Skyrmion at their positions. A half-Skyrmion or an anti-half-Skyrmion is bound to each hole and moves together. There are two ways to formulate the effect of half-Skyrmions on the doped holes. One way is to take a doped hole and the half-Skyrmion created by the hole as a composite particle. This approach is formulated in the next section, and we shall see that the theory leads to pseudogap behavior.

The other way is to include the effect of half-Skyrmions as fields mediating interaction between doped holes. By integrating out the half-Skyrmion degrees of freedom, we obtain the interaction between doped holes mediated by the half-Skyrmions. In this section, we take this approach and show that the interaction leads to a $d_{x^2-y^2}$-wave Cooper pairing between the doped holes.[37,38] An intuitive interpretation is also given about the origin of the pairing interaction based on a Berry phase.

The fact that each doped hole carries a half-Skyrmion is represented by

$$\nabla \times \alpha = \pi \sum_{s=\pm} s\psi_s^\dagger(r)\,\psi_s(r), \tag{13.5.10}$$

where $\alpha$ is the U(1) gauge field in the $CP^1$ model defined by Eq. (13.3.4). The index $s$ labels the sign of the topological charge. $s = +$ refers to a half-Skyrmion and $s = -$ refers to an anti-half-Skyrmion.

The interaction between the doped hole current and the gauge field is found as follows. Doped holes interact with the spins via a strong Kondo coupling,

$$H_K = J_K \sum_j S_j \cdot \left(c_j^\dagger \sigma c_j\right).$$

Meanwhile, the doped hole motion is described by

$$H_t = -t \sum_{\langle i,j \rangle} \left( c_i^\dagger c_j + h.c. \right).$$

From a perturbative calculation for the tight-binding model describing the $CuO_2$ plane, we find the Kondo coupling is $J_K \simeq 1eV$. Since $J_K$ is larger than $t \simeq 0.4eV$, we first diagonalize the Kondo coupling term. The diagonalization is carried out by the following unitary transformation,

$$c_j = U_j f_j,$$

where

$$U_j = \begin{pmatrix} z_{j\uparrow} & -z_{j\downarrow}^* \\ z_{j\downarrow} & z_{j\uparrow}^* \end{pmatrix}.$$

Under this transformation, the hopping term is

$$H_t = -t \sum_{\langle i,j \rangle} \left( f_i^\dagger U_i^\dagger U_j f_j + h.c. \right).$$

Extracting the terms including the gauge field $\alpha_\mu$, we find

$$H_{\text{int}} = it \sum_j \sum_{\delta=x,y} f_{j+\delta}^\dagger \alpha_\delta \sigma_z f_j + h.c.$$

(The effect of other terms is discussed in Ref. 39.) After Fourier transforming and taking the continuum limit, we obtain

$$H_{\text{int}} \simeq \frac{1}{m\Omega^{1/2}} \sum_{k,q} \sum_{\delta=x,y} \alpha_\delta\left(q\right) \cdot \left(k + \frac{q}{2}\right)_\delta f_{k+q}^\dagger \sigma_z f_k. \tag{13.5.11}$$

Here the effective mass $m$ is introduced by $t \simeq 1/2m$.

Now we derive the pairing interaction between the doped holes from Eqs. (13.5.10) and (13.5.11) by eliminating the gauge field. In order to fix the gauge, we take the Coulomb gauge. In wavevector space, we set $\alpha_x\left(q\right) = -\frac{iq_y}{q^2}\alpha\left(q\right)$ and $\alpha_y\left(q\right) = \frac{iq_x}{q^2}\alpha\left(q\right)$. From Eq. (13.5.10), we obtain

$$\alpha\left(q\right) = -\frac{\pi}{\Omega^{1/2}} \sum_{k,s} s f_{ks}^\dagger f_{k+q,s}.$$

Substituting this expression into Eq. (13.5.11), we obtain

$$H_{\text{int}} \simeq -\frac{i\pi}{m\Omega} \sum_{k,k',q} \sum_{s,s',\sigma,\sigma'} \frac{k_x q_y - k_y q_x}{q^2} s'\sigma f_{k's'\sigma'}^\dagger f_{k'+q,s'\sigma'} f_{k+q,s,\sigma}^\dagger f_{k,s,\sigma}.$$

Since we are interested in a Cooper pairing, we focus on terms with $k + k' + q = 0$. After symmetrizing the terms with respect to spin and half-Skyrmion indices, we obtain

$$H_{\text{int}} \simeq -\frac{i\pi}{m\Omega} \sum_{k \neq k'} \sum_{s,s',\sigma,\sigma'} \frac{k_x k_y' - k_y k_x'}{(k-k')^2} \left(s'\sigma + s\sigma'\right) f_{k's'\sigma'}^\dagger f_{-k',s,\sigma}^\dagger f_{-k,s,\sigma} f_{k,s'\sigma'}.$$

$$\tag{13.5.12}$$

This interaction term leads to a pairing of holes as shown below.

Following a standard procedure,[40] we apply the BCS mean field theory to the interaction (13.5.12). The mean field Hamiltonian reads,

$$
H = \sum_{k,s,\sigma} \left( f^\dagger_{k,s,\sigma} \; f^\dagger_{k,-s,-\sigma} \; f_{-k,s,\sigma} \; f_{-k,-s,-\sigma} \right)
$$

$$
\times \begin{pmatrix}
\xi_k & 0 & \Delta^{(+)}_{k,s,\sigma} & \Delta^{(-)}_{k,s,\sigma} \\
0 & \xi_k & \Delta^{(-)}_{k,-s,-\sigma} & \Delta^{(+)}_{k,-s,-\sigma} \\
\Delta^{(+)*}_{k,s,\sigma} & \Delta^{(-)*}_{k,-s,-\sigma} & -\xi_k & 0 \\
\Delta^{(-)*}_{k,s,\sigma} & \Delta^{(+)*}_{k,-s,-\sigma} & 0 & -\xi_k
\end{pmatrix}
\begin{pmatrix}
f_{k,s,\sigma} \\
f_{k,-s,-\sigma} \\
f^\dagger_{-k,s,\sigma} \\
f^\dagger_{-k,-s,-\sigma}
\end{pmatrix},
\qquad (13.5.13)
$$

The mean fields are defined by

$$
\Delta^{(+)}_{k,s,\sigma} = +\frac{2\pi i}{m\Omega} s\sigma \sum_{k(\neq k')} \frac{k_x k'_y - k_y k'_x}{(k - k')^2} \langle f_{-k',s,\sigma} f_{k',s,\sigma} \rangle,
\qquad (13.5.14)
$$

$$
\Delta^{(-)}_{k,s,c} = -\frac{2\pi i}{m\Omega} s\sigma \sum_{k(\neq k')} \frac{k_x k'_y - k_y k'_x}{(k - k')^2} \langle f_{-k',-s,-\sigma} f_{k',s,\sigma} \rangle.
\qquad (13.5.15)
$$

The mean field (13.5.14) is associated with the Cooper pairing between holes with the same spin and the same half-Skyrmion index. On the other hand, the mean field (13.5.15) is associated with the Cooper pairing between holes with the opposite spin and the opposite half-Skyrmion index.

If we consider the interaction (13.5.12) only, then the pairing states described by $\Delta^{(+)}_{k,s,\sigma}$ and $\Delta^{(-)}_{k,s,\sigma}$ are degenerate energetically. However, if we include the (anti-)half-Skyrmion-(anti-)half-Skyrmion interaction and the half-Skyrmion-anti-half-Skyrmion interaction, the Cooper pairing between holes with the opposite spin states and the opposite half-Skyrmion indices is favorable. Because the interaction between (anti-)half-Skyrmions is repulsive and the interaction between half-Skyrmions and anti-half-Skyrmions is attractive. So we may focus on the pairing correlations described by $\Delta^{(-)}_{k,s,\sigma}$.

At zero temperature, the BCS gap equation is

$$
\Delta^{(-)}_{k,s,\sigma} = -\frac{\pi i}{m\Omega} s\sigma \sum_{k(\neq k')} \frac{k_x k'_y - k_y k'_x}{(k - k')^2} \frac{\Delta^{(-)}_{k',s,\sigma}}{E_{k'}},
$$

where $E_k$ is the quasiparticle excitation energy. This gap equation is divided into two equations according to the relative sign between $s$ and $\sigma$. For $\Delta^{(-)}_{k,s,s} \equiv \Delta^{(-)}_{k,+}$, the gap equation is

$$
\Delta^{(-)}_{k,+} = -\frac{\pi i}{m\Omega} \sum_{k(\neq k')} \frac{k_x k'_y - k_y k'_x}{(k - k')^2} \frac{\Delta^{(-)}_{k',+}}{E_{k'}}.
\qquad (13.5.16)
$$

For $\Delta_{k,s,-s}^{(-)} \equiv \Delta_{k,-}^{(-)}$, the gap equation is

$$\Delta_{k,-}^{(-)} = +\frac{\pi i}{m\Omega} \sum_{k(\neq k')} \frac{k_x k_y' - k_y k_x'}{(k-k')^2} \frac{\Delta_{k',-}^{(-)}}{E_{k'}}. \tag{13.5.17}$$

Here we consider the case in which both of $\Delta_{k,+}^{(-)}$ and $\Delta_{k,-}^{(-)}$ describe the same pairing symmetry. Under this condition, we find $|\Delta_{k,+}^{(-)}| = |\Delta_{k,-}^{(-)}|$, and so $E_k$ is given by

$$E_k = \sqrt{\xi_k^2 + \left|\Delta_{k,+}^{(-)}\right|^2} = \sqrt{\xi_k^2 + \left|\Delta_{k,-}^{(-)}\right|^2}. \tag{13.5.18}$$

A similar gap equation was analyzed in Ref. 41 in the context of the fractional quantum Hall systems. Following Ref. 41, we introduce an ansatz,

$$\Delta_{k,+}^{(-)} = \Delta_k \exp\left(-i\ell\theta_k\right), \tag{13.5.19}$$

where $\ell$ is an integer and $\theta_k = \tan^{-1}(k_y/k_x)$. Substituting this expression into Eq. (13.5.16), the integration with respect to the angle $\theta_{k'}$ is carried out analytically. At this procedure, we find that there is no solution for $\ell = 0$. Therefore, there is no $s$-wave pairing state. Furthermore, the gap equation has the solutions only for $\ell > 0$. The gap $\Delta_k$ satisfies the following equation,

$$\Delta_k = \frac{1}{2m} \int_0^k dk' \frac{k'\Delta_{k'}}{E_{k'}} \left(\frac{k'}{k}\right)^\ell + \frac{1}{2m} \int_k^\infty dk' \frac{k'\Delta_{k'}}{E_{k'}} \left(\frac{k}{k'}\right)^\ell. \tag{13.5.20}$$

From the asymptotic forms in $k \to \infty$ and $k \to 0$, we assume the following form for $\Delta_k$,

$$\Delta_k/\varepsilon_F = \begin{cases} \Delta\left(k/k_F\right)^\ell, & (k < k_F), \\ \Delta\left(k_F/k\right)^\ell, & (k > k_F), \end{cases}$$

where $\epsilon_F$ is the Fermi energy of holes and $k_F$ is the Fermi wave number. The gap $\Delta$ is found numerically. The gap $\Delta$ decreases by increasing $\ell$. We find $\Delta = 0.916$ for $\ell = 1$ and $\Delta = 0.406$ for $\ell = 2$.

The gap equation (13.5.17) is analyzed similarly. However, because of the sign difference in the interaction the solution has the following form,

$$\Delta_k^{(-)} = \Delta_k \exp\left(i\ell\theta_k\right), \tag{13.5.21}$$

where $\Delta_k$ is the solution of Eq. (13.5.20). As a result, there are two types of Cooper pairs with opposite relative angular momentum.

For $p$-wave ($\ell = 1$) gap symmetry, the sum of $\Delta_{k,+}^{(-)}$ and $\Delta_{k,-}^{(-)}$ leads to a $p_x$-wave gap which is unstable in the bulk in the absence of symmetry breaking associated with spatial anisotropy. Since $s$-wave gap symmetry is ruled out as mentioned above, the lowest energy sate is obtained for $d_{x^2-y^2}$-wave gap symmetry.

The pairing mechanism based on half-Skyrmions is intuitively understood (Fig. 13.6). According to Eq. (13.5.10), a half-Skrymion, or a gauge flux, is induced around a hole. If another hole passes the gauge flux region at the Fermi velocity, a magnetic Lorentz-force-like interaction acts on that hole according to the interaction represented by Eq. (13.5.11). A similar pairing interaction is discussed at half-filled Landau levels.[41,42] In that system the gauge field is the Chern-Simons gauge field[43] whose gauge fluxes cancel the external magnetic field fluxes at the mean field level. Gauge field fluctuations give rise to a paring interaction between flux attached fermions.

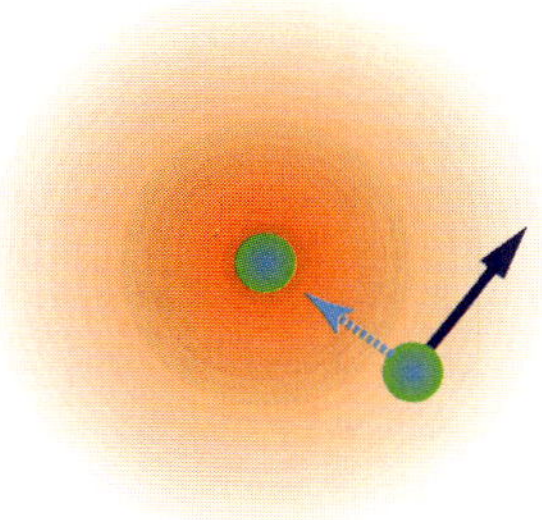

Fig. 13.6. Interaction between doped holes arising from a Berry phase effect associated with the gauge flux created by a half-Skyrmion.

## 13.6. Pseudogap

One of the most intriguing phenomena observed in high-temperature superconductors is the so-called pseudogap which is observed in various physical quantities.[5] Here we focus on the pseudogap behavior observed in ARPES. If the hole concentration is lower than the optimum hole concentration at which the transition temperature is the maximum, the Fermi surface is not a conventional Fermi surface expected from the band theory. Instead, a truncated, an ark-like Fermi surface is observed in ARPES.[6] In order to explain this Fermi arc, a standard approach is to consider a coupling with some boson modes, such as spin fluctuations or gauge field fluctuations associated with phase fluctuations of a mean field, expecting self-energy effects in the single body quasiparticle Green's function. However, it is not obvious that such a conventional analysis leads to qualitatively different physics.

The half-Skyrmion theory provides a completely different approach. To describe the doped hole dynamics which is associated with the spectral function observed by ARPES, we need to include the fact that each doped hole carries a half-Skyrmion. For that purpose, the direct way is to take a doped hole and the half-Skyrmion created by the hole as a composite particle so that the dynamics of the half-Skyrmion is included in the doped hole dynamics.[44]

The Hamiltonian describing the hole hopping is

$$H_{\mathrm{t}} = -\sum_{\langle i,j \rangle} \sum_{\sigma} t_{ij} c_{i\sigma}^{\dagger} c_{j\sigma} + h.c., \qquad (13.6.22)$$

where $t_{ij} = t$ for the nearest neighbor sites, $t_{ij} = t_1$ for the next nearest neighbor sites, and $t_{ij} = t_2$ for the third nearest neighbor sites. The parameters $t_1/t$ and $t_2/t$ are chosen so that the Fermi surface in the Fermi liquid phase is reproduced.[45]

Now let us include the half-Skyrmion dynamics. The half-Skyrmion dispersion is described by the Hamiltonian,

$$H_{\mathrm{hs}} = \sum_{k \in RBZ} \sum_{\sigma} \begin{pmatrix} c^{\dagger}_{ek\sigma} & c^{\dagger}_{ok\sigma} \end{pmatrix} \begin{pmatrix} 0 & \kappa_k \\ \kappa_k^* & 0 \end{pmatrix} \begin{pmatrix} c_{ek\sigma} \\ c_{ok\sigma} \end{pmatrix},$$

where the $k$-summation is taken over the reduced Brillouin zone, $|k_x \pm k_y| \leq \pi$, $c_{ek\sigma} = (c_{k\sigma} + c_{k+Q,\sigma})/\sqrt{2}$ and $c_{ok\sigma} = (c_{k\sigma} - c_{k+Q,\sigma})/\sqrt{2}$, and

$$\kappa_k = -v\left[(\cos k_x + \cos k_y) + i\left(\cos k_x - \cos k_y\right)\right].$$

The half-Skyrmion dispersion is given by $\pm|\kappa_k|$. This dispersion corresponds to Eq. (13.3.7) with $E_0 = 0$ and $c_{\mathrm{sw}} = v$. The half-Skyrmion creation energy $E_0$ is zero because it vanishes in the absence of the antiferromagnetic long-range order.[46] The spin-wave velocity for the doped system is denoted by $v$ which is different from $c_{\mathrm{sw}}$ in the undoped system. Here we use the same creation operators and the annihilation operators for doped holes and half-Skyrmions. This is because a doped hole and the half-Skyrmion carried by the hole is taken as a composite particle. Note that it is not necessary to distinguish between half-Skyrmions and anti-half-Skyrmions since their excitation spectra are the same.

The dispersion energy of the composite particle is calculated from $H = H_{\mathrm{t}} + H_{\mathrm{hs}}$ as,

$$E_k^{(\pm)} = \varepsilon_k^{(+)} \pm \left|\kappa_k + \varepsilon_k^{(-)}\right|,$$

with $\varepsilon_k^{(\pm)} = \left(\varepsilon_k \pm \varepsilon_{k+Q}\right)/2$ and $\varepsilon_k = -2t\left(\cos k_x + \cos k_y\right) - 4t_1 \cos k_x \cos k_y - 2t_2\left(\cos 2k_x + \cos 2k_y\right)$. The spectral function is calculated following a standard procedure.[47] The imaginary time Green's function for up-spin is defined as

$$G_{k\uparrow}(\tau) = -\left\langle T_\tau c_{k\uparrow}(\tau) c^{\dagger}_{k\uparrow}(0)\right\rangle,$$

where $c_{k\uparrow}(\tau) = \mathrm{e}^{\tau(H-\mu N)} c_{k\uparrow} \mathrm{e}^{-\tau(H-\mu N)}$ with $\mu$ the chemical potential and $N$ the number operator and $T_\tau$ is the imaginary time ($\tau$) ordering operator.

The Matsubara Green's function is obtained by

$$G_{k\uparrow}(i\omega_n) = \int_0^{(k_B T)^{-1}} d\tau e^{i\omega_n \tau} G_{k\uparrow}(\tau),$$

where $\omega_n = \pi(2n+1)k_B T$ $(n = 0, \pm 1, \pm 2, ...)$ is the fermion Matsubara frequency. By the analytic continuation, $i\omega_n \to \omega + i\delta$, with $\delta$ an infinitesimal positive number, the retarded Green's function is obtained. Thus, the spectral function is

$$A_k(\omega) = -\frac{1}{\pi} \mathrm{Im} G_{k\uparrow}(\omega + i\delta)$$

$$= \frac{1}{2}(1 + \zeta_k)\delta\left(\omega - E_k^{(+)} + \mu\right) + \frac{1}{2}(1 - \zeta_k)\delta\left(\omega - E_k^{(-)} + \mu\right), \quad (13.6.23)$$

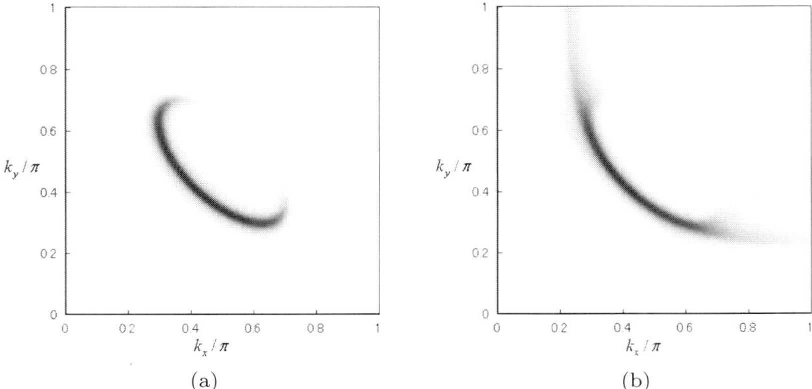



Fig. 13.7. (a) The intensity plot of the spectral function $A_k(\omega = 0)$ in a quadrant of the Brillouin zone. The doping concentration is $x = 0.10$. The other parameters are $t_1/t = -0.25$, $t_2/t = 0.10$, and $v/t = 1.0$. (b) The intensity plot of the spectral function $A_k(\omega = 0)$ with including the effect of the short-range antiferromagnetic correlation. The antiferromagnetic correlation length is taken as $\xi_{AF} = 10$.

where

$$\zeta_k = \frac{-\left(2t + v\right)\left(\cos k_x + \cos k_y\right)}{\sqrt{\left(2t + v\right)^2 \left(\cos k_x + \cos k_y\right)^2 + v^2 \left(\cos k_x - \cos k_y\right)^2}}.$$

In numerical computations, the parameter $\delta$ above is taken to be finite so that the $\delta$-function in the right hand side of Eq. (13.6.23) is replaced by the Lorentz function. Figure 13.7(a) shows $A_k(\omega = 0)$ with $\delta/t = 0.10$. The resulting Fermi surface is arc-like because the factors $1 \pm \zeta_k$ suppress the intensity in part of the Brillouin zone. This arc-like Fermi surface is consistent with the ARPES results.

However, there is some deviation from the experiment around the ends of the arc.[48] Basically the Fermi arc follows the underlying Fermi surface which appears at high-temperature above the characteristic pseudogap temperature, $T^*$. The deviation is suppressed by including the effect of the short-range antiferromagnetic correlation.[49,50] The effect is included by taking average over the wave vector change in the (incommensurate) antiferromagnetic correlation. The calculation is similar to that described in Ref. 49. Figure 13.7(b) shows how the spectral intensity is modified at the antiferromagnetic correlation length $\xi_{AF} = 10$.

Now we comment on the similarity to the d-density wave theory.[13] The calculation for the spectral function $A_k(\omega)$ is almost identical to that in the d-density wave theory.[51] However, the physical interpretation of $v$ is different. In d-density wave theory, $v$ is associated with d-density wave long-range order. But here $v$ is associated with the half-Skyrmions dynamics. Although the origin is different, the half-Skyrmion theory and the d-density wave theory may share some results about the pseudogap phenomenon because the Hamiltonian is almost the same.

## 13.7. Summary

In this chapter, we have reviewed the half-Skyrmion theory for high-temperature superconductivity. We have mainly focused on four topics. First, we have discussed the single hole doped system. The correlation of forming a singlet state between a doped hole spin and a copper spin has been investigated including the fact that the hole wave function extends over a space, and affected by the other copper site spins. A half-Skyrmion spin texture created by a doped hole has been described. It is shown that the half-Skyrmion excitation spectrum is in good agreement with the ARPES results in the undoped system.

The most important aspect of the half-Skyrmion is that the doped hole carries a topological charge which is represented by a gauge flux in the $CP^1$ representation. This property is in stark contrast to conventional spin polaron pictures. Because in that case interaction clouds arising from spin correlations are not characterized by a topological charge or a gauge flux.

Second, we have considered multi-half-Skymion configurations. In multi-half-Skyrmion configurations, antiferromagnetic long-range order is suppressed around the doping concentration, $x = 0.10$. The topological property of half-Skymions is kept for $x < 0.30$ as shown by numerical simulations. After suppression of antiferromagnetic long-range order, the spin correlation becomes incommensurate. Numerical simulations suggest that the origin of the incommensurate spin correlation is associated with stripe configurations of half-Skyrmions and anti-half-Skyrmions.

Third, we have discussed a mechanism of $d_{x^2-y^2}$-wave superconductivity. The gauge flux created by half-Skyrmions induces the interaction between doped holes. The interaction leads to a $d_{x^2-y^2}$-wave superconducting state of doped holes. The origin of the attractive interaction is a Lorentz force acting on a hole moving in a gauge flux created by another hole.

Finally, we have discussed a pseudogap phenomenon. We have considered a composite particle of a hole and half-Skyrmion. The pseudogap is associated with the excitation spectrum of the composite particle.

Although several aspects of the half-Skyrmion theory for high-temperature superconductivity are described in this review, we need further studies to establish the theory. In particular it is necessary to show the half-Skyrmion formation in the single hole doped system in a more convincing way to provide a sound starting point.


## Acknowledgements

I would like to thank T. Tohyama and G. Baskaran for helpful discussions. This work was supported by the Grant-in-Aid for Scientific Research from the Ministry of Education, Culture, Sports, Science and Technology (MEXT) of Japan, the Global COE Program "The Next Generation of Physics, Spun from Universality and Emergence," and Yukawa International Program for Quark-Hadron Sciences at

YITP. The numerical calculations were carried out in part on Altix3700 BX2 at YITP in Kyoto University.

# References


1. J.G. Bednorz and K.A. Müller, Possible high Tc superconductivity in the Ba-La-Cu-O system, *Z. Phys. B: Condens. Matt.* **64**(2) (1986) 189–193.
2. M. Imada, A. Fujimori and Y. Tokura, Metal-insulator transitions, *Rev. Mod. Phys.* **70**(4) (1998) 1039–1263.
3. D.J. Van Harlingen, Phase-sensitive tests of the symmetry of the pairing state in the high-temperature superconductors–Evidence for $d_{x^2-y^2}$ symmetry, *Rev. Mod. Phys.* **67**(2) (1995) 515–535.
4. E. Manousakis, The spin-$\frac{1}{2}$ Heisenberg antiferromagnet on a square lattice and its application to the cuprous oxides, *Rev. Mod. Phys.* **63**(1) (1991) 1–62.
5. T. Timusk and B.W. Statt, The pseudogap in high-temperature superconductors: an experimental survey, *Rep. Prog. Phys.* **62**(1) (1999) 61–122.
6. M.R. Norman, H. Ding, M. Randeria, J.C. Campuzano, T. Yokoya, T. Takeuchi, T. Takahashi, T. Mochiku, K. Kadowaki, P. Guptasarma and D.G. Hinks, Destruction of the fermi surface underdoped high-T-c superconductors, *Nature* **392** (1998) 157–160.
7. A. Damascelli, Z. Hussain and Z.X. Shen, Angle-resolved photoemission studies of the cuprate superconductors, *Rev. Mod. Phys.* **75**(2) (2003) 473.
8. C. Renner, B. Revaz, J.-Y. Genoud, K. Kadowaki and Ø. Fischer, Pseudogap precursor of the superconducting gap in under- and overdoped $Bi_2Sr_2CaCu_2O_{8+\delta}$, *Phys. Rev. Lett.* **80**(1) (1998) 149–152.
9. F.C. Zhang and T.M. Rice, Effective Hamiltonian for the superconducting Cu oxides, *Phys. Rev. B* **37**(7) (1988) 3759–3761.
10. P.A. Lee, N. Nagaosa and X.G. Wen, Doping a mott insulator: Physics of high-temperature superconductivity, *Rev. Mod. Phys.* **78**(1) (2006) 17–85.
11. P.W. Anderson, The resonating valence bond state in $La_2CuO_4$ and superconductivity, *Science* **235** (1987) 1196–1198.
12. S.A. Kivelson, I.P. Bindloss, E. Fradkin, V. Oganesyan, J.M. Tranquada, A. Kapitulnik and C. Howald, How to detect fluctuating stripes in the high-temperature superconductors, *Rev. Mod. Phys.* **75**(4) (2003) 1201–1241.
13. S. Chakravarty, R.B. Laughlin, D.K. Morr and C. Nayak, Hidden order in the cuprates, *Phys. Rev. B* **63**(9) (2001) 094503.
14. D.P. Arovas and A. Auerbach, Functional integral theories of low-dimensional quantum Heisenberg models, *Phys. Rev. B* **38**(1) (1988) 316–332.
15. J.E. Hirsch and S. Tang, Comment on a mean-field theory of quantum antiferromagnets, *Phys. Rev. B* **39**(4) (1989) 2850–2851.
16. D. Yoshioka, Boson mean field theory of the square lattice Heisenberg model, *J. Phys. Soc. Jpn.* **58** (1989) 3733–3745.
17. M. Raykin and A. Auerbach, $1/N$ expansion and long range antiferromagnetic order, *Phys. Rev. Lett.* **70**(24) (1993) 3808–3811.
18. A. Fetter and J. Walecka, *Quantum Theory of Many-Particle Systems* (Courier Dover Publications, 2003).
19. S. Chakravarty, B.I. Halperin and D.R. Nelson, Low-temperature behavior of two-dimensional quantum antiferromagnets, *Phys. Rev. Lett.* **60**(11) (1988) 1057–1060.

20. R.J. Gooding, Skyrmion ground states in the presence of localizing potentials in weakly doped $CuO_2$ planes, *Phys. Rev. Lett.* **66**(17) (1991) 2266–2269.
21. S. Haas, F.-C. Zhang, F. Mila and T.M. Rice, Spin and charge texture around in-plane charge centers in the $CuO_2$ planes, *Phys. Rev. Lett.* **77**(14) (1996) 3021–3024.
22. A.A. Belavin and A.M. Polyakov, Metastable states of two-dimensional isotropic ferromagnets, *JETP Lett.* **22**(10) (1975) 245–248.
23. R. Rajaraman, *Solitons and instantons* (North-Holland, 1987).
24. T.K. Ng, Topological spin excitations of Heisenberg antiferromagnets in two dimensions, *Phys. Rev. Lett.* **82**(17) (1999) 3504–3507.
25. T. Morinari, Half-skyrmion picture of a single-hole-doped $CuO_2$ plane, *Phys. Rev. B* **72** (2005) 104502.
26. B.O. Wells, Z.X. Shen, A. Matsuura, D.M. King, M.A. Kastner, M. Greven and R.J. Birgeneau, E versus k relations and many-body effects in the model insulating copper-oxide $Sr_2CuO_2Cl_2$, *Phys. Rev. Lett.* **74**(6) (1995) 964–967.
27. B.B. Beard, R.J. Birgeneau, M. Greven and U.-J. Wiese, Square-lattice Heisenberg antiferromagnet at very large correlation lengths, *Phys. Rev. Lett.* **80**(8) (1998) 1742–1745.
28. J.-K. Kim and M. Troyer, Low temperature behavior and crossovers of the square lattice quantum Heisenberg antiferromagnet, *Phys. Rev. Lett.* **80**(12) (1998) 2705–2708.
29. R.R.P. Singh, Thermodynamic parameters of the $T = 0$, spin-1/2 square-lattice Heisenberg antiferromagnet, *Phys. Rev. B* **39**(13) (1989) 9760–9763.
30. T. Morinari, Strong-coupling analysis of $QED_3$ for excitation spectrum broadening in the undoped high-temperature superconductors, *Phys. Rev. B* **77**(7) (2008) 075128.
31. M. Berciu and S. John, Magnetic structure factor in cuprate superconductors: Evidence for charged meron vortices, *Phys. Rev. B* **69**(22) (2004) 224515.
32. S.I. Belov and B.I. Kochelaev, Nuclear spin relaxation in two-dimensional Heisenberg antiferromagnet $S = 1/2$ with skyrmions, *Solid State Commun.* **106**(4) (1998) 207–210.
33. C. Timm and K.H. Bennemann, Doping dependence of the Néel temperature in Mott-Hubbard antiferromagnets: Effect of vortices, *Phys. Rev. Lett.* **84**(21) (2000) 4994–4997.
34. S.-W. Cheong, G. Aeppli, T.E. Mason, H. Mook, S.M. Hayden, P.C. Canfield, Z. Fisk, K.N. Clausen and J.L. Martinez, Incommensurate magnetic fluctuations in $La_{2-x}Sr_xCuO_4$, *Phys. Rev. Lett.* **67**(13) (1991) 1791–1794.
35. K. Yamada, C.H. Lee, K. Kurahashi, J. Wada, S. Wakimoto, S. Ueki, H. Kimura, Y. Endoh, S. Hosoya, G. Shirane, R.J. Birgeneau, M. Greven, M.A. Kastner and Y.J. Kim, Doping dependence of the spatially modulated dynamical spin correlations and the superconducting-transition temperature in $La_{2-x}Sr_xCuO_4$, *Phys. Rev. B* **57**(10) (1998) 6165–6172.
36. M. Matsuda, M. Fujita, K. Yamada, R.J. Birgeneau, M.A. Kastner, H. Hiraka, Y. Endoh, S. Wakimoto and G. Shirane, Static and dynamic spin correlations in the spin-glass phase of slightly doped $La_{2-x}Sr_xCuO_4$, *Phys. Rev. B* **62**(13) (2000) 9148–9154.
37. T. Morinari, Mechanism of $d_{x^2-y^2}$-wave superconductivity based on hole-doping-induced spin texture in high $T_c$ cuprates, *Phys. Rev. B* **73**(6) (2006) 064504.
38. T. Morinari, Mechanism of unconventional superconductivity induced by skyrmion excitations in two-dimensional strongly correlated electron systems, *Phys. Rev. B* **65** (6) (2002) 064513.
39. B.I. Shraiman and E.D. Siggia, Mobile vacancies in a quantum Heisenberg antiferromagnet, *Phys. Rev. Lett.* **61**(4) (1988) 467–470.

40. M. Sigrist and K. Ueda, Phenomenological theory of unconventional superconductivity, *Rev. Mod. Phys.* **63**(2) (1991) 239–311.
41. M. Greiter, X.G. Wen and F. Wilczek, Paired Hall states, *Nucl. Phys.* B **374**(3) (1991) 567–614.
42. T. Morinari, Composite fermion pairing theory in single-layer systems, *Phys. Rev. B* **62**(23) (2000) 15903–15912.
43. S.-C. Zhang, The Chern-Simons-Landau-Ginzburg theory of the fractional quantum Hall effect, *Int. J. Mod. Phys. B* **6**(1) (1992) 25–58.
44. T. Morinari, Fermi arc formation by chiral spin textures in high-temperature superconductors, *J. Phys. Chem. Solids* **69**(12) (2008) 2690–2692.
45. T. Tohyama and S. Maekawa, Angle-resolved photoemission in high $T_c$ cuprates from theoretical viewpoints, *Supercond. Sci. Technol.* **13**(4) (2000) R17–R32.
46. A. Auerbach, B.E. Larson and G.N. Murthy, Landau-level spin waves and Skyrmion energy in the two-dimensional Heisenberg antiferromagnet, *Phys. Rev. B* **43**(13) (1991) 11515–11518.
47. A.A. Abrikosov, L.P. Gor'kov and I. Dzyaloshinskii, *Methods of Quantum Field Theory in Statistical Physics* (Pergamon, New York, 1965).
48. M.R. Norman, A. Kanigel, M. Randeria, U. Chatterjee and J.C. Campuzano, Modeling the fermi arc in underdoped cuprates, *Phys. Rev. B* **76** (2007) 174501.
49. N. Harrison, R.D. McDonald and J. Singleton, Cuprate fermi orbits and fermi arcs: The effect of short-range antiferromagnetic order, *Phys. Rev. Lett.* **99**(20) (2007) 206406.
50. T. Morinari, Pseudogap and short-range antiferromagnetic correlation controlled fermi surface in underdoped cuprates: From fermi arc to electron pocket, arXiv: 0805.1977 (2008).
51. S. Chakravarty, C. Nayak and S. Tewari, Angle-resolved photoemission spectra in the cuprates from the d-density wave theory, *Phys. Rev. B* **68**(10) (2003) 100504(R).

# Chapter 14

# Deconfined Quantum Critical Points

T. Senthil,[1]* Ashvin Vishwanath,[1] Leon Balents,[2] Subir Sachdev,[3] Matthew P. A. Fisher[4]

The theory of second-order phase transitions is one of the foundations of modern statistical mechanics and condensed-matter theory. A central concept is the observable order parameter, whose nonzero average value characterizes one or more phases. At large distances and long times, fluctuations of the order parameter(s) are described by a continuum field theory, and these dominate the physics near such phase transitions. We show that near second-order quantum phase transitions, subtle quantum interference effects can invalidate this paradigm, and we present a theory of quantum critical points in a variety of experimentally relevant two-dimensional antiferromagnets. The critical points separate phases characterized by conventional "confining" order parameters. Nevertheless, the critical theory contains an emergent gauge field and "deconfined" degrees of freedom associated with fractionalization of the order parameters. We propose that this paradigm for quantum criticality may be the key to resolving a number of experimental puzzles in correlated electron systems and offer a new perspective on the properties of complex materials.

Much recent research in condensed-matter physics has focused on the behavior of matter near zero-temperature "quantum" phase transitions that are seen in several strongly correlated many-particle systems (*1*). Indeed, a popular view asribes many properties of correlated materials to the competition between qualitatively distinct ground states and the associated phase transitions. Examples of such materials include the cuprate high-temperature superconductors and the rare earth intermetallic compounds (known as the heavy fermion materials).

The traditional guiding principle behind the modern theory of critical phenomena is the association of the critical singularities with fluctuations of an order parameter that encapsulates the difference between the two phases on either side

[1]Department of Physics, Massachusetts Institute of Technology, Cambridge, MA 02139, USA.
[2]Department of Physics, University of California, Santa Barvara, CA 93106–4030, USA.
[3]Department of Physics, Yale University, P.O. Box 208120, New Haven, CT 06520–8120, USA.
[4]Kavli Institute for Theoretical Physics, University of California, Santa Barbara, CA 93106–4030, USA.

*To whom correspondence should be addressed. Email: senthil@mit.edu

of the critical point (a simple example is the average magnetic moment, which distinguishes ferromagnetic iron at room temperature from its high-temperature paramagnetic state). This idea, developed by Landau and Ginzburg (*2*), has been eminently successful in describing a wide variety of phase-transition phenomena. It culminated in the sophisticated renormalization group theory of Wilson (*3*), which gave a general prescription for understanding the critical singularities. Such an approach has been adapted to examine quantum critical phenomena as well and provides the generally accepted framework for theoretical descriptions of quantum transitions.

We present specific examples of quantum phase transitions that do not fit into this Landau-Ginzburg-Wilson (LGW) paradigm (*4*). The natural field theoretic description of their critical singularities is not in terms of the order parameter field(s) that describe the bulk phases, but in terms of new degrees of freedom specific to the critical point. In our examples, there is an emergent gauge field that mediates interactions between emergent particles that carry fractions of the quantum numbers of the underlying degrees of freedom. These fractional particles are not present (that is, are confined) at low energies on either side of the transition but appear naturally at the transition point. Laughlin has previously argued for fractionalization at quantum critical points on phenomenological grounds (*5*).

We present our examples using phase transitions in two-dimensional (2D) quantum magnetism, although other points of view are also possible (*6*). Consider a system of spin $S = 1/2$ moment $\vec{S}_r$ on the sites, $r$, of a 2D square lattice with the Hamiltonian

$$H = J \sum_{\langle rr' \rangle} \vec{S}_r \cdot \vec{S}_{r'} + \cdots \tag{1}$$

where $J > 0$ is the antiferromagnetic exchange interaction, and the ellipses represent other short-range interactions that may be tuned to drive various zero-temperature phase transitions.

Considerable progress has been made in elucidating the possible ground states of such a Hamiltonian. The Néel state has long-range magnetic order (Fig. 1A) and has been observed in a variety of insulators, including the prominent parent compound of the cuprates: $LA_2CuO_4$. Apart from such magnetic states, it is now recognized that models in the class of $H$ can exhibit a variety of quantum paramagnetic ground states. In such states, quantum fluctuations prevent the spins from developing magnetic long-range order. One class of paramagnetic states is the valence bond solids (VBS) (Fig. 1B). In such states, pairs of nearby spins form a singlet, resulting in an ordered pattern of valence bonds. Typically, such VBS states have an energy gap to spin-carrying excitations. Furthermore, for spin-1/2 systems on a square lattice, such states also necessarily break lattice translational symmetry. A second class of paramagnets has a liquid of valence bonds and need not break lattice translational symmetry, but we will not consider such states here. Our focus is on the nature of the phase transition between the ordered magnet and a VBS.

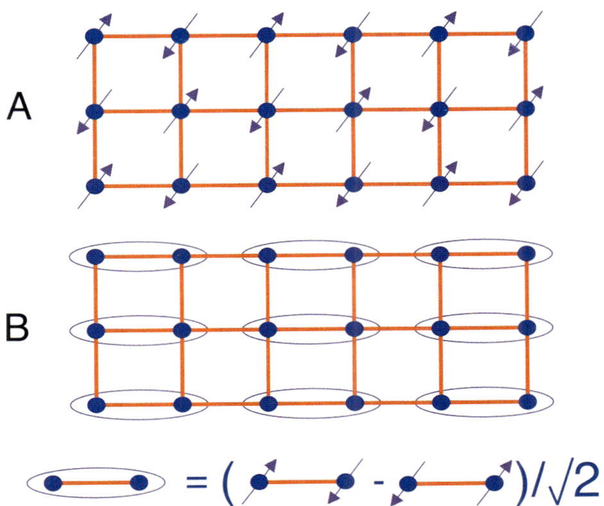

**Fig. 1. (A)** The magnetic Néel ground state of the Hamiltonian Eq. 1 on the square lattice. The spins, $\vec{S}_r$, fluctuate quantum-mechanically in the ground state, but they have a nonzero average magnetic moment, which is oriented along the directions shown. **(B)** A VBS quantum paramagnet. The spins are paired in singlet valence bonds, which resonate among the many different ways the spins can be paired up. The valence bonds crystallize, so that the pattern of bonds shown has a larger weight in the ground state wavefunction than its symmetry-related partners (obtained by 90° rotations of the above states about a site). This ground state is therefore fourfold degenerate.

We also restrict our discussion to the simplest kinds of ordered antiferromagnets: those with collinear order, where the order parameter is a single vector (the Néel vector).

Both the magnetic Néel state and the VBS are states of broken symmetry. The former breaks spin rotation symmetry and the latter that of lattice translations. The order parameters associated with these two different broken symmetries are very different. A LGW description of the competition between these two kinds of orders generically predicts either a first-order transition or an intermediate region of coexistence where both orders are simultaneously present. A direct second-order transition between these two broken symmetry phases requires fine-tuning to a multicritical point. Our central thesis is that for a variety of physically relevant quantum systems, such canonical predictions of LGW theory are incorrect. For $H$, we will show that a generic second-order transition is possible between the very different kinds of broken symmetry in the Néel and VBS phases. Our critical theory for this transition is, however, unusual and is not naturally described in terms of the order parameter fields of either phase. A picture related to the one developed here applies also to transitions between valence bond liquid and VBS states (*7*) and to transitions between different VBS states (*8*) in the quantum dimer model (*9, 10*).

**Field theory and topology of quantum antiferromagnets.** In the Néel phase or close to it, the fluctuations of the Néel order parameter are captured correctly by the well-known O(3) nonlinear sigma model field theory (*11–13*), with the following action $S_n$ in spacetime [we have promoted the lattice coordinate $r = (x, y)$ to a continuum spatial coordinate, and $\tau$ is imaginary time]:

$$S_n = \frac{1}{2g} \int d\tau \int d^2r \left[ \frac{1}{c^2} \left( \frac{\partial \hat{n}}{\partial \tau} \right)^2 + (\nabla_r \hat{n})^2 \right] + iS \sum_r (-1)^r A_r \tag{2}$$

Here $\hat{n} \propto (-1)^r \vec{S}_r$ is a unit three-component vector that represents the Néel order parameter [the factor $(-1)^r$ is $+1$ on one checker-board sublattice and $-1$ on the other]. The second term is the quantum-mechanical Berry phase of all the $S = 1/2$ spins: $A_r$ is the area enclosed by the path mapped by the time evolution of $\hat{n}_r$ on a unit sphere in spin space. These Berry phases play an unimportant role in the low-energy properties of the Néel phase (*12*) but are crucial in correctly describing the quantum paramagnetic phase (*14*). We show here that they also modify the quantum critical point between these phases, so that the exponents are distinct from those of the LGW theory without the Berry phases studied earlier (*12, 15*).

To describe the Berry phases, first note that in two spatial dimensions, smooth configurations of the Néel vector admit topological textures known as skyrmions (Fig. 2). The total skyrmion number associated with a configuration defines an integer topological quantum number $Q$

$$Q = \frac{1}{4\pi} \int d^2r \, \hat{n} \cdot \partial_x \hat{n} \times \partial_y \hat{n} \tag{3}$$

The sum over $r$ in Eq. 2 vanishes (*1, 11*) for all spin time histories with smooth equal-time configuratons, even if they contain skyrmions. For such smooth configurations, the total skyrmion number $Q$ is independent of time. However, the original microscopic model is defined on a lattice, and processes where $Q$ changes by some integer amount are allowed. Specifically, such a $Q$ changing event corresponds to a monopole (or hedge-hog) singularity of the Néel field $\hat{n}(r, \tau)$ in spacetime (a hedge-bog has $\hat{n}$ oriented radially outward in all spacetime directions away from its center).

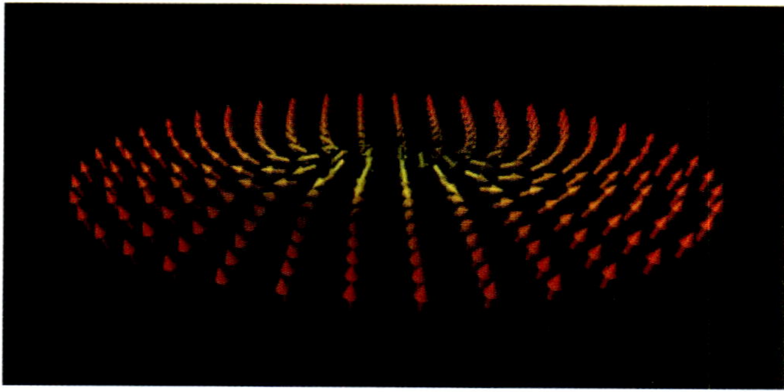

**Fig. 2.** A skyrmion configuration of the field $\hat{n}(r)$. Note that $\hat{n} = (-1)^r \vec{S}_r$, and so the underlying spins have a rapid sublattice oscillation, which is not shown. The skyrmion above has $\hat{n}(r = 0) = (0, 0, -1)$ and $\hat{n}(|r| \to \infty) = (0, 0, 1)$.

Haldance (*11*) showed that the sum over $r$ in Eq. 2 is nonvanishing in the presence of such monopole events. Precise calculation (*11*) gives a total Berry phase associated with each such $Q$ changing process, which oscillates rapidly on four sublattices of the dual lattice. This leads to destructive interference, which effectively suppresses all monopole events unless they are quadrupled (*11, 14*) (that is, they change $Q$ by 4).

The sigma model field theory augmented by these Berry phase terms is, in principle, powerful enough to correctly describe the quantum paramagent. Summing over the various monopole tunneling events shows that in the paramagnetic phase, the presence of the Berry phases leads to VBS order (*14*). Thus, $S_n$ contains within it the ingredients describing both the ordered phases of $H$. However, a description of the transition between these phases has so far proved elusive and will be provided here.

Our analysis of this critical point is aided by writing the Néel field $\hat{n}$ in the so-called CP$^1$ parametrization

$$\hat{n} = z^\dagger \vec{\sigma} z \tag{4}$$

with $\vec{\sigma}$ a vector of Pauli matrices. Here $z = z(r, \tau) = (z_1, z_2)$ is a two-component complex spinor of unit magnitude, which transforms under the spin-1/2 representation of the SU(2) group of spin rotations. The $z_{1,2}$ are the fractionalized "spinon" fields. To understand the monopoles in this representation, let us recall that the CP$^1$ representation has a U(1) gauge redundancy. Specifically, the local phase rotation

$$z \rightarrow e^{i\gamma(r,\tau)} z \tag{5}$$

leaves $\hat{n}$ invariant and hence is a gauge of degree of freedom. Thus, the spinon fields are coupled to a U(1) gauge field, $a_\mu$ [the space-time index $\mu = (r, \tau)$]. As is well known, the magnetic flux of $a_\mu$ is the topological charge density of $\hat{n}$ appearing in the integrand of Eq. 3. Specifically, configurations where the $a_\mu$ flux is $2\pi$ correspond to a full skyrmion (in the ordered Néel phase). Thus, the monopole events described above are spacetime magnetic monopoles (instantons) of $a_\mu$ at which $2\pi$ gauge flux can either disappear or be created. The fact that such instanton events are allowed means that the $a_\mu$ gauge field is to be regarded as compact.

We now state our key result for the critical theory between the Néel and VBS phases. We argue below that the Berry phase–induced quadrupling of monopole events renders monopoles irrelevant at the quantum critical point. So in the critical regime (but not away from it in the paramagnetic phase), we may neglect the compactness of $A_\mu$ and write down the simplest critical theory of the fractionalized spinons interacting with a noncompact U(1) gauge field with action $S_z = \int d^2 r d\tau L_z$ and

$$L_z = \sum_{\alpha=1}^{N} |(\partial_\mu - i a_\mu) z_\alpha|^2 + s|z|^2 + u(|z|^2)^2 + \kappa(\epsilon_{\mu\nu\kappa} \partial_\nu a_\kappa)^2. \tag{6}$$

Where $N = 2$ is the number of $z$ components, we have softened the length constaint on the spinons, with $|z|^2 \equiv \sum_{\alpha=1}^{N} |z_\alpha|^2$ allowed to fluctuate and the value of $s$ is to be tuned so that $L_z$ is at its scale-invariant critical point. The irrelevance of monopole tunneling events at the critical fixed point implies that the total gauge flux $\int d^2 r (\partial_x a_y - \partial_y a_x)$, or equivalently the skyrmion number $Q$, is asymptotically conserved. This emergent global topological conservation law provides precise meaning to the notion of deconfinement. It is important to note that the critical theory described by $L_z$ (*16*) is distinct from the LGW critical theory of the O(3) nonlinear sigma model obtained from Eq. 2 by dropping the Berry phases and tunning $g$ to a critical value (*17*). In particular, the latter model has a nonzero rate of monopole tunneling events at the transition, so that the global skyrmion number $Q$ is not conserved.

Among the important physical consequences of the theory $L_z$ (*7, 18*) are the presence of two diverging length scales upon approaching the critical point from the VBS side (the spin correlation length and a longer scale beyond which two spinons interact with a linear confining potential) and a large anomalous dimension for the Néel order parameter (because it is a composite of the critical spinons).

The critical theory $L_z$ is actually implied by existing results in the $N \to \infty$ limit (*18*). The following section illustrates the origin of $L_z$ by a physical derivation for the case of "easy-plane" anisotropy, when the spins prefer to lie in the $xy$ plane. Such arguments can be generalized to the isotropic case (*7, 18*).

**Duality transformations with easy-plane anisotropy.** For the easy-plane case, duality maps and an explicit derivation of a dual form of $L_z$ are already available in the literature (*6, 19*). Here we obtain this theory using simple physical arguments.

The easy-plane anisotropy reduces the continuous SU(2) spin rotational invariance to the U(1) subgroup of rotations about the $z$ axis of spin, along with a $Z_2$ (ising) spin reflection symmetry along the $z$ axis. With these symmetries, Eq. 2 allows an additional term $u_{ep} \int d\tau d^2 r (n^z)^2$, with $u_{ep} > 0$.

The classical Néel ground state of the easy-plane model $\hat{n}$ is independent of position and lies in the spin $xy$ plane. Topological defects above this ground state play an important role. These are vortices in the complex field $n^+ = n_x + in_y$, and along a large loop around the vortex the phase of $n^+$ winds by $2\pi m$, with $m$ an integer. In the core, the XY order is suppressed and the $\hat{n}$ vector will point along the $\pm \hat{z}$ direction. This corresponds to a nonzero staggered magnetization of the $z$ component of the spin in the core region. Thus, at the classical level, there are two kinds of vortices, often called merons, depending on the direction of the $\hat{n}$ vector at the core (Fig. 3). Either kind of vortex breaks the Ising-like $n^z \to -n^z$ symmetry at the core. Let us denote by $\Psi_1$ the quantum field that destroys a vortex whose core points in the up direction and by $\Psi_2$ the quantum field that destroys a vortex whose core points in the down direction.

Clearly, this breaking of the Ising symmetry is an artifact of the classical limit: Once quantum effects are included, the two broken symmetry cores will be able to

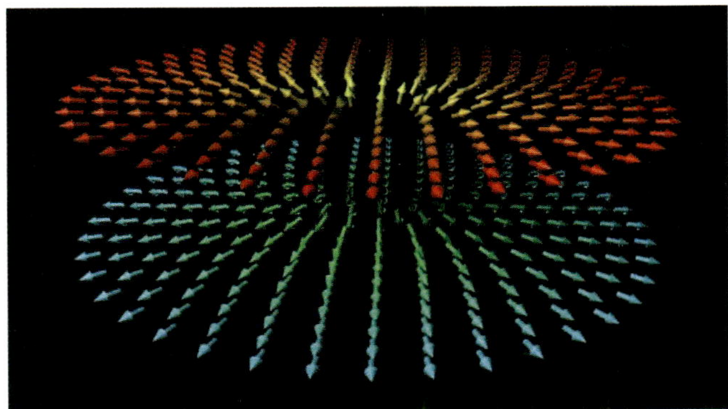

**Fig. 3.** The meron vortices $\Psi_1$ (above) and $\Psi_2$ in the easy-plane case. The $\Psi_1$ meron above has $\hat{n}(r=0)=(0,0,1)$ and $\hat{n}(|r|\to\infty)=(x,y,0)/|r|$; the $\Psi_2$ meron has $\hat{n}(r=0)=(0,0,-1)$ and the same limit as $|r|\to\infty$. Each meron above is half the skyrmion in Fig. 2: A composite of $\Psi_1$ and $\Psi_2^*$ makes one skyrmion.

tunnel into each other, and there will be no true broken Ising symmetry in the core. This tunneling is often called an "instanton" process that connects two classically degenerate states.

Surprisingly, such an instanton event is physically the easy-plane avatar of the space-time monopole described above for the fully isotropic model. This may be seen pictorially. Each classical vortex of Fig. 3 really represents half of the skyrmion configuration of Fig. 2. Now imagine the $\Psi_2$ meron at time $\tau\to-\infty$ and the $\Psi_1$ meron at time $\tau\to\infty$. These two configurations cannot be smoothly connected, and there must be a singularity in the $\hat{n}$ configuration, which we place at the origin of spacetime. A glance at Fig. 3 shows that the resulting configuration of $\hat{n}$ can be smoothly distorted into the radially symmetric hedgehog/monopole event. Thus, the tunneling process between the two merons is equivalent to creating a full skyrmion. This is precisely the monopole event. Hence, a skyrmion may be regarded as a composite of an "up" meron and a "down" antimeron, and the skyrmion number is the difference in the numbers of up and down merons.

The picture so far has not accounted for the Berry phases. The interference effect discussed above for isotropic antiferromagnets applies here too, leading to an effective cancellation of instanton tunneling events between single $\Psi_1$ and $\Psi_2$ merons. The only effective tunnelings are those in which four $\Psi_1$ merons come together and collectively flip their core spins to produce four $\Psi_2$ merons, or vice versa.

A different perspective on the $\Psi_{1,2}$ meron vortices is provided by the $\mathrm{CP}^1$ representation. Ordering in the $xy$ plane of spin space requires condensing the spinons

$$|\langle z_1\rangle| = |\langle z_2\rangle| \neq 0 \tag{7}$$

so that $n^+ = z_1^* z_2$ is ordered and there is no average value of $n^2 = |z_1|^2 - |z_2|^2$. Now, clearly, a full $2\pi$ vortex in $n^+$ can be achieved by either having a $2\pi$ vortex in

$z_1$ and not in $z_2$, or a $2\pi$ antivortex in $z_2$ and no vorticity in $z_1$. In the first choice, the amplitude of the $z_1$ condensate will be suppressed at the core, but $\langle z_2 \rangle$ will be unaffected. Consequently, $n^2 = |z_1|^2 - |z_2|^2$ will be nonzero and negative in the core, as in the $\Psi_2$ meron. The other choice also leads to nonzero $n^2$, which will now be positive, as in the $\Psi_1$ meron. Clearly, we may identify the $\Psi_2(\Psi_1)$ meron vortices with $2\pi$ vortices (antivortices) in the spinon fields $z_1(z_2)$. Note that in terms of the spinons, paramagnetic phases correspond to situations in which neither spinon field is condensed.

The above considerations and the general principles of boson duality in three spacetime dimensions (*20*) determine the form of the dual action $S_{\mathrm{dual}} = \int d\tau d^2 r L_{\mathrm{dual}}$ for $\Psi_{1,2}$ (*6, 19*)

$$L_{\mathrm{dual}} = \sum_{\alpha=1,2} |(\partial_\mu - iA_\mu)\Psi_\alpha|^2 + r_d|\Psi|^2 + u_d(|\Psi|^2)^2 + v_d|\Psi_1|^2|\Psi_2|^2$$

$$+ \kappa_d(\epsilon_{\mu\nu\kappa}\partial_\nu A_\kappa)^2 - \lambda[(\Psi_2^*\Psi_2)^4 + (\Psi_2^*\Psi_1)^4] \tag{8}$$

where

$$|\Psi|^2 = |\psi_2|^2 + |\Psi_2|^2$$

The correctness of this form may be argued as follows: First, from the usual boson-vortex duality transformation (*20*), the dual $\Psi_{1,2}$ vortex fields must be minimally coupled to a dual noncompact U(1) gauge field $A_\mu$. This dual gauge invariance is not related to Eq. 5 but is a consequence of the conservation of the total $S^z$: the "magnetic" flux $\varepsilon_{\mu\nu\kappa}\partial_\nu A_\kappa$ is the conserved $S^z$ current (*20*). Second, under the $Z_2$ reflection symmetry, the two vortices get interchanged; that is, $\Psi_1 \leftrightarrow \Psi_2$. The dual action must therefore be invariant under interchange of the 1 and 2 labels. Finally, if monopole events were to be disallowed by hand, the total skyrmion number — (the difference in the number of up and down meron vortices) would be conserved. This would imply a global U(1) symmetry [not to be confused with the U(1) spin symmetry about the $z$ axus] under which

$$\Psi_1 \to \Psi_1 \exp(i\phi); \quad \Psi_2 \to \Psi_2 \exp(-i\phi) \tag{9}$$

where $\phi$ is a constant. However, monopole events destroy the conservation of skyrmion number and hence this dual global U(1) symmetry. But because the monopoles are effectively quadrupled by cancellations from the Berry phases, skyrmion number is still conserved modulo 4. Thus, the symmetry in Eq. 9 must be broken down to the discrete cyclic group of four elements, $Z_4$.

The dual Lagrangian in Eq. 8 is the simplest one that is consistent with all these requirements. In particular, we note that in the absence of the $\lambda$ term, the dual global U(1) transformation in Eq. 9 leaves the Lagrangian invariant. The $\lambda$ term breaks this down to $Z_4$ as required. Thus, we may identify this term with the quadrupled monopole tunneling events. Berry phases are therefore explicitly included in $L_{\mathrm{dual}}$.

In this dual vortex theory, the XY ordered phase is simply characterized as a dual "paramagnet," where $\langle \Psi_{1,2} \rangle = 0$ and fluctuations of $\Psi_{1,2}$ are gapped. On the other hand, spin paramagnetic phases such as the VBS states correspond to condensates of the fields $\Psi_{1,2}$, which break the dual gauge symmetry. In particular, if both $\Psi_1$ and $\Psi_2$ condense with equal amplitude $|\langle \Psi_1 \rangle| = |\langle \Psi_2 \rangle| \neq 0$, then we obtain a paramagnetic phase where the global Ising symmetry is preserved. Note the remarkable complementarity between the description of the phases in this dual theory with that in terms of the spinon fields of the $CP^1$ representation: The descriptions map onto one another upon interchanging both $z_{1,2} \leftrightarrow \Psi_{1,2}$ and the role of the XY ordered and paramagnetic phases. This is a symptom of an exact duality between the two descriptions that obtains close to the transition (*17, 18*).

The combination $\Psi_1^* \Psi_2 \equiv |\Psi_1 \Psi_2| e^{i(\theta_1 - \theta_2)}$ actually serves as an order parameter for the translation symmetry — broken VBS ground state. This may be seen from the analysis of (*6, 19*). Alternatively, we may use the identification (*14*) of the skyrmion creation operator with the order parameter for translation symmetry breaking. Such a condensate of $\Psi_{1,2}$ breaks the global $Z_4$ symmetry of the action in Eq. 8. The preferred direction of the angle $\theta_1 - \theta_2$ depends on the sign of $\lambda$. The two sets of preferred directions correspond to columnar and plaquette patterns of translational symmetry breaking (Fig. 4). Also, the breaking of the dual $U(1)$ symmetry in Eq. 9 by $\lambda$ corresponds to a linear confinement of spinons in the paramagnet.

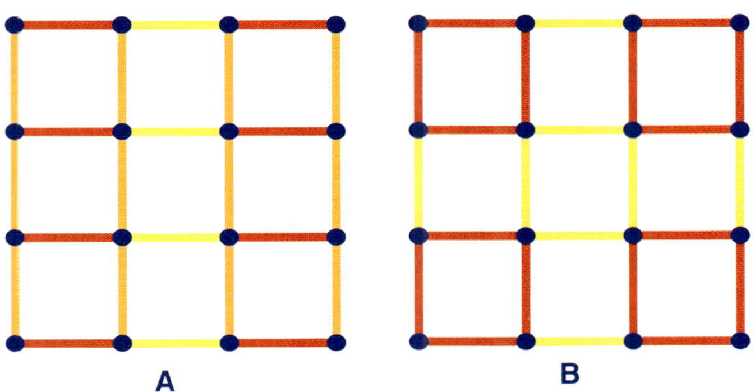



**Fig. 4.** Pattern of symmetry-breaking in the two possible VBS states (**A** and **B**) predicted by Eq. 8. The last term in Eq. 8 leads to a potential, $-\lambda \cos[4(\theta_1 - \theta_2)]$, and the sign of $\lambda$ chooses between the two states above. The distinct lines represent distinct values of $\langle \vec{S}_r \cdot \vec{S}_{r'} \rangle$ on each link. Note that the state in (A) is identical to that in Fig. 1B.

Despite its importance in the paramagnet, the $\lambda$ term is irrelevant at the critical point (*18*). In critical phenomena parlance, it is a dangerously irrelevant perturbation (*15*). Consequently the critical theory is deconfined, and the $z_{1,2}$ spinons (which are fractions of $n^+$), or in the dual description the $\Psi_{1,2}$ merons (which are fractions of a skyrmion), emerge as natural degrees of freedom right at the critical point. The spinons are confined in both adjacent phases, but the confinement length scale diverges on approaching the critical point. At a more sophisticated level, the critical fixed point is characterized by the emergence of an extra global U(1) symmetry in Eq. 9 that is not present in the microscopic Hamiltonian. This is associated with the conservation of skyrmion number and follows from the irrelevance of monopole tunneling events only at the critical point.

The absence of monopoles at the critical point, when generalized to the isotropic case (*18*), provides one of the justifications for the claimed critical theory in Eq. 6.

**Discussion.** Our results offer a new perspective on the phases of Mott insulators in two dimensions: Liquid resonating–valence-bond-like states, with gapless spinon excitations, can appear at isolated critical points between phases characterized by conventional confining orders. It appears probable that similar considerations apply to quantum critical points in doped Mott insulators, between phases with a variety of spin- and charge-density-wave orders and $d$-wave superconductivity. If so, the electronic properties in the quantum critical region of such critical points will be strongly non–Fermi-liquid-like, raising the prospect of understanding the phenomenology of the cuprate superconductors.

On the theoretical side, our results also illuminate studies of frustrated quantum antiferromagnets in two dimensions. A theory of the observed critical point between the Néel and VBS phases (*21*) is now available, and precise tests of the values of critical exponents should now be possible. A variety of other SU(2)-invariant antiferromagnets have been studied (*22*), and many of them exhibit VBS phases. It would be interesting to explore the characteristics of the quantum critical points adjacent to these phases and test our prediction of deconfinement at such points.

Our results also caricature interesting phenomena (*23, 24*) in the vicinity of the onset of magnetism in the heavy fermion metals. Remarkably, the Kondo coherence that characterizes the nonmagnetic heavy Fermi liquid seems to disappear at the same point at which magnetic long-range order sets in. Furthermore, strong deviations from Fermi liquid theory are seen in the vicinity of the quantum critical point. All of this is in contrast to naïve expectations based on the LGW paradigm for critical phenomena. However, this kind of exotic quantum criticality between two conventional phases is precisely the physics discussed in the present paper.

## References


1. S. Sachdev, *Quantum Phase Transition* (Cambridge Univ. Press, Cambridge, 1999).
2. L. D. Landau, E. M. Lifshitz, E. M. Pitaevskii, *Statistical Physics* (Butterworth-Heinemann, New York, 1999).
3. K. G. Wilson, J. Kogut, *Phys. Rep.* **12**, 75 (1974).
4. The Landau paradigm is also known to fail near 1D quantum critical points (or 2D classical critical points), such as the model considered by Haldane (*25*). This failure is caused by strong fluctuations in a low-dimensional system, a mechanism that does not generalize to higher dimensions.
5. R. B. Laughlin, *Adv. Phys.* **47**, 943 (1998).
6. C. Lannert, M. P. A. Fisher, T. Senthil, *Phys. Rev. B* **63**, 134510 (2001).

7. T. Senthil, L. Balents, S. Sachdev, A. Vishwanath, M. P. A. Fisher, http://xxx.lanl.gov/abs/cond-mat/0312617.

8. A. Vishwanath, L. Balents, T. Senthil, http://xxx.lanl.gov/abs/con-mat/0311085.

9. D. S. Rokhsar, S. A. Kivelson, *Phys. Rev. Lett.* **61**, 2376 (1988).

10. R. Moessner, S. L. Sondhi, E. Fradkin, *Phys. Rev. B* **65**, 024504 (2002).

11. F. D. M. Haldane, *Phys. Rev. Lett.* **61**, 1029 (1988).

12. S. Chakravarty, B. I. Halperin, D. R. Nelson, *Phys. Rev. B* **39**, 2344 (1989). These authors correctly noted that the Berry phases could at least be neglected in the Néel phase, but perhaps not beyond it; their critical theory applies to square lattice models with the spin $S$ an even integer (but not to $S$ half-odd-integer) and to dimerized or double-layer antiferromagnets with an even number of $S = 1/2$ spins per unit cell [such as the model in (*13*)].

13. M. Troyer, M. Imada, K. Ueda, *J. Phys. Soc. Jpn.* **66**, 2957 (1997).

14. N. Read, S. Sachdev, *Phys. Rev. Lett.* **62**, 1694 (1989).

15. A. V. Chubukov, S. Sachdev, J. Ye, *Phys. Rev. B* **49**, 11919 (1994). Speculations on the dangerous irrelevancy of Berry phase effects appeared here.

16. Halperin *et al.* (*26*) studied the critical theory of $L_z$ using expansions in $4 - D$ ($D$ is the dimension of spacetime) and in $1/N$. The former yielded a first-order transition and the latter second-order transition. Subsequent duality and numerical studies (*17, 20*) have shown the transition is second-order in $D = 3$ for $N = 1, 2$.

17. O. Motrunich, A. Vishwanath, http://xxx.lanl.gov/abs/cond-mat/0311222.

18. Scaling analyses, generalizations, and physical consequences appear in the supporting material on *Science* Online.

19. S. Sachdev. K. Park, *Ann. Phys. N.Y.* **298**, 58 (2002).

20. C. Dasgupta, B. I. Halperin, *Phys. Rev. Lett.* **47**, 1556 (1981).

21. A. W. Sandvik, S. Daul, R. R. P. Singh, D. J. Scalapino, *Phys. Rev. Lett.* **89**, 247201 (2002).

22. C. Lhuillier, G. Misguich, http://xxx.lanl.gov/abs/cond-mat/0109146.

23. P. Coleman, C. Pépin, Q. Si, R. Ramazashvili, *J. Phys. Condens. Matt.* **13**, 723 (2001).

24. Q. Si, S. Rabello, K. Ingersent, J. L. Smith, *Nature* **413**, 804 (2001).

25. F. D. M. Haldane, *Phys. Rev. B* **25**, 4925 (1982).

26. B. I. Halperin, T. C. Lubensky, S.-k. Ma, *Phys. Rev. Lett.* **32**, 292 (1974).

27. This research was generously supported by NSF under grants DMR-0213282, DMR-0308945 (T.S.), DMR-9985255 (L.B.), DMR-0098226 (S.S.), and DMR-0210790 and PHY-9907949 (M.P.A.F.). We also acknowledges funding from the NEC Corporation (T.S.), the Packard Foundation (L.B.), the Alfred P. Sloan Foundation (T.S. and L.B.), a Pappalardo Fellowship (A.V.), and an award from The Research Corporation (T.S.). We thank the Aspen Center for Physics for hospitality.


## Supporting Online Material

www.sciencemag.org/cgi/content/full/303/5663/1490/DC1

SOM Text

References and Notes

23 September 2003; accepted 7 January 2004

# PART 3
# String Theory

# Chapter 15

# Skyrmion and String Theory

Shigeki Sugimoto

*Institute for the Physics and Mathematics of the Universe,*
*The University of Tokyo, Chiba 277-8568, Japan*
*shigeki.sugimoto@ipmu.jp*

We review recent progress in baryon physics using gauge/string duality. Skyrme's idea to realize baryons as solitons is beautifully embedded in string theory.


## Contents


15.1 Introduction . . . . . . . . . . . . . . . . . . . . . . . . . . . . . . . . . . . . . 347
15.2 Holographic Description of QCD . . . . . . . . . . . . . . . . . . . . . . . . . 350
    15.2.1 Gauge/String duality . . . . . . . . . . . . . . . . . . . . . . . . . . . . 350
    15.2.2 Holographic description of Yang-Mills theory . . . . . . . . . . . . . . 351
    15.2.3 Adding quarks . . . . . . . . . . . . . . . . . . . . . . . . . . . . . . . . 352
    15.2.4 Hadrons in the model . . . . . . . . . . . . . . . . . . . . . . . . . . . . 353
15.3 Mesons from Open Strings . . . . . . . . . . . . . . . . . . . . . . . . . . . . . 354
    15.3.1 5 dim YM-CS theory as a theory of mesons . . . . . . . . . . . . . . . 354
    15.3.2 Mesons from 5 dim gauge field . . . . . . . . . . . . . . . . . . . . . . . 354
    15.3.3 Chiral symmetry . . . . . . . . . . . . . . . . . . . . . . . . . . . . . . . 355
    15.3.4 Skyrme model from 5 dim YM-CS theory . . . . . . . . . . . . . . . . 356
15.4 Baryons as Instantons in 5 Dim Gauge Theory . . . . . . . . . . . . . . . . . 357
    15.4.1 Connecting various descriptions of baryons . . . . . . . . . . . . . . . . 357
    15.4.2 Baryons as instantons . . . . . . . . . . . . . . . . . . . . . . . . . . . . 358
    15.4.3 Quantization . . . . . . . . . . . . . . . . . . . . . . . . . . . . . . . . . 360
    15.4.4 Currents . . . . . . . . . . . . . . . . . . . . . . . . . . . . . . . . . . . . 361
    15.4.5 Exploration . . . . . . . . . . . . . . . . . . . . . . . . . . . . . . . . . . 362
15.5 Conclusion . . . . . . . . . . . . . . . . . . . . . . . . . . . . . . . . . . . . . . 365
References . . . . . . . . . . . . . . . . . . . . . . . . . . . . . . . . . . . . . . . . . 365


## 15.1. Introduction

In the early 60's, even before quark model appeared, Skyrme proposed that baryons are realized as solitons, which are now called Skyrmions, in a pion effective action.[1] This great idea was developed further in the paper of Adkins, Nappi and Witten (ANW).[2] They calculated various quantities such as mean square radii, magnetic moments, axial radius, etc., for nucleons and found that they roughly agree with the experimental data. This remarkable result certainly suggest that the Skyrme model does catch the essential ingredients of the baryon physics. But it is not clear how it is related to the description of baryons in QCD, in which a baryon is described as a bound state of three quarks.

On the other hand, in the late 60's, string theory was born as a theory of hadrons. But, this proposal seemed to be less successful. People were discouraged by the features that did not appear to be close to our realistic world. For example, the space-time dimensions is higher than four, there exist massless hadrons with spin one and two in the spectrum, etc. Later, string theory evolved to a candidate of an ultimate unified theory that unifies all the elementary particles as well as interactions including quantum gravity, while QCD was recognized as the best candidate of the fundamental theory of hadrons.

The situation has been drastically changed, since the discovery of AdS/CFT correspondence, or more generally, gauge/string duality.[3–6] People have realized that a gauge theory can have a dual description based on string theory in a certain curved background. One of the surprising feature of this gauge/string duality is that the space-time dimensions of the two description are different. In general, the space-time dimensions in the string theory side is higher than that in the gauge theory side, and because of this fact, it is often called a holographic dual description. For example, four-dimensional $\mathcal{N} = 4$ supersymmetric Yang-Mills (SYM) theory is conjectured to be dual to type IIB string theory in $AdS_5 \times S^5$, which is ten-dimensional curved space-time. Although these two descriptions look completely different, they are conjectured to be equivalent, and there are numerous evidences supporting this conjecture.

What if we have a holographic dual of QCD? Suppose that there exists a string theory description that is equivalent to realistic four-dimensional QCD, what can we learn?

One of the nice points in QCD is that we can use the experimental data to check whether the gauge/string duality really works. Usually, in order to check the duality, it is inevitable to calculate some physical quantities in strongly coupled gauge theory to compare with the calculation in the string theory side. This is in general very difficult especially in the gauge theory without supersymmetry and conformal symmetry. But, in the case of QCD, we can skip all the calculations in the gauge theory side and simply compare the calculations in the string theory with the experimental data.

Once we accept the duality, it gives a new technology to analyze QCD. It enables us to calculate the meson effective theory including masses and couplings of various mesons. The calculation is actually very simple, powerful and fun, at least for the cases with large $N_c$ and large 't Hooft coupling, for which the string theory can be approximated by supergravity theory. Moreover, a lot of QCD phenomena, e.g. confinement, chiral symmetry breaking, origin of the hadron masses etc., can be understood quite easily from the topology of the background without getting into the detailed calculation.

Not only the practical usefulness, it provides more profound insight into particle physics. If the string theory description is completely equivalent to QCD, one cannot tell which one is more fundamental than the other as a theory of hadrons. It means that hadrons in our world can be described by string theory without using quarks and gluons. The concept of the "elementary particle" becomes ambiguous if there is a dual description. In the case of gauge/string duality, things are much more intricate than dualities in quantum field theory. The holographic dual of QCD is not even a theory of "particles", but a theory of strings living in a higher-dimensional curved space-time.

The gauge/sting duality suggests that the basic ideas of the string theory in the old days are essentially correct. The problems of string theory as a theory of hadrons can now be solved with the help of D-branes, curved space-time and holography. The ten-dimensional string theory can be dual to four-dimensional gauge theory and the massless spin one and two particles in ten-dimension correspond to massive mesons and glueballs in four-dimension.

The application of the gauge/string duality to QCD is proposed in Ref. 7. The holographic description of $U(N_c)$ QCD with $N_f$ massless quarks is obtained by putting $N_f$ probe D8-branes in a curved background corresponding to $N_c$ D4-branes in type IIA string theory. (See section 15.2 for a brief review of the model.) This system contains both open strings and closed strings. The closed strings are interpreted as glueballs and the open strings, which are attached on the D8-branes, are interpreted as mesons. The low energy effective theory of the open strings turns out to be a five-dimensional $U(N_f)$ Yang-Mills (YM) – Chern-Simons (CS) theory in a curved space-time. From this five-dimensional YM-CS theory, we can derive a four-dimensional meson effective action that contains infinitely many mesons, such as $\pi, \rho, a_1, \rho', a_1', \cdots$, and it was found that the masses and couplings of these mesons are roughly in agreement with the experimental data.

So far, we have only calculated the effective action up to the leading terms in the $1/N_c$ and $1/\lambda$ expansion. In the realistic QCD, we know $N_c = 3$, which may not be large enough, and we have to make $\lambda$ small in order to take a limit analogous to the continuum limit in lattice gauge theory. Therefore, the analysis is still very crude and the $1/N_c$ or $1/\lambda$ corrections may be large. But, believe it or not, the agreement with the experimental results turns out to be surprisingly better than what one would expect. See Ref. 7 for the details.

What about baryons? Now the Skyrme's idea plays a crucial role. As mentioned above, baryons are realized as solitons in Skyrme model. In a perfectly analogous way, the baryons in the holographic QCD are described as solitons in string theory. In the above model, a D4-brane wrapped on the non-trivial four-cycle in the background corresponds to a baryon. It can be shown that this wrapped D4-brane is equivalent to an instanton configuration localized in the four-dimensional space in the five-dimensional YM-CS theory. The instanton number is interpreted as the baryon number and it is directly related the Skyrmion in the Skyrme model. Moreover, it provides a new way to analyze properties of baryons, by applying the idea of ANW to this system. At the same time, as it was shown in Refs. 8 and 9, $N_c$ fundamental strings must be attached on the wrapped D4-brane because of the RR-flux in the background, and it can be viewed as a bound state of $N_c$ quarks. Therefore, the gauge/string duality provides a new description of baryons in string theory that connects the two old descriptions, namely, the Skyrmions and the bound states of $N_c$ quarks.

In this article, we mainly focus on the baryon physics. The main goal is to explain the basic idea and its consequences of the construction of baryons in the holographic description of QCD proposed in Ref. 7. In the next section, we will briefly review the construction of QCD in string theory. The holographic description of meson effective theory is reviewed in section 15.3 to the extent necessary for the analysis of baryons. Our main results for the baryons are given in section 15.4, which is based on our recent papers[10] and,[11] in which more details and further results can be found.[a]

## 15.2. Holographic Description of QCD

Here we briefly summarize the model.[7] If you accept the action (15.3.1) as our starting point, even if you are not familiar with string theory, you can just skip this section and go directly to section 15.3.

### 15.2.1. *Gauge/String duality*

The crucial step to obtain gauge theory in string theory was the discovery of D-branes.[15,16] Here a D$p$-brane is defined as a $(p+1)$-dimensional extended object, on which end points of open strings can be attached. By following the standard quantization procedure of the open strings attached on the D-brane, it can be shown that the massless spectrum of the open string contains a gauge particle. If we consider $N_c$ parallel D$p$-branes on top of each other, the low energy effective theory of the open strings attached on the D$p$-branes is a $(p+1)$-dimensional gauge theory with gauge group $U(N_c)$.

The basic idea of the gauge/string duality is that both gauge theory description and string theory description are realized as a certain limit of the same D-brane system and hence they should be equivalent. For example, the $\mathcal{N} = 4$ SYM theory with $U(N_c)$ gauge group can be realized on $N_c$ D3-branes in type IIB string theory, and $AdS_5 \times S^5$ space-time is the supergravity solution corresponding to this D3-brane system.[b]

---

[a]See also closely related works Refs. 12–14.

[b]To be more precise, we have to take a decoupling limit to pick up only massless open string degrees of freedom on the D3-brane. The $AdS_5 \times S^5$ geometry is obtained by taking the corresponding limit in the supergravity solution corresponding to the D3-branes. See Refs. 3 and 6 for more detail.

Roughly speaking, the loop expansion and $\alpha'$ expansion in string theory side correspond to the $1/N_c$ expansion and $1/\lambda$ expansion in the gauge theory side, respectively, where $\lambda$ is the 't Hooft coupling. Therefore, when $N_c$ and $\lambda$ are large, the string theory description can be approximated by supergravity, which is the low energy effective theory of the superstring at tree level. In other words, the strongly coupled gauge theory at the leading order in the $1/N_c$ expansion can be analyzed by the classical supergravity theory. This is why this duality is very powerful, but at the same time, this is why it is difficult to prove the duality.

### 15.2.2. *Holographic description of Yang-Mills theory*

The idea of the gauge/string duality explained above does not rely on the super-symmetry and conformal symmetry, although it becomes more difficult to analyze without these symmetries. Hence, it is natural to expect that it can be applied to more realistic gauge theories. The construction of four-dimensional pure Yang-Mills theory is proposed in Ref. 17. Let us consider $N_c$ D4-branes extended along $x^{0\sim4}$ directions in type IIA string theory and compactify the $x^4$ direction to $S^1$ of radius $M_{\mathrm{KK}}^{-1}$. The low energy effective theory of the open strings on the D4-brane world-volume is a $U(N_c)$ gauge theory with unwanted fermions and scalar fields in the adjoint representation of the gauge group. In order to break supersymmetry, we impose the anti-periodic boundary condition for all the fermions along the $S^1$. Then, the fermions become massive because of this boundary condition and scalar fields are also expected to acquire masses due to quantum corrections, since super-symmetry is completely broken. The only degree of freedom remaining massless is the gauge field and the low energy effective theory turns out to be four-dimensional $U(N_c)$ pure Yang-Mills theory.

Fortunately, the supergravity solution corresponding to this D4-brane system is explicitly known. The topology of this background is $\mathbb{R}^{1,3} \times \mathbb{R}^2 \times S^4$, where $\mathbb{R}^{1,3}$ is the four-dimensional space-time parametrized by $x^{0\sim3}$, the angular direction of the $\mathbb{R}^2$ is the $S^1$ parametrized by $x^4$, the radial direction of the $\mathbb{R}^2$ corresponds to the distance from the D4-brane, and the $S^4$ corresponds to the angular directions of the five-dimensional plane parametrized by $x^{5\sim9}$.

The type IIA string theory in this background is considered to be a holo-graphic description of pure Yang-Mills theory at low energy. A lot of quanti-ties have been calculated using this description within supergravity approxima-tion. They are roughly in good agreement with field theoretical results such as lattice gauge theory. (See Ref. 6 and references therein. See also Ref. 18 for the glueball spectrum.)

Note, however, that in the gauge theory description, the gluons live on the D4-brane world-volume wrapped on an $S^1$ of radius $M_{\mathrm{KK}}^{-1}$ and hence the system becomes five-dimensional if the energy scale is much higher than $M_{\mathrm{KK}}$. In order to get rid of the Kaluza-Klein modes associated with this $S^1$, we should consider a limit $M_{\mathrm{KK}} \to \infty$, keeping physical quantities, such as glueball mass, finite. This is analogous to the continuum limit in the lattice gauge theory. However, if we start with the string theory description and rely on the supergravity approximation, this limit is unfortunately not accessible. This is because the supergravity approximation is valid when the 't Hooft coupling is large, while the asymptotic freedom of Yang-Mills theory implies that the 't Hooft coupling becomes small when the "cut-off" scale $M_{\rm KK}$ becomes large. This means that we have to go beyond the supergravity approximation and deal with all the stringy corrections to take this limit. This is one of the long standing problems in this kind of approach. In this article, we use the supergravity approximation assuming that the 't Hooft coupling is large and do not attempt to take the "continuum limit" $M_{\rm KK} \to \infty$. Therefore, we should keep in mind that our "QCD" deviates from real QCD at the energy scale higher than $M_{\rm KK}$. As a consequence, in the supergravity description, there are a lot of particles with masses of order $M_{\rm KK}$ that cannot be interpreted as bound states of gluons. For example, the Kaluza-Klein modes associated with the $S^4$ in the background contains particles with non-trivial representations of the $SO(5)$ isometry, which acts as rotation of the $S^4$. But, since the Yang-Mills field in the gauge theory side is invariant under the $SO(5)$ symmetry, all the composites made by gluons should be $SO(5)$ invariant. The $SO(5)$ non-invariant particles are interpreted as bound states that involve massive Kaluza-Klein modes associated with the $S^1$, which are expected to decouple if we take the "continuum limit" $M_{\rm KK} \to \infty$. In the following, we simply neglect the effect of these artifacts and restrict our attention to the $SO(5)$ invariant sector.

### 15.2.3. *Adding quarks*

In order to add quarks to the system considered in the previous subsection, we add $N_f$ D8-$\overline{\rm D8}$ pairs to the D4-brane configuration.[7] Here, a D8-brane has two possible orientations, and in order to distinguish the two orientations, we call one of them as D8-brane and the other as $\overline{\rm D8}$-brane. D8-$\overline{\rm D8}$ pairs are extended along $x^{0\sim3}$ and $x^{5\sim9}$ directions. D8-branes are located at $x^4 = 0$ and $\overline{\rm D8}$-branes are located at $x^4 = \pi M_{\rm KK}^{-1}$.

As explained above, open strings attached on the D4-branes gives Yang-Mills field. In addition, there are open strings connecting between the D4-brane and the D8-brane ($\overline{\rm D8}$-brane), from which we find left (right) handed component of the quark field. In this way, it can be shown that the massless field content in this system is the four-dimensional $U(N_c)$ QCD with $N_f$ massless quarks. Note that the gauge symmetry for the D8-$\overline{\rm D8}$ pairs is $U(N_f)_{\rm D8} \times U(N_f)_{\overline{\rm D8}}$ and this is interpreted as the chiral symmetry $U(N_f)_L \times U(N_f)_R$ in massless QCD.

On the other hand, if the number of color $N_c$ and the 't Hooft coupling $\lambda$ are large, the system can be well-described by replacing D4-branes with the corresponding supergravity solution. Assuming $N_c \gg N_f$, the D8-$\overline{\rm D8}$ pairs are treated as probe D8-branes embedded in the background corresponding to the D4-branes. Now the topology of the D4-brane background is $\mathbb{R}^{1,3} \times \mathbb{R}^2 \times S^4$. In this background, the

D8-brane and $\overline{\text{D8}}$-brane should be smoothly connected and the system becomes a string theory with a single connected component of $N_f$ D8-branes embedded in the D4-brane background. As we will see more explicitly in section 15.3.3, this geometry induces the spontaneous chiral symmetry breaking in QCD. The D8-branes are extended along $\mathbb{R}^{1,3} \times S^4$ and one-dimensional subspace of $\mathbb{R}^2$ parametrized by $z \in (-\infty, +\infty)$.

Now, applying the idea of the gauge/string duality to the above construction, we conjecture that the four-dimensional QCD with $N_f$ massless quarks is dual to type IIA string theory in the D4-brane background with $N_f$ probe D8-branes (at low energy). The former gauge theory description is better when the 't Hooft coupling is small, while the latter string theory description is better when the 't Hooft coupling is large.

### 15.2.4. *Hadrons in the model*

How can we see hadrons in the holographic description? The objects that behave as point-like particles in four-dimensional space-time are closed strings in the bulk, open strings attached on the probe D8-branes, and D4-branes wrapped on $S^4$.[c] The closed strings are interpreted as glueballs as mentioned in section 15.2.2. Since each open string carries two flavor indices associated with the $N_f$ D8-branes, the open strings are interpreted as mesons. The D4-branes are more involved. They are five-dimensional objects, but since all the spatial directions are wrapped on the $S^4$, they behave as point-like particles. It was shown in[8,9] that $N_c$ units of electric charge are induced on a D4-brane wrapped on the $S^4$, because of the RR flux on the $S^4$, and $N_c$ fundamental strings must be attached to cancel the electric flux. (See Fig. 15.1.) This picture shows how this object is related to the bound state of $N_c$ quarks.

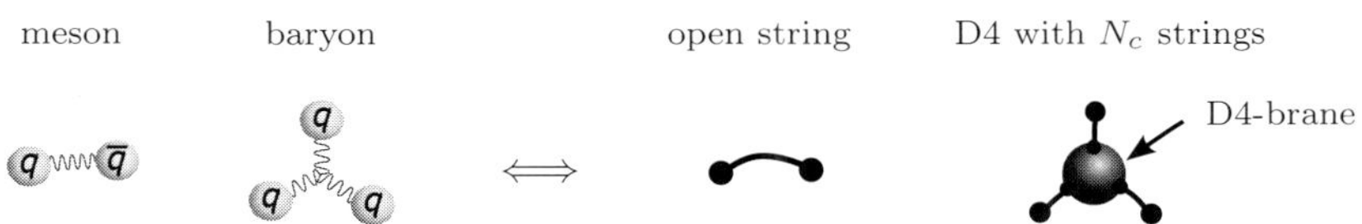


Fig. 15.1.     mesons and baryons in quark model and string theory

In the case without D8-branes, the fundamental strings attached on the D4-brane are extended to infinity, and this object is interpreted as an infinitely heavy baryon made by $N_c$ infinitely heavy external quarks. In our case, on the other hand, the other ends of the fundamental strings can be attached to the D8-branes and the mass of the baryon becomes finite. Actually, the mass of the baryon is roughly estimated as the tension of the D4-brane times the area of the $S^4$. Because the

---

[c]Do not confuse with the $N_c$ D4-branes used to construct QCD.

tension of the D4-brane is proportional to the inverse of the string coupling, the mass is of order $N_c$, which is in agreement with the well-known result in large $N_c$ QCD.[19] In this way, the origin of the baryon mass is understood geometrically.

## 15.3. Mesons from Open Strings

### 15.3.1. *5 dim YM-CS theory as a theory of mesons*

In the string theory description of QCD constructed in the previous section, the open strings attached on the probe D8-branes are interpreted as mesons. The low energy effective theory of the open strings on the D8-branes is given by a nine-dimensional $U(N_f)$ gauge theory. Here recall that there is an $SO(5)$ isometry that acts as rotation of the $S^4$ in the background. Since the gluon and quark fields are invariant under this $SO(5)$, we restrict our attention to the $SO(5)$ invariant sector, as in the case of Yang-Mills theory explain in section 15.2.2.

Then, the effective theory reduces to a five-dimensional $U(N_f)$ gauge theory. Using the supergravity solution of the D4-brane background, we can explicitly calculate the effective action which turns out to be the five-dimensional $U(N_f)$ YM-CS theory given by

$$S_{\text{5dim}} \simeq S_{\text{YM}} + S_{\text{CS}} ,$$

$$S_{\text{YM}} = -\kappa \int d^4x dz \, \text{Tr} \left( \frac{1}{2} h(z) F_{\mu\nu}^2 + k(z) F_{\mu z}^2 \right) ,$$

$$S_{\text{CS}} = \frac{N_c}{24\pi^2} \int_{\text{5dim}} \omega_5(A) , \tag{15.3.1}$$

where $\mu, \nu = 0, \cdots, 3$ are the Lorentz indices for the four-dimensional world and $z$ is the coordinate of the fifth dimension. The warp factors in the YM action are given by $k(z) = 1 + z^2$ and $h(z) = (1 + z^2)^{-1/3}$. $\kappa = a\lambda N_c$ ($a = 1/(216\pi^3)$) is a constant proportional to $N_c$ and the 't Hooft coupling $\lambda$, and $\omega_5(A)$ is the Chern-Simons five-form. Here and in the following, we mainly work in the $M_{\text{KK}} = 1$ unit. $M_{\text{KK}}$ dependence can easily be recovered by the dimensional analysis.

### 15.3.2. *Mesons from 5 dim gauge field*

In order to extract four-dimensional physics from the five-dimensional gauge theory, it is convenient to expand the five-dimensional gauge field as

$$A_\mu(x^\mu, z) = \sum_{n \geq 1} B_\mu^{(n)}(x^\mu)\psi_n(z) , \quad A_z(x^\mu, z) = \sum_{n \geq 0} \varphi^{(n)}(x^\mu)\phi_n(z) , \tag{15.3.2}$$

where $\{\psi_n(z)\}_{n \geq 1}$ and $\{\phi_n(z)\}_{n \geq 0}$ are complete sets of the (normalizable) functions of $z$. We choose the complete sets so that the kinetic terms and mass terms of the four-dimensional fields $B_\mu^{(n)}(x)$ and $\varphi^{(n)}(x)$ are diagonalized. The functions $\psi_n(z)$ ($n = 1, 2, 3, \cdots$) are eigenfunctions of the eigenequation

$$-h(z)^{-1}\partial_z(k(z)\partial_z\psi_n) = \lambda_n\psi_n , \tag{15.3.3}$$

where $\lambda_n$ are the eigenvalues, with the normalization condition

$$\kappa \int dz\, h(z)\psi_m \psi_n = \delta_{mn} \ . \qquad (15.3.4)$$

The functions $\phi_n(z)$ $(n = 0, 1, 2, \cdots)$ are given by $\phi_n(z) = \lambda_n^{-1/2}\partial_z \psi_n(z)$ for $n = 1, 2, 3, \cdots$ and $\phi_0(z) = (\kappa\pi)^{-1/2}k(z)^{-1}$.

Inserting this mode expansion (15.3.2) into the five-dimensional action (15.3.1), we obtain a four-dimensional action of mesons. It turns out that $B_\mu^{(n)}$ become massive vector and axial-vector meson fields for odd and even $n$, respectively, whose masses $m_n$ are related to the eigenvalues $\lambda_n$ by $m_n^2 = \lambda_n M_{\mathrm{KK}}^2$. The scalar fields $\varphi^{(n)}$ $(n = 1, 2, 3, \cdots)$ are eaten by $B_\mu^{(n)}$ to make them massive, while $\varphi^{(0)}(x)$ remain massless. We interpret $\varphi^{(0)}$, $B_\mu^{(1)}$, $B_\mu^{(2)}$, etc., as pion, $\rho$ meson, $a_1$ meson, etc., respectively. In this way, various meson fields $\pi$, $\rho$, $a_1$, etc. are beautifully unified in the five-dimensional gauge field. Furthermore, the structure of the interactions among these mesons reproduces various phenomenological models of meson effective theory such as Skyrme model, hidden local symmetry approach, vector meson dominance model, Gell-Mann–Sharp–Wagner model, etc. Even quantitatively, it was found that the masses and couplings of these mesons calculated from the five-dimensional YM-CS theory (15.3.1) are roughly in agreement with the experimental data. (See Ref. 7 for details.)

### 15.3.3. *Chiral symmetry*

As explained in section 15.2.3, the chiral symmetry $U(N_f)_L \times U(N_f)_R$ in QCD is realized as the gauge symmetry on the $N_f$ D8-$\overline{\mathrm{D8}}$ pairs. In terms of the five-dimensional gauge theory (15.3.1), the chiral symmetry correspond to the gauge symmetry at $z \to \pm\infty$, since the two boundaries at $z \to +\infty$ and $z \to -\infty$ correspond to the D8-branes and $\overline{\mathrm{D8}}$-branes, respectively, in the configuration before replacing the D4-branes with the corresponding supergravity background.

In the previous subsection, we implicitly assumed that the gauge field vanishes at $z \to \pm\infty$, since it was expanded by normalizable modes in (15.3.2). We could consider gauge field with non-zero boundary values as

$$A_{L\mu}(x^\mu) = \lim_{z \to +\infty} A_\mu(x^\mu, z) \ , \quad A_{R\mu}(x^\mu) = \lim_{z \to -\infty} A_\mu(x^\mu, z) \ . \qquad (15.3.5)$$

Then, the boundary values $(A_{L\mu}, A_{R\mu})$ are interpreted as the gauge fields associated with the chiral symmetry $U(N_f)_L \times U(N_f)_R$. However, since they correspond to non-normalizable modes, the coefficients of the kinetic terms of these gauge field diverge if we naively insert the five-dimensional gauge field with boundary condition (15.3.5) into the action (15.3.1). Therefore, the gauge fields $(A_{L\mu}, A_{R\mu})$ are not dynamical and should be considered as external fields, as expected from the fact that the chiral symmetry is a global symmetry in QCD.

Although the kinetic terms diverge, it is useful to introduce the external gauge fields associated with the chiral symmetry to read off the couplings between mesons and the external fields. We will use this to compute the currents in section 15.4.4.

Let us consider a $U(N_f)$-valued field defined by

$$U(x^\mu) \equiv e^{2i\Pi(x^\mu)/f_\pi} \equiv \mathrm{P}\exp\left[ i \int_{-\infty}^{\infty} dz\, A_z(x^\mu, z) \right] . \tag{15.3.6}$$

Under the five-dimensional gauge transformation

$$A_M \to A_M^g \equiv g A_M g^{-1} + i g \partial_M g^{-1} , \quad (M = 0, 1, 2, 3, z) , \tag{15.3.7}$$

this field transforms as

$$U \to g_L U g_R^{-1} , \tag{15.3.8}$$

where $g_L = \lim_{z \to +\infty} g$ and $g_R = \lim_{z \to -\infty} g$ . Since $(g_L, g_R)$ is interpreted as an element of the chiral symmetry $U(N_f)_L \times U(N_f)_R$, the transformation (15.3.8) is the same as that of the pion field in chiral Lagrangian. Therefore, we interpret the field defined in (15.3.6) as the pion field used in the chiral Lagrangian. It can be checked that the pion field $\varphi^{(0)}(x^\mu)$ in (15.3.2) and $\Pi(x^\mu)$ in (15.3.6) are identical up to the linear order. This explicitly shows that the chiral symmetry is spontaneously broken and the pion field appears as the Nambu-Goldstone mode.

### 15.3.4. *Skyrme model from 5 dim YM-CS theory*

In order to write down the effective action in terms of the pion field in (15.3.6), it is convenient to work in the $A_z = 0$ gauge. It can be achieved by the gauge transformation with

$$g^{-1}(x^\mu, z) = \mathrm{P}\exp\left[ i \int_{-\infty}^{z} dz'\, A_z(x^\mu, z') \right] , \tag{15.3.9}$$

which implies $A_z^g = 0$. In this gauge, the pion degrees of freedom appear in the boundary condition

$$A_\mu^g(x^\mu, z) \to \begin{cases} iU^{-1}\partial_\mu U & (z \to +\infty) \\ 0 & (z \to -\infty) \end{cases} , \tag{15.3.10}$$

where we have turned off the external fields $A_{L\mu} = A_{R\mu} = 0$. Then, the mode expansion (15.3.2) is modified to

$$A_\mu^g(x^\mu, z) = iU^{-1}\partial_\mu U(x^\mu)\psi_+(z) + \sum_{n \geq 1} B_\mu^{(n)}(x^\mu)\psi_n(z) , \tag{15.3.11}$$

where $\psi_+(z)$ is a function satisfying $\psi_+(-\infty) = 0$ and $\psi_+(+\infty) = 1$. A convenient choice of the function $\psi_+$ is

$$\psi_+(z) = \tfrac{1}{2}(1 + \psi_0(z)) , \quad \psi_0(z) \equiv \frac{2}{\pi} \arctan z , \tag{15.3.12}$$

which simplifies the calculation, because $\partial_z \psi_+$ is proportional to $\phi_0$, which is orthogonal to $\phi_n$ with $n \geq 1$.

Substituting (15.3.11) into the five-dimensional YM action in (15.3.1), we obtain

$$S_{\rm YM} = \int d^4x \left( \frac{f_\pi^2}{4} {\rm Tr} \left( U^{-1} \partial_\mu U \right)^2 + \frac{1}{32e_S^2} {\rm Tr} \left[ U^{-1} \partial_\mu U, U^{-1} \partial_\nu U \right]^2 + \cdots \right) ,$$

$$(15.3.13)$$

where '$\cdots$' are the terms including the vector/axial-vector mesons $B_\mu^{(n)}$. Here $f_\pi$ and $e_S$ are given by

$$f_\pi^2 = \frac{4}{\pi} \kappa M_{\rm KK}^2 , \quad e_S^{-2} = \kappa \int dz\, h(z)(1 - \psi_0^2)^2 \simeq 2.51 \cdot \kappa . \qquad (15.3.14)$$

The action (15.3.13) is precisely that of the Skyrme model. From (15.3.14), we find $f_\pi \sim \mathcal{O}(\sqrt{N_c})$ and $e_S \sim \mathcal{O}(1/\sqrt{N_c})$, which are in agreement with the results in large $N_c$ QCD. If we use the $\rho$-meson mass $m_\rho|_{\rm exp} \simeq 776$ MeV and the pion decay constant $f_\pi|_{\rm exp} \simeq 92.4$ MeV as inputs,[d] $M_{\rm KK}$ and $\kappa$ are fixed as

$$M_{\rm KK} \simeq 949 \text{ MeV} , \quad \kappa \simeq 0.00745 . \qquad (15.3.15)$$

Then, we obtain $e_S^{-2} \simeq 0.0187$ with which the Gasser-Leutwyler coefficients $L_1$, $L_2$ and $L_3$ calculated from the the effective action (15.3.13) are roughly consistent with the experimental values. Furthermore, it can be shown that the Wess-Zumino-Witten (WZW) term, including the terms with external fields ($A_{L\mu}, A_{R\mu}$), is correctly reproduced from the CS action in (15.3.1). The fifth coordinate introduced to write down the WZW term is now realized as the coordinate $z$ in our system. (See Ref. 7 for more details.)

## 15.4. Baryons as Instantons in 5 Dim Gauge Theory

### 15.4.1. *Connecting various descriptions of baryons*

Let us consider a static gauge configuration in the five-dimensional YM-CS theory (15.3.1) with non-trivial instanton number in the four-dimensional space parametrized by $x^M = (\vec{x}, z) \in \mathbb{R}^4$ ($M = 1, 2, 3, z$). Since the energy density is localized along the spatial directions, it behaves as a point-like particle and we interpret it as a baryon. Then, as one can easily guess, the baryon number $N_B$ is given by the instanton number as

$$N_B = \frac{1}{8\pi^2} \int {\rm Tr}\, F \wedge F . \qquad (15.4.16)$$

This relation can be checked by using the baryon number current which will be given in section 15.4.5.1.

Since we have obtained the Skyrme model in section 15.3.4, it is natural to expect that baryons can be constructed as Skyrmion. In fact, we can show that the

[d]This treatment should be considered with care. As explained in section 15.2.2, we should in principle take the "continuum limit" $M_{\rm KK} \to \infty$ and $\lambda \to 0$ with $m_\rho$ kept fixed at the experimental value, which is unfortunately not accessible within the supergravity approximation. If it were possible, the value of $f_\pi$ would be fixed and could not be used as an input.

instanton number is equal to the baryon number for the Skyrmions as follows. For our purpose, it is convenient to compactify three-dimensional space parametrized by $\{\vec{x}\}$ to $S^3$. The pion field $U(\vec{x})$ defines a map from the $S^3$ to $U(N_f)$, which is classified by its winding number in $\pi_3(U(N_f)) \simeq \mathbb{Z}$ and this winding number is interpreted as the baryon number in the Skyrme model. Let us consider the gauge field with the boundary condition (15.3.10). Using the relation

$$\mathrm{Tr}F \wedge F = d\omega_3(A) \,, \tag{15.4.17}$$

where $\omega_3(A)$ is the CS 3-form, and the Stokes' theorem, we obtain

$$\frac{1}{8\pi^2} \int_{S^3 \times \mathbb{R}} \mathrm{Tr}F \wedge F = \frac{1}{8\pi^2} \int_{S^3} \omega_3(A)\Big|_{z=+\infty} = -\frac{1}{24\pi^2} \int_{S^3} \mathrm{Tr}\left[(U^{-1}dU)^3\right] \,. \tag{15.4.18}$$

The last expression gives the winding number defined by the pion field.

It is interesting to note that the realization of baryon as an instanton configuration was introduced by Atiyah and Manton in 1989,[20] who proposed to use an instanton configuration to construct a Skyrmion via the relation (15.3.6). Our description of baryon naturally realizes their old idea and gives a definite physical interpretation of the fifth coordinate $z$ as one of the coordinates on the D8-brane world-volume.

On the other hand, as explained in section 15.2.4, D4-branes wrapped on the $S^4$ are also interpreted as baryons. Here, the D4-branes wrapped on the $S^4$ are embedded in the probe D8-branes. In general, it is known that D$p$-branes embedded within D$(p+4)$-branes are equivalent to the gauge configurations in the D$(p+4)$-brane world-volume gauge theory with non-trivial instanton number in the four-dimensional space transverse to the D$p$-brane.[21] Therefore, the D4-brane wrapped on the $S^4$ in our system is equivalent to the instanton configuration given above. The number of D4-branes is equal to the instanton number, which gives the baryon number.

As we have seen, the baryons can be described in a variety of ways, namely, bound states of $N_c$ quarks, solitons in the Skyrme model, D4-branes wrapped on $S^4$, and instanton in five-dimensional gauge theory. All of them are now connected to each other.

### 15.4.2. *Baryons as instantons*

For the rest of this article, we will outline the analysis of baryons using the instanton configurations in the five-dimensional YM-CS theory (15.3.1), following the papers.[10,11,e] From now we consider the case with $N_f = 2$.

Our strategy is the same as that given in ANW.[2] We first obtain the classical solution corresponding to the baryon, and use the moduli space approximation method to quantize it. Since our system includes a tower of massive vector and

---

[e]See also Refs. 12–14 for closely related papers.

axial-vector mesons, one can systematically incorporate the effect of these mesons by using the five-dimensional description. One of the advantages of this approach is that the action (15.3.1) is much simpler than traditional meson effective actions including $\rho$-meson and $a_1$-meson.

At first sight, one might think that the CS-term can be neglected since $\kappa$ in front of the YM action in (15.3.1) is proportional to $\lambda N_c$ and hence larger than the coefficient of the CS-term for large $\lambda$. However, one can easily show that if we ignore the CS-term, the instanton solution shrinks to zero size because of the warp factors $h(z)$ and $k(z)$. This is somewhat puzzling, since we have seen that the pion effective action contains the Skyrme term, which was originally introduced to stabilize the size of the Skyrmion. The tower of vector and axial-vector mesons have the effect of shrinking the size of the soliton.

On the other hand, the CS-term has an effect to make the size of the instanton larger. This is because the CS-term contains a term like

$$S_{\rm CS} \sim \frac{N_c}{8\pi^2} \int A^{U(1)} \wedge {\rm Tr} F \wedge F + \cdots , \tag{15.4.19}$$

where $A^{U(1)}$ is the $U(1)$ part of the $U(2)$ gauge field. This term acts as a source of the $U(1)$ charge for non-trivial instanton configuration. The repulsive force due to this $U(1)$ charge enlarge the size of the instanton and it is stabilized at a finite value. Note that the $U(1)$ part contains the $\omega$-meson and this mechanism is essentially the same as that proposed in Ref. 22, in which it was argued that the $\omega$-meson contributes to stabilize the size of the soliton even in the cases without Skyrme term.

The size of the instanton is estimated as $\mathcal{O}(\lambda^{-1/2})$ and hence it becomes small if $\lambda$ is large. If the instanton is small enough, the five-dimensional space-time can be approximated by the flat space-time, since we can use $h(z) \simeq k(z) \simeq 1$ for $|z| \ll 1$. Then, the leading order classical solution for large $\lambda$ is given by the instanton solution in the flat space-time. The $SU(2)$ part is given by

$$A_M^{\rm cl} = -i \frac{\xi^2}{\xi^2 + \rho^2} g \partial_M g^{-1} \tag{15.4.20}$$

with

$$g = \frac{(z - Z) - i(\vec{x} - \vec{X}) \cdot \vec{\tau}}{\xi} , \quad \xi = \sqrt{(\vec{x} - \vec{X})^2 + (z - Z)^2} . \tag{15.4.21}$$

and the $U(1)$ part is

$$A_0^{U(1)} = \frac{N_c}{8\pi^2 \kappa} \cdot \frac{1}{\xi^2} \left( 1 - \frac{\rho^4}{(\rho^2 + \xi^2)^2} \right) . \tag{15.4.22}$$

Here $\rho$ is the size of the instanton and $(\vec{X}, Z)$ represents the position of the instanton in the four-dimensional space.

### 15.4.3. *Quantization*

Next we consider a slowly moving (rotating) baryon configuration. We use the moduli space approximation method to quantize the system. The idea is to consider the classical solution (15.4.20) with the instanton moduli parameters promoted to time dependent variables and insert them into the action (15.3.1). Then, we obtain an action for the quantum mechanics on the instanton moduli space. The $SU(2)$ one instanton moduli space is parametrized by $\{(\vec{X}, Z, \rho, \boldsymbol{a})\} \equiv \{(X^\alpha)\}$ where $\boldsymbol{a} \in SU(2)$ represent the $SU(2)$ orientation of the instanton. Then the Lagrangian of the quantum mechanics obtained in the above procedure is

$$L_{\mathrm{QM}} = \frac{1}{2} G_{\alpha\beta} \dot{X}^\alpha \dot{X}^\beta - U(X^\alpha) \,, \tag{15.4.23}$$

where $G_{\alpha\beta}$ is the metric of the instanton moduli space and

$$U(X^\alpha) = 8\pi^2 \kappa \left( 1 + \frac{\rho^2}{6} + \frac{N_c^2}{5(8\pi^2\kappa)^2} \frac{1}{\rho^2} + \frac{Z^2}{3} + \cdots \right) \,. \tag{15.4.24}$$

The effect of the warp factors is taken into account perturbatively, which gives the non-trivial potential (15.4.24) for the parameters $\rho$ and $Z$. Neglecting the higher order terms denoted as '$\cdots$', the minimum of the potential (15.4.24) is given by

$$\rho^2 = \rho_{\mathrm{cl}}^2 \equiv \frac{N_c}{8\pi^2\kappa} \sqrt{\frac{6}{5}} \,, \quad Z = 0 \,. \tag{15.4.25}$$

Here $(\vec{X}, \boldsymbol{a})$ are genuine moduli that also appear in the analysis of ANW for the Skyrme model. On the other hand, $(\rho, Z)$ are not moduli parameters in the usual sense, since they have a non-trivial potential (15.4.24). We keep these new degrees of freedom, since one can show that they are light compared with the other massive modes for large $\lambda$.

Solving the Schrödinger equation for this quantum mechanics, we obtain the baryon states. For example, the wave function of the spin up proton state $|\, p \uparrow \rangle$ is given by

$$\psi(\vec{X}, \boldsymbol{a}, \rho, Z) \propto e^{i\vec{p}\cdot\vec{X}} R(\rho) \psi_Z(Z) T(\boldsymbol{a}) \tag{15.4.26}$$

where

$$R(\rho) = \rho^{-1 + 2\sqrt{1 + N_c^2/5}} e^{-\frac{8\pi^2\kappa}{\sqrt{6}} \rho^2} \,, \quad \psi_Z(Z) = e^{-\frac{8\pi^2\kappa}{\sqrt{6}} Z^2} \,, \quad T(\boldsymbol{a}) = a_1 + i a_2 \,. \tag{15.4.27}$$

Similarly, we can explicitly write down the wave functions for various baryon states including not only nucleon and $\Delta$, but also $N(1440)$, $N(1535)$, etc. The spectrum of the baryons obtained in this way is summarized in Fig. 15.2. As in the Skyrme model,[2] the isospin $I$ and spin $J$ of these baryon states turn out to be equal, and the baryons with $I = J$ found in the experiments are also listed. As the figure shows, the spectrum obtained in the model seems to catch the qualitative features of the observed one. However, it is less successful quantitatively. If we use the

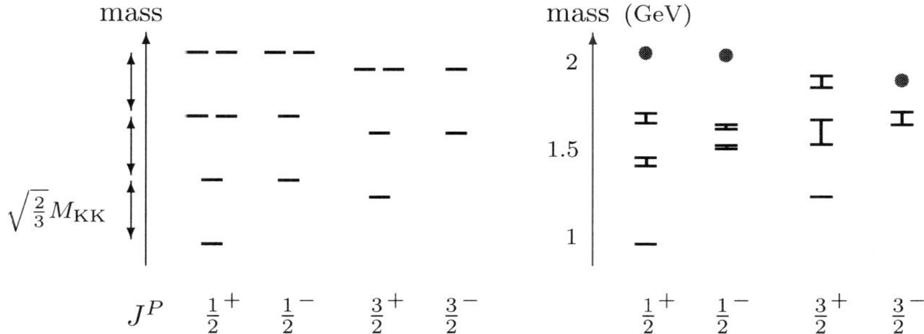

Fig. 15.2.  Left side is the baryon spectrum obtained in the model,[10] while the right side is the spectrum of the baryons with $I = J$ taken from PDG particle listings.[23]

value (15.3.15) for $M_{\mathrm{KK}}$, the mass differences are a bit too large compared to the experimental data.

### 15.4.4.  *Currents*

In order to extract the electromagnetic properties of baryons, we calculate the currents $(J_L^\mu, J_R^\mu)$ associated with the chiral symmetry $U(N_f)_L \times U(N_f)_R$. As explained in section 15.3.3, we can introduce the external gauge fields $(A_{L\mu}, A_{R\mu})$ for the chiral symmetry $U(N_f)_L \times U(N_f)_R$, considering the gauge field with non-trivial boundary values as in (15.3.5). The currents are obtained by inserting the gauge configuration with this boundary condition into the action (15.3.1) and picking up the terms linear with respect to the external gauge fields as

$$S_{\text{5dim}}\big|_{\mathcal{O}(A_L, A_R)} = -2 \int d^4x \, \mathrm{Tr}\left(A_{L\mu} J_L^\mu + A_{R\mu} J_R^\mu\right) . \tag{15.4.28}$$

As a result, we obtain

$$J_{L\mu} = -\kappa(k(z)F_{\mu z})\big|_{z=+\infty} \, , \quad J_{R\mu} = +\kappa(k(z)F_{\mu z})\big|_{z=-\infty} . \tag{15.4.29}$$

The vector and axial-vector currents are given by

$$J_{V\mu} = J_{L\mu} + J_{R\mu} = -\kappa\left[k(z)F_{\mu z}\right]_{z=-\infty}^{z=+\infty} ,$$

$$J_{A\mu} = J_{L\mu} - J_{R\mu} = -\kappa\left[\psi_0(z)k(z)F_{\mu z}\right]_{z=-\infty}^{z=+\infty} , \tag{15.4.30}$$

with $\psi_0(\pm\infty) = \pm 1$.

In order to calculate the current (15.4.29), we need to know how $F_{\mu z}$ behaves at $z \to \pm\infty$. For this purpose, we can no longer use the solution (15.4.20) and (15.4.22) which is valid only for $|z| \ll 1$. The key observation is that the gauge field (in a suitable gauge choice) becomes small when $\rho \ll |z|$ and the non-linear terms in the equations of motion can be neglected. Then, what we have to do is to solve the linearized equations of motion that agree with (15.4.20) in the intermediate region $\rho \ll |z| \ll 1$.

As an illustrative example, let us consider the time component of the $U(1)$ gauge field whose solution for $\xi \ll 1$ is given by (15.4.22). For $\rho \ll \xi \ll 1$, it can be approximated as

$$A_0^{U(1)} \simeq \frac{1}{8\pi^2 a\lambda} \cdot \frac{1}{\xi^2} \equiv -\frac{1}{2a\lambda} G^{\text{flat}}(\vec{x}, z; \vec{X}, Z) \qquad (15.4.31)$$

where $G^{\text{flat}}$ is the Green's function in the flat space satisfying

$$\partial_M \partial^M G^{\text{flat}} = \delta^3(\vec{x} - \vec{X})\delta(z - Z) . \qquad (15.4.32)$$

The solution for $\rho \ll \xi$ is then given by replacing flat space Green's function $G^{\text{flat}}$ with the Green's function for curved space-time $G$:

$$A_0^{U(1)} \simeq -\frac{1}{2a\lambda} G(\vec{x}, z; \vec{X}, Z) \qquad (15.4.33)$$

where $G$ is the Green's function in the curved space satisfying

$$\left(h(z)\partial_i^2 + \partial_z k(z)\partial_z\right) G = \delta^3(\vec{x} - \vec{X})\delta(z - Z) . \qquad (15.4.34)$$

Using (15.3.3), it can be shown that the Green's function $G$ can be written as

$$G = \kappa \sum_{n=1}^{\infty} \psi_n(z)\psi_n(Z) \left(-\frac{1}{4\pi} \frac{e^{-m_n r}}{r}\right) , \qquad (15.4.35)$$

where $r = |\vec{x} - \vec{X}|$.

### 15.4.5. *Exploration*

Now we are ready to calculate various physical quantities. We are going to show some numerical results that can be compared with experimental values. But, please keep in mind that you should not trust them too much, since our approximation is still very crude. For example, we have assumed $\lambda$ is large, but it may not be large enough in the realistic situation. Although we ignored the higher derivative terms in the action (15.3.1), they may also contribute. $N_c = 3$ is unfortunately not large enough. We know that our model deviates from real QCD at the energy scale higher than $M_{\text{KK}}$. We use (15.3.15) with which various quantities in the meson sector are consistent with the experimental values,[7] but we know the baryon mass differences are not in good agreement with the experimental data as we have seen in Fig. 15.2.

#### 15.4.5.1. *Baryon number (isoscalar) current*

Baryon number current is proportional to the $U(1)$ part of the vector current (15.4.30) and given by

$$J_B^\mu = -\frac{2}{N_c} \kappa \left[k(z)F_{U(1)}^{\mu z}\right]_{z=-\infty}^{z=+\infty} . \qquad (15.4.36)$$

Using the trick explained in section 15.4.4, we obtain

$$J_B^0 \simeq [k(z)\partial_z G]_{z=-\infty}^{z=+\infty} , \quad J_B^i \simeq -\frac{J^j}{16\pi^2\kappa}\epsilon^{ijk}\partial_k J_B^0 + \cdots , \qquad (15.4.37)$$

where $J^j = -i4\pi^2\kappa\rho^2\mathrm{Tr}(\tau^j \boldsymbol{a}^{-1}\dot{\boldsymbol{a}})$ is the spin operator. The terms in '$\cdots$' is the part irrelevant to the following calculation. It can be shown that the baryon number current is non-zero finite, since $k(z)$ and $\partial_z G$ behaves as $z^2$ and $1/z^2$ at $z \to \pm\infty$, respectively.

Using this expression, the isoscalar mean square radius for the nucleon is evaluated as

$$\langle r^2 \rangle_{I=0} = \int d^3x\, r^2 \langle J_B^0 \rangle \simeq (0.742 \text{ fm})^2 \,. \tag{15.4.38}$$

Here the expectation value $\langle J_B^0 \rangle$ is taken with respect to the nucleon wave function (15.4.27). This result is compared with the the experimental value $\langle r^2 \rangle_{I=0}^{1/2}|_{\exp} \simeq 0.806$ fm. The result of ANW is $\langle r^2 \rangle_{I=0}^{1/2}|_{\mathrm{ANW}} \simeq 0.59$ fm.

The isoscalar magnetic moment is given by

$$\mu_{I=0}^i = \frac{1}{2}\epsilon^{ijk} \int d^3x\, x^j J_B^k \simeq \frac{J^i}{16\pi^2\kappa} \,. \tag{15.4.39}$$

For a spin-up proton state $|\,p\uparrow\rangle$, we obtain

$$\langle p\uparrow|\mu_{I=0}^i|p\uparrow\rangle = \frac{\delta^{i3}}{32\pi^2\kappa} \equiv \frac{g_{I=0}}{4M_N}\delta^{i3} \,. \tag{15.4.40}$$

Here we have defined the isoscalar $g$-factor $g_{I=0}$ to compare with the experimental value in the unit of $1/(4M_N)$, where $M_N \simeq 940$ MeV is the nucleon mass. Using the values (15.3.15), the isoscalar $g$-factor is calculated as

$$g_{I=0} = \frac{M_N}{8\pi^2\kappa M_{\mathrm{KK}}} \simeq 1.68 \,. \tag{15.4.41}$$

The experimental value is $g_{I=0}|_{\exp} \simeq 1.76$ and the result of ANW is $g_{I=0}|_{\mathrm{ANW}} \simeq 1.11$.

### 15.4.5.2. *Isovector current*

The isovector current is obtained by applying the formula (15.4.30) to the $SU(2)$ part of the gauge field. The result is

$$J_V^{a\,0} \simeq I^a J_B^0 + \cdots \,, \quad J_V^{a\,i} \simeq (2\pi^2\kappa)\rho^2\mathrm{Tr}(\tau^a \boldsymbol{a}\tau^j \boldsymbol{a}^{-1})\epsilon^{ijk}\partial_k J_B^0 + \cdots \tag{15.4.42}$$

where $I^a = -i4\pi^2\kappa\rho^2\mathrm{Tr}(\tau^a \boldsymbol{a}\,\dot{\boldsymbol{a}}^{-1})$ is the isospin operator.

Then, the isovector magnetic moment is obtained as

$$\mu_{I=1}^i = \epsilon^{ijk} \int d^3x\, x^j J_V^{3\,k} \simeq -4\pi^2\kappa\rho^2\mathrm{Tr}(\tau^3 \boldsymbol{a}\tau^i \boldsymbol{a}^{-1}) \,. \tag{15.4.43}$$

Using the formula

$$\langle \mathrm{Tr}(\tau^a \boldsymbol{a}\tau^i \boldsymbol{a}^{-1}) \rangle = -\frac{2}{3}\langle \tau^a \sigma^i \rangle \,, \tag{15.4.44}$$

where $\tau^i$ and $\sigma^a$ in the right side are the Pauli matrices acting on the isospin and spin spaces, respectively. the expectation value with respect to the spin up proton state is obtained as

$$\langle p\uparrow|\mu^i_{I=1}|p\uparrow\rangle = \frac{8\pi^2\kappa}{3}\langle\rho^2\rangle\delta^{i3} \equiv \frac{g_{I=1}}{4M_N}\delta^{i3} \,. \tag{15.4.45}$$

If we approximate $\langle\rho^2\rangle$ by its classical value $\rho^2_{\rm cl}$ in (15.4.25), we obtain $g_{I=1}\simeq$ 4.34. If we evaluate $\langle\rho^2\rangle$ by using the nucleon wave function (15.4.27), we obtain

$$\langle\rho^2\rangle = \rho^2_{\rm cl}\left(\sqrt{1+\frac{5}{N_c^2}}+\frac{\sqrt{5}}{2N_c}\right) \simeq 1.62\,\rho^2_{\rm cl}\,, \tag{15.4.46}$$

which implies $g_{I=1}\simeq 7.03$. This value can be compared with $g_{I=1}|_{\rm exp}\simeq 9.41$ and $g_{I=1}|_{\rm ANW}\simeq 6.38$.

If we use the above results $g_{I=0}\simeq 1.68$ and $g_{I=1}\simeq 7.03$, the magnetic moments for proton and neutron (in the unit of nuclear magneton $\mu_N = 1/(2M_N)$) are evaluated as

$$\mu_p = \frac{1}{4}(g_{I=0}+g_{I=1}) \simeq 2.18\,, \quad \mu_n = \frac{1}{4}(g_{I=0}-g_{I=1}) \simeq -1.34\,, \tag{15.4.47}$$

which are compared with the experimental values and the results of ANW :

$$\mu_p|_{\rm exp}\simeq 2.79\,, \quad \mu_n|_{\rm exp}\simeq -1.91\,, \quad \mu_p|_{\rm ANW}\simeq 1.87\,, \quad \mu_n|_{\rm ANW}\simeq -1.31\,. \tag{15.4.48}$$

Note, however, that these values may not be meaningful, since $g_{I=0}=\mathcal{O}(N_c^0)$ and $g_{I=1}=\mathcal{O}(N_c^2)$.

### 15.4.5.3. *Summary of the result*

In a similar way, we can calculate various quantities for the baryons. Here we list some of the results obtained in Ref. 11.

|  | our result | experiment | ANW |
|---|---|---|---|
| $\langle r^2\rangle^{1/2}_{I=0}$ | 0.742 fm | 0.806 fm | 0.59 fm |
| $\langle r^2\rangle^{1/2}_{I=1}$ | 0.742 fm | 0.939 fm | $\infty$ |
| $\langle r^2\rangle^{1/2}_A$ | 0.537 fm | 0.674 fm | $-$ |
| $g_{I=0}$ | 1.68 | 1.76 | 1.11 |
| $g_{I=1}$ | 7.03 | 9.41 | 6.38 |
| $g_A$ | 0.734 | 1.27 | 0.61 |

$$\tag{15.4.49}$$

Here $g_A$ is axial coupling and $\langle r^2\rangle^{1/2}_A$ is the axial radius, which are obtained from the axial-vector current in (15.4.30). As a reference, we also listed the results in ANW,[2] though we should not compare them directly, since the way of fitting values in ANW is different from ours.

## 15.5. Conclusion

With the help of gauge/string duality, we have obtained a new description of hadrons. Baryons are described as solitons in string theory, revisiting Skyrme's pioneering idea. Gauge/string duality tells us how Skyrme's idea is connected to QCD and leads us to a new description of baryons as solitons in a five-dimensional YM-CS theory.

Generalizing the idea of ANW to our system, we analyzed static properties of baryons. The advantage of our model is that it automatically includes the contributions from various massive vector and axial-vector mesons in a reasonably simple action (15.3.1) Compared with the results in the Skyrme model (ANW), the agreement with the experimental values are improved in most of the cases. In this article, we only present our results for nucleons, but, other baryons, such as $\Delta$, $N(1440)$, $N(1535)$, etc., can also be treated in a similar way. (See Refs. 10 and 11 for details.)

## Acknowledgements

The author is especially grateful to K. Hashimoto, H. Hata, T. Sakai, S. Yamato for pleasant collaboration. This work is supported in part by JSPS Grant-in-Aid for Creative Scientific Research No. 19GS0219 and also by World Premier International Research Center Initiative (WPI Initiative), MEXT, Japan.

## References

1. T.H.R. Skyrme, "A Nonlinear field theory," *Proc. Roy. Soc. Lond. A* **260** (1961) 127; "Particle states of a quantized meson field," *Proc. Roy. Soc. Lond. A* **262** (1961) 237; "A Unified Field Theory Of Mesons And Baryons," *Nucl. Phys.* **31** (1962) 556.
2. G.S. Adkins, C.R. Nappi and E. Witten, "Static Properties Of Nucleons In The Skyrme Model," *Nucl. Phys. B* **228** (1983) 552.
3. J.M. Maldacena, "The large N limit of superconformal field theories and supergravity," *Adv. Theor. Math. Phys.* **2** (1998) 231 [*Int. J. Theor. Phys.* **38** (1999) 1113] [arXiv:hep-th/9711200].
4. S.S. Gubser, I.R. Klebanov and A.M. Polyakov, "Gauge theory correlators from non-critical string theory," *Phys. Lett. B* **428** (1998) 105 [arXiv:hep-th/9802109].
5. E. Witten, "Anti-de Sitter space and holography," *Adv. Theor. Math. Phys.* **2** (1998) 253 [arXiv:hep-th/9802150].
6. O. Aharony, S.S. Gubser, J.M. Maldacena, H. Ooguri and Y. Oz, "Large N field theories, string theory and gravity," *Phys. Rept.* **323** (2000) 183 [arXiv:hep-th/9905111].
7. T. Sakai and S. Sugimoto, "Low energy hadron physics in holographic QCD," *Prog. Theor. Phys.* **113** (2005) 843 [arXiv:hep-th/0412141]; "More on a holographic dual of QCD," *Prog. Theor. Phys.* **114** (2005) 1083 [arXiv:hep-th/0507073].
8. E. Witten, "Baryons and branes in anti de Sitter space," *JHEP* **9807** (1998) 006 [arXiv:hep-th/9805112].
9. D. J. Gross and H. Ooguri, "Aspects of large N gauge theory dynamics as seen by string theory," *Phys. Rev. D* **58**, 106002 (1998) [arXiv:hep-th/9805129].

10. H. Hata, T. Sakai, S. Sugimoto and S. Yamato, "Baryons from instantons in holographic QCD," *Prog. Theor. Phys.* **117** (2007) 1157 [arXiv:hep-th/0701280].
11. K. Hashimoto, T. Sakai and S. Sugimoto, "Holographic Baryons : Static Properties and Form Factors from Gauge/String Duality," *Prog. Theor. Phys.* **120** (2008) 1093 [arXiv:0806.3122 [hep-th]].
12. D.K. Hong, M. Rho, H.U. Yee and P. Yi, "Chiral dynamics of baryons from string theory," *Phys. Rev. D* **76** (2007) 061901 [arXiv:hep-th/0701276]; "Dynamics of Baryons from String Theory and Vector Dominance," *JHEP* **0709** (2007) 063 [arXiv:0705.2632 [hep-th]]; "Nucleon Form Factors and Hidden Symmetry in Holographic QCD," *Phys. Rev. D* **77** (2008) 014030 [arXiv:0710.4615 [hep-ph]].
13. H. Hata, M. Murata and S. Yamato, "Chiral currents and static properties of nucleons in holographic QCD," *Phys. Rev. D* **78** (2008) 086006 [arXiv:0803.0180 [hep-th]].
14. K.Y. Kim and I. Zahed, "Electromagnetic Baryon Form Factors from Holographic QCD," *JHEP* **0809** (2008) 007 [arXiv:0807.0033 [hep-th]].
15. J. Dai, R.G. Leigh and J. Polchinski, "New Connections Between String Theories," *Mod. Phys. Lett. A* **4** (1989) 2073.
16. J. Polchinski, "Dirichlet-Branes and Ramond-Ramond Charges," *Phys. Rev. Lett.* **75** (1995) 4724 [arXiv:hep-th/9510017].
17. E. Witten, "Anti-de Sitter space, thermal phase transition, and confinement in gauge theories," *Adv. Theor. Math. Phys.* **2** (1998) 505 [arXiv:hep-th/9803131].
18. R.C. Brower, S.D. Mathur and C.I. Tan, "Glueball Spectrum for QCD from AdS Supergravity Duality," *Nucl. Phys. B* **587** (2000) 249 [arXiv:hep-th/0003115].
19. E. Witten, "Baryons In The 1/N Expansion," *Nucl. Phys. B* **160** (1979) 57.
20. M.F. Atiyah and N.S. Manton, "Skyrmions from instantons," *Phys. Lett. B* **222** (1989) 438.
21. M.R. Douglas, "Branes within branes," arXiv:hep-th/9512077.
22. G.S. Adkins and C.R. Nappi, "Stabilization Of Chiral Solitons Via Vector Mesons," *Phys. Lett. B* **137** (1984) 251.
23. W.M. Yao *et al.* [Particle Data Group], "Review of particle physics," *J. of Phys. G* **33** (2006) 1.

# Chapter 16

# Holographic Baryons

Piljin Yi

*School of Physics, Korea Institute for Advanced Study, Seoul 130-722, Korea*

We review baryons in the D4-D8 holographic model of low energy QCD, with the large $N_c$ and the large 't Hooft coupling limit. The baryon is identified with a bulk soliton of a unit Pontryagin number, which from the four-dimensional viewpoint translates to a modified Skyrmion dressed by condensates of spin one mesons. We explore classical properties and find that the baryon in the holographic limit is amenable to an effective field theory description. We also present a simple method to capture all leading and subleading interactions in the $1/N_c$ and the derivative expansions. An infinitely predictive model of baryon-meson interactions is thus derived, although one may trust results only for low energy processes, given various approximations in the bulk. We showcase a few comparisons to experiments, such as the leading axial couplings to pions, the leading vector-like coupling, and a qualitative prediction of the electromagnetic vector dominance that involves the entire tower of vector mesons.[*]

## Contents

16.1 Low Energy QCD and Solitonic Baryons . . . . . . . . . . . . . . . . . . . 368
16.2 A Holographic QCD . . . . . . . . . . . . . . . . . . . . . . . . . . . . . . . 369
   16.2.1 Holographic pure QCD from D4 . . . . . . . . . . . . . . . . . . . 371
   16.2.2 Adding mesons via D4-D8 complex . . . . . . . . . . . . . . . . . . 372
16.3 Holographic Baryons . . . . . . . . . . . . . . . . . . . . . . . . . . . . . . . 375
   16.3.1 The instanton soliton . . . . . . . . . . . . . . . . . . . . . . . . . 375
   16.3.2 Quantum numbers . . . . . . . . . . . . . . . . . . . . . . . . . . . 378
16.4 Holographic Dynamics . . . . . . . . . . . . . . . . . . . . . . . . . . . . . . 379
   16.4.1 Dynamics of hairy solitons: generalities . . . . . . . . . . . . . . . 380
   16.4.2 The small size matters . . . . . . . . . . . . . . . . . . . . . . . . . 380
   16.4.3 Holographic dynamics of baryons . . . . . . . . . . . . . . . . . . . 381
   16.4.4 Basic identities and isospin-dependence . . . . . . . . . . . . . . . 383
16.5 Nucleons . . . . . . . . . . . . . . . . . . . . . . . . . . . . . . . . . . . . . 384
   16.5.1 Nucleon-meson effective actions . . . . . . . . . . . . . . . . . . . 385
   16.5.2 Numbers and comments . . . . . . . . . . . . . . . . . . . . . . . . 386
16.6 Electromagnetic Properties . . . . . . . . . . . . . . . . . . . . . . . . . . . 387
16.7 More Comments . . . . . . . . . . . . . . . . . . . . . . . . . . . . . . . . . 389
References . . . . . . . . . . . . . . . . . . . . . . . . . . . . . . . . . . . . . . . 391

---

[*]This note is an expanded version of a proceeding contribution to *"30 years of mathematical method in high energy physics,"* Kyoto 2008.

## 16.1. Low Energy QCD and Solitonic Baryons

QCD is a challenging theory. Its most interesting aspects, namely the confinement of color and the chiral symmetry breaking, have defied all analytical approaches. While there are now many data accumulated from the lattice gauge theory, the methodology falls well short of giving us insights on how one may understand these phenomena analytically, nor does it give us a systematic way of obtaining a low energy theory of QCD below the confinement scale.

A very useful approach in the conventional field theory language is the chiral perturbation theory.[1] It bypasses the question of how the confinement and the symmetry breaking occur but rather focuses on the implications. A quark bilinear condenses to break the chiral symmetry $U(N_F)_L \times U(N_F)_R$ to its diagonal subgroup $U(N_F)$, whereby $N_F^2$ Goldstone bosons appear, which we will refer to as pions. They are singled out as the lightest physical particles, and one guesses and constrains an effective Lagrangian for them. In the massless limit[a] of the bare quarks, the pions are packaged into a unitary matrix as

$$U(x) = e^{2i\pi(x)/f_\pi} , \qquad (16.1.2)$$

whose low energy action is written in a derivative expansion as

$$\int dx^4 \left( \frac{f_\pi^2}{4} \operatorname{tr} \left( U^{-1}\partial_\mu U \right)^2 + \frac{1}{32e_{Skyrme}^2} \operatorname{tr} \left[ U^{-1}\partial_\mu U, U^{-1}\partial_\nu U \right]^2 + \cdots \right) , \quad (16.1.3)$$

where the ellipsis denotes higher derivative terms as well as other possible quartic derivative terms. One can further add other massive mesons whose masses and interaction strengths are all left as free parameters to fit with data.

Another analytical approach is the large $N_c$ expansion.[2] Here, two different couplings $1/N_c$ and $\lambda = g_{YM}^2 N_c$ control the perturbation expansion, one counting the topology of the Feynman diagram and the other counting loops. An interesting question is how this large $N_c$ limit appears in the chiral Lagrangian approach. Since the pion fields (or any other meson fields that one can add) are already color-singlets, $N_c$ would enter only via the numerical coefficients of the various terms in the Lagrangian. Both terms shown in (16.1.3) can arise from planar diagrams of large $N_c$ expansion, and we expect

$$f_\pi^2 \sim N_c \sim \frac{1}{e_{Skyrme}^2} . \qquad (16.1.4)$$

Note that since $1/f_\pi^2$ and $1/(e_{Skyrme}^2 f_\pi^4)$ play the role of squared couplings for canonically normalized pions, the self-coupling of pions scales as $N_c^{-1/2}$.[3] In particular, this shows that baryons are qualitatively different than mesons in the large

---

[a]The effect of small bare masses for quarks can be incorporated by an explicit symmetry breaking term

$$\operatorname{tr}\left( MU + U^\dagger M^\dagger \right) \qquad (16.1.1)$$

with a matrix $M$, which in our holographic approach would be ignored.

$N_c$ chiral perturbation theory. Baryons involve $N_c$ number of quarks, so the mass is expected to grow linearly with $N_c$, or equivalently grows with the inverse square of pion self-couplings. In field theories, such a scaling behavior is a hallmark of nonperturbative solitons.

Indeed, it has been proposed early on that baryons are topological solitons, namely Skyrmions,[4] whose baryon number is cataloged by the third homotopy group of $U(N_F)$, $\pi_3(U(N_F)) = Z$. The topological winding is counted by how many times $U(x)$ covers a noncollapsible three-sphere in $U(N_F)$ manifold, as a function on $R^3$. Given such topological data, one must find a classical solution that minimizes the energy of the chiral Lagrangian. An order of magnitude estimate for the size $L_{Skyrmion}$ of a Skyrmion gives

$$L_{Skyrmion} \sim \frac{1}{f_\pi \, e_{Skyrme}} , \qquad (16.1.5)$$

which is independent of large $N_c$.

However, let us pose and consider whether this construction really makes sense. This solitonic picture says that baryons can be regarded as coherent states of Goldstone bosons of QCD. Although the latter are special due to the simple and universal origin and also due to the light mass, they are one of many varieties of bi-quark mesons. In particular, there are known and experimentally measured cubic couplings between pions and heavier spin one mesons, such $\rho$ mesons. A condensate of pions, as in a Skyrmion, would shows up as a source term for a $\rho$ meson equation of motion and $\rho$ must also have its own coherent state excited. In turn, this will disturb the conventional Skyrmion picture and modify it quantitatively. This is a clear signal that the usual Skyrmion picture of the baryon has to be modified significantly in the context of full QCD.

Perhaps because of this, and perhaps for other reasons, the picture of baryon as Skyrmion have produced mixed results when compared to experimental data. In this note, we will explore how this problem is partially cured, in a natural and simple manner without new unknown parameters, and how the resulting baryons look qualitatively and quantitatively different from that of Skyrmion. As we will see, the holographic picture naturally brings a gauge-principle in the bulk description of the flavor dynamics in such a way that all spin one mesons as well as pions would enter the construction of baryons on the equal footing. The basic concept of baryons as coherent states of mesons would remain unchanged, however. It is the purpose of this note to outline this new approach to baryons and to explore the consequences.

## 16.2. A Holographic QCD

A holographic QCD is similar to the chiral perturbation theory in the sense that we deal with exclusively gauge-invariant operators of the theory. The huge difference is, however, that this new approach tends to treat all gauge-invariant objects together. Not only the light meson fields like pions but also heavy vector mesons and baryons appear together, at least in principle. In other words, a holographic QCD deals with all color-singlets simultaneously, giving us a lot more predictive power. Later we will see examples of this more explicitly.

This new approach is motivated by the large $N_c$ limit of gauge theories[2] and in particular by the AdS/CFT correspondence.[5] One of the more interesting notion that emerged in this regard over the last three decades is the concept of *the master field*. The idea is that in the large $N$ limits of matrix theories with a gauge symmetry, the gauge-singlet observables behaves semiclassically in the large $N$ limit.[6] Probably the most astounding twist is the emergence of a new spatial direction in such a picture. As we learned from AdS/CFT, *the master fields* have to be thought of not as four-dimensional fields but at least five-dimensional, with the additional direction being labeled by energy scale. We refer to this new direction as the holographic direction.

The standard AdS/CFT duality gives us a precise equivalence between the large $N_c$ maximally supersymmetric Yang-Mills theories and the type IIB string theory or IIB supergravity in $AdS_5 \times S^5$. Here, *the master fields* are nothing but closed string fields such as the gravity multiplet and excited closed string fields. It is also believed that such a duality extends to other large $N$ field theories such as ordinary QCD which is neither supersymmetric nor conformal. The question is then how to find the right dual theory of the large $N_c$ QCD.

One set of ideas for this, dubbed bottom-up,[7] is similar in spirit to the chiral perturbation theory. One assumes that an approximate conformal symmetry exists for a wide range of energy scales and builds up a bulk gravity theory coupled to more bulk fields, as would be dictated by the AdS/CFT rules if QCD were conformal. The conformal symmetry is subsequently broken by cutting off the geometry at both the infrared and the ultraviolet and by introducing boundary conditions. Necessary degrees of freedoms, namely the master fields, are introduced as needed by construction, rather than derived, and in this sense the approach is similar to the conventional chiral perturbation theory.

The other approach is referred to as top-down, and here one tries to realize the QCD as a low energy limit of some open string theory on D-branes, from which a holographic model follows as the closed string theory dual. Arguably, the best model of this kind we know of is the D4-D8 system, where $U(N_c)$ D4 gauge theory compactified on a thermal circle provides large $N_c$ Yang-Mills sector. The $U(N_F)$ gauge theory on D8 brane, on the other hand, can be thought of bi-quark meson sector in the adjoint representation of the $U(N_F)$ flavor symmetry. A crucial aspect of this model expected from general AdS/CFT principles is that the vector-like flavor symmetry is promoted to a gauge theory in the bulk. This D4-D8 model was slowly developed over the years, starting with Witten's initial identification of the dual geometry for D4 branes wrapped on a thermal circle,[8] study of glueball mass spectra of pure QCD without matter,[9,10] the introduction of mesons via D8 branes,[11] and very recent study of baryons as solitonic objects[12–14] on D8 branes. In this section, we will review glueballs and mesons in this D4-D8 model.

### 16.2.1. *Holographic pure QCD from D4*

The story starts with a stack of D4 branes which is compactified on a circle. The circle here is sometimes called "thermal" in that one requires anti-periodic boundary condition on all fermions, just as one would for the Euclidean time circle when studying finite temperature field theory. The purpose of having a spatial "thermal" circle is to give mass to the fermionic superpartners and thus break supersymmetry. As is well known, the low energy theory on $N$ D$p$ branes is a maximally supersymmetric $U(N)$ Yang-Mills theory in $p+1$ dimensions, so putting $N_c$ D4 branes on a thermal circle, we obtain pure $U(N_c)$ Yang-Mills theory in the noncompact $3+1$ dimensions. We are interested in large $N_c$ limit, so the $U(1)$ part can be safely ignored, and we may pretend that we are studying $SU(N_c)$ theory instead. While the anti-periodic boundary condition generates massgap only to fermionic sector at tree level, scalar partners also become massive since there is no symmetry to prohibit their mass any more. Only the gauge multiplet is protected.

We then extrapolate the general idea of AdS/CFT to this non-conformal case, which states that, instead of studying strongly coupled large $N_c$ Yang-Mills theory, one may look at its dual closed string theory. The correct closed string background to use is nothing but the string background generated by the D4 branes in question. This geometry was first written down by Gibbons and Maeda[15] in the 1980's, and later reinterpreted by Witten in 1998 as the dual geometry for D4 branes on a thermal circle.[8] The metric is most conveniently written as

$$ds^2 = \left(\frac{U}{R}\right)^{3/2}\left(\eta_{\mu\nu}dx^\mu dx^\nu + f(U)d\tau^2\right) + \left(\frac{R}{U}\right)^{3/2}\left(\frac{dU^2}{f(U)} + U^2 d\Omega_4^2\right), \quad (16.2.6)$$

with $R^3 = \pi g_s N_c l_s^3$ and $f(U) = 1 - U_{KK}^3/U^3$. The topology of the spacetime is $R^{3+1} \times D \times S^4$, with the coordinate $\tau$ labeling the azimuthal angle of the disk $D$, with $\tau = \tau + \delta\tau$ and $\delta\tau = 4\pi R^{3/2}/(3U_{KK}^{1/2})$. The circle parameterized by $\tau$ is the thermal circle. The dilaton is

$$e^{-\Phi} = \frac{1}{g_s}\left(\frac{R}{U}\right)^{3/4}, \quad (16.2.7)$$

while the antisymmetric Ramond-Ramond background field $C_3$ is such that $dC_3$ carries $N_c$ unit of flux along $S^4$.

In the limit of large curvature radius, thus large $N_c$, and in the limit of large 't Hooft coupling $\lambda \equiv g_{YM}^2 N_c$, the duality collapse to a relationship between the theory of D4 branes to type IIA supergravity defined in this background. Given the lack of useful method of string theory quantization in curved background, this is the best we can do at the moment. Therefore, all computations in any of holographic QCD must assume such a limit and extrapolate to realistic regime at the end of the computation. This is also the route that we follow in this note.

Among remarkable works in early days of AdS/CFT is the study of glueball spectra in this background.[9,10] They considered small fluctuations of IIA gravity

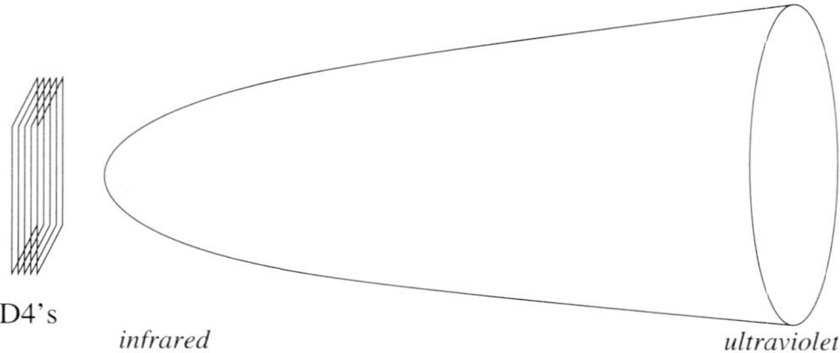



Fig. 16.1. A schematic diagram showing the dual geometry. A stack of D4's responsible for the dual geometry are shown for an illustrative purpose, although the actual spacetime does not include them. The manifold shown explicitly is spanned by the angle $\tau$ and the radial coordinate $U$. The thermal circle spanned by $\tau$ closes itself in the infrared end due to the strong interaction of QCD. Small excitations of metric (and its multiplet) at the infrared end correspond to glueballs.

multiplet in the above background, with the plane-wave like behavior along $x^\mu$ and $L^2$ normalizability along the remaining six directions. They identified each of such modes as glueballs up to spin 2, and computed their mass$^2$ eigenvalues as dictated by the linearized gravitational equation of motion.

This illustrates what is going on here. We can think of the duality here as a simple statement that the open string side and the closed string side is one and the same theory. The reason we have apparently more complicated description on the open string side is because there we started with a misleading and redundant set of elementary fields, namely the gauge field whose number scales as $N_c^2$, only to be off-set by the gauge symmetry. The closed string side, or its gravity limit, happens to be more smart about what are the right low energy degrees of freedom and encodes only gauge-invariant ones. For pure Yang-Mills theory like this, the only gauge-invariant objects are glueballs, so the dual gravitational side should compute the glueball physics.

The expectation that there exists a more intelligent theory consisting only of gauge-invariant objects in the large $N_c$ limit is thus realized via string theory in a somewhat surprising manner that *the master fields*, those truly physical degrees of freedom, actually live not in four dimensional Minkowskian world but in five or higher dimensional curved geometry. This is not however completely unanticipated, and was heralded in the celebrated work by Eguchi and Kawai in early 1980's[16] which is all the more remarkable in retrospect. For the rest of this note, we will continue this path and try to incorporate massless quarks to the story.

### 16.2.2. *Adding mesons via D4-D8 complex*

To add mesons, Sakai and Sugimoto introduced the $N_F$ D8 branes, which share the coordinates $x^\mu$ with the above D4 branes[11] and are transverse to the thermal circle

$\tau$. Before we trade off the $N_c$ D4 branes in favor of the dual gravity theory, this would have allowed massless quark as open strings ending on both the D4 and the D8 branes. As the D4's are replaced by the dual geometry, however, the 4-8 open strings have to be paired up into 8-8 open strings, which are naturally identified as bi-quark mesons. From the viewpoint of D8 branes, the lightest of such mesons belong to a $U(N_F)$ gauge field.

The $U(N_F)$ gauge theory on D8 branes has the action

$$-\frac{4\pi^2 l_s^4 \mu_8}{8} \int \sqrt{-h_{8+1}}\, e^{-\Phi}\, \mathrm{tr}\mathcal{F}^2 + \mu_8 \int C_3 \wedge \mathrm{tr}\, e^{2\pi\alpha'\mathcal{F}}\,, \tag{16.2.8}$$

where the contraction is via the induced metric of D8 and $\mu_p = 2\pi/(2\pi l_s)^{p+1}$ with $l_s^2 = \alpha'$. The induced metric on the D8 brane is

$$h_{8+1} = \frac{U^{3/2}(w)}{R^{3/2}}\left(dw^2 + \eta_{\mu\nu}dx^\mu dx^\nu\right) + \frac{R^{3/2}}{U^{1/2}(w)}d\Omega_4^2\,, \tag{16.2.9}$$

after we trade off the holographic (or radial) coordinate $U$ in favor of a conformal one $w$ as

$$w = \int_{U_{KK}}^{U} R^{3/2}dU'/\sqrt{U'^3 - U_{KK}^3}\,, \tag{16.2.10}$$

which resides in a finite interval of length $\sim O(1/M_{KK})$ where $M_{KK} \equiv 3U_{KK}^{1/2}/2R^{3/2}$. Thus, the topology of the D8 worldvolume is $R^{3+1} \times I \times S^4$. The nominal Yang-Mills coupling $g_{YM}^2$ is related to the other parameters as

$$g_{YM}^2 = 2\pi g_s M_{KK} l_s\,, \tag{16.2.11}$$

which is not, however, a physical parameter on its own. The low energy parameters of this holographic theory are $M_{KK}$ and $\lambda$, which together with $N_c$ sets all the physical scales such as the QCD scale and the pion decay constant.

In the low-energy limit, we ignore the $S^4$ direction on which D8's are completely wrapped, and find a five-dimensional Yang-Mills theory with a Chern-Simons term

$$-\frac{1}{4}\int_{4+1} \frac{1}{e(w)^2}\sqrt{-h_{4+1}}\,\mathrm{tr}\mathcal{F}^2 + \frac{N_c}{24\pi^2}\int_{4+1}\omega_5(\mathcal{A})\,, \tag{16.2.12}$$

where the position-dependent Yang-Mills coupling of this flavor gauge theory is

$$\frac{1}{e(w)^2} = \frac{e^{-\Phi}V_{S^4}}{2\pi(2\pi l_s)^5} = \frac{\lambda N_c}{108\pi^3}M_{KK}\frac{U(w)}{U_{KK}} \tag{16.2.13}$$

with $V_{S^4}$ the position-dependent volume of $S^4$. The Chern-Simons coupling with $d\omega_5(\mathcal{A}) = \mathrm{tr}\mathcal{F}^3$ arises because $\int_{S^4} dC_3 \sim N_c$.

As advertised, this by itself generates many of bi-quark mesons of QCD. More specifically, all of vector and axial-vector mesons and the pion multiplet are encoded in this five-dimensional $U(N_F)$ gauge field. The vector mesons and the axial vector mesons are more straightforward conceptually, since any "compactification" of five-dimensional Yang-Mills theory would lead to an infinite tower of four-dimensional

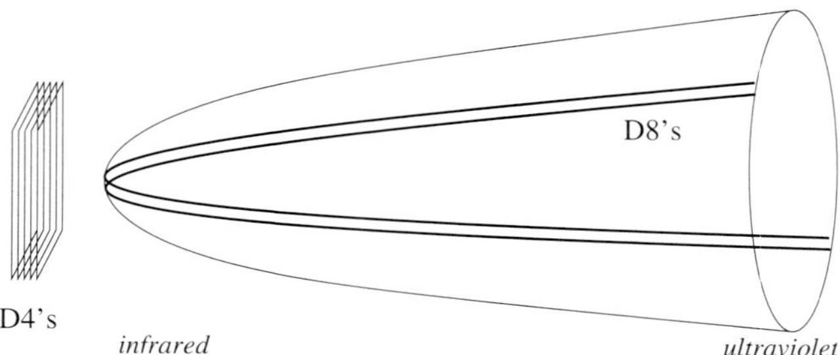



Fig. 16.2. The figure shows how D8's are added to the system. Low energy excitations (also located at the infrared end) of D8-D8 open strings are bi-quark mesons.

massive vector fields. Although the radial direction $w$ (or $U$) is infinite in terms of proper length, equation of motion is such that normalizable fields are strongly pushed away from the boundary, making it effectively a compact direction. The usual Kaluza-Klein reduction (in the somewhat illegal but convenient axial gauge $\mathcal{A}_w = 0$),

$$\mathcal{A}_\mu(x; w) = i\alpha_\mu(x)\psi_0(w) + i\beta_\mu(x) + \sum_n a_\mu^{(n)}(x)\psi_{(n)}(w) \tag{16.2.14}$$

contains an infinite number of vector fields, whose action can be derived explicitly as,

$$\int dx^4 \, \mathcal{L} = \int dx^4 \sum_n \mathrm{tr} \left\{ \frac{1}{2}\mathcal{F}_{\mu\nu}^{(n)}\mathcal{F}^{\mu\nu(n)} + m_{(n)}^2 a_\mu^{(n)} a^{\mu(n)} \right\} + \cdots, \tag{16.2.15}$$

with $\mathcal{F}_{\mu\nu}^{(n)} = \partial_\mu a_\nu^{(n)} - \partial_\nu a_\mu^{(n)}$. The ellipsis denotes zero mode part, to be discussed shortly, as well as infinite number of couplings among these infinite varieties of mesons, all of which come from the unique structure of the five-dimensional $U(N_F)$ Yang-Mills Lagrangian in (16.2.12). Because $\mathcal{A}$ has a specific parity, the parity of $a_n$'s are determined by the parity of the eigenfunctions $\psi_{(n)}(w)$ along the fifth direction. Since the parity of any one-dimensional eigenvalue system alternates, an alternating tower of vector and axial-vector fields emerge as the masses $m_{(n)}$ of the KK modes increase.

For each such eigenmode, a five-dimensional massless vector field has three degrees of freedom, so is natural for massive four-dimensional vector fields to appear. An exception to this naive counting, which is specific to the gauge theory, is the zero mode sector. In Eq. (16.2.14), we separated it out from the rest as $\alpha(x)$ and $\beta(x)$ terms. To understand this part, it is better to give up the axial gauge and consider the Wilson line,

$$U(x) = e^{i\int_w \mathcal{A}(x,w)}, \tag{16.2.16}$$

which, as the notation suggests, one identifies with the pion field $U(x) = e^{2i\pi(x)/f_\pi}$. Upon taking a singular gauge transformation back to $\mathcal{A}_w = 0$, one finds that it is related to $\alpha$ and $\beta$ as

$$\alpha_\mu(x) \equiv \{U^{-1/2}, \partial_\mu U^{1/2}\}\,, \qquad 2\beta_\mu(x) \equiv [U^{-1/2}, \partial_\mu U^{1/2}]\,. \qquad (16.2.17)$$

Truncating to this zero mode sector reproduces a Skyrme Lagrangian of pions[4] as a dimensional reduction of the five-dimensional Yang-Mills action,

$$\int dx^4 \left( \frac{f_\pi^2}{4} \mathrm{tr}\,\left(U^{-1}\partial_\mu U\right)^2 + \frac{1}{32 e_{Skyrme}^2} \mathrm{tr}\,\left[U^{-1}\partial_\mu U, U^{-1}\partial_\nu U\right]^2 \right)\,, \qquad (16.2.18)$$

with $f_\pi^2 = (g_{YM}^2 N_c)N_c M_{KK}^2/54\pi^4$ and $1/e_{Skyrme}^2 \simeq 61(g_{YM}^2 N_c)N_c/54\pi^7$. No other quartic term arises, nor do we find higher order terms in derivative, although we do recover the Wess-Zumino-Witten term from the Chern-Simons term.[11] To compare against actual QCD, we must fix $\lambda = g_{YM}^2 N_c \simeq 17$ and $M_{KK} \simeq 0.94\,GeV$ to fit both the pion decay constant $f_\pi$ and the mass of the first vector meson. After this fitting, all other infinite number of masses and coupling constants are fixed. This version of the holographic QCD is extremely predictive.

Let us emphasize that the meson system here comes with a qualification. Note that we treated D8 branes differently than D4 branes. The latter are replaced by the dual geometry while the former are kept as branes. This has to be because we are interested in objects charged under $U(N_F)$, whereas we are only interested in singlets under $U(N_c)$. However, we not only treated D8 as branes but also as probe branes, meaning that the backreaction of D8 to the dual geometry of D4's is ignored. In terms of field theory language, we effectively ignored Feynman diagrams involving quarks in the internal lines, resulting in the quenched approximation.

## 16.3. Holographic Baryons

The baryon can be naturally regarded as a coherent state of mesons in the large $N_c$. In the conventional chiral Lagrangian approach, is the Skyrmion made from pions, which we argued cannot be the full picture. In D4-D8 model of holographic QCD above, especially, pions are only the zero mode part of a holographic flavor theory, and infinite towers of vector and axial-vector mesons are packaged together with pions into a single five-dimensional $U(N_F)$ gauge field. This suggests that the picture of baryon as a soliton must be lifted to a five-dimensional soliton of this $U(N_F)$ gauge theory in the bulk, in such a manner that spin one mesons contribute to the construction of baryons as well. In this section, we explore classical and quantum properties of this holographic and new version of Skyrmion.

### 16.3.1. *The instanton soliton*

The five-dimensional effective action for the $U(N_F)$ gauge field in Eq. (16.2.12) admits solitons which carry a Pontryagin number

$$\frac{1}{8\pi^2} \int_{R^3 \times I} \mathrm{tr} F \wedge F = k\,, \qquad (16.3.19)$$

with integral $k$. We denoted by $F$ the non-Abelian part of $\mathcal{F}$ (and similarly later, $A$ for non-Abelian part of $\mathcal{A}$). The smallest unit with $k = 1$ turns out to carry quantum numbers of the baryon. The easiest way to see this identification is to relate it to the Skyrmion[4] of the chiral perturbation theory .

Recall that both instantons and Skyrmions are labeled by the third homotopy group $\pi_3$ of a group manifold, which is the integer for any semi-simple Lie group manifold $G$. For the Skyrmion, the winding number shows up in the classification of maps

$$U(x) : R^3 \to SU(N_F) . \tag{16.3.20}$$

For the instanton whose asymptotic form is required to be pure gauge,

$$A(x, w \to \pm\infty) = ig_\pm(x)^\dagger dg_\pm(x) , \tag{16.3.21}$$

the winding number is in the classification of the map

$$g_-(x)^\dagger g_+(x) : R^3 \to SU(N_F) . \tag{16.3.22}$$

The relationship between the two types of the soliton is immediate.[18] Recall that the $U$ field of chiral perturbation theory is obtained in our holographic picture as the open and infinite Wilson line along $w$ direction. On the other hand, the Wilson line computes nothing but $g_-(x)^\dagger g_+(x)$, so we find that

$$U_k(x) = e^{i \int_w A^{(k)}(x,w)} \tag{16.3.23}$$

carries $k$ Skyrmion number exactly when $A^{(k)}$ carries $k$ Pontryagin number. Therefore, the instanton soliton in five dimensions is the holographic image of the Skyrmions in four dimensions. We will call it the instanton soliton.

Normal instantons on a conformally flat four-manifold are well studied, and the counting of zero modes says that for a $k$ instanton in $U(N_F)$ theory, there are $4kN_F$ collective coordinates. For the minimal case with $k = 1$ and $N_F = 2$, there are eight collective coordinates. They are four translations, one overall size, and three gauge rotations. For our instanton solitons, this counting does not hold any more.

Unlike the usual Yang-Mills theory in trivial $R^4$ background, the effective action has a position-dependent inverse Yang-Mills coupling $1/e(w)^2$ which is a monotonically increasing function of $|w|$. Since the Pontryagin density contributes to action as multiplied by $1/e(w)^2$, this tends to position the soliton near $w = 0$ and also shrink it for the same reason. The $F^2$ energy of a trial configuration with size $\rho$ can be estimated easily in the small $\rho$ limit,[b]

$$E_{\text{Pontryagin}} = \frac{\lambda N_c}{27\pi} M_{KK} \times \left(1 + \frac{1}{6} M_{KK}^2 \rho^2 + \cdots\right) , \tag{16.3.24}$$

which clearly shows that the energy from the kinetic term increases with $\rho$. This by itself would collapse the soliton to a point-like one, making further analysis impossible.

---

[b]The estimate of energy here takes into account the spread of the instanton density $D(x^i, w) \sim \rho^4/(r^2+w^2+\rho^2)^4$, but ignores the deviation from the flat geometry along the four spatial directions.

A second difference comes from the presence of the additional Chern-Simons term $\sim \mathrm{tr}\mathcal{A} \wedge \mathcal{F} \wedge \mathcal{F}$, whereby the Pontryagin density $F \wedge F$ sources some of the gauge field $\mathcal{A}$ minimally. This electric charge density costs the Coulombic energy

$$E_{\mathrm{Coulomb}} \simeq \frac{1}{2} \times \frac{e(0)^2 N_c^2}{10\pi^2 \rho^2} + \cdots , \qquad (16.3.25)$$

again in the limit of $\rho M_{KK} \ll 1$. This Coulombic energy tends to favor larger soliton size, which competes against the shrinking force due to $E_{Pontryagin}$.

The combined energy is minimized at[12-14]

$$\rho_{baryon} \simeq \frac{(2 \cdot 3^7 \cdot \pi^2/5)^{1/4}}{M_{KK}\sqrt{\lambda}} , \qquad (16.3.26)$$

and the classical mass of the stabilized soliton is

$$m_B^{classical} = (E_{\mathrm{Pontryagin}} + E_{\mathrm{Coulomb}})\Big|_{\mathrm{minimum}}$$
$$= \frac{\lambda N_c}{27\pi} M_{KK} \times \left(1 + \frac{\sqrt{2 \cdot 3^5 \cdot \pi^2/5}}{\lambda} + \cdots \right) . \qquad (16.3.27)$$

As was mentioned above, the size $\rho_{baryon}$ is significantly smaller than $\sim 1/M_{KK}$. We have a classical soliton whose size is a lot smaller than the fundamental scale of the effective theory. On the other hand, this small soliton size is still much larger than its own Compton size $1/m_B^{classical} \simeq 27\pi/(M_{KK}\lambda N_c)$, justifying our assertion that this is indeed a soliton.

Note that the instanton soliton size is much smaller than the Skyrmion size when the 't Hooft coupling is large.[c] We already saw that the Skyrmion size is determined by the ratio of the two dimensionful couplings in the chiral Lagrangian. Using the values of these coupling derived from our D4-D8 model, the would-be Skyrmion size is

$$L_{Skyrmion} \sim \frac{1}{f_\pi e_{Skyrme}} \sim \frac{1}{M_{KK}} . \qquad (16.3.28)$$

On the other hand, the size of the holographic baryon is

$$\rho_{baryon} \sim \frac{1}{M_{KK}\sqrt{\lambda}} . \qquad (16.3.29)$$

The difference is substantial in the large 't Hooft coupling limit where this holographic QCD makes sense. Why is this?

Simply put, the Skyrmion solution of size $\sim 1/M_{KK}$ is a bad approximation, because it solves the chiral Lagrangian which neglects all other spin one mesons. This truncation can be justified for processes involving low energy pions. The baryon is, however, a heavy object and contains highly excited modes of pions, and will excite

---

[c]One must not confuse these solitonic sizes with the electromagnetic size of baryons. The latter is dictated by how photons interact with the baryon, and in the holographic QCD with $\lambda \gg 1$ is determined at $\rho$ meson scale and independent of $\lambda$, due to the vector dominance. One may think of these solitonic sizes as being hadronic.

relatively light vector mesons as well since $U$ is coupled to vector and axial-vector mesons nontrivially at cubic level. Therefore, the truncation to the pion sector is not a good approximation as far as solitonic baryons are concerned, especially for large 't Hooft coupling constant.[d] We emphasize this difference because many of existing computation of the baryon physics based on the Skyrmion picture must be thus rethought in terms of the new instanton soliton picture. We will consider implication of this new picture of the baryon in next sections.

Our solitonic picture of the baryon has a close tie to the usual AdS/CFT picture of baryons as wrapped D-branes. A D4 brane wrapped along the compact $S^4$ corresponds to a baryon vertex on the five-dimensional spacetime,[11] as follows from an argument originally due to Witten.[20] To distinguish them from the D4 branes supporting QCD, let us call them D4$'$. On the D4$'$ worldvolume we have again a Chern-Simons coupling of the form,

$$\mu_4 \int C_3 \wedge 2\pi\alpha' d\mathcal{A}' \tag{16.3.30}$$

with D4$'$ gauge field $\mathcal{A}'$, which can be evaluated over $S^4$ as

$$2\pi\alpha'\mu_4 \int_{S^4 \times R} dC_3 \wedge \mathcal{A}' = N_c \int_R \mathcal{A}' , \tag{16.3.31}$$

where $R$ denotes the worldline in the noncompact part of the spacetime. This shows that the background $dC_3$ flux over $S^4$ induces $N_c$ unit of the electric charge. On the other hand, the Gauss constraint for $\mathcal{A}'$ demands that the net charge should be zero, so the wrapped D4$'$ can exist only if $N_c$ end points of fundamental strings are attached to D4$'$ to cancel this charge. In turn, the other ends of the fundamental strings must go somewhere, and the only place it can go is D8 branes. One can think of these strings as individual quarks that constitute the baryon. Also, because of these fundamental strings, the wrapped D4$'$ cannot be separated from D8's without a lot of energy cost. The lowest energy state would be one where D4$'$ is on top of D8's, which then would smear out as an instanton. The latter is exactly the instanton soliton of ours.

### 16.3.2. *Quantum numbers*

For the sake of simplicity, and also because the quarks in this model have no bare mass, we will take $N_F = 2$ for the rest of the note. A unit instanton soliton in question comes with six collective coordinates. Three correspond to the position in $R^3$, and three correspond to the gauge angles in $SU(N_F = 2)$. If the soliton is small enough ($\rho M_{KK} \ll 1$), there exists approximate symmetries $SO(4) = SU(2)_+ \times SU(2)_-$ at $w = 0$, so the total rotational symmetry of a small solution at origin

---

[d]There were previous studies that incorporated the effect of coupling a single vector meson, namely the lightest $\rho$ meson, on the Skyrmion which showed a slight shrinkage of the soliton[19] as we would have expected in retrospect.

is $SU(N_F = 2) \times SU(2)_+ \times SU(2)_-$. Let us first see how the quantized instanton soliton fits into representations of this approximate symmetry group.

The instanton can be rotated by an conjugate $SU(2)$ action as,

$$F \quad \to \quad S^\dagger F S \,, \tag{16.3.32}$$

with any $2 \times 2$ special unitary matrices $S$ which span $\boldsymbol{S}^3$.[e] Then, the quantization of the soliton is a matter of finding eigenstates of free and nonrelativistic nonlinear sigma-model onto $\boldsymbol{S}^3$.[21] $S$ itself admits an $SO(4)$ symmetry of its own,

$$S \quad \to \quad U S V^\dagger \,. \tag{16.3.33}$$

Because of the way the spatial indices are locked with the gauge indices, these two rotations are each identified as the gauge rotation, $SU(N_F = 2)$, and half of the spatial rotations, say, $SU(2)_+$. Eigenstates on $\boldsymbol{S}^3$ are then nothing but the familiar angular momentum eigenfunctions of three Euler angles, conventionally denoted as

$$|s : p, q\rangle \,. \tag{16.3.34}$$

Recall that the quadratic Casimirs of the two $SU(2)$'s (associated with $U$ and $V$ rotations) always coincide to be $s(s+1)$. One can proceed exactly in the same manner for anti-instantons, where $SU(2)_+$ is replaced by $SU(2)_-$.

Therefore, under $SU(N_F = 2) \times SU(2)_+ \times SU(2)_-$, the quantized instantons are in[23]

$$(2s + 1; 2s + 1; 1) \,, \tag{16.3.35}$$

while the quantized anti-instantons are in

$$(2s + 1; 1; 2s + 1) \,. \tag{16.3.36}$$

Possible values for $s$ are integers and half-integers. However, we are eventually interested in $N_c = 3$, in which case spins and isospins are naturally half-integral. Thus we will subsequently consider the case of fermionic states only. Exciting these isospin come at energy cost. See Hata et.al.[13] for mass spectra of some excited instanton solitons.

## 16.4. Holographic Dynamics

The solitonic baryon is a coherent object which is made up of pions as well as of vector and axial-vector mesons. This implies that the structure of the soliton itself contains all the information on how the baryon interacts with these infinite tower of mesons. This sort of approach has been also used[17] in the Skyrmion picture of old days, where, for instance, the leading axial coupling for a nucleon emitting a soft pion was computed following such thoughts. The difference here is that, instead of just pions, all spin one mesons enter this holographic construction of the baryon, and this enables us to compute all low energy meson-hadron vertices simultaneously.

[e]Since $S$ and $-S$ rotates the solution the same way the moduli space is naively $\boldsymbol{S}^3/Z_2$. However at quantum level, we must consider states odd under this $Z_2$ as well, so the moduli space is $\boldsymbol{S}^3$.

### 16.4.1. *Dynamics of hairy solitons: generalities*

First, we would like to illustrate the point by considering another kind of solitons. The magnetic monopoles[22] appear as solitons in non-Abelian Yang-Mills theories spontaneously broken to a subgroup containing a $U(1)$ factor, such as in $SU(2) \to U(1)$, and carries a magnetic charge. Usually it is a big and fluffy object and must be treated as a classical object. However, if we push the electric Yang-Mills coupling to be large enough, so that the magnetic monopole size becomes comparable or even smaller than the symmetry breaking scale, we have no choice but to treat it as a point-like object. The effective action for this monopole field $\mathcal{M}$ (spinless for example) should contain at least,

$$\left| \left( \partial_\mu + i \frac{4\pi}{e} \tilde{A}_\mu \right) \mathcal{M} \right|^2 , \tag{16.4.37}$$

where $\tilde{A}$ is the dual photon of the unbroken $U(1)$ gauge field. We know this coupling exists simply because the monopole has the magnetic charge $4\pi/e$. But how do we know the latter fact? Because the soliton solution itself exhibits a long range magnetic Coulomb tail of the form

$$F^{monopole} \sim \frac{4\pi}{e} \frac{1}{r^2} . \tag{16.4.38}$$

If we replace the solitonic monopole by the quanta of the field $\mathcal{M}$ but do not couple to the dual photon field as above, we would end up with a local excitation. However, a magnetic monopole (or an electrically charged particle) is not really a local object. Creating one always induces the corresponding long range magnetic (electric) Coulomb field. To ensure that the effective field theory represent the magnetic monopole accurately, we must make sure that creating a quanta of $\mathcal{M}$ is always followed by creation of the necessary magnetic Coulomb field. This is achieved by coupling the local field $\mathcal{M}$ to the gauge field $\tilde{A}$ at an appropriate strength. This is a somewhat unconventional way to understand the origin of the minimal coupling of the monopole to the dual gauge field $\tilde{A}$.

### 16.4.2. *The small size matters*

Before going further, let us briefly pose and ask about the validity of such an approach for our solitonic baryon. The key to this is a set of inequalities among three natural scales that enter the baryon physics, which are

$$\frac{1}{M_{KK}} \gg \frac{1}{M_{KK}\sqrt{\lambda}} \gg \frac{1}{M_{KK}N_c\lambda} . \tag{16.4.39}$$

They hold in the large $N_c$ and large $\lambda$ limit. The first is the length scale of mesons, the second is the classical size of the solitonic baryon, and the third is the Compton wavelength of the baryon since its mass is $\sim M_{KK}N_c\lambda$.

The first inequality tells us that the baryon tends to be much smaller than mesons and thus can be regarded almost pointlike when interacting with mesons.

This justifies the effective field theory approach where we think of each baryon as small excitation of a field. One does this precisely when the object in question can be treated as if it has no internal structure other than quantum numbers like spins.

The second inequality tells us that the quantum uncertainty associated with the baryon is far smaller than the classical core size of the soliton. This is important because, otherwise, one may not be able to trust anything about the classical features of the soliton at quantum level. When the second inequality holds, it enables us to make use of the classical shape of the soliton and to extract information about how mesons interact with the baryon. The fact we have a small soliton size and an even smaller Compton size of that soliton is very fortunate.

### 16.4.3. *Holographic dynamics of baryons*

As with the small magnetic monopole case, we wish to trade off the (quantized) instanton soliton in favor of local baryon field(s) and make sure to encode the long-range tails of the soliton in how the baryon field(s) interacts with the low energy gauge fields. Our instanton soliton has two types of distinct but related long-range field. The first is due to the Pontryagin density and goes like

$$F_{mn} \sim \frac{\rho^2_{baryon}}{(r^2 + w^2)^2} , \qquad (16.4.40)$$

while the second is the Coulomb field due to the Chern-Simons coupling between $\mathcal{A}$ and $F \wedge F$,

$$\mathcal{F}_{0n} \sim \frac{e(w)^2 N_c}{(r^2 + w^2)^{3/2}} . \qquad (16.4.41)$$

The latter is the five-dimensional analog of the electric Coulomb tail.

Apart from the fact that we have two kinds of long-range fields, there is another important difference from the monopole case. As we saw in section 3.2, the solitonic baryon has $S^3$ worth of internal moduli, quantization of which gave us the various spin/isospin baryons. Since the gauge direction of the magnetic long range field is determined by coordinate on $S^3$, the field strengths associated with the Pontryagin density should be smeared out by quantum fluctuation along the moduli space. It is crucial for our purpose that what we mean by long-range fields of the instanton soliton are actually these quantum counterpart, not the naive classical one. Basic features of the smearing out effect and relevant identities can be found in next subsection.

The electric Coulomb tail should be encoded in a minimal coupling to the Abelian part of $\mathcal{A}$. For a spin/isospin half baryon, $\mathcal{B}$, we anticipate a minimal term of the form

$$\bar{\mathcal{B}}(N_c \mathcal{A}^{U(1)}_m + A_\mu)\gamma^m \mathcal{B} . \qquad (16.4.42)$$

This is uniquely fixed by the Coulomb charge $N_c$ and the $SU(N_F = 2)$ representation of the quantized instanton. The purely magnetic tail of the soliton is more subtle to deal with. From the simple power counting, it is obvious that the coupling responsible for such a tail must have one higher dimension than the minimal coupling, hinting at the field strength $F$ of the $SU(N_F = 2)$ part coupling directly to a baryon bilinear, such as

$$\bar{\mathcal{B}} F_{mn} \gamma^{mn} \mathcal{B} \,. \tag{16.4.43}$$

It turns out that this is precisely the right structure to mimic the long-range magnetic fields of quantized instantons and anti-instantons.[f]

To show that the latter vertex is indeed precisely the right one, one must consider the following points. (1) Is this the unique term that can reproduce the correct quantum-smeared long-range instanton and anti-instanton tail? (2) If so, how do we fix the coefficient, taking into account the quantum effects. (3) And is the estimate reliable? The answers are long and technical. We refer the readers to literatures[12,14,23] for precise answers to these questions, but here state that the answers are all affirmative and that the effective action of mesons and baryons is uniquely determined by this simple consideration. This is true at least in the large $N_c$ and the large $\lambda$ limit.

This leads to the following five-dimensional effective action,

$$\int d^4x dw \left[ -i\bar{\mathcal{B}}\gamma^m D_m \mathcal{B} - i m_{\mathcal{B}}(w)\bar{\mathcal{B}}\mathcal{B} + \frac{2\pi^2 \rho_{baryon}^2}{3e^2(w)} \bar{\mathcal{B}}\gamma^{mn} F_{mn}\mathcal{B} \right]$$
$$- \int d^4x dw \frac{1}{4e^2(w)} \, \text{tr} \, \mathcal{F}_{mn}\mathcal{F}^{mn} \,, \tag{16.4.44}$$

with the covariant derivative given as $D_m = \partial_m - i(N_c \mathcal{A}_m^{U(1)} + A_m)$ with $A_m$ in the fundamental representation of $SU(N_F = 2)$.

The position-dependent mass $m_{\mathcal{B}}(w) \sim 1/e(w)^2$ is a very sharp increasing function of $|w|$, such that in the large $N_c$ and large $\lambda$ limit, the baryon wavefunction is effectively localized at $w = 0$. This is the limit where the above effective action is trustworthy. We find

$$\frac{2\pi^2 \rho_{baryon}^2}{3e^2(0)} = \frac{N_c}{\sqrt{30}} \cdot \frac{1}{M_{KK}} \,, \tag{16.4.45}$$

so the last term involving baryons can be actually dominant over the minimal coupling, despite that it looks subleading in the derivative expansion. As it turns out, this term is dominant for cubic vertex processes involving pions or axial vector mesons, whereas the minimal coupling dominates for those involving vector mesons.[14]

---

[f]In fact, a prototype of this simple method makes a brief appearance in the landmark work on Skyrmion by Adkins, Nappi, and Witten.[17] In their case, however, this gives only the pion-baryon interactions, forcing them to a related but somewhat different formulation. In our case, this method generates all meson-baryon interactions, however.

### 16.4.4. *Basic identities and isospin-dependence*

We have discussed general ideas behind the effective action approach and given the explicit results for isospin 1/2 case. The only term that is not obvious is the coupling between baryons and the field strength $F$, with the coefficient $2\pi^2 \rho^2_{baryon}/3e^2(0)$, and we would like to spend a little more time on its origin. Apart from convincing readers that the derivation of the effective action is actually rigorous, this would also allow us to outline how the result generalizes for higher isospin baryons, such $\Delta$ particles, as well.

Each and every quantum of the baryon field $\mathcal{B}$ is supposed to represent a quantized (anti-)instanton soliton. Let us recall that the quantization of the soliton involves finding wavefunctions on the moduli space of the soliton, which is $S^3$. Since the moduli encode the gauge direction of the instanton soliton, the classical gauge field is quantum mechanically smeared and should be replaced by its expectation values as

$$F \;\rightarrow\; \langle\langle S^\dagger F S \rangle\rangle = \langle\langle \Sigma_{ab} \rangle\rangle F^b \,, \tag{16.4.46}$$

with $2\Sigma_{ab} \equiv \mathrm{tr}\left[\tau_a S^\dagger \tau_b S\right]$. $\langle\langle \cdots \rangle\rangle$ means taking expectation value on wavefunctions on the moduli space of the soliton, and computes the quantum smearing effect.

The effective action (16.4.44) would make sense if and only if each quanta of the baryon field $\mathcal{B}$ is equipped with precisely the right smeared-out gauge field of this type. How is this possible? For the simplest case of isospin 1/2, the relevant identity that shows this reads[g]

$$\langle\langle 1/2 : p', q' | \Sigma_{ab} | 1/2 : p, q \rangle\rangle = -\frac{1}{3}(\mathcal{U}(1/2 : p', q')^{\epsilon'}_{\beta'})^* \sigma_a^{\beta'\beta} \tau_b^{\epsilon'\epsilon} \mathcal{U}(1/2 : p, q)^{\epsilon}_{\beta} \tag{16.4.47}$$

where $\mathcal{U}(1/2 : p, q)$ is the two-component spinor/isospinor of $J_3 = p$, $I_3 = q$, and $J^2 = I^2 = 3/4$. Identifying the two-component spinor $\mathcal{U}$ as the upper half of the four-component spinor $\mathcal{B}$ representing positive energy states, one can show that the equation of motion for the gauge field coupled to $\mathcal{B}$ is

$$(\nabla \cdot F)^a_m \sim \nabla_n \left(\bar{\eta}^b_{nm} \mathcal{U}^\dagger (\sigma_b \tau^a) \mathcal{U}\right) + \cdots \,, \tag{16.4.48}$$

which shows, via (16.4.47), that the quanta $\mathcal{U}$ of $\mathcal{B}$ would be accompanied by the correctly smeared long range tail of gauge field of type (16.4.46). The right hand side comes from the coupling of type

$$\bar{\mathcal{B}} F \mathcal{B} \tag{16.4.49}$$

in (16.4.44). A similar match can be shown for negative energy states, where the 't Hooft symbol $\bar{\eta}$ is replaced by $\eta$ and $\mathcal{U}$ by its anti-particle counterpart $\mathcal{V}$. A careful check of the normalization leads us to the coefficient $2\pi^2 \rho^2_{baryon}/3e^2(0)$, where the number 3 in the denominator came from the factor 1/3 in Eq. (16.4.47).

---

[g]This identity for $s = 1/2$ is originally due to Adkins, Nappi, and Witten, who obtained it in the context of the Skyrmion. The moduli space of a Skyrmion and that of our instanton soliton coincide, so the same identity holds.

It turns out that this goes beyond $s = 1/2$. The identity (16.4.47) is generalized to arbitrary half-integral $s$ as[23]

$$\langle\langle s : p', q' | \Sigma_{ab} | s : p, q \rangle\rangle$$
$$= -\frac{s}{s+1} \cdot (\mathcal{U}(s : p', q')^{\epsilon' \epsilon_2 \cdots \epsilon_{2s}}_{\beta' \alpha_2 \cdots \alpha_{2s}})^* \sigma_a^{\beta' \beta} \tau_b^{\epsilon' \epsilon} \mathcal{U}(s : p, q)^{\epsilon \epsilon_2 \cdots \epsilon_{2s}}_{\beta \alpha_2 \cdots \alpha_{2s}} , \qquad (16.4.50)$$

where the left-hand-side is again evaluated as wavefunction-overlap integral on the moduli space $S^3$ of the instanton soliton. $\mathcal{U}$ is now that of higher spin/isospin field with symmetrized multi-spinor/multi-isospinor indices. As with $\mathcal{U}(1/2)$, $\mathcal{U}(s)$'s are positive energy spinors with each index taking values 1 and 2. This implies a cubic interaction term of type

$$\bar{\mathcal{B}}_s F \mathcal{B}_s \qquad (16.4.51)$$

where $\mathcal{B}_s$ denotes a local baryon field of isospin $s$ and $SO(4) = SU(2)_+ \times SU(2)_-$ angular momentum $[s]_+ \otimes [0]_- \oplus [0]_+ \otimes [s]_-$. Relative to the isospin $1/2$ case, the coefficient is increased from $1/3$ to $s/(s+1)$, which reflects the obvious fact that higher angular momentum states would be less and less smeared.

Finally, with $s > 1/2$ baryons included, there are one more type of processes allowed where a baryon changes its own isospin by emitting isospin 1 mesons. The relevant identities for these processes are

$$\langle\langle s : p', q' | \Sigma_{ab} | s+1 : p, q \rangle\rangle$$
$$= -\frac{1}{2}\sqrt{\frac{2s+1}{2s+3}} \cdot \left[ \mathcal{U}(s : p', q')^\dagger \mathcal{U}(s+1 : p, q)_{ab} \right] , \qquad (16.4.52)$$

where $3 \times 3$ spin/isospin $s$ wavefunctions $\mathcal{U}(s+1 : p, q)_{ab}$ are

$$(\mathcal{U}(s+1 : p, q)_{ab})^{\epsilon_1 \cdots \epsilon_{2s}}_{\alpha_1 \cdots \alpha_{2s}} \equiv (\sigma_2 \sigma_a)^{\beta \beta'} (\tau_2 \tau_b)_{\epsilon \epsilon'} \mathcal{U}(s+1 : p, q)^{\epsilon \epsilon' \epsilon_1 \cdots \epsilon_{2s}}_{\beta \beta' \alpha_1 \cdots \alpha_{2s}} . \qquad (16.4.53)$$

This shows up in the effective action of baryon as a coupling of type

$$\bar{\mathcal{B}}_{s+1} F \mathcal{B}_s . \qquad (16.4.54)$$

The complete effective action of baryons with such arbitrary half-integer isospins was given in Ref. 23. For the rest of the note, we will confine ourselves to isospin $1/2$ case.

## 16.5. Nucleons

Nucleons are the lowest lying baryons with isospin and spin $1/2$. As such, they arise from the isospin $1/2$ holographic baryon field $\mathcal{B}$ whose effective action is given explicitly above. This effective action contains interaction terms between currents of $\mathcal{B}$ with the $U(N_F)$ gauge field of five dimensions, and thus contain an infinite number of interaction terms between nucleons and mesons, specifically all cubic couplings involving nucleons emitting pion, vector mesons, or axial-vector mesons. Extracting four-dimensional amplitudes of interests is a simple matter of dimensional reduction from $R^{3+1} \times I$ to $R^{3+1}$. In this section, we show this procedure, showcase some of the simplest examples for comparisons, and comment on how the results should be taken in view of various approximation schemes we relied on.

### 16.5.1. *Nucleon-meson effective actions*

The effective action for the four-dimensional nucleons is derived from this, by identifying the lowest eigenmode of $\mathcal{B}$ upon the KK reduction along $w$ direction as the proton and the neutron. Higher KK modes would be also isospin-half baryons, but the gap between the ground state and excited state is very large in the holographic limit, so we consider only the ground state. We mode expand $\mathcal{B}_\pm(x^\mu, w) = \mathcal{N}_\pm(x^\mu) f_\pm(w)$, where $\pm$ refers to the chirality along $w$ direction, and reconsitute a four-dimensional spinor $\mathcal{N}$ with $\gamma^5 \mathcal{N}_\pm = \pm \mathcal{N}_\pm$ as its chiral and anti-chiral components. The lowest KK eigenmodes $f_\pm(w)$ solve

$$\left[ -\partial_w^2 \mp \partial_w m_\mathcal{B}(w) + (m_\mathcal{B}(w))^2 \right] f_\pm(w) = m_\mathcal{N}^2 f_\pm(w) \,, \qquad (16.5.55)$$

with some minimum eigenvalue $m_\mathcal{N} > m_\mathcal{B}(0) = m_\mathcal{B}^{classical}$. This nucleon mass $m_\mathcal{N}$ will generally differ from the five-dimensional soliton mass $m_\mathcal{B}^{classical}$, due to quantization of light modes such as spread of the wavefunction $f_{L,R}$ along the fifth direction.

Inserting this into the action (16.4.44), we find the following structure of the four-dimensional nucleon action

$$\int dx^4 \, \mathcal{L}_4 = \int dx^4 \left( -i\bar{\mathcal{N}}\gamma^\mu \partial_\mu \mathcal{N} - i m_\mathcal{N} \bar{\mathcal{N}}\mathcal{N} + \mathcal{L}_{\text{vector}} + \mathcal{L}_{\text{axial}} \right) \,, \qquad (16.5.56)$$

where we have, schematically, the vector-like couplings

$$\mathcal{L}_{\text{vector}} = -i\bar{\mathcal{N}}\gamma^\mu \beta_\mu \mathcal{N} - \sum_{k \geq 0} g_V^{(k)} \bar{\mathcal{N}}\gamma^\mu a_\mu^{(2k+1)} \mathcal{N} \,, \qquad (16.5.57)$$

and the axial couplings to axial mesons,

$$\mathcal{L}_{\text{axial}} = -\frac{i g_A}{2} \bar{\mathcal{N}}\gamma^\mu \gamma^5 \alpha_\mu \mathcal{N} - \sum_{k \geq 1} g_A^{(k)} \bar{\mathcal{N}}\gamma^\mu \gamma^5 a_\mu^{(2k)} \mathcal{N} \,. \qquad (16.5.58)$$

All the coupling constants $g_{V,A}^{(k)}$ and $g_A$ are calculated by suitable wave-function overlap integrals involving $f_\pm$ and $\psi_{(n)}$'s.

Although we did not write so explicitly, isospin triplet mesons and singlet mesons have different coupling strengths to the nucleons, so there are actually two sets of couplings $(g_A, g_A^{(k)}, g_V^{(k)})$, one for isosinglet mesons, such as $\omega$ and $\eta'$, and the other for isotriplet mesons, such as $\rho$ and $\pi$. The leading contribution to axial couplings in the isospin triplet channel arise from the direct coupling to $F_{mn}$, and are all proportional to $\rho_{baryon}^2$. All the rest are dominated by terms from the five-dimensional minimal coupling to $\mathcal{A}_m$. We refer interested readers to Refs. 12 and 14 for explicit form of these coupling constants.

### 16.5.2. *Numbers and comments*

To showcase typical predictions from the above setup, let us quote two notable examples for the nucleons.[14] The first is the cubic coupling of the lightest vector mesons to the nucleon, to be denoted as $g_{\rho\mathcal{N}\mathcal{N}}$ for the isotriplet meson $\rho$ and $g_{\omega\mathcal{N}\mathcal{N}}$ for the iso-singlet meson $\omega$. In the above effective action, these two are denoted collectively as $g_V^{(0)}$. An interesting prediction of this holographic effective action of nucleons is that

$$\frac{g_{\omega\mathcal{N}\mathcal{N}}}{g_{\rho\mathcal{N}\mathcal{N}}} = N_c + \delta , \tag{16.5.59}$$

where the leading $N_c$ is a consequence from the five-dimensional minimal coupling to $\mathcal{A}$ while the subleading correction $\delta$ arises from the direct coupling to the field strength $F$. With $N_c = 3$ and $\lambda \simeq 17$ (the latter is required by fitting $f_\pi$ and $M_{KK}$ to the pion decay constant and the vector meson masses to actual QCD), we find

$$\frac{g_{\omega\mathcal{N}\mathcal{N}}}{g_{\rho\mathcal{N}\mathcal{N}}} \simeq 3 + 0.6 = 3.6 . \tag{16.5.60}$$

Extracting ratios like this from experimental data is somewhat model-dependent, with no obvious consensus, but the ratio is believed to be larger than 3 and numbers around 4-5 are typically found. Given the crude nature of our approximation and that there is no tunable parameter other than the QCD scale and $f_\pi$, the agreement is uncanny. A more complete list of various cubic couplings between spin one mesons and nucleons has been worked out in Ref. 14 and further elaborated recently in Ref. 24.

The leading axial coupling to pions, $g_A$, is somewhat better measured at $\simeq 1.26$. Our prediction is[12]

$$g_A = \frac{2\lambda N_c (\rho_{baryon} M_{KK})^2}{81\pi^2} + \cdots = \left(\frac{24}{5\pi^2}\right)^{1/2} \times \frac{N_c}{3} + \cdots , \tag{16.5.61}$$

where the leading term arise from the direct coupling to the field strength $F$ and the ellipsis denotes the subleading and higher correction. While this does not look too good, we must remember that this holographic model is effectively a quenched QCD, missing out on possible $O(1)$ corrections. From old studies of large $N_c$ constituent models, a group theoretical $O(1)$ correction has been proposed for this type of operators, which states that the next leading correction would amount to $N_c \rightarrow N_c + 2$.[h] So, in a more realistic version where we take into account of the backreaction of D8 branes on the dual geometry, we may anticipate for $N_c = 3$

$$g_A \simeq \left(\frac{24}{5\pi^2}\right)^{1/2} \times \frac{N_c + 2}{3} + O(1/N_c) \simeq 1.16 + O(1/N_c) . \tag{16.5.62}$$

Finally, $O(1/N_c)$ is partly captured by the minimal coupling term in our quenched model, which turns out to give roughly a 10% positive correction, making the total very close to the measured quantity 1.26.

---

[h]See Ref. 14 for more explanations and references.

These two illustrate nicely what kind of predictions can be made and how accurate their predictions can be when compared to experimental data. Much richer array of predictions exist, such as other cubic couplings between mesons and baryons, anomalous magnetic moment,[12] complete vector dominance of electromagnetic form factors,[14] and detailed prediction on momentum dependence of such form factors.[25,26]

However, one should be a bit more cautious. The model, as an approximation to real QCD, has many potential defects. The main problem is that all of this is in the context of large $N_c$ and that any prediction, such as above two, has to involve an extensive extrapolation procedure. Many ambiguities can be found in such a procedure, and we chose a particular strategy of computing all quantities and analytically continuing the final expressions for the amplitudes to realistic QCD regime. The fact it works remarkably well does not really support its validity in any rigorous sense. Also the D4-D8 model we employed include many massive fields which are not part of ordinary four-dimensional QCD, and one should be cautious in using the holographic QCD for physics other than simple low energy processes.

Despite such worries, the D4-D8 holographic QCD turned out to be far better than one may have anticipated. We have shown how it accommodates not only the (vector) meson sector but the baryon sector very competently.[i] Whether or not the holographic QCD can be elevated to a controlled and justifiable approximation to real QCD remains to be seen, depending crucially on having a better understanding of the string theory in the curved spacetime. Nevertheless, it is fair to say that we finally have a rough grasp of the physics that controls *the master fields*, and perhaps this insight by itself will lead to a better and more practical formulation of the QCD in the future.

## 16.6. Electromagnetic Properties

Holographic baryons and their effective action in the bulk also encode how baryons, and in particular, nucleons would interact with electromagnetism. For this, one follows the usual procedure of AdS/CFT where operators in the field theory are matched up with non-normalizable modes of bulk fields. Operationally, one simply introduces the boundary photon field $\mathcal{V}$ as a nonnormalizable mode, which adds to $\beta$-term in the expansion of $\mathcal{A}$,

$$A_\mu(x;w) = i\alpha_\mu(x)\psi_0(w) + \mathcal{V}_\mu(x) + i\beta_\mu(x) + \sum_n a_\mu^{(n)}(x)\psi_{(n)}(w) , \qquad (16.6.63)$$

and repeat the dimensional reduction to the four dimensions. For instance, computation of anomalous magnetic moments of proton and neutron can be done with relative ease, and gives remarkably good agreement with measured values.[14]

---

[i]One of the acutely missing parts is how the spinless mesons (except Goldstone bosons) would fit in the story. Initial investigation of this gave a possibly disappointing result, although it may have more to do with how the lightest scalar mesons are rather complicated objects and may not be a bi-quark meson of conventional kind.[27]

For more detailed accounts of electromagnetic properties of baryon, we refer the readers to Refs. 14, 25 and 26. Here we will only consider the most notable feature of the electromagnetic properties, namely the complete vector dominance, whereby all electromagnetic interactions are entirely mediated by the infinite tower of vector mesons. This also illustrates well how the holographic QCD can give a sweeping and qualitative prediction and also where it could fail.

The vector dominance means that there is no point-like charge, which, in view of the minimal coupling between $\mathcal{A}$ and $\mathcal{B}$ in (16.4.44), sounds pretty odd. To understand what's going on, one must consider quadratic structures in the vector meson sector. Defining

$$\zeta_k = \int dw \frac{1}{2e(w)^2}\, \psi_{(2k+1)}(w)\,, \qquad (16.6.64)$$

for parity even eigenfunctions $\psi_{(2k+1)}$'s, the quadratic part of the vector meson is[11]

$$\sum_k \mathrm{tr}\left[-\frac{1}{2}|\,dv^{(k)}|^2 - m^2_{(2k+1)}|\,v^{(k)} - \zeta_k(\mathcal{V}+i\beta)|^2\right]\,, \qquad (16.6.65)$$

where we introduced the shifted vector fields

$$v^{(k)} = a^{(2k+1)} + \zeta_k(\mathcal{V}+i\beta)\,. \qquad (16.6.66)$$

This mixing of vector mesons and photon is at the heart of the vector dominance. (The axial-vector mesons, $a^{(2k)}$'s, do not mix with photon because of the parity.)

Now let us see how this mixing of vector fields enters the coupling of baryons with electromagnetic vector field $\mathcal{V}$. Taking the minimal coupling, we find

$$\int dw\, \bar{\mathcal{B}}\gamma^m A_m \mathcal{B} = \bar{\mathcal{B}}\gamma^\mu \mathcal{V}_\mu B + \sum_k g^{(k)}_{V,min}\bar{\mathcal{B}}\gamma^\mu a^{(2k+1)}_\mu B + \cdots\,, \qquad (16.6.67)$$

where the ellipsis denotes axial couplings to axial vectors as well as coupling to pions via $\alpha_\mu$ and $\beta_\mu$. $g^{(k)}_{V,min}$ is the cubic coupling between $k$-th vector meson and the baryon, or more precisely its leading contribution coming from the minimal coupling to $\mathcal{A}$. Again, the presence of the direct minimal coupling to the photon $\mathcal{V}$ seems to contradict the notion of vector dominance. However, it is advantageous to employ the canonically normalized vector fields $v^{(k)}$ in place of $a^{(k)}$, upon which this becomes

$$\bar{B}\gamma^\mu \mathcal{V}_\mu B + \sum_k g^{(k)}_{V,min}\bar{B}\gamma^\mu(v^{(k)}_\mu - \zeta_k \mathcal{V}_\mu)B + \cdots\,. \qquad (16.6.68)$$

On the other hand,

$$\sum_k g^{(k)}_{V,min}\zeta_k = \sum_k \int dw'\, |f_+(w')|^2 \psi_{(2k+1)}(w') \times \int dw\, \frac{1}{2e(w)^2}\psi_{(2k+1)}(w)$$

$$= \int dw'\, |f_+(w')|^2 \times \int dw\, \delta(w-w') = 1\,, \qquad (16.6.69)$$

where we made use of the definite parities of $1/e(w)^2$ and $\psi_{(n)}$'s and also of the completeness of $\psi_{(n)}$'s. This sum rule $\sum_k g_{V,min}^{(k)} \zeta_k = 1$ forces

$$\bar{B}\gamma^\mu \mathcal{V}_\mu B + \sum_k g_{V,min}^{(k)} \bar{B}\gamma^\mu (v_\mu^{(k)} - \zeta_k \mathcal{V}_\mu) B + \cdots = \sum_k g_{V,min}^{(k)} \bar{B}\gamma^\mu v_\mu^{(k)} B + \cdots \quad (16.6.70)$$

and the baryon couples to the photon field $\mathcal{V}$ only via $v^{(k)}$'s which mixes with $\mathcal{V}$ in their mass terms.

This choice of basis is only for the sake of clarity. Regardless of the basis, the above shows that no coupling between $\mathcal{V}$ and $\mathcal{B}$ can occur in the infinite momentum limit. This statement is clear in the $\{\mathcal{V}; v^{(k)}\}$ basis which is diagonal if the mass term is negligible. Alternatively, we can ask for the invariant amplitude of the charge form factor, to which the minimal coupling contributes[14]

$$F_{1,min}(q^2) = 1 - \sum_k \frac{g_{V,min}^{(k)} \zeta_k q^2}{q^2 + m_{(2k+1)}^2} = \sum_k \frac{g_{V,min}^{(k)} \zeta_k m_{(2k+1)}^2}{q^2 + m_{(2k+1)}^2} \quad (16.6.71)$$

with the momentum transfer $q$. For small momentum transfer, the first few light vector mesons dominate the form factors by mediating betwee the baryon and the photon. This end fit with experimental data pretty well. Similar computation can be done for the magnetic form factor, from which one also finds the (anomalous) magnetic moment that fits the data pretty well.[12,14]

However, for large momentum transfer, the form factor decays as $1/q^2$ which is actually too slow for real QCD baryons. Estimates based on the parton picture say that the decay should be $\sim 1/q^{2(N_c-1)}$. This dramatic failure of the form factor for large momentum regime should not be a big surprise. The theory we started with is a low energy limit of D4-D8 complex compactified (with warp factors) on $S^1 \times S^4$. As such, one has to truncate infinite number of massive modes in order to reach a QCD-like theory in the boundary and must stay away from that cut-off scale to be safe from this procedure. For large momentum transfers, say larger than $M_{KK}$, the computation we relied on has no real rationale. This should caution readers that the holographic QCD, at least in the limited forms that are available now, is not a fix for everything. One really must view it as a vastly improved version of the chiral Lagrangian approach, with many hidden symmetries now manifest, but still suitable only for low energy physics.

## 16.7. More Comments

D4-D8 holographic model of QCD is the most successful model of its kind known. It reproduces in particular detailed particle physics of mesons and baryons. One reason for its success can be found in the fact that it builds on the the meson sector, the lightest of which is lighter than the natural cut-off scale $M_{KK}$. Apart from $1/N_c$ and $1/\lambda$ expansions imposed by general AdS/CFT ideas, one also must be careful with low energy expansion as well, because, as we stated before, the model includes many more massive Kaluza-Klein modes and even string modes that are not part of ordinary QCD. For low energy processes, nevertheless, one would hope that these extra massive states (above $M_{KK}$) do not contribute too much, which seems to be the case for low lying meson sector.[11]

Our solitonic and holographic model of baryons elevates the classic Skyrme picture based on pions to a unified model involving all spin one mesons in addition to pions. This is why the picture is extremely predictive. As we saw in this note, for low momentum processes, such as soft pion processes, soft rho meson exchanges, and soft elastic scattering of photons, the model's predictions compare extremely well with experimental data. It is somewhat mysterious that the baryon sector works out almost as well as the meson sector, since baryons are much heavier than $M_{KK}$ in the large $N_c$ and the large $\lambda$ limit.

Note that the soliton underlying the baryon is nearly self-dual in the large $\lambda$ limit. For instance, Eq. (16.3.27) shows that the leading, would-be BPS, mass is dominant over the rest by a factor of $\lambda$. There must be a sense in which the soliton is approximately supersymmetric with respect to the underlying IIA string theory, even though the background itself breaks all supersymmetry at scale $M_{KK}$. One may argue that even though there are many KK modes and even stringy modes lying between the naive cut-off scale $M_{KK}$ and the baryon mass scale $M_{KK}N_c\lambda$, these non-QCD degrees of freedom would be paired into approximate supermultiplets, reducing their potentially destructive effect, especially because the baryon itself is roughly BPS. Whether or not one can actually quantify such an idea for the model we have is unclear, but if possible it would be an important step toward rigorously validating holographic approaches to baryons in this D4-D8 set-up.

There are more works to be done. One important direction is to perform more refined comparisons against experiments. In particular, extracting coupling constants from raw data seems quite dependent on theoretical models, and it is important to compute directly measurable amplitudes starting from the effective action of ours. Nucleon-nucleon scattering amplitudes or more importantly the nucleon-nucleon potential would be a good place to start.[24,28,29] Another profitable path would be to consider dense system such as neutron stars as well as physics of light nuclei, where our model with far less tunable parameters would give unambiguous predictions. This will in turn further test the model as well.


## Acknowledgements

This note is based on a set of collaborative works with D.K. Hong, J. Park, M. Rho, and H.-U. Yee. The author wishes to thank SITP of Stanford University, Aspen Center for Physics, and also organizers of the conference "30 years of mathematical method in high energy physics" for hospitality. This work is supported in part by the Science Research Center Program of KOSEF (CQUeST, R11-2005-021), the Korea Research Foundation (KRF-2007-314-C00052), and by the Stanford Institute for Theoretical Physics (SITP Quantum Gravity visitor fund).

# References


1. C.G. Callan, S.R. Coleman, J. Wess and B. Zumino, "Structure of phenomenological Lagrangians. 2," *Phys. Rev.* **177** (1969) 2247; S.R. Coleman, J. Wess and B. Zumino, "Structure cf phenomenological Lagrangians. 1," *Phys. Rev.* **177** (1969) 2239.
2. G. 't Hooft, "A planar diagram theory for strong interactions," *Nucl. Phys. B* **72** (1974) 461.
3. E. Witten, 'Baryons In The 1/N Expansion," *Nucl. Phys. B* **160** (1979) 57.
4. T.H.R. Skyrme, "A unified field theory of mesons and baryons," *Nucl. Phys.* **31** (1962) 556.
5. J.M. Maldacena, "The large N limit of superconformal field theories and supergravity," *Adv. Theor. Math. Phys.* **2** (1998) 231 [arXiv:hep-th/9711200]; S.S. Gubser, I.R. Klebanov and A.M. Polyakov, "Gauge theory correlators from non-critical string theory," *Phys. Lett. B* **428** (1998) 105 [arXiv:hep-th/9802109]; E. Witten, "Anti-de Sitter space and holography," *Adv. Theor. Math. Phys.* **2** (1998) 253 [arXiv:hep-th/9802150].
6. See, for example, R. Gopakumar and D.J. Gross, "Mastering the master field," *Nucl. Phys. B* **451** (1995) 379 [arXiv:hep-th/9411021].
7. J. Babington, J. Erdmenger, N.J. Evans, Z. Guralnik and I. Kirsch, "Chiral symmetry breaking and pions in non-supersymmetric gauge/gravity duals," *Phys. Rev. D* **69** (2004) 066007 [arXiv:hep-th/0306018]; J. Erlich, E. Katz, D.T. Son and M.A. Stephanov, Phys. Rev. Lett. **95** (2005) 261602 [arXiv:hep-ph/0501128].
8. E. Witten, 'Anti-de Sitter space, thermal phase transition, and confinement in gauge theories," *Adv. Theor. Math. Phys.* **2** (1998) 505 [arXiv:hep-th/9803131].
9. C. Csaki, H. Ooguri, Y. Oz and J. Terning, "Glueball mass spectrum from supergravity," *JHEP* **9901** (1999) 017 [arXiv:hep-th/9806021].
10. R.C. Brower, S.D. Mathur and C.I. Tan, "Glueball spectrum for QCD from AdS supergravity duality," *Nucl. Phys. B* **587** (2000) 249 [arXiv:hep-th/0003115].
11. T. Sakai and S. Sugimoto, "Low energy hadron physics in holographic QCD," *Prog. Theor. Phys.* **113** (2005) 843 [arXiv:hep-th/0412141]; "More on a holographic dual of QCD," *Prog. Theor. Phys.* **114** (2006) 1083 [arXiv:hep-th/0507073].
12. D.K. Hong, M. Rho, H.U. Yee and P. Yi, "Chiral dynamics of baryons from string theory," *Phys. Rev. D* **76** (2007) 061901 [arXiv:hep-th/0701276].
13. H. Hata, T. Sakai, S. Sugimoto and S. Yamato, "Baryons from instantons in holographic QCD," [arXiv:hep-th/0701280].
14. D.K. Hong, M. Rho, H.U. Yee and P. Yi, "Dynamics of Baryons from String Theory and Vector Dominance," *JHEP* **0709** (2007) 063 [arXiv:0705.2632 [hep-th]].
15. G.W. Gibbons and K.I. Maeda, "Black Holes And Membranes In Higher Dimensional Theories With Dilaton Fields," *Nucl. Phys. B* **298** (1988) 741.
16. T. Eguchi and H. Kawai, "Reduction Of Dynamical Degrees Of Freedom In The Large N Gauge Theory," *Phys. Rev. Lett.* **48** (1982) 1063.
17. G.S. Adkins, C.R. Nappi and E. Witten, "Static properties of nucleons in the Skyrme model," *Nucl. Phys. B* **228** (1983) 552.
18. M.F. Atiyah and N.S. Manton, "Skyrmions from instantons," *Phys. Lett. B* **222** (1989) 438.
19. T. Fujiwara et al, " An effective Lagrangian for pions, $\rho$ mesons and skyrmions," *Prog. Theor. Phys.* **74** (1985) 128; U.-G. Meissner, N. Kaiser, A. Wirzba and W. Weise, "Skyrmions with $\rho$ and $\omega$ mesons as dynamical gauge bosons," *Phys. Rev. Lett.* **57** (1986) 1676; U.G. Meissner and I. Zahed, "Skyrmions in the presence of vector mesons," *Phys. Rev. Lett.* **56** (1986) 1035; K. Nawa, H. Suganuma and T. Kojo, "Baryons in Holographic QCD," *Phys. Rev. D* **75** (2007) 086003 [arXiv:hep-th/0612187].

20. E. Witten, "Baryons and branes in anti de Sitter space," *JHEP* **9807** (1998) 006 [arXiv:hep-th/9805112].
21. D. Finkelstein and J. Rubinstein, "Connection between spin, statistics, and kinks," *J. Math. Phys.* **9** (1968) 1762.
22. G. 't Hooft, "Magnetic monopoles in unified gauge theories," *Nucl. Phys.* B **79** (1974) 276.
23. J. Park and P. Yi, "A Holographic QCD and Excited Baryons from String Theory," *JHEP* **0806** (2008) 011 [arXiv:0804.2926 [hep-th]].
24. Y. Kim, S. Lee and P. Yi, "Holographic Deuteron and Nucleon-Nucleon Potential," arXiv:0902.4048 [hep-th].
25. D.K. Hong, M. Rho, H.U. Yee and P. Yi, "Nucleon Form Factors and Hidden Symmetry in Holographic QCD," *Phys. Rev.* D **77** (2008) 014030 [arXiv:0710.4615 [hep-ph]].
26. K. Hashimoto, T. Sakai and S. Sugimoto, "Holographic Baryons: Static Properties and Form Factors from Gauge/String Duality," arXiv:0806.3122 [hep-th].
27. S. Sugimoto, private communication.
28. K.Y. Kim and I. Zahed, "Nucleon-Nucleon Potential from Holography," arXiv:0901.0012 [hep-th].
29. K. Hashimoto, T. Sakai and S. Sugimoto, "Nuclear Force from String Theory," arXiv:0901.4449 [hep-th].

# Chapter 17

# The Cheshire Cat Principle from Holography

Holger Bech Nielsen[*] and Ismail Zahed[†]

*Niels Bohr Institute, 17 Blegdamsvej, Copenhagen, Denmark
†Department of Physics and Astronomy, SUNY Stony-Brook, NY 11794



The Cheshire cat principle states that hadronic observables at low energy do not distinguish between hard (quark) or soft (meson) constituents. As a result, the delineation between hard/soft (bag radius) is like the Cheshire cat smile in Alice in Wonderland. This principle reemerges from current holographic descriptions of chiral baryons whereby the smile appears in the holographic direction. We illustrate this point for the baryonic form factor.


## Contents


17.1 Introduction . . . . . . . . . . . . . . . . . . . . . . . . . . . . . . . . . . . . 393
17.2 The Principle and Holography . . . . . . . . . . . . . . . . . . . . . . . . . 394
17.3 The Holographic Model . . . . . . . . . . . . . . . . . . . . . . . . . . . . . 396
17.4 The Baryon Current . . . . . . . . . . . . . . . . . . . . . . . . . . . . . . . 397
17.5 Baryonic Form Factor . . . . . . . . . . . . . . . . . . . . . . . . . . . . . . 398
17.6 Conclusions . . . . . . . . . . . . . . . . . . . . . . . . . . . . . . . . . . . . 400
References . . . . . . . . . . . . . . . . . . . . . . . . . . . . . . . . . . . . . . 400


## 17.1. Introduction

Back in the eighties, quark bag models were proposed as models for hadrons that capture the essentials of asymptotic freedom through weakly interacting quarks and gluons within a bag, and the tenets of nuclear physics through strongly interacting mesons at the boundary. The delineation or bag radius was considered as a fundamental and physically measurable scale that separates ultraviolet from infrared QCD.[1]

The Cheshire cat principle[2] suggested that this delineation was unphysical in low energy physics, whereby fermion and color degrees of freedom could readily leak through the bag radius, making the latter immaterial. In a way, the bag radius was like the smile of the Cheshire cat in Alice in Wonderland. The leakage of the fundamental charges was the result of quantum effects or anomalies.[3]

In 1+1 dimensions exact bosonization shows that a fermion can translate to a boson and vice-versa making the separation between a fermionic or quark and a bosonic or meson degree of freedom arbitrary. In 3+1 dimensions there is no known

393

exact bozonization transcription, but in large $N_c$ the Skyrme model has shown that baryons can be decently described by topological mesons. The Skyrmion is the ultimate topological bag model with zero size bag radius,[4] lending further credence to the Cheshire cat principle.

The Skyrme model was recently seen to emerge from holographic QCD once chiral symmetry is enforced in bulk.[5] In holography, the Skyrmion is dual to a flavor instanton in bulk at large $N_c$ and strong t'Hooft coupling $\lambda = g^2 N_c$.[5,6] The chiral Skyrme field is just the holonomy of the instanton in the conformal direction. This construction shows how a flavor instanton with instanton number one in bulk, transmutes to a baryon with fermion number one at the boundary.

Of course, QCD is not yet in a true correspondence with a known string theory, as $N = 4$ SYM happens to be according to Maldacena's conjecture.[7] Perhaps, one way to achieve this is through the bottom-up string approach advocated in Ref. 8. Throughout, we will assume that the correspondence when established will result in a model perhaps like the one suggested in Ref. 5 for the light mesons and to which we refer to as holographic QCD.

With this in mind, holographic QCD provides a simple realization of the Cheshire cat principle at strong coupling. In section 2, we review briefly the holonomy construction for the Skyrmion in holography and illustrate the Cheshire cat principle. In section 3 we outline the holographic model. In section 4 we construct the baryonic current. In section 5 we derive the baryonic form factor. Many of the points presented in this review are borrowed from recent arguments in Ref. 9.

## 17.2. The Principle and Holography

In holographic QCD, a baryon is initially described as a flavor instanton in the holographic Z-direction. The latter is warped by gravity. For large $Z$, the warped instanton configuration is not known. However, at large $\lambda = g^2 N_c$ the warped instanton configuration is forced to $Z \sim 1/\sqrt{\lambda}$ due to the high cost in gravitational energy. As a result, the instanton in leading order is just the ADHM configuration with an additional U(1) barynonic field, with gauge components[5]

$$\widehat{\mathbb{A}}_0 = -\frac{1}{8\pi^2 a\lambda} \frac{2\rho^2 + \xi^2}{(\rho^2 + \xi^2)^2} \ , \qquad \mathbb{A}_M = \eta_{iMN} \frac{\sigma_i}{2} \frac{2x_N}{\xi^2 + \rho^2} \ , \qquad (17.2.1)$$

with all other gauge components zero. The size is $\rho \sim 1/\sqrt{\lambda}$. We refer to[5] (last reference) for more details on the relevance of this configuration for baryons. The ADHM configuration has maximal spherical symmetry and satisfies

$$(\mathbb{R}\mathbb{A})_Z = \mathbb{A}_Z(\mathbb{R}\vec{x}) \ , \qquad (\mathbb{R}^{ab}\mathbb{A}^b)_i = \mathbb{R}^T_{ij}\mathbb{A}^a_j(\mathbb{R}\vec{x}) \ , \qquad (17.2.2)$$

with $\mathbb{R}^{ab}\tau^b = \Lambda^+ \tau^a \Lambda$ a rigid SO(3) rotation, and $\Lambda$ is SU(2) analogue.

The holographic baryon is just the holonomy of (17.2.1) along the gravity bearing and conformal direction $Z$,

$$U^{\mathbb{R}}(x) = \Lambda P \exp\left(-i \int_{-\infty}^{+\infty} dZ\, \mathbb{A}_Z\right) \Lambda^+ . \qquad (17.2.3)$$

The corresponding Skyrmion in large $N_c$ and leading order in the strong coupling $\lambda$ is $U(\vec{x}) = e^{i\vec{\tau}\cdot\hat{x}F(\vec{x})}$ with the profile

$$F(\vec{x}) = \frac{\pi|\vec{x}|}{\sqrt{\vec{x}^2 + \rho^2}} . \qquad (17.2.4)$$

In a way, the holonomy (17.2.3) is just the fermion propagator for an infinitly heavy flavored quark with the conformal direction playing the role of *time*. (17.2.3) is the bosonization of this *conformal* quark in 3+1 dimensions.

The ADHM configuration in bulk acts as a point-like Skyrmion on the boundary. The baryon emerges from a semiclassical organization of the quantum fluctuations around the point-like source (17.2.3). To achieve this, we define

$$\begin{aligned} A_M(t, x, Z) = \mathbb{R}(t)(&\mathbb{A}_M(x - X_0(t), Z - Z_0(t)) \\ &+ C_M(t, x - X_0(t), Z - Z_0(t))) , \end{aligned} \qquad (17.2.5)$$

The collective coordinates $\mathbb{R}, X_0, Z_0, \rho$ and the fluctuations $C$ in (17.2.5) form a redundant set. The redundancy is lifted by constraining the fluctuations to be orthogonal to the zero modes. This can be achieved either rigidly[10] or non-rigidly.[11] We choose the latter as it is causality friendly. For the collective iso-rotations the non-rigid constraint reads

$$\int_{x=Z=0} d\hat{\xi} C\, G^B \mathbb{A}_M , \qquad (17.2.6)$$

with $(G^B)^{ab} = \epsilon^{aBb}$ the real generators of $\mathbb{R}$.

For $Z$ and $\rho$ the non-rigid constraints are more natural to implement since these modes are only soft near the origin at large $\lambda$. The vector fluctuations at the origin linearize through the modes

$$d^2\psi_n/dZ^2 = -\lambda_n\psi_n , \qquad (17.2.7)$$

with $\psi_n(Z) \sim e^{-i\sqrt{\lambda_n}Z}$. In the spin-isospin 1 channel they are easily confused with $\partial_Z\mathbb{A}_i$ near the origin as we show in Fig. 17.1. Using the non-rigid constraint, the double counting is removed by removing the origin from the vector mode functions

$$\psi'_n(Z) = \theta(|Z| - Z_C)\psi_n(Z) , \qquad (17.2.8)$$

with $Z_C \sim \rho \sim 1/\sqrt{\lambda}$ which becomes the origin for large $\lambda$. In the non-rigid semiclassical framework, the baryon at small $\xi < |Z_C|$ is described by a flat or uncurved instanton located at the origin of $R^4$ and rattling in the vicinity of $Z_C$. At large $\xi > |Z_C|$, the rattling instanton sources the vector meson fields described

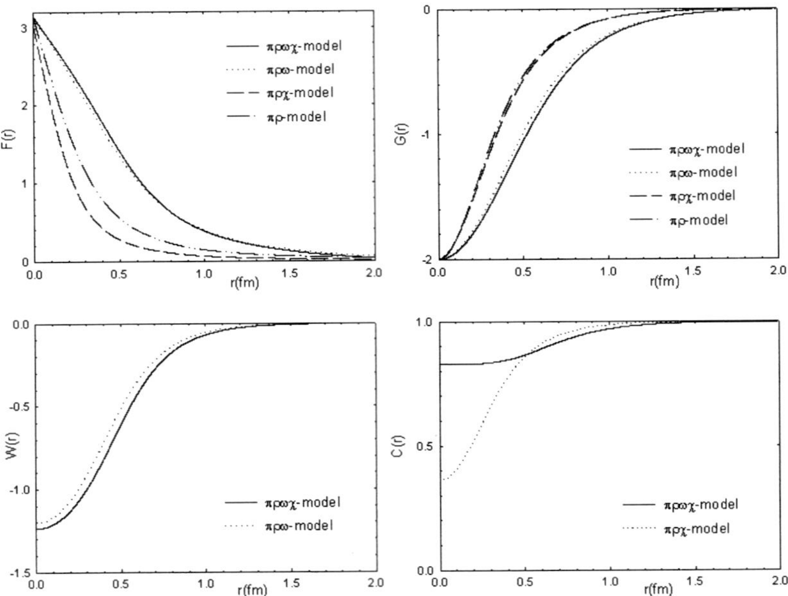

Fig. 17.1. The Z-mode in the non-rigid gauge vs $\partial_Z \mathbb{A}_i$.

by a semi-classical expansion with non-rigid Dirac constraints. Changes in $Z_C$ (the core boundary) are reabsorbed by a residual gauge transformation on the core instanton. This is a holographic realization of the Cheshire cat principle[2] where $Z_C$ plays the role of the Cheshire cat smile.

## 17.3. The Holographic Model

To illustrate the Cheshire cat mechanism more quantitatively, we now summarize the holographic Yang-Mills-Chern-Simons action in 5D curved background. This is the leading term in a $1/\lambda$ expansion of the D-brane Born-Infeld (DBI) action on D8,[5]

$$S = S_{YM} + S_{CS} \,, \tag{17.3.9}$$

$$S_{YM} = -\kappa \int d^4x dZ \, \mathrm{Tr} \left[ \frac{1}{2} K^{-1/3} \mathcal{F}_{\mu\nu}^2 + M_{\mathrm{KK}}^2 K \mathcal{F}_{\mu Z}^2 \right] \,, \tag{17.3.10}$$

$$S_{CS} = \frac{N_c}{24\pi^2} \int_{M^4 \times R} \omega_5^{U(N_f)}(\mathcal{A}) \,, \tag{17.3.11}$$

where $\mu, \nu = 0, 1, 2, 3$ are 4D indices and the fifth(internal) coordinate $Z$ is dimensionless. There are three things which are inherited by the holographic dual gravity theory: $M_{\mathrm{KK}}, \kappa$, and $K$. $M_{\mathrm{KK}}$ is the Kaluza-Klein scale and we will set $M_{\mathrm{KK}} = 1$

as our unit. $\kappa$ and $K$ are defined as

$$\kappa = \lambda N_c \frac{1}{216\pi^3} \equiv \lambda N_c a \, , \qquad K = 1 + Z^2 \, . \tag{17.3.12}$$

$\mathcal{A}$ is the 5D $U(N_f)$ 1-form gauge field and $\mathcal{F}_{\mu\nu}$ and $\mathcal{F}_{\mu Z}$ are the components of the 2-form field strength $\mathcal{F} = \mathrm{d}\mathcal{A} - i\mathcal{A} \wedge \mathcal{A}$. $\omega_5^{U(N_f)}(\mathcal{A})$ is the Chern-Simons 5-form for the $U(N_f)$ gauge field

$$\omega_5^{U(N_f)}(\mathcal{A}) = \mathrm{Tr}\left( \mathcal{A}\mathcal{F}^2 + \frac{i}{2}\mathcal{A}^3\mathcal{F} - \frac{1}{10}\mathcal{A}^5 \right) \, , \tag{17.3.13}$$

We note that $S_{YM}$ is of order $\lambda$, while $S_{CS}$ is of order $\lambda^0$. These terms are sufficient to carry a semiclassical expansion around the holonomy (17.2.3) with $\hbar = 1/\kappa$ as we now illustrate it for the baryon current.

## 17.4. The Baryon Current

To extract the baryon current, we source the reduced action with $\hat{\mathcal{V}}_\mu(x)$ a $U(1)_V$ flavor field on the boundary in the presence of the vector fluctuations $(C = \hat{v})$. The effective action for the $U(1)_V$ source to order $\hbar^0$ reads

$$S_{\mathrm{eff}}[\hat{\mathcal{V}}_\mu] = \sum_{n=1}^{\infty} \int d^4x \left[ -\frac{1}{4}\left( \partial_\mu \hat{v}_\nu^n - \partial_\nu \hat{v}_\mu^n \right)^2 - \frac{1}{2}m_{\hat{v}^n}^2 (\hat{v}_\mu^n)^2 \right.$$

$$\left. -\kappa K \widehat{\mathbb{F}}^{Z\mu} \widehat{\mathcal{V}}_\mu (1 - \alpha_{v^n}\psi_{2n-1}) \Big|_{Z=B} \right.$$

$$\left. + a_{v^n} m_{\hat{v}^n}^2 \hat{v}_\mu^n \widehat{\mathcal{V}}^\mu - \kappa K \widehat{\mathbb{F}}^{Z\mu} \hat{v}_\mu^n \psi_{2n-1} \Big|_{Z=B} \right] \, , \tag{17.4.14}$$

The first line is the free action of the massive vector meson which is

$$\Delta_{\mu\nu}^{mn}(x) = \int \frac{d^4p}{(2\pi)^4} e^{-ipx} \left[ \frac{-g_{\mu\nu} - p_\mu p_\nu/m_{v^n}^2}{p^2 + m_{v^n}^2}\delta^{mn} \right] \, , \tag{17.4.15}$$

in Lorentz gauge. The second line is the direct coupling between the core instanton and the $U(1)_V$ source as displayed in Fig. 2(a) while the last line corresponds to the vector omega, omega$'$, ... mediated couplings (VMD) as displayed in Fig. 2(b). These couplings are

$$\kappa K \widehat{\mathbb{F}}^{Z\mu} \hat{v}_\mu^n \psi_{2n-1} \, , \tag{17.4.16}$$

which are large and of order $1/\sqrt{\hbar}$ since $\psi_{2n-1} \sim \sqrt{\hbar}$. When $\rho$ is set to $1/\sqrt{\lambda}$ after the book-keeping noted above, the coupling scales like $\lambda\sqrt{N_c}$, or $\sqrt{N_c}$ in the large $N_c$ limit taken first.

The direct coupling drops by the sum rule

$$\sum_{n=1}^{\infty} \alpha_{v^n}\psi_{2n-1} = 1 \, , \tag{17.4.17}$$

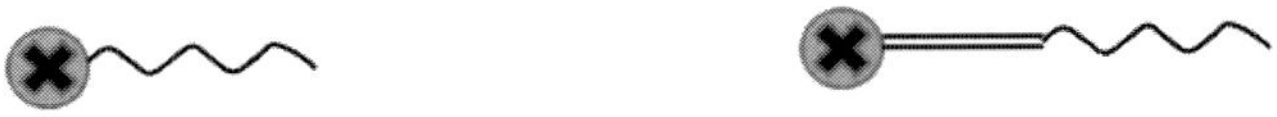

Fig. 17.2.   (a) Direct coupling; (b) VMD coupling.

following from closure in curved space

$$\delta(Z - Z') = \sum_{n=1}^{\infty} \kappa \psi_{2n-1}(Z)\psi_{2n-1}(Z')K^{-1/3}(Z') \ . \tag{17.4.18}$$

in complete analogy with VMD for the pion in holography.[5]

Baryonic VMD is exact in holography provided that an infinite tower of radial omega's are included in the mediation of the $U(1)_V$ current. To order $\hbar^0$ the baryon current is

$$J_B^\mu(x) = -\sum_{n,m} m_{v^n}^2 a_{v^n} \psi_{2m-1} \int d^4y \ \kappa K \widehat{\mathbb{F}}_{Z\nu}(y,Z)\Delta_{mn}^{\nu\mu}(y-x)\Big|_{Z=B} \ . \tag{17.4.19}$$

This point is in agreement with the effective holographic approach described in Ref. 13. The static baryon charge distribution is

$$J_B^0(\vec{x}) = -\sum_{n} \int d\vec{y} \ \frac{2}{N_c}\kappa K \widehat{\mathbb{F}}_{Z0}(\vec{y},Z) \ \Delta_n(\vec{y}-\vec{x}) \ a_{v^n} m_{v^n}^2 \psi_{2n-1} \Big|_{Z=B} \ , \tag{17.4.20}$$

with

$$\Delta_n(\vec{y}-\vec{x}) \equiv \int \frac{d\vec{p}}{(2\pi)^3} \frac{e^{-i\vec{p}\cdot(\vec{y}-\vec{x})}}{\vec{p}^2 + m_{v^n}^2} \ . \tag{17.4.21}$$

The extra $2/N_c$ follows the normalization $\widehat{\mathcal{V}}_\mu = \delta_{\mu 0}\frac{\sqrt{2N_f}}{N_c}\widehat{B}_0(\vec{x})$ for the baryon number source.

## 17.5.  Baryonic Form Factor

The static baryon form factor is a purely surface contribution from

$$J_B^0(\vec{q}) = \int d\vec{x}\,e^{i\vec{q}\cdot\vec{x}}J_B^0(x)$$

$$= -\sum_{n} \int dZ\partial_Z \left[\left(\int d\vec{x}\,e^{i\vec{q}\cdot\vec{x}}\mathcal{Q}_0(x,Z)\right)\psi_{2n-1}\right]\frac{a_{v^n} m_{v^n}^2}{\vec{q}^{\,2} + m_{v^n}^2} \tag{17.5.22}$$

$$= \int d\vec{x}\,e^{i\vec{q}\cdot\vec{x}}\sum_{n} \frac{a_{v^n} m_{v^n}^2}{\vec{q}^{\,2} + m_{v^n}^2} \ \psi_{2n-1}(Z_C)2\mathcal{Q}_0(\vec{x},Z_C) \ , \tag{17.5.23}$$

with

$$Q_0(x, Z) \equiv \frac{1}{N_c} \kappa K \widehat{\mathbb{F}}_{Z0}(x, Z) \ . \tag{17.5.24}$$

The boundary contribution at $Z = \infty$ vanishes since $\psi_{2n-1} \sim 1/Z$ for large $Z$. In the limit $q \to 0$ we pick the baryon charge

$$\int d\vec{x}\, e^{i\vec{q}\cdot\vec{x}}\, 2Q_0(\vec{x}, Z_C) \ , \tag{17.5.25}$$

due to the sum rule (17.4.17), with the limits $\lim_{q\to 0} \lim_{Z\to 0}$ understood sequentially.

The surface density follows from the U(1) bulk equation

$$\frac{4}{N_c} \kappa K \widehat{\mathbb{F}}_{Z0}(Z_c) = \int_{-Z_C}^{Z_C} dZ \frac{1}{32\pi^2} \epsilon_{MNPQ} \left( \mathrm{Tr}(\mathbb{F}_{MN}\mathbb{F}_{PQ}) + \frac{1}{2}\widehat{\mathbb{F}}_{MN}\widehat{\mathbb{F}}_{PQ} \right)$$

$$+ \frac{2}{N_c} \int_{-Z_C}^{Z_C} dZ \kappa K^{-1/3} \partial^i \widehat{\mathbb{F}}_{0i} \ , \tag{17.5.26}$$

The baryon number density lodged in $|Z| < Z_c$ integrates to 1 since

$$B = \int d\vec{x}\, J_B^0(\vec{x}) = \int d\vec{x}\, 2Q_0(\vec{x}, Z_c)$$

$$= \int d\vec{x} \int_{-Z_C}^{Z_C} dZ \frac{1}{32\pi^2} \epsilon_{MNPQ} \mathrm{Tr}(\mathbb{F}_{MN}\mathbb{F}_{PQ}) = 1 \ , \tag{17.5.27}$$

as the spatial flux vanishes on $R_X^3$ is zero for a sufficiently localized SU(2) instanton in $R_X^3 \times R_Z$.

The isoscalar charge radius, can be read from the $q^2$ terms of the form factor

$$\langle r^2 \rangle_0 = \frac{3}{2} \frac{Z_c \rho^2}{\sqrt{Z_c^2 + \rho^2}} + \int dZ\, \Delta_C(Z, Z_c) \tag{17.5.28}$$

with $r \equiv \sqrt{(\vec{x})^2}$. The first contribution is from the core and of order $1/\lambda$,

$$\int d\vec{x}\, r^2\, 2Q_0(\vec{x}, Z_c) = \frac{3}{2}\rho^2 \frac{Z_c}{\sqrt{Z_c^2 + \rho^2}} \to \frac{3}{2}\rho^2 \ . \tag{17.5.29}$$

The second contribution is from the cloud and of order $\lambda^0$,

$$\sum_{n=1}^{\infty} \frac{a_{v^n} \psi_{2n-1}(Z_c)}{m_n^2} = \int dZ \Delta_C(Z, Z_c) \tag{17.5.30}$$

with $\Delta_C = \Box_C^{-1} \equiv -\partial_Z^{-1} K^{-1} \partial_Z^{-1} K^{-1/3}$ the inverse vector meson propagator in bulk.

The results presented in this section were derived in Ref. 9 using the Cheshire cat descriptive. They were independently arrived at in Ref. 12 using the strong coupling source quantization. They also support, the effective 5-dimensional nucleon approach described in Ref. 13 using the heavy nucleon expansion.

## 17.6. Conclusions

The holography model presented here provides a simple realization of the Cheshire principle, whereby a zero size Skyrmion emerges to order $1/\hbar = \kappa$ through a holonomy in 5 dimensions. The latter is a bosonized form of a heavy quark sitting still in the conformal direction viewed as *time*. The baryon has zero size.

To order $\hbar^0$, the core Skyrmion is dressed by an infinite tower of vector mesons which couple in the holographic direction a distance $Z_C$ away from the core. The emergence of $Z_C$ follows from a non-rigid semiclassical quantization constraint to prevent double counting. $Z_C$ divides the holographic direction into a core dominated by an instanton and a cloud described by vector mesons.

Observables are $Z_C$ independent provided that the curvature in both the core and the cloud is correctly accounted for. This is the Cheshire cat mechanism in holography with $Z_C$ playing the role of the Cheshire cat smile. We have illustrated this point using the baryon form factor, where $Z_C$ was taken to zero using the uncurved or flat ADHM instanton. The curved instanton is not known. Most of these observations carry to other baryonic observables[9,12] and baryonic matter[14] (and references therein).

## Acknowledgments

IZ thanks Keun-Young Kim for his collaboration on numerous aspects of holographic QCD. This work was supported in part by US-DOE grants DE-FG02-88ER40388 and DE-FG03-97ER4014.

## References

1. A. Chodos, R. Jaffe, K. Johnson and C. Thorn, "Baryon Structure in the Bag Theory", *Phys. Rev. D* **10** (1974) 2599; G.E. Brown and M. Rho, "The Little Bag," *Phys. Lett. B* **82** (1079) 177.
2. S. Nadkarni, H.B. Nielsen and I. Zahed, "Bosonization Relations As Bag Boundary Conditions," *Nucl. Phys. B* **253** (1985) 308.
3. H.B. Nielsen, M. Rho, A. Wirzba and I. Zahed, "Color Anomaly in Hybrid Bag Model," *Phys. Lett. B* **269** (1991) 389; M. Rho, "The Cheshire Cat Hadrons Revisited," *Phys. Rep.* **240** (1994) 1.
4. I. Zahed and G.E. Brown, "The Skyrme Model", *Phys. Rep.* **142** (1986) 1.
5. T. Sakai and S. Sugimoto, "Low energy hadron physics in holographic QCD," *Prog. Theor. Phys.* **113** (2005) 843; T. Sakai and S. Sugimoto, "More on a holographic dual of QCD," *Prog. Theor. Phys.* **114** (2006) 1083; H. Hata, T. Sakai, S. Sugimoto and S. Yamato, "Baryons from instantons in holographic QCD," arXiv:hep-th/0701280.
6. D. Hong, M. Rho, H. Yee and P. Yi, "Chiral Dynamics of Baryons from String Theory" *Phys. Rev. D* **76** (2007) 061901.
7. J. Maldacena, "The Large N limit of Superconformal Field Theories and Supergravity" *Adv. Theor. Math. Phys.* **2** (1998) 231.
8. C. Csaki and M. Reece, "Toward a Systematic Holographic QCD: A Braneless Approach", arXiv:hep-th/0608266.

9. K.Y. Kim and I. Zahed, "Electromagnetic Baryon Form Factors from Holographic QCD." *JHEP* **0809** (2008) 007.

10. C. Adami and I. Zahed, "Soliton quantization in chiral models with vector mesons," *Phys. Lett. B* **215** (1988) 387.

11. H. Verschelde and H. Verbeke, "Nonrigid quantization of the skyrmion," *Nucl. Phys. A* **495** (1989) 523.

12. K. Hashimoto, T. Sakai and S. Sugimoto, "Holographic Baryons: Static Properties and Form Factors from Gauge/String Duality," arXiv:0806.3122 [hep-th].

13. D.K. Hong, M. Rho, H.U. Yee and P. Yi, "Nucleon Form Factors and Hidden Symmetry in Holographic QCD," arXiv:0710.4615 [hep-ph]; M. Rho, "Baryons and Vector Dominance in Holographic Dual QCD," arXiv:0805.3342 [hep-ph].

14. K.Y. Kim, S.J. Sin and I. Zahed, "Dense Holographic QCD in the Wigner Seitz Approximation," *JHEP* **0809** (2008) 001.

## Chapter 18

# Baryon Physics in a Five-Dimensional Model of Hadrons


Alex Pomarol[*,‡] and Andrea Wulzer[†,§]

*Departament de Física, Universitat Autònoma de Barcelona,
08193 Bellaterra, Barcelona
†Institut de Théorie des Phénomènes Physiques, EPFL,
CH–1015 Lausanne, Switzerland



We review the procedure to calculate baryonic properties using a recently proposed five-dimensional approach to QCD. We show that this method gives predictions to baryon observables that agree reasonably well with the experimental data.


## Contents


18.1 Introduction . . . . . . . . . . . . . . . . . . . . . . . . . . . . . . . . . . 403
18.2 A Five-Dimensional Model for QCD Mesons . . . . . . . . . . . . . . . . . 405
    18.2.1 Meson physics and calculability . . . . . . . . . . . . . . . . . . . 408
18.3 Baryons from 5D Skyrmions . . . . . . . . . . . . . . . . . . . . . . . . . . 411
    18.3.1 4D Skyrmions from 5D Solitons . . . . . . . . . . . . . . . . . . . 411
    18.3.2 The static solution . . . . . . . . . . . . . . . . . . . . . . . . . . 412
    18.3.3 Zero-mode fluctuations . . . . . . . . . . . . . . . . . . . . . . . . 413
    18.3.4 The Lagrangian of collective coordinates . . . . . . . . . . . . . . 416
    18.3.5 Skyrmion quantization . . . . . . . . . . . . . . . . . . . . . . . . 417
    18.3.6 The nucleon form factors . . . . . . . . . . . . . . . . . . . . . . . 419
18.4 Properties of Baryons: Results . . . . . . . . . . . . . . . . . . . . . . . . . 421
18.5 Conclusions and Outlook . . . . . . . . . . . . . . . . . . . . . . . . . . . . 427
A.1 Numerical Methods . . . . . . . . . . . . . . . . . . . . . . . . . . . . . . . 428
    A.1.1 COMSOL implementation . . . . . . . . . . . . . . . . . . . . . . . 430
References . . . . . . . . . . . . . . . . . . . . . . . . . . . . . . . . . . . . . . . 431


## 18.1. Introduction

In 1973 Gerard 't Hooft proposed, in a seminal article,[1] a dual description for QCD. He showed that in the limit of large number of colors ($N_c$) strongly-interacting gauge theories could be described in terms of a weakly-interacting theory of mesons. It was later recognized[2] that, in this dual description, baryons appeared as solitons

---


‡alex.pomarol@uab.cat

§andrea.wulzer@epfl.ch

made of meson fields, as Skyrme had pointed out long before.[3] These solitonic states were therefore referred to as skyrmions.

Skyrmions have been widely studied in the literature, with some phenomenological successes.[4] Nevertheless, since the full theory of QCD mesons is not known, these studies have been carried out in truncated low-energy models either incorporating only pions[2,3] or few resonances.[4] It is unclear whether these approaches capture the physics needed to fully describe the baryons, since the stabilization of the baryon size is very sensitive to resonances around the GeV. In the original Skyrme model with only pions, for instance, the inverse skyrmion size $\rho_s^{-1}$ equals the chiral perturbation theory cut-off $\Lambda_{\chi PT} \sim 4\pi F_\pi$ (as it should be, since this is the only scale of the model), rendering baryon physics completely incalculable. Other examples are models with the $\rho$-meson[5] or the $\omega$-meson[6] which were shown to have a stable skyrmion solution. The inverse size, also in this case, is of order $m_\rho \sim \Lambda_{\chi PT}$, which is clearly not far from the mass of the next resonances. Including the latter could affect strongly the physics of the skyrmion, or even destabilize it.

In the last ten years the string/gauge duality[7–9] has allowed us to gain new insights into the problem of strongly-coupled gauge theories. This duality has been able to relate certain strongly-coupled gauge theories with string theories living in more than four dimensions. A crucial ingredient in these realizations is a (compact) warped extra dimension that plays the role of the energy scale in the strongly-coupled 4D theory. This has suggested that the QCD dual theory of mesons proposed by 'tHooft[1] must be a theory formulated in more than 4 dimensions.

Inspired by this duality, a five-dimensional field theory has been proposed in Refs. 10 and 11 to describe the properties of mesons in QCD. This 5D theory has a cut-off scale $\Lambda_5$ which is above the lowest-resonance mass $m_\rho$. The gap among these two scales, which ensures calculability in the meson sector, is related to the number of colors $N_c$ of QCD. In the large $N_c$-limit, one has $\Lambda_5/m_\rho \to \infty$ and the 5D model describes a theory of infinite mesonic resonances, corresponding to the Kaluza-Klein (KK) spectrum. This 5D model has provided a quite accurate description of meson physics in terms of a very limited number of parameters.

Further studies, boosted by this success, have recently shown that the 5D model can also successfully describe baryon physics.[12–14] As Skyrme proposed,[3] baryons must appear in this 5D theory as solitons. These 5D skyrmion-like solitons have been numerically obtained (see Fig. 18.1) and their properties have been studied. Their inverse size $\rho_s^{-1} \sim m_\rho$ have been found to be smaller than the cut-off scale $\Lambda_5$, showing then that, contrary to the 4D case, they can be consistently studied in 5D effective theories. Indeed, the expansion parameter which ensures calculability is provided by $1/(\rho_s \Lambda_5) \ll 1$.

In this article we will review the properties of baryons obtained in Refs. 12–14 using the five-dimensional model of QCD of Refs. 10, 11 and 15. We will show how the calculation of the static properties of the nucleons, such as masses, radii and

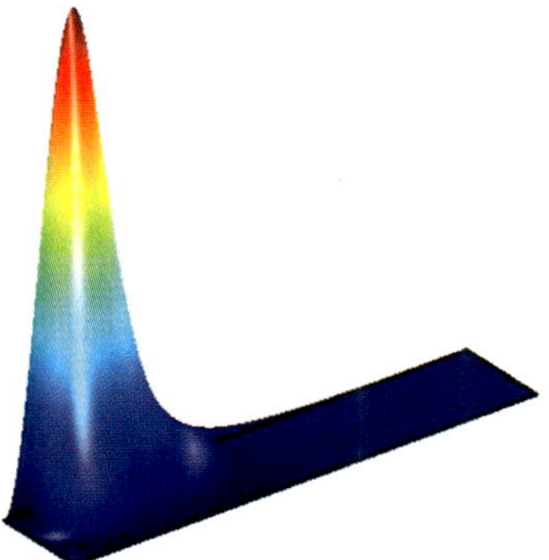

Fig. 18.1.   Energy density, in the plane of the 4D radial and the extra fifth coordinate, of the skyrmion in a 5D model for QCD.

form factors, are performed, and will compare the predictions of the model with experiments. As we will see, these predictions show a reasonably good agreement with the data.

There have been alternative studies to baryon physics using 5D models. Nevertheless, these studies have encountered several problems. For example, the first approaches[16] truncated the 5D theory and only considered the effects of the first resonances. This leads to skyrmions whose size is of the order of the inverse of the truncation scale, and therefore sensitive to the discarded heavier resonances. Later studies[17–19] were performed within the Sakai-Sugimoto model.[20]   It was shown, however, that baryons are not calculable in this framework as their inverse size is of the order of the string scale which corresponds to the cut-off of the theory.[17]

## 18.2.  A Five-Dimensional Model for QCD Mesons

The 5D model that we will consider to describe mesons in two massless flavor QCD is the following. This is a $U(2)_L \times U(2)_R$ gauge theory with metric $ds^2 = a(z)^2 \left( \eta_{\mu\nu} dx^\mu dx^\nu - dz^2 \right)$, where $x^\mu$ represent the usual 4 coordinates and $z$, which runs in the interval $[z_{\mathrm{UV}}, z_{\mathrm{IR}}]$, denotes the extra dimension. We will work in $\mathrm{AdS}_5$ where the warp factor $a(z)$ is

$$a(z) = \frac{z_{\mathrm{IR}}}{z} \,, \tag{18.2.1}$$

and $z_{\mathrm{UV}} \to 0$ to be taken at the end of the calculations. In this limit $z_{\mathrm{IR}}$ coincides with the AdS curvature and the conformal length

$$L = \int_{z_{\mathrm{UV}}}^{z_{\mathrm{IR}}} dz \,. \tag{18.2.2}$$

The $U(2)_L$ and $U(2)_R$ gauge connections, denoted respectively by $L_M$ and $R_M$ ($M = \{\mu, 5\}$), are parametrized by $L_M = L_M^a \sigma_a/2 + \widehat{L}_M \mathbb{1}/2$ and $R_M = R_M^a \sigma_a/2 +$

$\widehat{R}_M \mathbf{1}/2$, where $\sigma_a$ are the Pauli matrices. This chiral gauge symmetry is broken by the conditions on the boundary at $z = z_{\text{IR}}$ (IR-boundary), which read

$$(L_\mu - R_\mu)\,|_{z=z_{\text{IR}}} = 0\,, \qquad (L_{\mu 5} + R_{\mu 5})\,|_{z=z_{\text{IR}}} = 0\,, \tag{18.2.3}$$

where the 5D field strength is defined as $L_{MN} = \partial_M L_N - \partial_N L_M - i[L_M, L_N]$, and analogously for $R_{MN}$. At the other boundary, the UV one, we can consider generalized Dirichlet conditions for all the fields:

$$L_\mu\,|_{z=z_{\text{UV}}} = l_\mu\,, \qquad R_\mu\,|_{z=z_{\text{UV}}} = r_\mu\,. \tag{18.2.4}$$

The 4D fields $l_\mu$ and $r_\mu$ are arbitrary but fixed and they can be interpreted, as we will now discuss, as external sources for the QCD global currents. We will eventually be interested in taking the sources to vanish.

We can now, inspired by the "holographic" formulation of the AdS/CFT correspondence,[7–9] try to interpret the above 5D model in terms of a 4D QCD-like theory, whose fields we will generically denote by $\Psi(x)$ and its action by $S_4$. This is a strongly coupled 4D theory that possesses an $U(2)_L \times U(2)_R$ global symmetry with associated Noether currents $j^\mu_{L,R}$. If the 4D theory were precisely massless QCD with two flavors, the currents would be given by the usual quark bilinear, $\left(j^\mu_{L,R}\right)_{ij} = \overline{Q}^j_{L,R}\gamma^\mu Q^i_{L,R}$. Defining $Z[l_\mu, r_\mu]$ as the generating functional of current correlators, we state our correspondence as

$$Z\,[l_\mu, r_\mu] \equiv \int \mathcal{D}\Psi \exp\left[iS_4\,[\Psi] + i\int d^4x \,\mathrm{Tr}\,(j^\mu_L l_\mu + j^\mu_R r_\mu)\right]$$

$$= \int \mathcal{D}L_M \mathcal{D}R_M \exp\left[iS_5\,[L, R]\right]\,, \tag{18.2.5}$$

where the 5D partition function depends on the sources $l_\mu$, $r_\mu$ through the UV-boundary conditions in Eq. (18.2.4).

Equation (18.2.5) leads to the following implication. Under local chiral transformations, $Z$ receives a contribution from the $U(2)^3$ anomaly, which is known in QCD.[a] This implies[9,21,22] that the 5D action must contain a Chern-Simons (CS) term

$$S_{CS} = -i\frac{N_c}{24\pi^2}\int [\omega_5(L) - \omega_5(R)]\,, \tag{18.2.6}$$

whose variation under 5D local transformations which does not reduce to the identity at the UV exactly reproduces the anomaly. The CS coefficient will be fixed to $N_c = 3$ when matching QCD. The CS 5-form, defining $A = -iA_M dx^M$, is

$$\omega_5(A) = \mathrm{Tr}\left[A(dA)^2 + \frac{3}{2}A^3(dA) + \frac{3}{5}A^5\right]\,. \tag{18.2.7}$$

---

[a]The 5D semiclassical expansion we perform in our model corresponds, as we will explain in the following, to the large-$N_c$ expansion. This is why we are ignoring the $U(1)$-$SU(N_c)^2$ anomaly of QCD, which is subleading at large-$N_c$. Being this anomaly responsible for the $\eta'$ mass, our model will contain a massless $\eta'$.

When $A$ is the connection of an $U(2)$ group, as in our case, one can use the fact that $SU(2)$ is an anomaly-free group to write $\omega_5$ as

$$\omega_5(A) = \frac{3}{2}\widehat{A}\mathrm{Tr}\left[F^2\right] + \frac{1}{4}\widehat{A}\left(d\widehat{A}\right)^2 + d\,\mathrm{Tr}\left[\widehat{A}AF - \frac{1}{4}\widehat{A}A^3\right], \qquad (18.2.8)$$

where $A = A + \widehat{A}\mathbb{1}/2$ and $A$ is the $SU(2)$ connection. The total derivative part of the above equation can be dropped, since it only adds to $S_{CS}$ an UV-boundary term for the sources.

The full 5D action will be given by $S_5 = S_g + S_{CS}$, where $S_g$ is made of locally gauge invariant terms. $S_g$ is also invariant under transformations which do not reduce to the identity at the UV-boundary, and for this reason it does not contribute to the anomalous variation of the partition function. Taking the operators of the lowest dimensionality, we have

$$S_g = -\int d^4x \int_{z_{\mathrm{UV}}}^{z_{\mathrm{IR}}} dz\, a(z)\,\frac{M_5}{2}\left\{\mathrm{Tr}\left[L_{MN}L^{MN}\right] + \frac{\alpha^2}{2}\widehat{L}_{MN}\widehat{L}^{MN} + \{L \leftrightarrow R\}\right\}.$$
$$(18.2.9)$$

We have imposed on the 5D theory invariance under the combined $\{x \to -x, L \leftrightarrow R\}$, where $x$ denotes ordinary 3-space coordinates. This symmetry, under which $S_{CS}$ is also invariant, corresponds to the usual parity on the 4D side. We have normalized differently the kinetic term of the $SU(2)$ and $U(1)$ gauge bosons, since we do not have any symmetry reason to put them equal. In the large-$N_c$ limit of QCD, however, the Zweig's rule leads to equal couplings (and masses) for the $\rho$ and $\omega$ mesons, implying $\alpha = 1$ in our 5D model. Since this well-known feature of large-$N_c$ QCD does not arise automatically in our 5D framework (as, for instance, the equality of the $\rho$ and $\omega$ masses does), we will keep $\alpha$ as a free parameter. The CS term, written in component notation, will be given by

$$S_{CS} = \frac{N_c}{16\pi^2}\int d^5x\left\{\frac{1}{4}\epsilon^{MNOPQ}\widehat{L}_M\mathrm{Tr}\left[L_{NO}L_{PQ}\right]\right.$$
$$\left. + \frac{1}{24}\epsilon^{MNOPQ}\widehat{L}_M\widehat{L}_{NO}\widehat{L}_{PQ} - \{L \leftrightarrow R\}\right\}. \qquad (18.2.10)$$

The 5D theory defined above has only 3 independent parameters: $M_5$, $L$ and $\alpha$.

Let us make again use of Eq. (18.2.5) to determine the current operators through which the theory couples to the external EW bosons. These currents are obtained by varying Eq. (18.2.5) with respect to $l_\mu$ (exactly the same would be true for $r_\mu$) and then taking $l_\mu = r_\mu = 0$. The variation of the l.h.s. of Eq. (18.2.5) simply gives the current correlator of the 4D theory, while in the r.h.s. this corresponds to a variation of the UV-boundary conditions. The effect of this latter can be calculated in the following way. We perform a field redefinition $L_\mu \to L_\mu + \delta L_\mu$ where $\delta L_\mu(x, z)$ is chosen to respect the IR-boundary conditions and fulfill $\delta L_\mu(x, z_{\mathrm{UV}}) = \delta l_\mu$. This redefinition removes the original variation of the UV-boundary conditions, but leads a new term in the 5D action, $\delta S_5$. One then has

$$i \int d^4x \, \text{Tr}\left[\langle j_L^\mu(x)\rangle \delta l_\mu(x)\right] = i \int \mathcal{D}L_M \mathcal{D}R_M \, \delta S_5\left[L, R\right] \exp\left[iS_5\left[L, R\right]\right], \quad (18.2.11)$$

where the 5D path integral is now performed by taking $l_\mu = r_\mu = 0$, *i.e.* normal Dirichlet conditions. The explicit value of $\delta S_5$ is given by

$$\delta S_5 = \int d^4x \, \text{Tr}\left[J_L^\mu(x)\delta l_\mu(x)\right] + \int d^5x (\text{EOM}) \cdot \delta L, \quad (18.2.12)$$

where $J_{L\,\mu} = J_{L\,\mu}^a \sigma^a + \widehat{J}_{L\,\mu} \mathbb{1}$ and

$$J_{L\,\mu}^a = M_5\left(a(z)L_{\mu\,5}^a\right)|_{z=z_{\rm UV}}, \qquad \widehat{J}_{L\,\mu} = \alpha^2 M_5\left(a(z)\widehat{L}_{\mu\,5}\right)|_{z=z_{\rm UV}}. \quad (18.2.13)$$

The last term of Eq. (18.2.12) corresponds to the 5D "bulk" part of the variation, which leads to the equations of motion (EOM). Remembering that the EOM always have zero expectation value,[b] we find that we can identify $J_L^\mu$ of Eq. (18.2.13) with the current operator on the 5D side: $\langle j_L^\mu\rangle_{\rm 4D} = \langle J_L^\mu\rangle_{\rm 5D}$. Notice that the CS term has not contributed to Eq. (18.2.12) due to the fact that each term in $S_{CS}$ which contains a $\partial_z$ derivative (and therefore could lead to a UV-boundary term) also contains $L_\mu$ or $R_\mu$ fields; these fields on the UV-boundary are the sources $l_\mu$ and $r_\mu$ that must be put to zero.

### 18.2.1. *Meson physics and calculability*

The phenomenological implications for the lightest mesons of 5D models like the one described above have been extensively studied in the literature. Let us briefly summarize the main results here. If rewritten in 4D terms, the theory contains massless Goldstone bosons that parametrize the $U(2)_L \times U(2)_R/U(2)_V$ coset and describe the pion triplet and a massless $\eta'$. The pion decay constant is given by

$$F_\pi^2 = 2M_5 \left(\int \frac{dz}{a(z)}\right)^{-1} = \frac{4M_5}{L}. \quad (18.2.14)$$

The massive spectrum consists of infinite towers of vector and axial-vector spin-one KK resonances. Among the vectors we have an isospin triplet, the $\rho^{(n)}$, and a singlet $\omega^{(n)}$. The axial-vectors are again a triplet $a_1^{(n)}$ and a singlet $f_1^{(n)}$. We want to interpret, as our terminology already suggests, the lightest states of each tower as the $\rho(770)$, $\omega(782)$, $a_1(1260)$ and $f_1(1285)$ resonances, respectively. The model predicts at leading order, *i.e.* at tree-level,

$$m_\rho = m_\omega \simeq \frac{3\pi}{4L}, \qquad m_{a_1} = m_{f_1} \simeq \frac{5\pi}{4L}, \quad (18.2.15)$$

compatibly with observations. The model also predicts the decay constants $F_i$ and couplings $g_i$ for the mesons as a function of $M_5$, $L$ and $\alpha$ that can be found in

---

[b]We have actually shown this here; notice that $\delta L_\mu$ was completely arbitrary in the bulk, but the variation of the functional integral can only depend on $\delta l_\mu = \delta L_\mu(x, z_{\rm UV})$.

Table 18.1. Global fit to mesonic physical quantities. Masses, decay constants and widths are given in MeV. Physical masses have been used in the kinematic factors of the partial decay widths.

|  | Experiment | AdS$_5$ | Deviation |
|---|---|---|---|
| $m_\rho$ | 775 | 824 | $+6\%$ |
| $m_{a_1}$ | 1230 | 1347 | $+10\%$ |
| $m_\omega$ | 782 | 824 | $+5\%$ |
| $F_\rho$ | 153 | 169 | $+11\%$ |
| $F_\omega/F_\rho$ | 0.88 | 0.94 | $+7\%$ |
| $F_\pi$ | 87 | 88 | $+1\%$ |
| $g_{\rho\pi\pi}$ | 6.0 | 5.4 | $-10\%$ |
| $L_9$ | $6.9 \cdot 10^{-3}$ | $6.2 \cdot 10^{-3}$ | $-10\%$ |
| $L_{10}$ | $-5.2 \cdot 10^{-3}$ | $-6.2 \cdot 10^{-3}$ | $-12\%$ |
| $\Gamma(\omega \to \pi\gamma)$ | 0.75 | 0.81 | $+8\%$ |
| $\Gamma(\omega \to 3\pi)$ | 7.5 | 6.7 | $-11\%$ |
| $\Gamma(\rho \to \pi\gamma)$ | 0.068 | 0.077 | $+13\%$ |
| $\Gamma(\omega \to \pi\mu\mu)$ | $8.2 \cdot 10^{-4}$ | $7.3 \cdot 10^{-4}$ | $-10\%$ |
| $\Gamma(\omega \to \pi ee)$ | $6.5 \cdot 10^{-3}$ | $7.3 \cdot 10^{-3}$ | $+12\%$ |

Refs. 10–12 and 15; here we only notice, for later use, their scaling with the 5D coupling:

$$F_i \sim \sqrt{M_5} \,, \qquad g_i \sim \frac{1}{\sqrt{M_5}} \,, \tag{18.2.16}$$

while the masses, as shown above, do not depend on $M_5$. In Table 18.1 we show a fit to 14 meson quantities. The best fit is obtained for the values of $1/L = 343$ MeV, $M_5 L = 0.0165$ and $\alpha = 0.94$ for the three parameters of our model. The minimum Root Mean Square Error (RMSE) corresponding to those values is found to be 11% and the relative deviation of each single prediction is below around 15%.

Concerning the choice of the meson observables, some remarks are in order. First of all, we are only considering the lowest state of each KK tower because we expect the masses and couplings of the heavier mesons to receive large quantum corrections. Our model is indeed, as we will explain below, an effective theory valid up to a cut-off $\Lambda_5 \sim 2$ GeV and our tree-level calculations only correspond to the leading term of an $E/\Lambda_5$ expansion. Apart from this restriction, we must include in our fit observables with an experimental accuracy better than 10%. This is because we want to neglect the experimental error in order to obtain an estimate of the accuracy of our theoretical predictions. Much more observables can be computed, once the best-fit value of the parameters are obtained, and several of them have already been considered in the literature. For instance, one can study the other low-energy constants of the chiral lagrangian, the physics of the $f_1$ resonance or the pseudo–scalar resonances which arise when the explicit breaking of the chiral symmetry is taken into account.[11] It would also be interesting to compute the $a_1 \to \pi\gamma$ decay, which is absent in our model at tree-level and only proceeds via loop effects or higher-dimensional terms of our 5D effective lagrangian.[c]

---

[c]Higher order contributions will also change our tree-level prediction $L_9 + L_{10} = 0$, which is again related with the absence of the $a_1$–$\pi$–$\gamma$ vertex.

As discussed in the Introduction, the semiclassical expansion in the 5D model should correspond to the large-$N_c$ expansion on the 4D side. The results presented above provide a confirmation of this interpretation: at large-$N_c$ meson masses are expected to scale like $N_c^0$, while meson couplings and decay constants scale like $g_i, 1/F_i \sim 1/\sqrt{N_c}$. These scalings agree with Eq. (18.2.15) and (18.2.16) if the parameters $\alpha$, $L$ and $M_5$ are taken to scale like[d]

$$\alpha \sim N_c^0 \,, \quad L \sim N_c^0 \,, \quad M_5 \sim N_c \,. \tag{18.2.17}$$

This leads us to define the adimensional $N_c$-invariant parameter

$$\gamma \equiv \frac{N_c}{16\pi^2 M_5 L \alpha} \,, \tag{18.2.18}$$

whose experimental value is $\gamma = 1.23$ and will be useful later on. We will also show in the following that the assumed scaling of the 5D parameters leads to the correct $N_c$ scaling in the baryon sector as well.

Other descriptions of vector mesons in terms of massive vector fields, *i.e.* models with Hidden Local Symmetry (HLS)[23,24] or two-form fields,[25] also correctly reproduce the meson physical properties. Nevertheless, we believe that 5D models, as the one discussed here, present more advantages.[e] First of all, they contain less parameters. In the models of Refs. 23–25, for example, the mass and the couplings of each meson are independent parameters; also anomalous processes, those involving an odd number of pions, depend on several operators with unknown coefficients which arise at the same order, while in our case they all arise from a single operator, the 5D CS term. Finally, Vector Meson Dominance is automatic in our scenario, while it needs to be imposed "by hand" in the case of HLS.

Moreover, and perhaps more importantly, 5D models are calculable effective field theory in which higher-dimensional operators are suppressed by the cut-off of the theory $\Lambda_5$. Calculations can be organized as an expansion in $E/\Lambda_5$, where $E$ is the typical scale of the process under consideration. Given that the cut-off is parametrically bigger than the mass of the lightest mesons, reliable calculations of masses and couplings can be performed.

Let us now use naive dimensional arguments to estimate the maximal value of our cut-off $\Lambda_5$. This is determined by the scale at which loops are of order of tree-level effects. Computing loop corrections to the $F^2$ operator, which arise from the $F^2$ term itself, one gets $\Lambda_5 \sim 24\pi^3 M_5$. Nevertheless, one gets a lower value for $\Lambda_5$ from the CS term due to the $N_c$ dependence of its coefficient. Indeed, at the one-loop level, the CS term gives a contribution of order $M_5$ to the $F^2$ operator for $\Lambda_5 \sim 24\pi^3 M_5/N_c^{2/3}$. Even though the cut-off scale lowers due to the presence of

---

[d]This scaling can also be obtained from the AdS/CFT correspondence.

[e]It must be possible, generalizing what was done in Ref. 26, to rewrite our model as a 4D HLS with infinitely many $U(2)$ hidden symmetry groups. The comparison with HLS models that we perform in this section only applies, therefore, to the standard case of a finite number of hidden symmetries.

the CS term, we can still have, in the large-$N_c$ limit, a 5D weakly coupled theory where higher-dimensional operators are suppressed. The cut-off can be rewritten as

$$\Lambda_5 \sim \frac{3\pi}{2} \frac{N_c^{1/3}}{\gamma \alpha L} \sim 2 \text{ GeV} \,,$$

where we have used the best-fit value of our parameters.

The power of calculability of our 5D model makes it very suitable for studying baryon physics. Indeed, the typical size of the 5D skyrmion solution will be of order $\rho_s \sim 1/m_\rho$, guaranteeing that effects from higher-dimensional operators will be suppressed by $m_\rho/\Lambda_5 \sim 0.4$. This is therefore, we believe, the first fully consistent approach towards baryon physics.

## 18.3. Baryons from 5D Skyrmions

### 18.3.1. *4D Skyrmions from 5D Solitons*

Time-independent configurations of our 5D fields, which correspond to allowed initial $(t \to -\infty)$ and final $(t \to +\infty)$ states of the time evolution, are labeled by the topological charge

$$B = \frac{1}{32\pi^2} \int d^3x \int_{z_{\mathrm{UV}}}^{z_{\mathrm{IR}}} dz\, \epsilon_{\hat\mu\hat\nu\hat\rho\hat\sigma} \mathrm{Tr} \left[ L^{\hat\mu\hat\nu} L^{\hat\rho\hat\sigma} - R^{\hat\mu\hat\nu} R^{\hat\rho\hat\sigma} \right] \,, \qquad (18.3.19)$$

where the indeces $\hat\mu, \hat\nu, \ldots$ run over the 4 spatial coordinates, but they are raised with Euclidean metric. We will now show that $B$ can only assume integer values, which ensures that it cannot be changed by the time evolution. This makes $B$ a topologically conserved charge which we identify with the baryon number. In order to show this, and with the aim of making the relation with the skyrmion more precise, it is convenient to go to the axial gauge $L_5 = R_5 = 0$. The latter can be easily reached, starting from a generic gauge field configuration, by means of a Wilson-line transformation. In the axial gauge both boundary conditions Eqs. (18.2.3) and (18.2.4) (in which we take now $l = r = 0$) cannot be simultaneously satisfied. Let us then keep Eq. (18.2.3) but modify the UV-boundary condition to

$$\widetilde{L}_i \,|_{z=z_{\mathrm{UV}}} = i\, U(\boldsymbol{x}) \partial_i U(\boldsymbol{x})^\dagger \,, \qquad \widetilde{R}_i \,|_{z=z_{\mathrm{UV}}} = 0 \,, \qquad (18.3.20)$$

where $\widetilde{L}_i$ and $\widetilde{R}_i$ are the gauge fields in the axial gauge and $i$ runs over the 3 ordinary space coordinates. The field $U(\boldsymbol{x})$ in the equation above precisely corresponds to the Goldstone field in the 4D interpretation of the model.[22] Remembering that $F \wedge F = d\omega_3$, where $\omega_3$ is the third CS form, the 4D integral in Eq. (18.3.19) can be rewritten as an integral on the 3D boundary of the space:

$$B = \frac{1}{8\pi^2} \int_{3D} \left[ \omega_3(\widetilde{L}) - \omega_3(\widetilde{R}) \right] \,. \qquad (18.3.21)$$

The contribution to $B$ coming from the IR-boundary vanishes as the $L$ and $R$ terms in Eq. (18.3.21) cancel each other due to Eq. (18.2.3). This is crucial for $B$ to be quantized and it is the reason why we have to choose the relative minus sign among the $L$ and $R$ instanton charges in the definition of $B$. At the $\boldsymbol{x}^2 \to \infty$ boundary, the contribution to $B$ also vanishes since in the axial gauge $\partial_5 A_i = 0$ (in order to have $F_{5i} = 0$). We are then left with the UV-boundary which we can topologically regard as the 3-sphere $S_3$. Therefore, we find

$$
\begin{aligned}
B &= -\frac{1}{8\pi^2} \int_{\mathrm{UV}} \omega_3 \left[ \widetilde{L}_i \left( = i\, U \partial_i U^\dagger \right) \right] \\
&= \frac{1}{24\pi^2} \int d^3 x \, \epsilon^{ijk} \mathrm{Tr} \left[ U \partial_i U^\dagger \, U \partial_j U^\dagger \, U \partial_k U^\dagger \right] \ \in \mathbb{Z}.
\end{aligned}
\tag{18.3.22}
$$

The charge $B$ is equal to the Cartan-Maurer integral invariant for $SU(2)$ which is an integer.

In the next section we will discuss regular static solutions with nonzero $B$. If they exist, they cannot trivially correspond to a pure gauge configuration. Moreover, the particles associated to solitons with $B = \pm 1$ will be stable given that they have minimal charge. Eq. (18.3.22) also makes the relation with 4D skyrmions explicit: topologically non-trivial 5D configurations are those for which the corresponding pion matrix $U(\boldsymbol{x})$ is also non-trivial. The latter corresponds to a 4D skyrmion with baryon number $B$. In a general gauge, the skyrimion configuration $U(\boldsymbol{x})$ will be given by

$$
U(\boldsymbol{x}) = P \left\{ \exp\left[ -i \int_{z_{\mathrm{UV}}}^{z_{\mathrm{IR}}} dz' \, R_5(\boldsymbol{x}, z') \right] \right\} \cdot P \left\{ \exp\left[ i \int_{z_{\mathrm{UV}}}^{z_{\mathrm{IR}}} dz' \, L_5(\boldsymbol{x}, z') \right] \right\},
\tag{18.3.23}
$$

where $P$ indicates path ordering. From a 4D perspective, the 5D soliton that we are looking for can be considered to be a 4D skyrmion made of Goldstone bosons and the massive tower of KK gauge bosons.

### 18.3.2. *The static solution*

In order to obtain the static soliton solution of the 5D EOM of our theory it is crucial to specify a correct Ansatz, which is best constructed by exploiting the symmetries of our problem. Let us impose, first of all, our solution to be invariant under time-reversal $t \to -t$ combined with $\widehat{L} \to -\widehat{L}$ and $\widehat{R} \to -\widehat{R}$, under which also the CS term is invariant. This transformation reduces, in static configurations, to a sign change of the temporal component of $L$ and $R$ and of the spatial components of $\widehat{L}$ and $\widehat{R}$. We can therefore consistently put them to zero. We also use parity invariance ($\{L \leftrightarrow R, x \leftrightarrow -x\}$) to restrict to configurations for which $L_i(x, z, t) = -R_i(-x, z, t)$, $L_{5,0}(x, z, t) = R_{5,0}(-x, z, t)$ and analogously for $\widehat{L}, \widehat{R}$. We impose, finally, invariance under "cylindrical" transformations,[27] *i.e.* the simultaneous action of 3D space rotations $x_a \sigma^a \to \theta^\dagger x_a \sigma^a \theta$, with $\theta \in SU(2)$, and vector $SU(2)$ global transformations $L, R \to \theta (L, R) \theta^\dagger$. An equivalent way to state the invariance is that a 3D rotation with $\theta$ acts on the solution exactly as an $SU(2)$ vector one in the opposite direction (*i.e.* with $\theta^\dagger$) would do. The resulting Ansatz for the static solution (which we denote by "barred" fields) is entirely specified 4 real 2D fields

$$
\begin{cases}
\overline{R}_j^a(x,z) = A_1(r,z)\widehat{x}_a\widehat{x}_j + \dfrac{1}{r}\varepsilon_{ajk}\widehat{x}_k - \dfrac{\phi_{(x)}}{r}\varepsilon^{(x,y)}\Delta^{(y),aj} \,, \\[2mm]
\overline{R}_5^a(x,z) = A_2(r,z)\widehat{x}^a \,, \\[2mm]
\alpha\widehat{\overline{R}}_0{}'(x,z) = \dfrac{s(r,z)}{r} \,,
\end{cases}
\tag{18.3.24}
$$

where $r^2 = \sum_i x^i x^i$, $\widehat{x}^i = x^i/r$, $\varepsilon^{(x,y)}$ is the antisymmetric tensor with $\varepsilon^{(1,2)} = 1$ and the "doublet" tensors $\Delta^{(1,2)}$ are

$$
\Delta^{(x),ab} = \begin{bmatrix} \epsilon^{abc}\widehat{x}^c \\ \widehat{x}^a\widehat{x}^b - \delta^{ab} \end{bmatrix}.
\tag{18.3.25}
$$

Substituting the Ansatz in the topological charge Eq. (18.3.19) we find

$$
B = \frac{1}{2\pi} \int_0^\infty dr \int_{z_{\mathrm{UV}}}^{z_{\mathrm{IR}}} dz\, \epsilon^{\bar\mu\bar\nu} \left[ \partial_{\bar\mu}(-i\phi^* D_{\bar\nu}\phi + h.c.) + A_{\bar\mu\bar\nu} \right],
\tag{18.3.26}
$$

where $x^{\bar\mu} = \{r,z\}$, $A_{\bar\mu} = \{A_1, A_2\}$, $A_{\bar\mu\bar\nu}$ its field-strenght, $\phi = \phi_1 + i\phi_2$ and the covariant derivative will be defined in Eq. (18.3.35). The charge can be written, as it should, as an integral over the 1D boundary of the 2D space. Finite-energy regular solutions with $B = 1$ which obey Eqs. (18.2.3) and (18.2.4) must respect the following boundary conditions:

$$
z = z_{\mathrm{IR}} : \quad
\begin{cases}
\phi_1 = 0 \\
\partial_2\phi_2 = 0 \\
A_1 = 0 \\
\partial_2 s = 0
\end{cases},
\qquad
z = z_{\mathrm{UV}} : \quad
\begin{cases}
\phi_1 = 0 \\
\phi_2 = -1 \\
A_1 = 0 \\
s = 0
\end{cases},
\tag{18.3.27}
$$

and

$$
r = 0 : \quad
\begin{cases}
\phi_1/r \to A_1 \\
(1+\phi_2)/r \to 0 \\
A_2 = 0 \\
s = 0
\end{cases}
\qquad
r = \infty : \quad
\begin{cases}
\phi = -ie^{i\pi z/L} \\
A_2 = \frac{\pi}{L} \\
s = 0
\end{cases}.
\tag{18.3.28}
$$

Solutions of the EOM with the required boundary conditions exist, and have been obtained numerically in Ref. 13 using the COMSOL package[28] (see Appendix for details). The 2D energy density of this solution is given in Fig. 18.1.

### 18.3.3. *Zero-mode fluctuations*

Let us now consider time-dependent infinitesimal deformations of the static solutions. Among these, the zero-mode (*i.e.* zero frequency) fluctuations are particularly important as they will describe single-baryon states. Zero-modes can be defined as directions in the field space in which uniform and slow motion is permitted by the classical dynamics and they are associated with the global symmetries of the problem, which are in our case $U(2)_V$ and 3-space rotations plus 3-space translations. The latter would describe baryons moving with uniform velocity and therefore can be ignored in the computation of static properties like the form factors. Of course, the global $U(1)_V$ acts trivially on all our fields and the global $SU(2)_V$ has the same effect as 3-space rotations on the static solution (18.3.24) because of the cylindrical symmetry. The space of static solutions which are of interest for us is therefore parametrized by 3 real coordinates –denoted as collective coordinates– which define an $SU(2)$ matrix $U$.

To construct zero-modes fluctuations we consider collective coordinates with general time dependence, *i.e.* we perform a global $SU(2)_V$ transformation on the static solution

$$R_{\hat{\mu}}(x,z;U) = U\,\overline{R}_{\hat{\mu}}(x,z)\,U^\dagger\,, \qquad \widehat{R}_0(x,z;U) = \widehat{\overline{R}}_0(x,z)\,, \tag{18.3.29}$$

but we allow $U = U(t)$ to depend on time. It is only for constant $U$ that Eq. (18.3.29) is a solution of the time-dependent EOM. For infinitesimal but non-zero rotational velocity

$$K = k_a \sigma^a/2 = -iU^\dagger dU/dt\,,$$

Eq. (18.3.29) becomes an infinitesimal deformation of the static solution. Along the zero-mode direction uniform and slow motion is classically allowed, for this reason our fluctuations should fulfill the time-dependent EOM at linear order in $K$ provided that $dK/dt = 0$.

From the action (18.2.9) and (18.2.10) the following EOM are derived

$$\begin{cases} D_{\hat{\nu}}\left(a(z)R^{\hat{\nu}}_0\right) + \dfrac{\gamma\alpha L}{4}\epsilon^{\hat{\nu}\hat{\omega}\hat{\rho}\hat{\sigma}} R_{\hat{\nu}\hat{\omega}}\widehat{R}_{\hat{\rho}\hat{\sigma}} = 0 \\[2mm] \alpha\partial_{\hat{\nu}}\left(a(z)\widehat{R}^{\hat{\nu}}{}_0\right) + \dfrac{\gamma L}{4}\epsilon^{\hat{\nu}\hat{\omega}\hat{\rho}\hat{\sigma}}\left[\mathrm{Tr}\left(R_{\hat{\nu}\hat{\omega}}R_{\hat{\rho}\hat{\sigma}}\right) + \dfrac{1}{2}\widehat{R}_{\hat{\nu}\hat{\omega}}\widehat{R}_{\hat{\rho}\hat{\sigma}}\right] = 0 \\[2mm] D_{\hat{\nu}}\left(a(z)R^{\hat{\nu}\hat{\mu}}\right) - a(z)D_0 R_0{}^{\hat{\mu}} - \dfrac{\gamma\alpha L}{2}\epsilon^{\hat{\mu}\hat{\nu}\hat{\rho}\hat{\sigma}}\left[R_{\hat{\nu}0}\widehat{R}_{\hat{\rho}\hat{\sigma}} + R_{\hat{\nu}\hat{\rho}}\widehat{R}_{\hat{\sigma}0}\right] = 0 \\[2mm] \alpha\partial_{\hat{\nu}}\left(a(z)\widehat{R}^{\hat{\nu}\hat{\mu}}\right) - \alpha a(z)\widehat{\partial}_0\widehat{R}_0{}^{\hat{\mu}} - \gamma L\epsilon^{\hat{\mu}\hat{\nu}\hat{\rho}\hat{\sigma}}\left[\mathrm{Tr}\left(R_{\hat{\nu}0}R_{\hat{\rho}\hat{\sigma}}\right) + \dfrac{1}{2}\widehat{R}_{\hat{\nu}0}\widehat{R}_{\hat{\rho}\hat{\sigma}}\right] = 0 \end{cases}$$
$$\tag{18.3.30}$$

We only need to specify the EOM for one chirality since we are considering, as explained in the previous section, a parity invariant Ansatz. We would like to find solutions of Eq. (18.3.30) for which $R_{\hat{\mu}}$ and $\widehat{R}_0$ are of the form (18.3.29); it is easy to see that the time-dependence of $U$ in Eq. (18.3.29) acts as a source for the components $R_0$ and $\widehat{R}_{\hat{\mu}}$, which therefore cannot be put to zero as in the static case. Notice that the same happens in the case of the 4D skyrmion,[4] in which the temporal and spatial components of the $\rho$ and $\omega$ mesons are turned on in the rotating skyrmion solution. Also, it can be shown that Eq. (18.3.30) can be solved, to linear order in $K$ and for $dK/dt = 0$, by the Ansatz in Eq. (18.3.29) if the fields $R_0$ and $\widehat{R}_{\hat{\mu}}$ are chosen to be linear in $K$. Even though $K$ must be constant for the EOM to be solved, it should be clear that this does not imply any constraint on the allowed form of the collective coordinate matrix $U(t)$ in Eq. (18.3.29), which can have an arbitrary dependence on time. What we actually want to do here is to find an appropriate functional dependence of the fields on $U(t)$ such that the time-dependent EOM would be solved if and only if the rotational velocity $K = -iU^\dagger dU/dt$ was constant.

In order to solve the time-dependent equations (18.3.30) we will consider a 2D Ansatz obtained by a generalization of the cylindrical symmetry of the static case. The Ansatz for $R_{\hat\mu}$ and $\widehat{R}_0$ is specified by Eq. (18.3.29) in which the static fields are given by Eq. (18.3.24). Due to the cylindrical symmetry of the static solution the fields in Eq. (18.3.29) are invariant under 3D space rotations $x_a\sigma^a \to \theta^\dagger x_a\sigma^a\theta$ combined with vector $SU(2)$ global transformations $L, R \to \theta\,(L,R)\,\theta^\dagger$ if $U$ also transforms as $U \to \theta^\dagger U\theta$. We are therefore led to consider a generalized cylindrical symmetry under which $k_a$ also rotates as the space coordinates do. Compatibly with this symmetry and with the fact that $R_0$ and $\widehat{R}_{\hat\mu}$ must be linear in $K$ we write the Ansatz as

$$R_0(x,z;U) = U\,\overline{R}_0(x,z;K)\,U^\dagger + i\,U\partial_0 U^\dagger, \qquad \widehat{R}_{\hat\mu}(x,z;U) = \widehat{\overline{R}}_{\hat\mu}(x,z;K),$$

$$(18.3.31)$$

where

$$
\begin{cases}
\overline{R}_0^a(x,z;K) = \chi_{(x)}(r,z)k_b\Delta^{(x),ab} + v(r,z)(k\cdot\widehat{x})\widehat{x}^a \\[2mm]
\alpha\widehat{\overline{R}}_i(x,z;K) = \dfrac{\rho(r,z)}{r}\left(k^i - (k\cdot\widehat{x})\widehat{x}^i\right) + B_1(r,z)(k\cdot\widehat{x})\widehat{x}^i + Q(r,z)\epsilon^{ibc}k_b\widehat{x}_c \\[2mm]
\alpha\widehat{\overline{R}}_5(x,z;K) = B_2(r,z)(k\cdot\widehat{x})
\end{cases}.
$$

$$(18.3.32)$$

It must be observed that our Ansatz has not fixed the 5D gauge freedom completely; its form is indeed preserved by chiral $SU(2)_{L,R}$ gauge transformations of the form $g_R = U(t)\cdot g\cdot U^\dagger(t)$ and $g_L = U(t)\cdot g^\dagger\cdot U^\dagger(t)$ with

$$g = \exp[i\alpha(r,z)x^a\sigma_a/(2r)],$$

$$(18.3.33)$$

under which the 2D fields $\phi_{(x)}$ and $\chi_{(x)}$ defined respectively in Eq. (18.3.24) and (18.3.32) transform as charged complex scalars. The fields $A_{\hat\mu}$ transform as gauge fields. There is also a second residual $U(1)$ associated with chiral $U(1)_{L,R}$ 5D transformations of the form $\widehat{g}_R = \widehat{g}$ and $\widehat{g}_L = \widehat{g}^\dagger$ with

$$\widehat{g} = \exp\left[i\beta(r,z)\frac{(k\cdot\widehat{x})}{\alpha}\right].$$

$$(18.3.34)$$

Under this second residual $U(1)$ only $B_{\bar\mu} = \{B_1, B_2\}$ and $\rho$ transform non trivially; $B_{\bar\mu}$ is a gauge field and $\rho$ a Goldstone. In order to make manifest the residual gauge invariance of the observables we will compute we introduce gauge covariant derivatives for the $\phi$, $\chi$ and $\rho$ fields

$$
\begin{cases}
(D_{\bar\mu}\phi)_{(x)} = \partial_{\bar\mu}\phi_{(x)} + \epsilon^{(xy)}A_{\bar\mu}\phi_{(y)} \\[2mm]
(D_{\bar\mu}\chi)_{(x)} = \partial_{\bar\mu}\chi_{(x)} + \epsilon^{(xy)}A_{\bar\mu}\chi_{(y)} \\[2mm]
D_{\bar\mu}\rho = \partial_{\bar\mu}\rho - B_{\bar\mu}
\end{cases}.
$$

$$(18.3.35)$$

At this point it is straightforward to find the zero-mode solution. The EOM for the 2D fields can be obtained by plugging the Ansatz in Eq. (18.3.30), while the conditions at the IR and UV boundaries are derived from Eq. (18.2.3) and (18.2.4), respectively. The boundary conditions at $r = 0$ are obtained by imposing the regularity of the Ansatz, while those for $r \to \infty$ come from requiring the energy of the solution to be finite and $B = 1$. Also in this case, numerical solutions can be obtained with the methods discussed in the appendix. The reader not interested in detail can simply accept that a solution of Eq. (18.3.30) exists and is given by our Ansatz for some particular functional form of the 2D fields which we are able to determine numerically. In the rest of the paper the 2D fields will always denote this numerical solution of the 2D equations.

### 18.3.4. *The Lagrangian of collective coordinates*

The collective coordinate matrix $U(t)$ will be associated with static baryons. The classical dynamics of the collective coordinates is obtained by plugging Eqs. (18.3.29) and (18.3.31) in the 5D action. One finds $S[U] = \int dt L$ where

$$L = -M + \frac{\lambda}{2} k_a k^a \,. \tag{18.3.36}$$

The mass $M$ and the moment of inertia $\lambda$ are given respectively by

$$M = 8\pi M_5 \int_0^\infty dr \int_{z_{\rm UV}}^{z_{\rm IR}} dz \left\{ a(z) \left[ |D_{\bar\mu}\phi|^2 + \frac{1}{4} r^2 A_{\bar\mu\bar\nu}^2 + \frac{1}{2r^2} \left(1 - |\phi|^2\right)^2 - \frac{1}{2} \left(\partial_{\bar\mu}s\right)^2 \right] \right.$$
$$\left. - \frac{\gamma L}{2} \frac{s}{r} \epsilon^{\bar\mu\bar\nu} \left[ \partial_{\bar\mu}(-i\phi^* D_{\bar\nu}\phi + h.c.) + A_{\bar\mu\bar\nu} \right] \right\} \,, \tag{18.3.37}$$

and

$$\lambda = 16\pi M_5 \frac{1}{3} \int_0^\infty dr \int_{z_{\rm UV}}^{z_{\rm IR}} dz \left\{ a(z) \left[ -(D_{\bar\mu}\rho)^2 - r^2 \left(\partial_{\bar\mu}Q\right)^2 - 2Q^2 - \frac{r^2}{4} B_{\bar\mu\bar\nu}B_{\bar\mu\bar\nu} \right. \right.$$
$$\left. + r^2 (D_{\bar\mu}\chi)^2 + \frac{r^2}{2} \left(\partial_{\bar\mu}v\right)^2 + \left(\chi_{(x)}\chi_{(x)} + v^2\right) \left(1 + \phi_{(x)}\phi_{(x)}\right) - 4v\phi_{(x)}\chi_{(x)} \right]$$
$$+ \gamma L \left[ -2\epsilon^{\bar\mu\bar\nu} D_{\bar\mu}\rho \, \chi_{(x)} \left(D_{\bar\nu}\phi\right)_{(x)} + 2\epsilon^{\bar\mu\bar\nu} \partial_{\bar\mu}\left(rQ\right) \chi_{(x)} \epsilon^{(xy)} \left(D_{\bar\nu}\phi\right)_{(y)} \right.$$
$$\left. \left. - v \left( \frac{1}{2} \epsilon^{\bar\mu\bar\nu} B_{\bar\mu\bar\nu} \left(\phi_{(x)}\phi_{(x)} - 1\right) + r Q \epsilon^{\bar\mu\bar\nu} A_{\bar\mu\bar\nu} \right) + \frac{2rQ}{\alpha^2} \epsilon^{\bar\mu\bar\nu} D_{\bar\mu}\rho \partial_{\bar\nu}\left(\frac{s}{r}\right) \right] \right\} \,. \tag{18.3.38}$$

The numerical values of $M$ and $\lambda$ are easily computed, once the numerical solution for the 2D fields is known. Using the best-fit values of the parameters we find $M = 1132$ MeV and $1/\lambda = 227$ MeV.

Let us give some more detail on this theory. For now we proceed at the classical level and we will discuss the quantization in the next section. Our lagrangian can be rewritten as

$$L = -M + \lambda \text{Tr} \left[ \dot{U}^\dagger \dot{U} \right] = -M + 2\lambda \sum_i \dot{u}_i^2 \,, \tag{18.3.39}$$

where we have parametrized the collective coordinates matrix $U$ as $U = u_0 \mathbb{1} + i\, u_i \sigma^i$, with $\sum_i u_i^2 = 1$. The lagrangian (18.3.39) is the one of the classical spherical rigid rotor. The variables $\{u_0, u_i\}$ are restricted to the unitary sphere $S^3$, which is conveniently parametrized by the coordinates $q^\alpha \equiv \{x, \phi_1 \phi_2\}$ –which run in the $x \in [-1, 1]$, $\phi_1 \in [0, 2\pi)$ and $\phi_2 \in [0, 2\pi)$ domains– as

$$u_1 + i\, u_2 \equiv z_1 = \sqrt{\frac{1-x}{2}}\, e^{i\,\phi_1}\,, \qquad u_0 + i\, u_3 \equiv z_2 = \sqrt{\frac{1+x}{2}}\, e^{i\,\phi_2}\,, \qquad (18.3.40)$$

where we also introduced the two complex coordinates $z_{1,2}$. We can now rewrite the Lagrangian as

$$L = -M + 2\lambda\, g_{\alpha\beta}\dot{q}^\alpha\dot{q}^\beta\,, \qquad (18.3.41)$$

where $g$ is the metric of $S^3$ which reads in our coordinates

$$ds^2 = g_{\alpha\beta}dq^\alpha dq^\beta = \frac{1}{4}\frac{1}{1-x^2}\, dx^2 + \frac{1-x}{2}\, d\phi_1{}^2 + \frac{1+x}{2}\, d\phi_2{}^2\,. \qquad (18.3.42)$$

The conjugate momenta are $p_\alpha = \partial L/\partial \dot{q}^\alpha = 4\lambda g_{\alpha\beta}\dot{q}^\beta$ and therefore the classical Hamiltonian is

$$H_c = M + \frac{1}{8\lambda}p_\alpha g^{\alpha\beta}(q)p_\beta\,. \qquad (18.3.43)$$

It should be noted that the points $U$ and $-U$ in what we denoted as the space of collective coordinates actually describe the same field configuration (see Eqs. (18.3.29) and (18.3.31)). The $SU(2) = S_3$ manifold we are considering is actually the universal covering of the collective coordinate space which is given by $S_3/Z_2$. This will be relevant when we will discuss the quantization.

### 18.3.5. *Skyrmion quantization*

We should now quantize the classical theory described above, by replacing as usual the classical momenta $p_\alpha$ with the differential operator $-i\partial/\partial q^\alpha$ acting on the wave functions $f(q)$. Given that the metric depends on $q$, however, there is an ambiguity in how to extract the quantum hamiltonian $H_q$ from the classical one in Eq. (18.3.43). This ambiguity is resolved by requiring the quantum theory to have the same symmetries that the classical one had. At the classical level, we have an $SO(4) \simeq SU(2) \times SU(2)$ symmetry under $U \to U \cdot \theta^\dagger$ and $U \to g \cdot U$ with $\theta, g \in SU(2)$. These correspond, respectively, to rotations in space and to isospin (*i.e.* global vector) transformations, as one can see from the Ansatz in Eqs. (18.3.29) and (18.3.31). This is because $K$ is invariant under left multiplication by $g$, and that the Ansatz is left unchanged by performing a rotation $x_a\sigma^a \to \theta^\dagger x_a\sigma^a\theta$ and simultaneously sending $U \to U \cdot \theta$. The spin and isospin operators must be given, in the quantum theory, by the generators of these transformations on the space of wave functions $f(q)$ which are defined by

$$[S^a, U] = U\sigma^a/(2)\,, \qquad [I^a, U] = -\sigma^a/(2)U\,. \qquad (18.3.44)$$

After a straightforward calculation one finds

$$
\begin{cases}
S^3 = -\dfrac{i}{2}\left(\partial_{\phi_1} + \partial_{\phi_2}\right) \\[2mm]
S^+ = \dfrac{1}{\sqrt{2}}e^{i(\phi_1+\phi_2)}\left[i\sqrt{1-x^2}\,\partial_x + \dfrac{1}{2}\sqrt{\dfrac{1+x}{1-x}}\,\partial_{\phi_1} - \dfrac{1}{2}\sqrt{\dfrac{1-x}{1+x}}\,\partial_{\phi_2}\right] \\[2mm]
S^- = \dfrac{1}{\sqrt{2}}e^{-i(\phi_1+\phi_2)}\left[i\sqrt{1-x^2}\,\partial_x - \dfrac{1}{2}\sqrt{\dfrac{1+x}{1-x}}\,\partial_{\phi_1} + \dfrac{1}{2}\sqrt{\dfrac{1-x}{1+x}}\,\partial_{\phi_2}\right]
\end{cases}
$$

$$
\begin{cases}
I^3 = -\dfrac{i}{2}\left(\partial_{\phi_1} - \partial_{\phi_2}\right) \\[2mm]
I^+ = -\dfrac{1}{\sqrt{2}}e^{i(\phi_1-\phi_2)}\left[i\sqrt{1-x^2}\,\partial_x + \dfrac{1}{2}\sqrt{\dfrac{1+x}{1-x}}\,\partial_{\phi_1} + \dfrac{1}{2}\sqrt{\dfrac{1-x}{1+x}}\,\partial_{\phi_2}\right] \\[2mm]
I^- = -\dfrac{1}{\sqrt{2}}e^{-i(\phi_1-\phi_2)}\left[i\sqrt{1-x^2}\,\partial_x - \dfrac{1}{2}\sqrt{\dfrac{1+x}{1-x}}\,\partial_{\phi_1} - \dfrac{1}{2}\sqrt{\dfrac{1-x}{1+x}}\,\partial_{\phi_2}\right]
\end{cases}
\tag{18.3.45}
$$

where the raising/lowering combinations are $S^{\pm} = (S^1 \pm iS^2)/\sqrt{2}$.

The operators in Eq. (18.3.45) should obey the Hermiticity conditions $\left(S^3\right)^{\dagger} = S^3$, $\left(S^+\right)^{\dagger} = S^-$, and analogously for the isospin. In order for the Hermiticity conditions to hold we choose the scalar product to be

$$
\langle A|B\rangle \equiv \int d^3q\,\sqrt{g}\,f_A^{\dagger}(q)f_B(q)\,,
\tag{18.3.46}
$$

where $\sqrt{g} = 1/4$ in our parametrization of $S_3$. The reason why this choice of the scalar product gives the correct Hermiticity conditions is that $S^a$ and $I^a$ (where $a = 1, 2, 3$) can be written as $X^{\alpha}\partial_{\alpha}$ with $X^{\alpha}$ Killing vectors of the appropriate $S_3$ isometries. The Killing equation $\nabla_{\alpha}X_{\beta} + \nabla_{\beta}X_{\alpha} = 0$ ensures the generators to be Hermitian with respect to the scalar product (18.3.46).

Knowing that the scalar product must be given by Eq. (18.3.46) greatly helps in guessing what the quantum Hamiltonian, which has to be Hermitian, should be. We can multiply and divide by $\sqrt{g}$ the kinetic term of $H_c$ and move one $\sqrt{g}$ factor to the left of $p_{\alpha}$. Then we apply the quantization rules and find [f]

$$
H_q = M - \frac{1}{8\lambda}\frac{1}{\sqrt{g}}\partial_{\alpha}\left(\sqrt{g}g^{\alpha\beta}\partial_{\beta}\right) = M - \frac{1}{8\lambda}\nabla_{\alpha}\nabla^{\alpha}\,,
\tag{18.3.47}
$$

which is clearly Hermitian. We can immediately show that $H_q$ commutes with spin and isospin, so that the quantum theory is really symmetric as required: a straightforward calculation gives indeed

$$
H_q = M + \frac{1}{2\lambda}S^2 = M + \frac{1}{2\lambda}I^2\,.
\tag{18.3.48}
$$

It would not be difficult to solve the eigenvalue problem for the Hamiltonian (18.3.47), but in order to find the nucleon wave functions it is enough to note that

---

[f]The last equality holds because $H_q$ is supposed to be acting on the wave functions, which are scalar functions.

the versor of $n$-dimensional Euclidean space provides the $n$ representation of the $SO(n)$ isometry group. In our case, $n = 4 = (2, 2)$, which is exactly the spin/isospin representation in which nucleons live. It is immediately seen that $z_1$, as defined in Eq. (18.3.40), has $S^3 = I^3 = 1/2$. Acting with the lowering operators we easily find the wave functions

$$|p \uparrow\rangle = \frac{1}{\pi} z_1 \,, \qquad |n \uparrow\rangle = \frac{i}{\pi} z_2 \,,$$
$$|p \downarrow\rangle = -\frac{i}{\pi} \bar{z}_2 \,, \qquad |n \downarrow\rangle = -\frac{1}{\pi} \bar{z}_1 \,, \tag{18.3.49}$$

which are of course normalized with the scalar product (18.3.46). The mass of the nucleons is therefore $E = M + 3/(8\lambda)$.

Notice that the nucleon wave functions are odd under $U \to -U$, meaning that they are double-valued on the genuine collective coordinate space $S_3/Z_2$. This corresponds, following,[29] to quantize the skyrmion as a fermion and explains how we could get spin-$1/2$ states after a seemingly bosonic quantization without violating spin-statistics.

Let us now summarize some useful identities which will be used in our calculation. First of all, it is not hard to check that, after the quantization is performed the rotational velocity becomes

$$k^a = -i \operatorname{Tr}\left[U^\dagger \dot{U} \sigma^a\right] = \frac{1}{\lambda} S^a \,, \tag{18.3.50}$$

and analogously

$$i \operatorname{Tr}\left[\dot{U} U^\dagger \sigma^a\right] = \frac{1}{\lambda} I^a \,. \tag{18.3.51}$$

Other identities which we will use in our calculations are

$$\langle \operatorname{Tr}\left[U \sigma^b U^\dagger \sigma^a\right] = -\frac{8}{3} S^b I^a \rangle \,,$$
$$\langle \operatorname{Tr}\left[U \sigma^b \hat{x}_b (k \cdot \hat{x}) U^\dagger \sigma^a\right] = -\frac{2}{3\lambda} I^a \rangle \,, \tag{18.3.52}$$

where the VEV symbols $\langle ... \rangle$ mean that those are not operatorial identities, but they only hold when the operators act on the subspace of nucleon states. Notice that the second equation in (18.3.52) is implied by the first one if one also uses the commutation relation (18.3.44), Eq. (18.3.51) and the fact that, on nucleon states, $\langle \{S^a, S^i\} = \delta^{ai}/2 \rangle$.

### 18.3.6. *The nucleon form factors*

The nucleon form factors parametrize the matrix element of the currents on two nucleon states. For the isoscalar and isovector currents we have

$$\langle N_f(p')|J_S^\mu(0)|N_i(p)\rangle = \bar{u}_f(p') \left[ F_1^S(q^2)\gamma^\mu + \frac{i F_2^S(q^2)}{2M_N}\sigma^{\mu\nu} q_\nu \right] u_i(p),$$

$$\langle N_f(p')|J_V^{\mu a}(0)|N_i(p)\rangle = \bar{u}_f(p') \left[ F_1^V(q^2)\gamma^\mu + \frac{i F_2^V(q^2)}{2M_N}\sigma^{\mu\nu} q_\nu \right] (2I^a) \, u_i(p), \tag{18.3.53}$$

where the currents are defined as $J_V^a = J_R^a + J_L^a$ and $J_S = 1/3\left(\widehat{J}_R + \widehat{J}_L\right)$ in terms of the chiral ones. In the equation above $q \equiv p' - p$ is the 4-momentum transfer, $N_i$ and $N_f$ are the initial and final nucleon states and $u_i(p)$, $\bar{u}_f(p')$ their wave functions, $I^a = \sigma^a/2$ is the isospin generators and $\sigma^{\mu\nu} \equiv i/2[\gamma^\mu, \gamma^\nu]$. For the axial current $J_A^a = J_R^a - J_L^a$ we have

$$\langle N_f(p')|J_{A\mu}^a(0)|N_i(p)\rangle = \bar{u}_f(p')G_A(q^2)\left[\gamma_\mu - \frac{2M_N}{q^2}q^\mu\right]\gamma^5 I^a u_f(p)\,. \qquad (18.3.54)$$

Exact axial and isospin symmetries, which hold in our model, have been assumed in the definitions above.

In our non-relativistic model the current correlators will be computed in the Breit frame in which the initial nucleon has 3-momentum $-\vec{q}/2$ and the final $+\vec{q}/2$ (i.e. $p^\mu = (E, -\vec{q}/2)$ and $p'^\mu = (E, \vec{q}/2)$, and $q^2 = -\vec{q}^2$, with $E = \sqrt{M_N^2 + \vec{q}^2/4}$). Notice that the textbook definitions in Eqs. (18.3.53) and (18.3.54) involve nucleon states which are normalized with $\sqrt{2E}$; in order to match with our non-relativistic normalization we have to divide all correlators by $2M_N$. The vector currents become

$$\langle N_f(\vec{q}/2)|J_S^0(0)|N_i(-\vec{q}/2)\rangle = G_E^S(\vec{q}^2)\chi_f^\dagger \chi_i\,,$$

$$\langle N_f(\vec{q}/2)|J_S^i(0)|N_i(-\vec{q}/2)\rangle = i\frac{G_M^S(\vec{q}^2)}{2M_N}\chi_f^\dagger 2(\vec{S}\times\vec{q})^i\chi_i\,,$$

$$\langle N_f(\vec{q}/2)|J_V^{0a}(0)|N_i(-\vec{q}/2)\rangle = G_E^V(\vec{q}^2)\chi_f^\dagger(2I^a)\chi_i\,,$$

$$\langle N_f(\vec{q}/2)|J_V^{ia}(0)|N_i(-\vec{q}/2)\rangle = i\frac{G_M^V(\vec{q}^2)}{2M_N}\chi_f^\dagger 2(\vec{S}\times\vec{q})^i(2I^a)\chi_i\,, \qquad (18.3.55)$$

where we defined

$$G_E^{S,V}(-q^2) = F_1^{S,V}(q^2) + \frac{q^2}{4M_N^2}F_2^{S,V}(q^2)\,, \qquad G_M^{S,V}(-q^2) = F_1^{S,V}(q^2) + F_2^{S,V}(q^2)\,, \qquad (18.3.56)$$

and used the definition $(\vec{S}\times\vec{q})^i \equiv \varepsilon^{ijk}S^j q^k$. The nucleon spin/isospin vectors of state $\chi_{i,f}$ are normalized to $\chi^\dagger\chi = 1$. For the axial current we find

$$\langle N_f(\vec{q}/2)|J_A^{i,a}(0)|N_i(-\vec{q}/2)\rangle = \chi_f^\dagger\frac{E}{M_N}G_A(\vec{q}^2)2S_T^i I^a\chi_i\,,$$

$$\langle N_f(\vec{q}/2)|J_A^{0,a}(0)|N_i(-\vec{q}/2)\rangle = 0 \qquad (18.3.57)$$

where $\vec{S}_T \equiv \vec{S} - \hat{q}\,\vec{S}\cdot\hat{q}$ is the transverse component of the spin operator.

It is straightforward to compute the matrix elements of the currents in position space on static nucleon states. Plugging the Ansatz (18.3.24), (18.3.29), (18.3.32) and (18.3.31) in the definition of the currents (18.2.13) and performing the quantization one obtains quantum mechanical operators acting on the nucleons. The matrix elements are easily computed using the results of sect. 3.1. We finally obtain the form factors by taking the Fourier transform and comparing with Eqs. (18.3.55) and (18.3.57). We have[g]

$$
\begin{aligned}
G_E^S &= -\frac{N_c}{6\pi\gamma L} \int dr\, r\, j_0(qr)\, (a(z)\partial_z s)_{UV} \\
G_E^V &= \frac{4\pi M_5}{3\lambda} \int dr\, r^2\, j_0(qr) \left[ a(z)\left(\partial_z v - 2\,(D_z\chi)_{(2)}\right)\right]_{UV} \\
G_M^S &= \frac{8\pi M_N M_5 \alpha}{3\lambda} \int dr\, r^3\, \frac{j_1(qr)}{qr}\, (a(z)\partial_z Q)_{UV} \\
G_M^V &= \frac{M_N N_c}{3\pi L\gamma\alpha} \int dr\, r^2\, \frac{j_1(qr)}{qr} \left( a(z)\,(D_z\phi)_{(2)}\right)_{UV} \\
G_A &= \frac{N_c}{3\pi\alpha\gamma L} \int dr\, r \left[ a(z)\frac{j_1(qr)}{qr}\left((D_z\phi)_{(1)} - r\,A_{zr}\right) - a(z)\,(D_z\phi)_{(1)}\, j_0(qr)\right]_{UV}
\end{aligned}
$$

$$(18.3.58)$$

where $j_n$ are spherical Bessel functions which arise because of the Fourier transform.

## 18.4. Properties of Baryons: Results

In this section we will present our results. After discussing some qualitative features, such as the large-$N_c$ scaling of the form factors and the divergences of the isovector radii due to exact chiral symmetry, we extrapolate to the physically relevant case of $N_c = 3$ and perform a quantitative comparison with the experimental data. Consistently with our working hypothesis that the 5D model really describes large-$N_c$ QCD we find a 30% relative discrepancy.

*Large-$N_c$ scaling*

As explained in sect. 2.1, all the three parameters $\alpha$, $\gamma$ and $L$ of our 5D model should scale like $N_c^0$, Eq. (18.2.17), in order for the large-$N_c$ scaling of meson couplings and masses to be correctly reproduced. This implies the following scaling for the baryon observables. First, we notice that the solitonic solution is independent of $N_c$ given that $M_5$ factorizes out of the action and does not appear in the EOM. This implies that the radii of the soliton does not scale with $N_c$, while the classical mass $M$ and the moment of inertia $\lambda$ scale like $N_c$. Using this we can read the $N_c$-scaling of the electric and magnetic form factors from Eq. (18.3.58):

$$G_E^S \sim N_c\,, \quad G_E^V \sim N_c^0\,, \quad \frac{G_M^S}{M_N} \sim N_c^0\,, \quad \frac{G_M^V}{M_N} \sim N_c\,. \tag{18.4.59}$$

In large-$N_c$ QCD the baryon masses scale like $N_c$,[31] as in our model. The matrix elements of the currents on nucleon states are also expected to scale like $N_c$, even

---

[g]It is quite intuitive that the form factors can be computed in this way. Given that solitons are infinitely heavy at small coupling, in the Breit frame they are almost static during the process of scattering with the current. To check this, however, we should perform the quantization of the collective coordinates associated with the center-of-mass motion, as it was done in Ref. 30 for the original 4D Skyrme model.

though cancellations are possible.[32] The radii, therefore, must scale like $N_c^0$ as we find and, looking at the definition (18.3.55), $G_E^{S,V}$ and $G_M^{S,V}/M_N$ should both scale like $N_c$ up to cancellations. It is very simple to understand why, both in QCD and in our model, there must be a cancellation in $G_E^V$. Remembering that the temporal component of the current at zero momentum gives the conserved charge and looking at the definitions (18.3.55), one immediately obtains $G_E^V(0) = 1/2$ because the skyrmion, as the nucleon, is in the $1/2$ representation of isospin. This condition is respected by our model as it is implied by the EOM, and fulfilled to great accuracy (0.1%) by the numerical solution. Similarly we find at zero momentum $G_E^S(0) = N_c/6$ as required for a bound-state made of $N_c$ quarks of $U(1)_V$ charge $1/6$ each (in our conventions). Also this condition is implied by the EOM and verified by the numerical solution.

Concerning the second cancellation, *i.e.* $G_M^S/M_N \sim N_c^0$, we are not able to prove that it must take place in large-$N_c$ QCD as it does in our model. We can, however, check that it occurs in the naive quark model, or better in its generalization for arbitrary odd $N_c = 2k + 1$.[33] In this non-relativistic model the nucleon wave function is made of $2k + 1$ quark states $q_i$, $2k$ of which are collected into $k$ bilinear spin/isospin singlets while the last one has free indices which give to the nucleon its spin/isospin quantum numbers. Of course, the wave function is symmetrized in flavor and spin given that the color indices are contracted with the antisymmetric tensor and the spatial wave function is assumed to be symmetric. The current operator is the sum of the currents for the $2k + 1$ quarks, each of which will assume by symmetry the same form as in Eq. (18.3.55). If $S_{1,2}$ and $I_{1,2}$ represent the spin and isospin operators on the quarks $q_{1,2}$ the operators $S_1 + S_2$ and $I_1 + I_2$ will vanish on the singlet combination of the two quarks, but $S_1 I_1 + S_2 I_2$ will not. The $k$ singlets will therefore only contribute to $G_E^S$, $G_M^V$ and $G_A$, which will have the naive scaling, while for the others we find cancellations.

A detailed calculation can be found in[34] where, among other things, the proton and neutron magnetic moments and the axial coupling are computed in the naive quark model. The magnetic moments are related to the form factor at zero momentum as $\mu_V/\mu_N = G_M^V(0)$ and $\mu_S/\mu_N = G_M^S(0)$ where $\mu_N = 1/(2M_N)$ is the nuclear magneton and $2\mu_V = \mu_p - \mu_n$, $2\mu_S = \mu_p + \mu_n$. In accordance with the previous discussion, the results in the naive quark model are $2\mu_S = \mu_u + \mu_d$ and $2\mu_V = 2k/3(\mu_u - \mu_d)$, where $\mu_{u,d}$ are the quark magnetic moments, while for the axial coupling one finds $g_A = G_A(0) = 2k/3 + 1$ which scales like $N_c$ as expected:

$$g_A = \frac{N_c}{3} + \frac{2}{3}. \tag{18.4.60}$$

Notice that for $N_c = 3$ the subleading term in the $1/N_c$-expansion represents a 60% correction. We have of course no reason to believe that such big corrections should persist in the true large-$N_c$ QCD; this remark simply suggests that "large" $1/N_c$ corrections to the form factors are not excluded.

### Divergences in the chiral limit

It is well known that in QCD the isovector electric $\langle r^2_{E,V} \rangle$ and magnetic $\langle r^2_{M,V} \rangle$ radii which are proportional, respectively, to the $q^2$ derivative of $G^V_E$ and $G^V_M$ at zero momentum, diverge in the chiral limit.[35] In our model, as in the Skyrme model, divergences in the integrals of Eq. (18.3.58) which define the form factors are due, as in QCD, to the massless pions. If all the fields were massive, indeed, any solution to the EOM would fall down exponentially at large $r$ while in the present case power-like behaviors can appear. These power-like terms in the large-$r$ expansion of the solution can be derived analytically by performing a Taylor expansion of the fields around infinity $(1/r = 0)$, substituting into the EOM and solving order by order in $1/r$. The exponentially suppressed part of the solution will never contribute to the expansion. This procedure allows us to determine the asymptotic expansion of the solution completely, up to an integration constant $\beta$. Substituting the expansion into the definitions of the form factors (18.3.58) one gets

$$
\begin{cases}
G^S_E \propto \beta^3 \displaystyle\int dr \, \frac{1}{r^7} j_0(qr) + \ldots \\[2mm]
G^V_E \propto \beta^2 \displaystyle\int dr \, \frac{1}{r^2} j_0(qr) + \ldots \\[2mm]
G^S_M \propto \beta^3 \displaystyle\int dr \, \frac{1}{r^5} \frac{j_1(qr)}{qr} + \ldots \\[2mm]
G^V_M \propto \beta^2 \displaystyle\int dr \, \frac{1}{r^2} \frac{j_1(qr)}{qr} + \ldots
\end{cases}
\tag{18.4.61}
$$

All the form factors are finite for any $q$, including $q = 0$. The electric and magnetic radii, however, are defined as

$$
\langle r^2_{E,M} \rangle = -\frac{6}{G_{E,M}(\vec{q}^2 = 0)} \left. \frac{dG_{E,M}(\vec{q}^2)}{d\vec{q}^2} \right|_{\vec{q}^2 = 0},
\tag{18.4.62}
$$

and taking a $q^2$ derivative of Eqs. (18.4.61) makes one more power of $r^2$ appear in the integral. It is easy to see that the scalar radii are finite, while the vector ones are divergent. For the axial form factor $G_A$ we find

$$
G_A \propto \int dr \left[ \left( \frac{3}{r}\beta - \frac{1}{r^5}\beta^3 \right) \frac{j_1(qr)}{qr} + \left( -\frac{1}{r}\beta + \frac{5}{7r^5}\beta^3 \right) j_0(qr) + \ldots \right],
\tag{18.4.63}
$$

The integral in Eq. (18.4.63) is convergent for any $q \neq 0$ but, however, it is not uniformly convergent for $q \to 0$. The leading $1/r$ term in Eq. (18.4.63) is indeed given by $I(q) = \beta \int_0^\infty dr \, (1/r) \, (3j_1(qr)/(qr) - j_0(qr))$, which is independent of $q$ and equal to $\beta/3$, while the argument of the integral vanishes for $q \to 0$ so that exchanging the limit and integral operations would give the wrong result $I(0) = 0$. To restore uniform convergence and obtain an analytic formula for $g_A$ one can subtract the $I(q)$ term from the expression in Eq. (18.3.58) for $G_A$ and replace it with $\beta/3$. Rewriting the axial form factor in this way is also useful to establish that the axial radius, which seems divergent if looking at Eq. (18.4.63), is on the contrary finite. The $I(q)$ term, indeed, does not contribute to the $q^2$ derivative and the ones which are left in Eq. (18.4.63) give a finite contribution.

We have found, compatibly with the QCD expectation, that all the form factors and radii are finite but the isovector ones. Notice that the structure of the divergences is completely determined by the asymptotic large-$r$ behaviour of the solution, and not by its detailed form (*i.e.*, for instance, by the actual value of the integration constant $\beta$ which depends on the entire solution). Our model coincides, in the IR, with the Skyrme model, therefore the asymptotic behaviour of the current densities is expected to be the same in the two cases. This explains why we obtained the same divergences as in the Skyrme model.

*Pion form factor and Goldberger-Treiman relation*

It is of some interest to define and compute the pion-nucleon form factor which parametrizes the matrix element on Nucleon states of the pion field. In the Breit frame (for normalized nucleon states) it is

$$\langle N_f(\vec{q}/2)|\pi^a(0)|N_i(-\vec{q}/2)\rangle = -\frac{i}{2M_N\vec{q}^2}G_{NN\pi}(\vec{q}^2)\chi_f^\dagger(2S^i)q_i(2I^a)\chi_i\,, \quad (18.4.64)$$

where $\pi^a(x)$ is the normalized and "canonical" pion field operator. The field is canonical in the sense that its quadratic effective lagrangian only contains the canonical kinetic term $\mathcal{L}_2 = 1/2(\partial\pi_a)^2$, or equivalently that its propagator is the canonical one, without a non-trivial form factor. With this definition, $G_{NN\pi}$ is the vertex form factor of the meson-exchange model for nucleon-nucleon interactions[36] and corresponds to an interaction[h]

$$\mathcal{L}_{NN\pi} = i\,(G_{NN\pi}(\Box)\pi_a)\overline{N}\gamma^\mu\gamma_5(2I^a)N\,. \quad (18.4.65)$$

On-shell, the form factor reduces to the pion-nucleon coupling constant, $G_{NN\pi}(0) = g_{NN\pi}$, whose experimental value is $g_{NN\pi} = 13.5 \pm 0.1$.

The pion field which matches the requirements above is given by the zero-mode of the KK decomposition. In the unitary gauge $\partial_z(a(z)A_5) = 0$, where $A_M \equiv (L_M - R_M)/2$, and for AdS$_5$ space, one has

$$A_5^{(un)}(x,z) = \frac{1}{F_\pi L}\frac{1}{a(z)}\pi^a(x)\sigma_a\,, \quad (18.4.66)$$

where $F_\pi$ is given in Eq. (18.2.14). Gauge-transforming back to the gauge in which our numerical solution is provided and using the Ansatz in Eqs. (18.3.24,18.3.29) we find the pion field

$$\pi^a = -\frac{F_\pi}{2}\int_{z_{\rm UV}}^{z_{\rm IR}} dz\,A_2(r,z)\hat{x}^b\mathrm{Tr}\left[U\sigma_b U^\dagger\sigma^a\right]\,. \quad (18.4.67)$$

---

[h]Nucleon scattering, in our model, is a soliton scattering process and we have no reason to believe that it can be described by meson-exchange, *i.e.* that contact terms are suppressed. Therefore, we will not attempt any comparison of our form factor with the one used in meson-exchange models.

Table 18.2. Prediction of the nucleon observables with the microscopic parameters fixed by a fit on the mesonic observables. The deviation from the empirical data is computed using the expression $(th - exp)/\min(|th|, |exp|)$, where $th$ and $exp$ denote, respectively, the prediction of our model and the experimental result.

| | Experiment | AdS$_5$ | Deviation |
|---|---|---|---|
| $M_N$ | 940 MeV | 1130 MeV | $+20\%$ |
| $\mu_S$ | 0.44 | 0.34 | $-30\%$ |
| $\mu_V$ | 2.35 | 1.79 | $-31\%$ |
| $g_A$ | 1.25 | 0.70 | $-79\%$ |
| $\sqrt{\langle r_{E,S}^2 \rangle}$ | 0.79 fm | 0.88 fm | $+11\%$ |
| $\sqrt{\langle r_{E,V}^2 \rangle}$ | 0.93 fm | $\infty$ | |
| $\sqrt{\langle r_{M,S}^2 \rangle}$ | 0.82 fm | 0.92 fm | $+12\%$ |
| $\sqrt{\langle r_{M,V}^2 \rangle}$ | 0.87 fm | $\infty$ | |
| $\sqrt{\langle r_A^2 \rangle}$ | 0.68 fm | 0.76 fm | $+12\%$ |
| $\mu_p/\mu_n$ | $-1.461$ | $-1.459$ | $+0.1\%$ |

Taking the matrix element of the above expression and comparing with Eq. (18.4.64) one obtains

$$G_{NN\tau}(q^2) = -\frac{8\pi}{3} M_N F_\pi q \int_0^\infty dr\, j_1(qr) \int dz\, r^2 A_2(r,z)\,. \qquad (18.4.68)$$

At $q \to 0$ the form factor $G_{NN\pi}$ is completely determined by the large-$r$ behavior of the field $A_2$, given by $A_2 \to \beta/r^2$. We then find

$$g_{NN\pi} = -\frac{32\pi}{3} M_N F_\pi \beta L^2\,. \qquad (18.4.69)$$

By using Eqs. (33,34) of Ref. 2 which show that also $g_A$ is determined by the asymptotic behavior of the axial current, one finds

$$g_A = -\frac{32\pi}{3} F_\pi^2 \beta L^2\,, \qquad (18.4.70)$$

that, together with Eq. (18.4.69), leads to the famous Goldberger-Treiman relation $F_\pi g_{\pi NN} = M_N g_A$. This relation, which is a consequence of having exact chiral symmetry, has been numerically verified to 0.01%.

*Comparison with experiments*

Let us now compare our results with real-world QCD. We therefore fix the number of colors $N_c = 3$ and choose our microscopic parameters to be those that gave the best fit to the mesonic quantities: $1/L \simeq 343$ MeV, $M_5 L \simeq 0.0165$ and $\alpha \simeq 0.94$ ($\gamma \simeq 1.23$). The numerical results of our analysis and the deviation with respect to the experimental data are reported in Table 18.2. We find a fair agreement with the experiments, a 36% total RMSE which is compatible with the expected size of $1/N_c$ corrections. The axial charge $g_A$ is the one which shows the larger (80%)

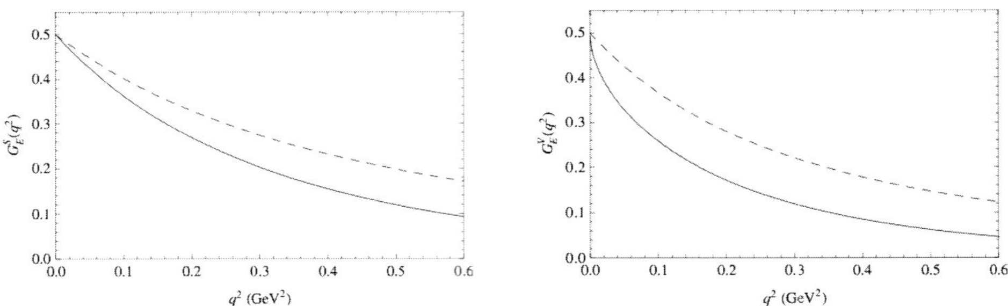

Fig. 18.2.   Scalar (left) and vector (right) electric form factors. We compare the results with the empirical dipole fit (dashed line).[4]

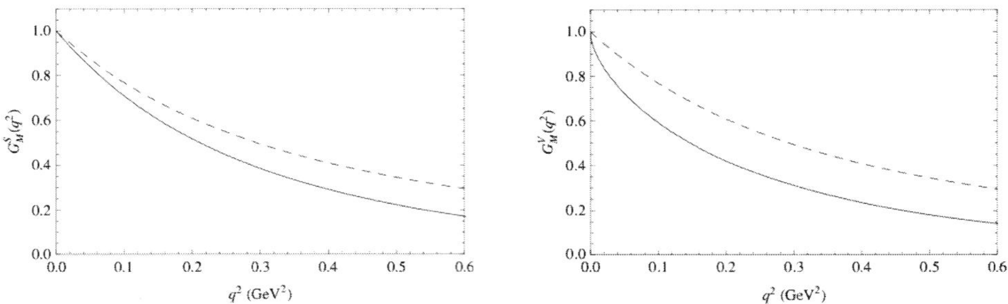

Fig. 18.3.   Normalized scalar (left) and vector (right) magnetic form factors. We compare the results with the empirical dipole fit (dashed line).[4]

deviation, and indeed removing this observable the RMSE decreases to 21%. We cannot exclude that, in a theory in which the naive expansion parameter is $1/3$, enhanced 80% corrections to few observables might appear at the next-to-leading order. Nevertheless, we think that this result could be very sensitive to the pion mass and therefore could be substantially improved in 5D models that incorporate explicit chiral breaking. The reason for this is that $g_A$ is strongly sensitive to the large-$r$ behavior of the solution (see the discussion following Eq. (18.4.63)) which is in turn heavily affected by the presence of the pion mass. Notice that a larger value, $g_A \simeq 0.99$, is obtained in the "complete" model described in Ref. 4, a model with similar features to our 5D scenario and which includes a nonzero pion mass. This expectation, however, fails in the original Skyrme model, where the addition of the pion mass does not affect $g_A$ significantly[37] and one finds $g_A \simeq 0.65$.

Table 18.2 also shows the proton-neutron magnetic moment ratio, $\mu_p/\mu_n$, which is in perfect agreement with the experimental value. This observable is the only one in the list that includes two orders of the $1/N_c$ expansion. Indeed, due to the scaling $\mu_V \sim N_c$ and $\mu_S \sim N_c^0$, we have $\mu_p/\mu_n = -(\mu_V + \mu_S)/(\mu_V - \mu_S) \simeq -1 - 2\mu_S/\mu_V$.

In Figs. 18.2, 18.3 and 18.4 we compare the normalized nucleon form factors at $q^2 \neq 0$ with the dipole fit of the experimental data. The shape of the scalar and

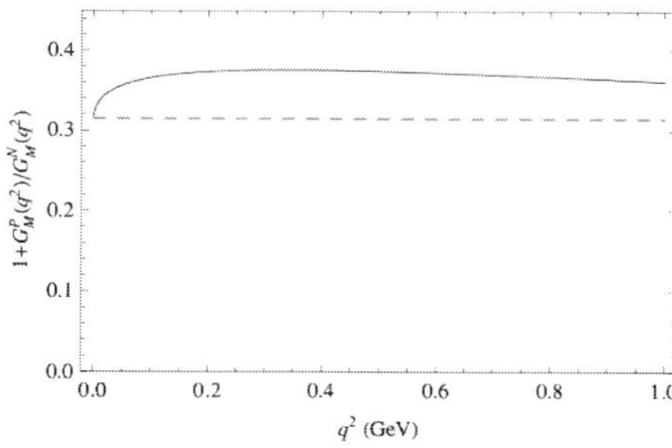
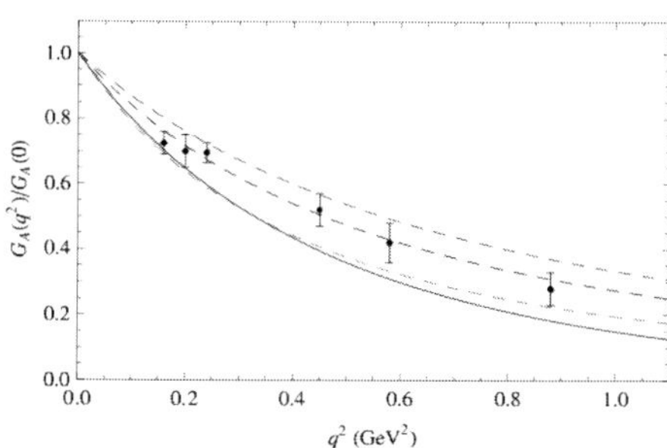

Fig. 18.4. Left: deviation of the ratio of proton and neutron magnetic form factors from the large $N_c$ value (solid line), compared with the dipole fit of the experimental data (dashed line). Right: normalized axial form factor (solid line) compared with the empirical dipole fit (dashed lines)[4] and with the experimental data taken from.[38,39]

axial form factors is of the dipole type, the discrepancy is mainly due to the error in the radii. The shape of vector form factors is of course not of the dipole type for small $q^2$, but this is due to the divergence of the derivative at $q^2 = 0$. Including the pion mass will for sure improve the situation given that it will render finite the slope at zero momentum; it would be interesting to see if the dipole shape of these form factors is recovered in the presence of the pion mass. We also plot in the left panel of Fig. 18.4 the deviation of ratio of the proton and neutron magnetic form factors from the large $N_c$ value which is given, due to the the different large-$N_c$ scaling of the isoscalar and isovector components, by $G_M^P(q)/G_M^N(q) = -1$. Not only do we find that this quantity is quite well predicted, with an error $\lesssim 15\%$, but also see that its shape, in agreement with observations, is nearly constant away from $q^2 = 0$. Also in this case corrections from the pion mass are expected to go in the right direction.

## 18.5. Conclusions and Outlook

We have shown that five-dimensional models, used to describe meson properties of QCD, can also be considered to study baryon physics. Baryons appear in these theories as soliton of sizes of order $1/m_\rho$ stabilized by the presence of the CS term. We have reviewed the procedure to calculate the static properties of the nucleons that have shown to be in reasonable good agreement with the experimental data. This shows, once again, that 5D models provide an alternative and very promising tool to study properties of QCD in certain regimes.

There are further issues that deserve to be analyzed. The most urgent one is the inclusion of a nonzero pion mass. As we have pointed out above, this will be crucial to calculate the isovector radii and, maybe, improve the prediction for $g_A$. For this purpose we need to use a 5D model along the lines of Refs. 10 and 11 where an explicit breaking of the chiral symmetry, corresponding to the quark masses, is introduced. We can also use this approach to study systems with high baryon densities, analyze possible phase transitions or study the properties of nuclear matter.

## Acknowledgments

A. Wulzer thanks G. Panico for the many useful discussions. The work of AP was partly supported by the Research Projects CICYT-FEDER-FPA2005-02211, SGR2005-00916 and "UniverseNet" (MRTN-CT-2006-035863).

## A.1. Numerical Methods

In this technical appendix we explain how the numerical determination of the soliton solution is performed.

### *Equations of motion and boundary conditions*

Let us first of all write down the EOM for the 2D fields which characterize our Ansatz in Eqs. (18.3.24), (18.3.31), (18.3.29) and (18.3.32). These can be obtained by plugging the Ansatz either directly in the 5D equations (18.3.30) or in the 5D action in Eqs. (18.2.9) and (18.2.10). In the second case one gets the 2D action specified by Eqs. (18.3.36), (18.3.37) and (18.3.38) and the EOM are obtained by performing the variation. In both cases one gets

$$
\left\{
\begin{aligned}
&D^{\bar{\mu}}\left(a(z)D_{\bar{\mu}}\phi\right) + \frac{a(z)}{r^2}\phi(1-|\phi|^2) + i\gamma L\epsilon^{\bar{\mu}\bar{\nu}}\partial_{\bar{\mu}}\left(\frac{s}{r}\right)D_{\bar{\nu}}\phi = 0 \\
&\partial^{\bar{\mu}}\left(r^2 a(z)A_{\bar{\mu}\bar{\nu}}\right) - a(z)\left(i\phi^{\dagger}D_{\bar{\nu}}\phi + h.c.\right) + \gamma L\epsilon^{\bar{\mu}\bar{\nu}}\partial_{\bar{\mu}}\left(\frac{s}{r}\right)(|\phi|^2-1) = 0 \quad , \qquad \text{(A.1)}\\
&\partial_{\bar{\mu}}\left(a(z)\partial^{\bar{\mu}}s\right) - \frac{\gamma L}{2r}\epsilon^{\bar{\mu}\bar{\nu}}\left[\partial_{\bar{\mu}}(-i\phi^{\dagger}D_{\bar{\nu}}\phi + h.c.) + A_{\bar{\mu}\bar{\nu}}\right] = 0
\end{aligned}
\right.
$$

for the fields which are already "turned on" in the static case. For the "new" fields which appear in the rotating skyrmion solution we have

$$
\left\{
\begin{aligned}
&\partial^{\bar{\mu}}(r^2 a(z)\partial_{\bar{\mu}}v) - 2a(z)\left[v(1+|\phi|^2) - \chi\phi^{\dagger} - \phi\chi^{\dagger}\right] \\
&\qquad\qquad\qquad\qquad + \gamma L\epsilon^{\bar{\mu}\bar{\nu}}\left[\frac{1}{2}(|\phi|^2-1)B_{\bar{\mu}\bar{\nu}} + rQA_{\bar{\mu}\bar{\nu}}\right] = 0 \\
&D^{\bar{\mu}}(r^2 a(z)D_{\bar{\mu}}\chi) + a(z)\left[2v\phi - (1+|\phi|^2)\chi\right] - \gamma L\epsilon^{\bar{\mu}\bar{\nu}}(D_{\bar{\mu}}\phi)\left[i\partial_{\bar{\nu}}(rQ) + D_{\bar{\nu}}\rho\right] = 0 \\
&\frac{1}{r}\partial^{\bar{\mu}}(r^2 a(z)\partial_{\bar{\mu}}Q) - \frac{2}{r}a(z)Q - \frac{\gamma L}{2}\epsilon^{\bar{\mu}\bar{\nu}}\left[(iD_{\bar{\mu}}\phi(D_{\bar{\nu}}\chi)^{\dagger} + h.c.)\right.\\
&\qquad\qquad\qquad\left. + \frac{1}{2}A_{\bar{\mu}\bar{\nu}}(2v - \chi\phi^{\dagger} - \phi\chi^{\dagger}) - \frac{2}{\alpha^2}D_{\bar{\mu}}\rho\,\partial_{\bar{\nu}}\left(\frac{s}{r}\right)\right] = 0 \qquad \text{(A.2)}\\
&\partial_{\bar{\mu}}(a(z)D_{\bar{\mu}}\rho) - \frac{\gamma L}{2}\epsilon^{\bar{\mu}\bar{\nu}}\left[(D_{\bar{\mu}}\phi(D_{\bar{\nu}}\chi)^{\dagger} + h.c.) + \frac{i}{2}A_{\bar{\mu}\bar{\nu}}(\phi\chi^{\dagger} - \chi\phi^{\dagger})\right.\\
&\qquad\qquad\qquad\qquad\left. + \frac{2}{\alpha^2}\partial_{\bar{\mu}}(rQ)\partial_{\bar{\nu}}\left(\frac{s}{r}\right)\right] = 0 \\
&\partial^{\bar{\nu}}\left(r^2 a(z)B_{\bar{\nu}\bar{\mu}}\right) + 2a(z)D_{\bar{\mu}}\rho \\
&\qquad + \gamma L\epsilon^{\bar{\mu}\bar{\nu}}\left\{\left[(\chi - v\phi)(D_{\bar{\nu}}\phi)^{\dagger} + h.c.\right] + (1-|\phi|^2)\partial_{\bar{\nu}}v - \frac{2r}{\alpha^2}Q\,\partial_{\bar{\nu}}\left(\frac{s}{r}\right)\right\} = 0
\end{aligned}
\right.
$$

In order to solve numerically the EOM, they must be rewritten as a system of elliptic partial differential equations. This can be achieved by choosing a 2D Lorentz gauge condition for the residual $U(1)$ gauge fields

$$\partial^{\bar{\mu}} A_{\bar{\mu}} = 0 , \qquad \partial^{\bar{\mu}} B_{\bar{\mu}} = 0 . \tag{A.3}$$

The equations for $A_{\bar{\nu}}$ become $J^{\bar{\nu}} = \partial_{\bar{\mu}} \left( r^2 a A^{\bar{\mu}\bar{\nu}} \right) = r^2 a \partial_{\bar{\mu}} \partial^{\bar{\mu}} A^{\bar{\nu}} + \partial_{\bar{\mu}} (r^2 a) A^{\bar{\mu}\bar{\nu}}$ which is an elliptic equation and a similar result is obtained for $B_{\bar{\mu}}$.

The gauge condition needs only to be imposed at the boundaries, while in the bulk one can just solve the "gauge-fixed" EOM treating the two gauge field components as independent. The fact that the currents are conserved, $\partial_{\bar{\nu}} J^{\bar{\nu}} = 0$, implies indeed an elliptic equation for $\partial^{\bar{\mu}} A_{\bar{\mu}}$ which has a unique solution once the boundary conditions are specified. If imposed on the boundary, therefore, the gauge conditions are maintained also in the bulk.

The IR and UV boundary conditions on the 2D fields follow from Eq. (18.2.3) and Eq. (18.2.4) and from the gauge choice in Eq. (A.3). They are given explicitly by

$$z = z_{\text{IR}} \; : \quad \begin{cases} \phi_1 = 0 \\ \partial_2 \phi_2 = 0 \\ A_1 = 0 \\ \partial_2 A_2 = 0 \\ \partial_2 s = 0 \end{cases} \quad \begin{cases} \chi_1 = 0 \\ \partial_2 \chi_2 = 0 \\ \partial_2 v = 0 \\ \partial_2 Q = 0 \end{cases} \quad \begin{cases} \rho = 0 \\ B_1 = 0 \\ \partial_2 B_2 = 0 \end{cases} , \tag{A.4}$$

and

$$z = z_{\text{UV}} \; : \quad \begin{cases} \phi_1 = 0 \\ \phi_2 = -1 \\ A_1 = 0 \\ \partial_2 A_2 = 0 \\ s = 0 \end{cases} \quad \begin{cases} \chi_1 = 0 \\ \chi_2 = -1 \\ v = -1 \\ Q = 0 \end{cases} \quad \begin{cases} \rho = 0 \\ B_1 = 0 \\ \partial_2 B_2 = 0 \end{cases} . \tag{A.5}$$

The boundary conditions at $r = \infty$ have to ensure that the energy of the solution is finite; this means that the fields should approach a pure-gauge configuration. At the same time one has to require that the solution is non-trivial and its topological charge (Eq. (18.3.19)) is equal to one. We have

$$r = \infty \; : \quad \begin{cases} \phi = -i e^{i\pi z/L} \\ \partial_1 A_1 = 0 \\ A_2 = \frac{\pi}{L} \\ s = 0 \end{cases} \quad \begin{cases} \chi = i e^{i\pi z/L} \\ v = -1 \\ Q = 0 \end{cases} \quad \begin{cases} \rho = 0 \\ \partial_1 B_1 = 0 \\ B_2 = 0 \end{cases} . \tag{A.6}$$

The $r = 0$ boundary of our domain requires an ad hoc treatment, given that the EOM become singular there. Of course this boundary is not a true boundary of our 5D space, but it represents some internal points. Thus we must require the 2D solution to give rise to regular 5D vector fields at $r = 0$ and we must also require the gauge choice to be fulfilled. These conditions are

$$
r = 0 \; : \quad
\begin{cases}
\phi_1/r \to A_1 \\
(1 + \phi_2)/r \to 0 \\
A_2 = 0 \\
\partial_1 A_1 = 0 \\
s = 0
\end{cases}
\qquad
\begin{cases}
\chi_1 = 0 \\
\chi_2 = -v \\
\partial_1 \chi_2 = 0 \\
Q = 0
\end{cases}
\qquad
\begin{cases}
\rho/r \to B_1 \\
\partial_1 B_1 = 0 \\
B_2 = 0
\end{cases}
\quad . \quad \text{(A.7)}
$$

### A.1.1. *COMSOL implementation*

To obtain the numerical solution of the EOM we used the COMSOL 3.4 package,[28] which permits to solve a generic system of differential elliptic equations by the finite elements method. A nice feature of this software is that it allows us to extend the domain up to boundaries where the EOM are singular (*i.e.* the $r = 0$ line), because it does not use the bulk equations on the boundaries, but, instead, it imposes the boundary conditions.

In order to improve the convergence of the program and the numerical accuracy, one is forced to perform a coordinate and a field redefinition. The former is needed to include the $r = \infty$ boundary in the domain in which the numerical solution is computed. The advantage of this procedure is the fact that in this way one can correctly enforce the right behaviour of the fields at infinity by imposing the $r = \infty$ boundary conditions. A convenient coordinate change is given by

$$
x = c \arctan\left(\frac{r}{c}\right) , \qquad \text{(A.8)}
$$

where $x$ is the new coordinate used in the program and $c$ is an arbitrary constant. The domain in the $x$ direction is now reduced to the interval $[0, c\pi/2]$. The parameter $c$ has been introduced to improve the numerical convergence of the solution. A good choice for $c$ is $c \sim 10$, which allows to have a reasonable domain for $x$ and, at the same time, does not compress the solution towards $x = 0$.

A field redefinition is needed to impose the regularity conditions at $r = 0$ (Eq. (A.7)). For this purpose we use the rescaled fields

$$
\begin{cases}
\phi_1 = x\psi_1 \\
\phi_2 = -1 + x\psi_2 \\
\rho = x\tau
\end{cases}
\quad . \qquad \text{(A.9)}
$$

With these redefinitions, in the new coordinates, the $r = 0$ boundary conditions read as

$$
r = 0 \; : \quad
\begin{cases}
\psi_1 - A_1 = 0 \\
\psi_2 = 0 \\
A_2 = 0 \\
\partial_x A_1 = 0
\end{cases}
\qquad
\begin{cases}
\chi_1 = 0 \\
\partial_x \chi_2 = 0 \\
v = -\chi_2 \\
Q = 0
\end{cases}
\qquad
\begin{cases}
\tau - B_1 = 0 \\
\partial_x B_1 = 0 \\
B_2 = 0
\end{cases}
\quad . \quad \text{(A.10)}
$$

In order to ensure the convergence of the program another modification is needed. As already discussed, to obtain a soliton solution with non-vanishing topological charge we have to impose non-trivial boundary conditions for the 2D fields at $r = \infty$ (Eq. (A.6)). It turns out that if imposing such conditions the program is not able to reach a regular solution. This is so because the $r = \infty$ boundary is singular and imposing non-trivial (though gauge-equivalent to the trivial ones) boundary conditions at a singular point spoils the regularity of the numerical solution; the same would happen if the topological twist was located at $r = 0$. To fix this problem we have to perform a gauge transformation which reduces the $r = \infty$ conditions to trivial ones and preserves the ones at $r = 0$ at the cost of introducing a "twist" on the UV boundary. For this, we use a transformation of the residual $U(1)$ chiral gauge symmetry associated to $SU(2)_{L,R}$ (Eq. (18.3.33)) with

$$\alpha(r, z) = (1 - z/L)f(r), \tag{A.11}$$

where $f(r)$ can be an arbitrary function which respects the conditions

$$\begin{cases} f(0) = 0 \\ f(\infty) \to \pi \end{cases} \quad \text{and} \quad \begin{cases} f''(0) = 0 \\ f''(\infty) \to 0 \end{cases}. \tag{A.12}$$

For $c \sim 10$ it turns out that a good choice for $f(r)$ is $f(r) = 2\arctan r$. The gauge-fixing condition for $A_{\bar{\mu}}$ is now modified as

$$\partial_r A_1 + \partial_z A_2 - (1 - z/L)f''(r) = 0, \tag{A.13}$$

the UV boundary conditions are given by

$$z = z_{\mathrm{UV}} : \quad \begin{cases} x\psi_1 = \sin f(r) \\ (-1 + x\psi_2) = -\cos f(r) \\ A_1 = f'(r) \\ \partial_z A_2 = 0 \\ s = 0 \end{cases} \quad \begin{cases} \chi_1 = -\sin f(r) \\ \chi_2 = \cos f(r) \\ v = -1 \\ Q = 0 \end{cases} \quad \begin{cases} \tau = 0 \\ B_1 = 0 \\ \partial_z B_2 = 0 \end{cases}, \tag{A.14}$$

and the $r = \infty$ constraints are now trivial

$$r = \infty : \quad \begin{cases} \psi_1 = 0 \\ (-1 + x\psi_2) = 1 \\ \partial_x A_1 = 0 \\ A_2 = 0 \\ s = 0 \end{cases} \quad \begin{cases} \chi = -i \\ v = -1 \\ Q = 0 \end{cases} \quad \begin{cases} \tau = 0 \\ \partial_x B_1 = 0 \\ B_2 = 0 \end{cases}, \tag{A.15}$$

whereas the $r = 0$ and the IR boundary conditions are left unchanged. Notice that in the new gauge the EOM for $A_{\bar{\mu}}$ are modified in accord to Eq. (A.13), however they are still in the form of elliptic equations.

## References


1. G. 't Hooft, "A planar diagram theory for strong interactions," *Nucl. Phys. B* **72** (1974) 461.
2. G.S. Adkins, C.R. Nappi and E. Witten, "Static Properties Of Nucleons In The Skyrme Model," *Nucl. Phys. B* **228** (1983) 552.
3. T.H.R. Skyrme, "A Nonlinear field theory," *Proc. Roy. Soc. Lond. A* **260** (1961) 127.

4. For a review see U. G. Meissner, "Low-Energy Hadron Physics From Effective Chiral Lagrangians With Vector Mesons," *Phys. Rept.* **161** (1988) 213.

5. Y. Igarashi, M. Johmura, A. Kobayashi, H. Otsu, T. Sato and S. Sawada, "Stabilization Of Skyrmions Via Rho Mesons," *Nucl. Phys.* B **259** (1985) 721.

6. G.S. Adkins and C.R. Nappi, "Stabilization Of Chiral Solitons Via Vector Mesons," *Phys. Lett.* B **137** (1984) 251.

7. J.M. Maldacena, "The large N limit of superconformal field theories and supergravity," *Adv. Theor. Math. Phys.* **2** (1998) 231.

8. S.S. Gubser, I.R. Klebanov and A.M. Polyakov, "Gauge theory correlators from non-critical string theory," *Phys. Lett.* B **428** (1998) 105.

9. E. Witten, "Anti-de Sitter space and holography," *Adv. Theor. Math. Phys.* **2** (1998) 253.

10. J. Erlich, E. Katz, D.T. Son and M.A. Stephanov, "QCD and a holographic model of hadrons," *Phys. Rev. Lett.* **95** (2005) 261602.

11. L. Da Rold and A. Pomarol, "Chiral symmetry breaking from five dimensional spaces," *Nucl. Phys.* B **721** (2005) 79; *JHEP* **0601** (2006) 157.

12. A. Pomarol and A. Wulzer, "Stable skyrmions from extra dimensions," *JHEP* **0803** (2008) 051.

13. A. Pomarol and A. Wulzer, "Baryon Physics in Holographic QCD," *Nucl. Phys.* B **809** (2009) 347.

14. G. Panico and A. Wulzer, "Nucleon Form Factors from 5D Skyrmions," arXiv:0811.2211 [hep-ph].

15. J. Hirn and V. Sanz, "Interpolating between low and high energy QCD via a 5D Yang-Mills model," *JHEP* **0512** (2005) 030.

16. K. Nawa, H. Suganuma and T. Kojo, "Baryons in Holographic QCD," *Phys. Rev. D* **75** (2007) 086003; "Brane-induced Skyrmions: Baryons in holographic QCD," *Prog. Theor. Phys. Suppl.* **168** (2007) 231.

17. H. Hata, T. Sakai, S. Sugimoto and S. Yamato, "Baryons from instantons in holographic QCD," arXiv:hep-th/0701280.

18. D.K. Hong, T. Inami and H.U. Yee, "Baryons in AdS/QCD," *Phys. Lett.* B **646** (2007) 165; D.K. Hong, M. Rho, H.U. Yee and P. Yi, "Chiral dynamics of baryons from string theory," *Phys. Rev. D* **76** (2007) 061901; "Nucleon Form Factors and Hidden Symmetry in Holographic QCD," *Phys. Rev. D* **77** (2008) 014030.

19. H. Hata, M. Murata and S. Yamato, "Chiral currents and static properties of nucleons in holographic QCD," arXiv:0803.0180 [hep-th]; K. Hashimoto, T. Sakai and S. Sugimoto, "Holographic Baryons: Static Properties and Form Factors from Gauge/String Duality," arXiv:0806.3122 [hep-th].

20. T. Sakai and S. Sugimoto, "Low energy hadron physics in holographic QCD," *Prog. Theor. Phys.* **113** (2005) 843.

21. C.T. Hill, "Exact equivalence of the D = 4 gauged Wess-Zumino-Witten term and the D = 5 Yang-Mills Chern-Simons term," *Phys. Rev. D* **73** (2006) 126009.

22. G. Panico and A. Wulzer, "Effective Action and Holography in 5D Gauge Theories," *JHEP* **0705** (2007) 060.

23. M. Bando, T. Kugo and K. Yamawaki, "Nonlinear Realization and Hidden Local Symmetries," *Phys. Rept.* **164** (1988) 217.

24. H. Georgi, "Vector Realization Of Chiral Symmetry," *Nucl. Phys.* B **331** (1990) 311.

25. G. Ecker, J. Gasser, H. Leutwyler, A. Pich and E. de Rafael, "Chiral Lagrangians for Massive Spin 1 Fields," *Phys. Lett.* B **223** (1989) 425; G. Ecker, J. Gasser, A. Pich and E. de Rafael, "The Role Of Resonances In Chiral Perturbation Theory," *Nucl. Phys.* B **321** (1989) 311.

26. D.T. Son and M.A. Stephanov, "QCD and dimensional deconstruction," *Phys. Rev. D* **69** (2004) 065020.

27. E. Witten, "Some exact multipseudoparticle solutions of classical Yang-Mills theory," *Phys. Rev. Lett.* **38** (1977) 121.

28. See http://www.comsol.com.

29. D. Finkelstein and J. Rubinstein, "Connection between spin, statistics, and kinks," *J. Math. Phys.* **9** (1968) 1762.

30. E. Braaten, S.M. Tse and C. Willcox, "Electromagnetic Form-Factors In The Skyrme Model," *Phys. Rev. Lett.* **56** (1986) 2008.

31. E. Witten, "Baryons In The 1/N Expansion," *Nucl. Phys. B* **160** (1979) 57.

32. See, for example, A.V. Manohar, "Large N QCD," arXiv:hep-ph/9802419.

33. E. Witten, "Current Algebra, Baryons, And Quark Confinement," *Nucl. Phys. B* **223** (1983) 433.

34. G. Karl and J.E. Paton, "Naive Quark Model For An Arbitrary Number Of Colors," *Phys. Rev. D* **30** (1984) 238.

35. M.A.B. Beg and A. Zepeda, "Pion radius and isovector nucleon radii in the limit of small pion mass," *Phys. Rev. D* **6** (1972) 2912.

36. For a review see R. Machleidt, K. Holinde and C. Elster, "The Bonn Meson Exchange Model for the Nucleon Nucleon Interaction," *Phys. Rept.* **149** (1987) 1.

37. G.S. Adkins and C.R. Nappi, "The Skyrme Model With Pion Masses," *Nucl. Phys. B* **233**, 109 (1984).

38. E. Amaldi *et al.*, "Axial-vector form-factor of the nucleon from a coincidence experiment on electroproduction at threshold," *Phys. Lett. B* **41** (1972) 216.

39. A. Del Guerra *et al.*, "Threshold $\pi^+$ electroproduction at high momentum transfer: a determination of the nucleon axial vector form-factor," *Nucl. Phys. B* **107** (1976) 65.

# Author Index


Balents, L., 333
Battye, R.A., 3
Brey, L., 291

Diakonov, D., 57

Ezawa, Z.F., 233

Fertig, H.A., 291
Fisher, M.P.A., 333

Girvin, S.M., 217

Hen, I., 179
Holzwarth, G., 41
Hong, D.K., 165

Karliner, M., 179

Lee, H.K., 147

Manton, N.S., 3
Moon, K., 269

Nielsen, H.B., 393

Park, B.-Y., 115
Petrov, V., 57
Pomarol, A., 403

Rho, M., 147

Sachdev, S., 333
Scoccola, N.N., 91
Senthil, T., 333
Sugimoto, S., 347
Sutcliffe, P.M., 3

Tsitsishvili, G., 233

Vento, V., 115
Vishwanath, A., 333

Wulzer, S, 403

Yi, P., 367

Zahed, I., 393

$\omega$-meson problem
  in dense skyrmion, 140
  resolution, 140, 154

alpha ($\alpha$) particles as skyrmions, 15
Atiyah-Manton ansatz, 122, 358

baby skyrmions, 180, 259
  half skyrmions, 186
  lattice structure, 185
  relation to 3D skyrmions, 199
bilayer quantum Hall systems, 223, 253, 269, 299

Callan-Klebanov model, 108
Chern-Simons term, 359, 373, 377, 406
Cheshire Cat
  from holography, 393
  holographic model, 396
  principle, 394
chiral $\pi$-$\rho$-$\omega$-meson model, 44
chiral quark soliton model, 60
coherence network model, 303
color flavor locked (CFL)
  phase, 159, 168
  quark matter, 167

d-wave superconductivity
  role of half skyrmions, 321
deconfined quantum critical phenomenon, 333
  half skyrmions/merons, 339

magnetic Néel-VBS transition, 335
dense skyrmion matter, 115
  chiral symmetry restoration, 128
  pion flucturations, 130
  with vector mesons, 136
dilatons
  in hidden local symmetry, 150
  scale invariance, 125

entanglement
  spin-charge, 271

finite nuclei as skyrmions, 3
five-dimensional model of hadrons
  AdS/QCD, 403
    large $N_c$ scaling, 421
  baryons, 411
  mesons, 405
  skyrmions, 411
form factors
  in chiral soliton models, 41
    electric proton, 43, 47
    magnetic proton, 47
    time-like $Q^2$, 51
  nucleons in AdS/QCD, 419
  nucleons in holographic QCD, 389, 398
  pions in AdS/QCD, 424

gapless collective spin mode, 297
gapped quark, 170
gauge/string duality, 348–350, 370, 404

Goldberger-Treiman relation
  for nucleons in AdS/QCD, 424
  for pentaquark $\Theta^+$, 83

hadronic freedom, 149, 156
half skyrmions
  anomaly matching, 149
  for high temperature superconductivity, 311
  hole doped system, 314
  in baby skyrmions, 186
  in dense hadronic matter, 129, 147
  in Néel-VBS transition, 339
  in quarkyonic phase, 149
  multi configurations, 318
  pseudogap, 129
heavy-meson-soliton binding, 94
heavy-quark skyrmions, 91
heavy-quark symmetry, 93
hidden local symmetry (HLS), 109, 118, 136, 151
holographic dynamics
  baryons, 375
  mesons, 372
  pure QCD, 371
holographic QCD, 350, 369
  chiral symmetry, 355
holographic Yang-Mills theory, 351

instantons
  holographic baryons, 357, 367
  in magnetic Néel-VBS transition, 337, 339

kaon-skyrmion scattering, 71

merons, 228, 280
  confinement-deconfinement transition, 287
  electron-like, 262
  hole-like, 262
  magnetic Néel-VBS transition, 339
  states, 262

noncommutative geometry, 236
nuclear energy levels
  Skyrme model, 32

pentaquark $\Theta^+$, 63
  Goldberger-Treiman relation, 83
  narrow width, 76
pseudogap
  in dense hadronic matter, 125
  in high T superconductivity, 325
pseudospin, 229, 271, 291
  QH ferromagnet, 253

quantum Hall effect, 217, 291
  ferromagnetism, 219
  systems, 233

rational map, 11, 198

scale invariance
  dilatons, 125
    in hidden local symmetry, 150
skyrmion crystal, 294
  BCC, 120
  FCC, 138
skyrmion texture
  NMR shifts, 221
spin texture, 291
  half skyrmion, 314
spontaneously broken rotational symmetry (SBRS)
  in baby skyrmions, 202
superqualitons, 159, 165
  $Q$-matter, 175

vector dominance, 387
  with infinite tower, 388
vector limit, 155
vector manifestation (VM), 142
vector symmetry, 155